Green Energy and Technology

Climate change, environmental impact and the limited natural resources urge scientific research and novel technical solutions. The monograph series Green Energy and Technology serves as a publishing platform for scientific and technological approaches to “green”—i.e. environmentally friendly and sustainable—technologies. While a focus lies on energy and power supply, it also covers "green" solutions in industrial engineering and engineering design. Green Energy and Technology addresses researchers, advanced students, technical consultants as well as decision makers in industries and politics. Hence, the level of presentation spans from instructional to highly technical.

Indexed in Scopus

Indexed in Ei Compendex

Israel Sunday Dunmade
Michael Olawale Daramola
Samuel Ayodele Iwarere
Editors

Sustainable Engineering

Concepts and Practices

Springer

Editors
Israel Sunday Dunmade
Department of Earth and Environmental Sciences
Mount Royal University
Calgary, AB, Canada

Samuel Ayodele Iwarere
Department of Chemical Engineering, Faculty of Engineering, Built Environment and Information Technology
University of Pretoria
Pretoria, South Africa

Michael Olawale Daramola
Department of Chemical Engineering, Faculty of Engineering, Built Environment and Information Technology
University of Pretoria
Pretoria, South Africa

ISSN 1865-3529 ISSN 1865-3537 (electronic)
Green Energy and Technology
ISBN 978-3-031-47214-5 ISBN 978-3-031-47215-2 (eBook)
https://doi.org/10.1007/978-3-031-47215-2

This Springer imprint is published by the registered company Springer Nature Switzerland AG
The registered company address is: Gewerbestrasse 11, 6330 Cham, Switzerland

Paper in this product is recyclable.

Foreword

We are living in a time when all academic disciplines and professions are being called upon to examine their practices with respect to environmental and economic sustainability and consider the ethical impact on people and cultures. The innovations of the past have come with serious impacts on the present and our shared future. By drawing together a wide range of topics such as food, water and energy and emerging areas such as artificial intelligence and nanotechnology, the authors, from Africa, Asia, Europe, North America and Oceania, provide invaluable context for practice both in the West and in developing nations. Indeed, Prof Dunmade, Prof Daramola and Dr Iwarere bring unique perspectives from Environmental Science and Chemical Engineering from Canada and South Africa, and their biographies attest to that.

The focus of this book is sustainable engineering and the interrelation of products, processes and systems, but it highlights two important realities: (1) we must always recognise that the context of our work shapes our practice, and (2) applying theory is often challenging and we need to adapt to meet real-world situations. The diverse range of authors from around the world brings the needed context and the lessons they have learned in addressing sustainability. Engineers, allied professionals, governmental agencies and graduate students will benefit from the broad experiences reflected in the chapters.

By reading this work, you will gain a deeper understanding of sustainable engineering, its practices, principles and methods in a variety of fields. You will gain insights into making informed decisions that will contribute to sustainability with particular emphasis on waste reduction.

President & Vice-Chancellor, Timothy Rahilly
Professor Health, Community, & Education
Mount Royal University,
Calgary, Alberta, Canada

Preface

Engineering as a discipline has contributed significantly to the economic growth of countries globally. Indeed, the first and second industrial revolutions have helped several countries in the Western world to achieve significant development while the other countries are following suit. However, the developments we have witnessed have come at a cost to our society, with a strong social and environmental impact. Hence, the abundance of waste of resources at private (industrial) and governmental facilities in both developed and developing countries triggered a strong passion in us to contribute to the elimination/minimisation of these wastes. Doing this would not only be good for the environment but would also be socially and economically beneficial. In addition, we also saw the need to equip students in the tertiary institutions and professionals with skills that would enable them to develop systems that would meet the needs of people in an environmentally sustainable manner. Therefore, this book is seen as one of the ways to achieve these goals. The book is born out of a passionate desire for a text that inspires professionals, managers and researchers, particularly in the field of engineering, to incorporate sustainability into what they create (products, processes and systems), and into how they develop and manage them throughout the lifecycle of such creations.

In today's growing global environmentally conscious and competitive economy, it is becoming increasingly important for role players in our economic activities to incorporate sustainability consideration into the design and lifecycle management of their products, processes, systems and services. Engineering and other professionals that are involved in the development and operational management of these products, systems and services need to have adequate instruments, in terms of educational training and technology, to foster the sustainability of our systems and services.

Moreover, in the last few decades, engineers and other professionals have been taking progressive steps towards the development and management of sustainable engineering products, processes, systems and services. There is a need for a body of work that documents experiences of some of these colleagues. This would inspire others to learn from their success and avoid their mistakes. It would also serve as a useful resource for education and training for those whose choices and decisions would affect the future of our world. This edited book would be found to be one of such useful resources.

Today's challenge is complex and diverse, there is therefore a need for a sustainable engineering book that does not only cover a broad spectrum of traditional engineering areas but various aspects of our economic activities and emerging areas that relate with sustainability of engineering activities with the aim of reaching a broader audience, hence this edited book is unique in its spread. The process of developing such a sustainability engineering book revealed the need to approach the project as an edited book with chapters contributed by academics and professionals from various branches of engineering and allied professions.

This book has a large spread in topics coverage ranging from basic needs such as food, water and energy to new and emerging areas like artificial intelligence and nanotechnology. There is also a diversity in the categories of contributors. The chapters were contributed by young and emerging researchers, professors, and professionals from corporate, non-governmental and governmental organisations. In addition, contributors have widespread academic and professional backgrounds ranging from various kinds of engineering disciplines to non-engineering disciplines like biology, arts and gerontology. The geographical spread of contributors is also global, ranging from Canada and the USA in North America to Australia and New Zealand in Oceania. We also have submissions from such countries as Denmark, Finland, France, Germany, The Netherlands, Spain, Turkey and the United Kingdom in Europe; Nigeria and South Africa in Africa; and China, India, Indonesia and Malaysia in Asia. A number of the chapters were produced through collaboration by scholars from different universities and even from different continents. The book is truly broad and comprehensive in perspective and viewpoints.

The book consists of 28 chapters divided into five parts. Part I entitled Food-Water-Energy Nexus and Sustainable Engineering consists of six chapters that are essentially focused on some of the current and unsustainable practices that affects food, water and energy and how a more sustainable approach can be introduced. Part II entitled Sustainability in Materials Recovery and Processing contains five chapters that discuss how different tools such as life cycle analysis could be employed as a sustainable approach for materials recovery and processing. Readers would find eight chapters in Part III entitled Sustainability in Manufacturing, Infrastructure, and Engineering Design that discussed various issues relating to the incorporation of sustainability into engineering design, engineering infrastructure and manufacturing. Three chapters with in-depth discussions on sustainable biotechnology and sustainability in nanotechnology will be found in Part IV entitled Sustainable Engineering in Biotechnology and Nanotechnology, while Part V entitled Artificial Intelligence and Socio-economic Impact in Sustainable Engineering contains six chapters with a focus on sustainability, artificial intelligence and socio-economic impact.

Thus, readers of this edited book would benefit from diverse and holistic viewpoints and insight into how academics and professionals from various branches of engineering and allied fields all over the world are incorporating and fostering sustainability in their research works and professional activities at a national and international level of engagement. The content of this book should be found helpful by students in tertiary education, engineers, planners, managers and regulators in taking sustainability steps that would make our world a better place.

Calgary, AB, Canada
Pretoria, South Africa

Israel Sunday Dunmade
Michael Olawale Daramola
Samuel Ayodele Iwarere

Contents

About the Editors

Israel Sunday Dunmade (PhD, PEng, PMP, EP) is a Professor of Sustainable Engineering in the Department of Earth and Environmental Sciences, Faculty of Science and Technology, Mount Royal University, Calgary, Canada. His research interest is in the field of lifecycle engineering – sustainable (water, energy, health and transport) infrastructure, sustainable design, sustainable manufacturing, lifecycle analysis, circular economy, campus sustainability and rural sustainability. He is the Alberta Regional Director of the Canadian Society for Bioengineering (CSBE) and a member of the Board of Directors of the Canadian Society for Bioengineering Foundation. Furthermore, Dunmade is a member of the International Network for the Science and Technology of Sustainability. In addition, he was a DAAD scholar, the Past President of the African Network of the American Society of Agricultural and Biological Engineers and the former Chair of the International Affairs Division of the American Society for Agricultural and Biological Engineers. He is a recipient of several research grants and awards of recognition. Dunmade is a Professional Engineer, Project Management Professional and Environmental Professional. He has published two books, book chapters and authored/co-authored several peer-reviewed publications and conference presentations.

Michael Olawale Daramola (PhD, FIChemE, FSAICHE, PGDip (HE), CEng) is a Professor of Chemical Engineering and Head of the Department of Chemical Engineering at the University of Pretoria. He holds a BScEng (with Hons.), an MScEng and a PhD degree in Chemical Engineering. He also holds an MSc degree in Biotechnology (Process Technology option), a PGDip in Higher Education and a Certificate in Academic Leadership. In addition, Prof Daramola holds a certificate in Research Management and Leadership from IREX (affiliated to Michigan State University and George Mason University in the USA) and a certificate in Postgraduate Supervision from Rhodes University (in collaboration with Nuffic, The Netherlands). He is a global scholar with training, teaching and research experiences from The Netherlands, France, the USA, South Africa, Singapore, Ghana and Nigeria. He is a chartered engineer with the Engineering Council of the UK and a COREN-registered engineer with the Council for Regulation of Engineering in

Nigeria. Prof Daramola's primary research focuses on sustainable energy and the environment, with major activities in developing and applying nanomaterials, waste treatment and valorisation. The research provides understanding and develops concepts towards providing solutions in sustainable energy and environment. Prof Daramola has been among the world's top 2% of scientists since 2021. He was a finalist in the South African Oscar Science award in 2022 (in TW Kambule-NSTF Award: Researcher category and in Engineering Research Capacity Development Award category), a Carnegie Corporation of New York's Fellow in Research Leadership, a Nuffic Fellow, etc. He serves on the editorial boards of international journals such as *ACS Sustainable Chemistry and Engineering, Materials Today Proceedings, Groundwater for Sustainable Development, Membranes, International Journal of Oil, Gas and Coal Technology (IJOGT)* and *Frontiers in Chemical Engineering*. He was the Associate Editor of *Chemical Engineering Communicatio*ns from 2016 to 2020. He was a finalist in the 2021/2022 NSTF-South32 Awards. He is the national panel member and vice chairperson of the South African Bureau of Standard's (SABS) Technical Committee on carbon dioxide capture, transportation and geological storage (SABS/TC 0265 or ISO/TC 265). He is also a national ISO/TC 323 'Circular economy' panel member. Prof Daramola has produced 55 graduate students (22 doctoral and 33 master's) and mentored more than 15 early career researchers and academics (postdoctoral and junior lecturers). He has also contributed enormously to knowledge creation in engineering with publications in excess of 240 publications including one granted patent, 15 book chapters and two edited books. These publications have attracted more than 4200 citations.

Samuel Ayodele Iwarere (PhD, MScEng, BSc Hons.) is currently a Senior Lecturer in the Department of Chemical Engineering at the University of Pretoria in South Africa. He previously worked as a Senior Researcher in the same Department from July 2020 to December 2022 and a Royal Society FLAIR Fellow from May 2020 to January 2023. His expertise is in the field of Plasma Science and Engineering. He is a recipient of the prestigious Future Leaders – African Independent Research (FLAIR) Fellowship, a partnership between the African Academy of Sciences and the Royal Society that is funded by the UK government as part of the Global Challenges Research Fund (GCRF). He graduated with a Bachelor of Science Second Class Honours (Upper Division) in Industrial Chemistry from the University of Ilorin, Nigeria, and MScEng (Chemical) cum laude from the University of KwaZulu-Natal (UKZN) in South Africa. He obtained a PhD in Chemical Engineering with a specialisation in Plasma Technology from UKZN in collaboration with the Plasma Group at Mines ParisTech, France, under the supervision of Prof Deresh Ramjugernath and Prof Laurent Fulcheri. Dr Iwarere completed three years (2014 to 2016) as a Postdoctoral Research Fellow at UKZN in the Thermodynamics Research Unit. From 2017 to June 2020, he worked full-time at UKZN as the principal investigator on a Water Research Commission-funded project while with the Thermodynamics Research Unit. In addition, he holds a CREST certificate in Doctoral Supervision issued by Stellenbosch University, South Africa in 2020, from October 2019 to March 2020 with the Centre for Research and Evaluation, Science and Technology. His current research activities involve the understanding of plasma-liquid interactions in water and wastewater treatment using experimental and computational approaches and other plasma environmental-related applications, including solid waste valorisation. He works in collaboration with a multidisciplinary team of researchers.

Part I
Food-Water-Energy Nexus and Sustainable Engineering

Digitalization for Sustainable Agriculture: Enabling Farm Digitalization Through Decentralized Control and Ownership

Alvaro Romera, Glenn Parry, James Turner, Martin Espig, Michael Rogerson, and Munir Shah

1 Introduction

Agricultural digitalization is often promoted in terms of enabling intensified on-farm production, improved eco-efficiency, and profitability gains, but digital technologies also have the potential to support the broader sustainable development of agri-food systems. Multiple authors have described the various ways in which digitalization could benefit efforts to sustainably transform farming practices (Mondejar et al. 2021). Even though we do not automatically equate digitalization to promoting sustainable agriculture (see, e.g., Fleming et al. 2018, 2021; Jakku et al. 2023; Lioutas et al. 2021), we recognize the role digital technologies could play in realizing relevant sustainable development goals. Among the promises of smart agriculture or Agriculture 4.0 is a future in which robots autonomously perform most of the farm tasks (see, e.g., da Silveira et al. 2021). Farmers would manage their farms from the comfort of their office, with a great deal of decision-making done by artificial intelligence (AI) algorithms. Such algorithms would operate using big data

The original version of the chapter has been revised. A correction to this chapter can be found at https://doi.org/10.1007/978-3-031-47215-2_30

A. Romera (✉) · J. Turner
AgResearch Ltd, Ruakura Agricultural Centre, Hamilton, New Zealand
e-mail: alvaro.romera@agresearch.co.nz

G. Parry
Surrey Business School, University of Surrey, Guildford, Surrey, UK

M. Espig · M. Shah
AgResearch Ltd, Lincoln Research Centre, Christchurch, New Zealand

M. Rogerson
University of Sussex Business School, University of Sussex, Brighton, UK

I. S. Dunmade et al. (eds.), *Sustainable Engineering*, Green Energy and Technology, https://doi.org/10.1007/978-3-031-47215-2_1

generated by sensors, Internet of Things (IoT) devices, satellites, and drones, which monitor all aspects of the farm system. The much greater precision and accuracy that these systems afford could make farming more environmentally sustainable, efficient, and profitable (Lajoie-O'Malley et al. 2020; Shepherd et al. 2020). Rural communities could be more prosperous as a result. This imagined future has the potential to generate a lot of data about a farm and the supply chains within which it is located (Wolfert et al. 2017). This provides an opportunity for farmers to respond effectively to the increasing pressure to collect data for supporting on-farm decision-making, regulatory compliance, and reporting to downstream supply chain partners.

Consumers also benefit in this imagined future. Food would be safer due to the more precise and targeted use of agrichemicals and the reduction of food safety risks. Data would flow along supply chains, enabling the accurate fulfillment of consumer needs. Waste would then be virtually eliminated, which together with higher yields that these technologies enable may help to address some global food security problems. Full transparency of the production system would give consumers greater certainty about the way their food is produced, allowing them to make informed choices related to the sustainability and health qualities of the products they buy. Or at least that is the promised future.

This chapter will use illustrative case examples from New Zealand farms to critically reflect on these envisioned futures and discuss how the adoption of digital innovation in farm systems may be accelerated in ways that equally create value for all actors across the agri-food supply chain, in particular farmers. We identify key barriers that need to be addressed in doing so. New Zealand offers a useful context as it is considered to be a digital leader (Chakravorti and Chaturvedi 2017) and is actively seeking to advance the agricultural technology sector (e.g., AgriTech Industry Transformation Plan (Ministry of Business, Innovation and Employment 2020)).

2 Future Vision

Digital technologies are increasingly incorporated across modern farms around the world. This includes software and hardware. An increasing number of farms are connected to the Internet, particularly in developed countries, but also in developing nations. Farmers use a variety of software products daily to help them with data recording and to assist in their decision-making. For that, they use computers, tablets, or mobile phones. Farm machinery, like tractors, harvesters, and spreaders, are equipped with increasingly sophisticated technologies, such as vehicle health and usage monitoring systems (HUMS) or ground radar and global positioning systems (GPS), and they generate high volumes of data. Virtual fences can be used to herd cows on the field and move them around the farm, all from a computer or smartphone. Sensors in milk vats monitor temperature, volume, and milking times, and the information is useful to farmers as well as processors who collect the milk. All of these data offer potential value to not only the technology providers and users but multiple decision-makers connected to on-farm operations, including farm managers, input suppliers, agricultural processors, marketers, retailers, consumers, and regulatory and policy actors (van der Burg et al. 2019; Wolfert et al. 2017).

For benefits to be realized by wider decision-makers, ideally all these digital devices would be interoperable and data shared among decision-makers (Wolfert et al. 2023). That is, data collected or generated by any of them could be readily utilized by any other. Those who need the information could be given appropriate access to it without double handling of data or requiring complex information handling procedures. Whenever possible, data processing will be automated through efficient pipelines, for example, to generate reports or timely insights to support decision-making. Furthermore, data collected by many farmers, perhaps from an entire region, nation, or even globally, could be integrated, shared, and utilized for public as well as private good. In sum, as in other sectors of the economy, agricultural data would become a valuable commodity because it generates value beyond the person or firm who collected it originally.

3 The Current State of Digitalization in New Zealand Farm Systems

The future vision portrayed in the introduction has not yet arrived, at least not for open-field farm systems, which produce by far most of the food consumed worldwide today. Some of the propositions may not be viable, or even desirable. Indeed, while many authors claim that digital agriculture will deliver all those benefits, it is worth noting that some authors have critically reflected on the potential risks and benefits of Agriculture 4.0 (e.g. Eastwood et al. 2023; Jakku et al. 2023). So, what would be a more realistic expectation for a nearer term?

A high degree of on-farm digitalization, with seamless interoperability between technologies, could enable farmers to extract maximum benefits from data. Well-functioning digital systems might also assist them in responding to growing societal demands for greater sustainability, food safety, and accountability. Such techno-optimistic framings dominate the discourses around the envisioned futures, and emerging realities, of agricultural digitalization and corresponding venture capital models of investment (Barrett and Rose 2022; Espig et al. 2022; Fielke et al. 2022; Klerkx and Rose 2020). At the same time, however, the value proposition of digital agricultural technologies has been juxtaposed to potential negative impacts associated with on-farm practice changes and far-reaching societal transformations. This has prompted the notion of "digital disruptions" to describe the substantial change processes associated with digital innovation (Cook et al. 2022; Eastwood et al. 2019; Jakku et al. 2023).

Despite gradual steps toward digitalization and its promising possibilities, digital technologies often still only support some aspects of on-farm operations and supply chain integration, with varying degrees of interconnectivity and interoperability. This partial application and fragmented data flows between systems limit the perceived and actual value offered by digital technologies. Concurrently, ongoing debates concerning data ownership, trust, competing business models, and equitable distribution of benefits have hampered digitalization efforts (Fleming et al. 2018; Jakku et al. 2019; Shepherd et al. 2020). Building on these insights, we outline the current situation of three distinct New Zealand farm systems. We aim to demonstrate that, rather than simply being failures to realize the possibilities of digitalization, the partial utilization of digital technologies tends to be rooted in practical challenges and broader sociocultural adoption barriers. These need to be addressed to unlock the proposed value of digital transformations in a responsible and socially beneficial manner (Bronson 2019; Eastwood et al. 2023; Espig et al. 2022; Jakku et al. 2023).

3.1 Farm 1: Small Sheep Farm

Derrick[1] is in his 60s and a livestock farmer as well as quality assurance auditor with over 45 years' experience in the meat sector. While still employed as an auditor, he operates a 40-hectare sheep farm on New Zealand's South Island, largely by himself, with occasional help from his wife. They run around 300 breeding ewes and finish their own lambs. The farm is stocked conservatively to limit the workload as Derrick got older. He acknowledges that the farm operation is relatively low-tech and uncomplicated, with a simple animal health plan and only modest agrichemical inputs that are mainly recorded in a manual diary. Having no employees, contractors are brought in for cropping their winter feed, fertilizer applications, and spraying of crop protection chemicals.

In some cases, contractors share data recorded on their tractors via a flash drive, but Derrick often notes applied agrichemicals manually in his diary. He is aware that this is a rather traditional way of recording farm data:

> I'm still a bit of a diary man. I've kept a diary forever and referred back to those notes through the diary, year in and year out ... [A]ll of that stuff that's required through the recording and the assurance program is in those diaries. I'm just old-school in the way that I'm recording and doing things here, because it's quite a simple operation.

Continuing to use diaries makes Derrick feel that he can easily keep track of farm operations, although he sees the potential benefits of available recording and decision support software. For example, proof of placement for applied agrichemicals can be recorded through platforms such as FarmIQ, a farm management software, or TracMap, a GPS guidance and precision management system. In his view, these software tools will have to be used by all New Zealand livestock farmers in the future.

To maximize the value of digital technologies for his business, Derrick would like to have an easy way to capture key livestock data in one place, such as linking batch numbers, expiry dates on agrichemicals, withholding periods, safe dates for

[1] All names are pseudonyms.

slaughter, or the dates an animal is drenched or receives treatments. This would reduce the likelihood of crucial information getting lost.

Suitable software could also assist with agrichemical records. At the moment, Derrick manually monitors on-farm agrichemical stocks and replenishes them as needed, with current quality assurance criteria only requiring farmers to keep a register of original product containers and note when these have been used. Having a system to digitalize this process and syncing records in a centralized location would be a game changer. As Derrick notes:

> [J]ust a way of linking that data from the devices or whatever back to a central location or a PC, ... it doesn't always happen. But if it's some sort of an automated system whereby they [farmers] enter in that program on their smartphone, it's automatically going to the home PC or the tablet, as soon as they've coverage, then that works quite well. But otherwise, it tends to sit on their devices and then come audit time, they're all trying to drag it back out.

However, he acknowledges that fully realizing the potential value of such software would require good training and support for farmers. For smaller farms like him, associated capital and operational expenditures are also relatively large compared to their operating budgets.

3.2 Farm 2: Mixed Crop and Livestock Farm

Rod is a 57-year-old farmer who together with his wife operates a 450-hectare, fully irrigated mixed crop and livestock farm in the Canterbury region on New Zealand's South Island. Like many others in the area, they grow a diverse mixture of cash crops, ranging from cereals as their mainstay to seed crops (ryegrass, clover, brassicas, etc.), and various vegetables. Crop rotations are relatively flexible, allowing them to respond to changing market trends. Going into the winter months, they also bring in up to 6000 lambs for grazing. The livestock fattening program accounts for about 30% of farm income.

Rod uses a range of digital technologies and software tools to support his operations. He programmed tractor computers to record the seed varieties sown on each paddock. While the tractors run on software by the American company Trimble, a sprayer he bought secondhand only supports Topcon software. He decided to keep both setups rather than configure a new spraying unit or tractor software. Additionally, his irrigation wells contain data loggers to record water usage as required in his irrigation permit consent. To record core cropping production data, such as agrichemical inputs, and activities, Rod uses the ProductionWise software by the Foundation for Arable Research, a levy-funded industry good body. The software also allows him to meet quality assurance requirements, for instance, by showing proof of placement of crop protection chemicals to supply chain partners, and it features an automated inventory system that removes agrichemical stocks once applications are recorded. The same software is used for the family's farm planning, for instance, to quickly calculate gross margins and review historical crop performance data. ProductionWise is Rod's main recording and decision support software, but he further records nutrient inputs (e.g., nitrogen fertilizers) in the OverseerFM software for regulatory compliance and auditing.

While these digital technologies and tools certainly support Rod's operations, they also create new pressures and practical challenges that hinder realizing their full value. He notes the large number of data entries required to record the diverse activities across their farm, with up to two dozen entries per day. As he explains:

> I want to record every irrigation, every cultivation, every application of spray that we would do to a paddock and every single action, particularly if we want to get accurate gross margins coming out. … [Q]uite [a number] of activities can go on, … we've got nine irrigators on the farm and they're all on set paddocks, so that's nine entries … and we've got our own sprayer, the contractor doing fertiliser. And while he is … going more into digital recording for proof of placement, it's not apparent to have that capability linked into our ProductionWise. … [A] frustration might be around dual activity or repetitive services of getting all the information to the one platform.

Rod acknowledges that they could improve their real-time recording via mobile phone, but he sees the applications' user interfaces as cumbersome and prefers data entry on his computer in the evenings, which creates time delays and relies partly on paper records collected throughout the day. One frustrating consequence of using multiple software tools is lacking interoperability, causing the need to duplicate data entries. In many cases, data inputs and outputs have to be transferred manually from written record or across platforms. With some of this recordkeeping solely done for retrospective compliance purposes, Rod also does not see much value added to his business from those records, for example, for the family's farm planning activities.

At the moment, the capital and operational cost for these digital technologies still only constitutes a small percentage of their overall expenses. As digitalization across the sector accelerates, however, Rod feels that related hard- and software costs will become significant to their operation. Further challenges emerge from new hardware often requiring updated software packages, and vice versa. These changes not only create growing costs but also mean that increasingly specialized skills are required to set up and maintain new digital technologies, with Rod highlighting their diminishing ability to fix equipment directly on-farm and after-sale services often not being great.

3.3 Farm 3: Corporate Dairy Business

Erin is employed by a company that manages almost two dozen dairy farms across New Zealand's South Island. Now in her later 30s, she shifted into a support position following several years in on-farm roles. Much of her time is spent implementing technologies and innovation projects across individual farms. Given the structure of their operations, these are rather complex tasks. The company has several contract milkers and leasing agreements, with some involving equity holdings. Most farms are overseen by a business manager, while on-farm operations are run by a farm manager and three to six staff. The company's portfolio covers a wide range of farm types, from low- to higher-input systems (e.g., imported feed) and herd sizes ranging from 550 through to 1200 cows. Almost all farms are irrigated.

Technologies used on farm are equally as diverse. The Protrack software is used in about two-thirds of milking shed, which, among other features, offers automated livestock drafting and herd management functions. All farms also have in-shed feeding systems and computer terminals to monitor operations. Different systems are used across their farms to record amounts and locations of fertilization applications. Similarly, water usage is monitored for regulatory reporting, with processes and systems varying between different catchment irrigation schemes and resource consents. A pasture metering tool is used to record yield data and guide required fertilizer usage. The company is also experimenting with a virtual fencing system by the New Zealand-based company on one farm, which works well after some initial setup problems.

Farm management and regulatory compliance is supported by several software platforms. OverseerFM is used to estimate nitrogen leaching (e.g., from animal urine) and greenhouse gas emissions generated on farm (e.g., from ruminant animals) and embodied emissions in products brought onto farm. Records on animal health and reproduction, milk production and milk quality are recorded in MINDA. Additionally, FarmIQ is trialed as another management tool on some farms. To assist accounting and have easy access to product receipts required for quality assurance reporting, the company uses the integrated accounting software Greentree.

Working across these different platforms means that data entries and outputs frequently need to be handled and transferred multiple times. Achieving interconnectivity and interoperability thus remains challenging. As Erin remarks:

> none of them talk to each other... Most farming operations couldn't do this, but because we are a group, ... we pool all of our information into a system called Patika and then we display that in Power BI so that we can actually measure our farms' performance against one another. And we haven't got everything pooling in there because there are so many systems.

Not having all information readily accessible in one place not only complicates timely decision-making but also makes auditing processes time-consuming. Moreover, hard copy records, such as feed declaration forms, need to be digitized before they can be shared.

While some of these practical challenges are technological and related to the on-the-ground complexities of farming, Erin is cognizant that cultural and social aspects also play a role. This involves questions around trust, ownership, privacy, and commercial interests:

> [I]n agriculture, we've been notoriously bad at wanting to hoard data, because we think that data is what holds power, and we don't want to open up. ... So, you still own that information and that data yourself, because ... it's our farm data, we're producing it. ... [W]e don't just want it free-sourced so that anyone can access it. ... [I]t's our competitive advantage in New Zealand, our information and our data and the way that we farm. And the way that we do things is what gives us a competitive advantage. So, we've got to be careful at protecting that.

More mundane obstacles she sees are many farmers still preferring to use hard copies and physical files. In other instances, human errors in copy-and-pasting or manually reentering data can cause issues and additional costs. With many rural areas in New Zealand having patchy mobile reception, Erin notes that Internet connectivity is another key constraining factor for accelerating on-farm digitalization. The company installed signal boosters in several places, but some digital tools are still hard to utilize to their fullest potential. In her view, connectivity issues will likely limit widespread uptake of existing digital technologies and prevent potential innovation to some degree.

3.4 Synthesis of Current State

These brief accounts of three New Zealand farm systems demonstrate that the uptake of digital technologies remains partial and uneven between different operations and business models (see Table 1). Much of this uneven adoption is driven by a range of technical, farm system, economic, and sociocultural factors, all of which affect the perceived and actual value digital technologies might add to commercial farm operations against a background of their implementation challenges. For small-scale farmers like Derrick, digital technologies and software platforms offer limited value, given his relatively simple operation, while their adoption requires considerable efforts and costs compared to his overall operating budget. Conversely, digital technologies can create substantial value in the type of large-scale commercial operations Erin supports, given their significant data requirements for regulatory reporting and informed decision-making. Limiting factors in these farm systems manifest, in large parts, around lacking interconnectivity and interoperability between digital systems.

Whether digitalization can yield on-farm benefits, and what associated implementation challenges must be navigated, thus varies between farm systems. Therefore, current levels of farm digitalization in New Zealand differ significantly. Our anecdotal insights correspond to those presented by other New Zealand researchers and industry bodies (e.g., AgriTechNZ 2022; Eastwood et al. 2023; Shepherd et al. 2020). This work illustrates that digitalization by itself is not resulting automatically in net benefits for all farmers equally. This emphasizes the need to carefully examine the value proposition of digital agricultural innovation, with due consideration of socio-technical contexts and the specificities of each farm system.

Table 1 Three contrasting farm systems in New Zealand

Criteria	Farm system 1	Farm system 2	Farm system 3
Type of enterprise	Sheep farm	Mixed crop and livestock farm (cereals, seeds, vegetables, and lamb finishing)	Corporate dairy business
Size	40 ha	450 ha	More than 20 farms, ranging from 550 to 1200 cows
Technology on farm	Low-tech operation with most on-farm work undertaken by contractors, such as agrichemical application	Several irrigators Topcon equipment on sprayerTrimble on tractors Data loggers on irrigation wells	Pivots irrigators. Protrack automatic drafting gates. Fertilizer spreading capturing data. Pasture metering, recoding place and yield. Halter collars on cows: automate herd movements, remote heat and health detection
Software	Mostly paper based and happy with where things are Farm does not require Farm Environmental Plan auditing	ProductionWise for data recording, calculations of production, and financial indicatorsOverseer for estimation of nitrogen losses for Farm Environmental Plan auditing (irrigation resource consent)	Overseer for estimation of nitrogen losses for Farm Environmental Plan auditing (irrigation resource consent). MINDA: animal health records. FarmIQ: farm management and recording software. Greentree: accounting and quality assurance reporting. Patika: data warehouse to bring information together Power BI: to display information
Issues identified around ICT	Recording software is too expensive and not fully justified for small operations, yet some of the capabilities are necessary	Purchase and use of agrichemicals not linked to recording systems and different providers. Irrigators (multiple) record water application and location, but data must be manually transferred to the recording system. Multiple simultaneous operations that need recording. Contractors collect information that has to be transferred manually. Underutilized capability on smart phones to record data on the go. Duplication of data entry. Farm Environment Plans (compulsory auditing) and Overseer require much of the same data. Retrospective information for compliance not well used for planning forwards. Data recording for compliance is a small part of costs, but it is growing fast. Data recording for quality assurance not utilized for obtaining better prices. Less than optimal post-sale service of equipment and software. Operating sophisticated equipment, but not feeling confident with data handling	Data double handling (copy-and-paste or reentering data). Systems do not talk to each other. Getting all the information in one place for auditing, with records sitting in many different places. IT providers seen to hoard data that farmers feel they own, as they generated it. Concerns about data protection to ensure New Zealand farmers' competitive advantage. Questions on how to ensure data is accurate. Variable connectivity and Internet access between farms

4 Future Imperfect and Possible Pathways

As the case studies described in the previous section testify, the big Agriculture 4.0 dream is not here yet, not even in the more modest form discussed in Sect. 2. The challenges described by farmers in the cases above are not isolated examples. A literature review by Eastwood et al. (2023) explained in detail a series of barriers for digitalization in agriculture. They concluded that barriers are not simply technical. Rather, they occur at several interconnected levels, including at the farm, societal, innovation system, and policy level. Similarly, da Silveira et al.'s (2021) systematic literature review classified the barriers for the development of Agriculture 4.0 into technological, economic, political, social, and environmental. The same set of barriers was identified in a recent study specific for New Zealand (AgriTechNZ 2022).

While summarizing all these barriers and their causes is beyond the present scope, it is clear that some of the underlying causes lay in dominant business models associated with agricultural technologies and particularly how value is extracted from data beyond the direct use for which it was collected. We focus here on some of the key barriers associated with these business models.

First, there are no clear data ownership policies and technical tools to implement multi-stakeholder data management in practice. The agricultural data ecosystem is very complex, with often multiple parties in the data value chain. As mentioned in Sect. 3, even in simpler farm scenario, some data is collected and managed by farmers themselves and some by multiple third parties. Factors to consider while attributing data ownership in these contexts are (1) on-farm hardware cost, (2) data system and management cost (including cybersecurity), and (3) data integration and analysis cost. For example, let's consider a scenario where farmers pay a contractor to apply fertilizer on a crop and to a service company to collect related data generated in the process. The service company is responsible for data management, integration, and analysis. However, it is the contractors' work that initially generates the data within the context of the farmers' businesses. Three actors can thus understandably claim some ownership of the data.

Second, different actors in the agricultural data value chain can have different attitudes toward data sharing. For example, some farmers express scepticism about the benefits of data sharing, with arguments such as the following: (1) regulators may use it against me, (2) data is very valuable for me and if I share them I will lose competitive advantage, and (3) most value from shared data is extracted by downstream actors in the value chains (i.e., by processors and retailers) with only limited value for me (see, e.g., Jakku et al. 2019). These arguments are understandable, and different strategies are required to address related concerns.

Third, current digital systems in agriculture often lack interoperability despite the value of data increasing significantly when integrating it from multiple components within a farm or across multiple farms (e.g., across a region or an entire sector). Integration is currently often cumbersome and expensive. One can distinguish device interoperability and software interoperability on farms. In the case of device interoperability, on-farm devices and sensors are relatively new developments, and

ISO standards regarding data collection, including metadata, are often nonexistent. Each company therefore collects data with their own standards. In terms of software interoperability, multiple companies provide services to farmers and collect data using their own proprietary software and user interfaces. These companies are often competitors, which means they are unwilling to provide easy ways for interoperability with other software.

Fourth, even if a perfect formula to attribute data ownership can be designed and farmers are willing to share data, there is currently a lack of digital tools to implement and monetize the necessary exchanges. To create a more dynamic data economy that supports the full value propositions of agricultural digitalization, a methodology to attribute the right value of the data needs to be designed. Key issues that need answering in this regard include the following: (1) what is the cost of data collection, storage, management, integration, and analysis? (2) Who is benefiting from it? (3) The fact that data is a non-rival good, i.e., data can be shared and used countless times (Romer 1990; Samuelson 1954).

Given these barriers to realizing the value of agricultural digitalization, the next section focuses on the problems and tensions in business models through what one may call a "Web3.0 lens."

4.1 Web3.0 and Its Components

The World Wide Web has undergone a number of iterations (Fig. 1). Its initial form, now described as Web1.0, comprised web pages navigated via hyperlinks and was largely static and "read only." Web2.0 offered users the ability to enter data and

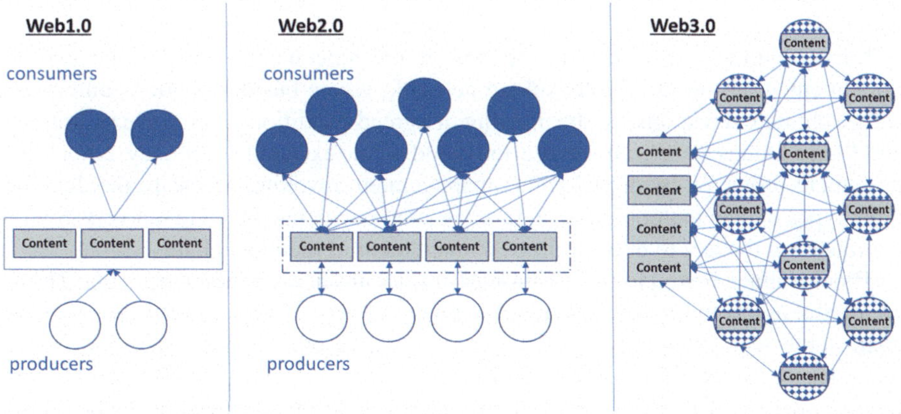

Fig. 1 Three stages of the World Wide Web so far. Web 1 is the read web – still quite small. Web 2 is the read/write web and, though larger, dominated by a number of large content producers who hold the data on centralized servers. Web 3 still has some centralized servers but is dominated by producer/consumers who also hold and directly share content

interact with websites. The "read-write" web enabled webpages to become interactive and facilitate personalized experience and targeted advertising, providing the foundation for Internet business model development.

Web3.0 refers to the third generation of the World Wide Web, with its primary vision that the Web is owned by users and creators rather than by Big Tech. The defining properties of Web3.0 are that it is (1) open (it runs on systems developed with open-source standards, (2) trustless (users can interact privately or publicly without the need for central authority, thus avoiding the risk of intermediary mishandling their data or their privacy), (3) permissionless (it provides tools and protocols for seamless interoperability and peer-to-peer networks without the need for consent from a controlling third party organization), and (4) decentralized (data is stored and managed in a decentralized manner via peer-to-peer interconnection). Key technologies enabling the Web3.0 vision are blockchain, smart contracts, non-fungible tokens, artificial intelligence, and edge computing.

Blockchain is the key enabling technology for the creation of decentralized applications and services where data is maintained. In a blockchain environment, data is managed across numerous computers (nodes) in a distributed manner. These computers are linked together in a peer-to-peer network. Data is structured into chunks (blocks) that are strung together forming a chain, and verification of the new block creation in the network is done through consensus algorithms. Each block is timestamped and cannot be deleted or changed once created, making it immutable (Rogerson and Parry 2020). Immutability of the data begins to solve the problem of trust in economic activities (Abou Jaoude and Saade 2019). Also, each node (computer) in the blockchain network stores a copy of data synchronously, which helps to increase interoperability in a data economy (Rogerson and Parry 2020). The immutable nature of data in blockchains removes the plasticity of Web2.0. Transactions, exchanges, and claims are assigned to an individual, or individuals, and locked to a point in time. These properties enable digital "smart" contracting, which facilitates the attribution of ownership and rights over digital assets.

Smart contracts are codified versions of the agreements between the parties implemented on top of the blockchain network, where business logic is automatically executed in response to the meeting of agreed conditions (Fernández-Caramés and Fraga-Lamas 2018). This execution could be the exchange of money, unlocking of content protected by digital rights management, revoking access to the data, or other types of data manipulation, e.g., automated payments linked to delivery. Smart contracts are the key to decentralized management of the blockchains.

Blockchain is linked to the creation of cryptocurrencies in popular media. These are digital assets in the form of fungible tokens: a type of digital asset that is interchangeable with other tokens of the same type. In other words, each individual token has the same value as any other token of the same type. As such, cryptocurrencies prove useful within blockchain systems as forms of payment. If they have stable value, they can be used to rapidly pay partners anywhere in the world via smart contracts. However, the value of many tokens (Bitcoin, Ethereum, etc.) has been the subject of much speculation, leading to highly volatile cryptocurrency markets. This has limited their utility and damaged user confidence. However, the development of stablecoins (those that do not rapidly change value) may lead to cryptocurrencies find extensive utility in global payment systems in the longer term.

In contrast to fungible tokens, non-fungible tokens (NFTs) have unique identification codes and metadata to distinguish them from other tokens. These cryptographic tokens may contain data that either represent or point toward digital or physical assets. Each NFT essentially provides the certificate of ownership of assets. "Tokenizing" assets makes buying, selling, and trading more efficient. Trading on blockchain helps these exchanges happen in a trusted manner, reducing the need for central authorities. As with the fungible "crypto" tokens, NFTs have been through a media hype cycle, mainly linking them with digital art. There are also issues to address, not least the legal standing, or lack thereof, for NFTs (Moringiello and Odinet 2021). However, longer term they may be usefully linked to physical assets (machinery, land, livestock, seed, etc.) and to aid efficient trade.

Automation of services based on artificial intelligence (AI) is critical for successful implementation of Web3.0-based applications and services. AI technologies, such as machine learning, deep learning, big data, and natural language processing, help empower more intelligent and adaptive applications, such as smart autonomous farm vehicles that navigate to a field and execute tasks as appropriate to conditions.

Web3.0 embraces the Internet of Things (IoT), the network of physical objects, such as devices, vehicles, and buildings, that are equipped with sensors, software, and other technologies for the purpose of collecting and exchanging data. This is expanded to what is called "Industry 4.0," which envisions the development of IoT by industry, creating cyber-physical systems (CPS) that integrate information, communication, and industrial technologies (Afy-Shararah and Rich 2018; Lu 2017). Digital and intelligent systems combine to provide data-driven and customizable processes that support organizations. Industry 4.0 further enables the creation of new flexible production models that provide firm-specific products and services (Zhou et al. 2015; Kache and Seuring 2017). Connecting equipment with sensors removes people from the data input process. Such automation provides supply chains with potentially more accurate and timely information. Such data can reveal the patterns of use and consumption of products and services, allowing for process monitoring, forecasting, and better management of reverse supply processes for recycling, waste management, and waste reduction (Parry et al. 2016). Such advanced information technologies can create competitive advantage through production planning, flexibility, improved efficiency, and reduced cost (Ho 1996; Lei and Goldhar 1991).

Web3.0 may also help to overcome the centralization of power through the use of edge computing. Most computing devices such as mobile phones, laptops, appliances, sensors, and IoTs are capable of processing large volumes of data. Rather than the centralized data storage and manipulation seen in Web2.0, Web3.0 utilizes these devices as nodes in the network. Computation happens not at the center, but at the edge, and data storage and processing occurs in a distributed fashion. This approach potentially reduces the power of platform providers, hence acting as key enablers for the implementation of distributed, and potentially more democratic and user empowered, Web3.0 applications and services.

4.2 No Panaceas

While often still only beginning to be realized on-farm, the promises of Web3.0 are already a reality elsewhere in the food industry. Blockchain has been successful in enhancing visibility of processes along supply chains, reducing human error, increasing efficiency, and enhancing trust in product veracity (Rogerson and Parry 2020). However, as with any novel solution to intractable problems, practical challenges remain. These may be resolved as use of such technologies matures, but they may also include challenges that simply cannot be resolved by technological improvements alone.

First, a fundamental problem we see in the digitization literature is the *risk of error or fraud* in the entry of data – the nexus of the physical and digital worlds. Data might be more trusted if they are retained in blockchain or what are considered advanced digital systems, but such systems may create a "halo of trust." The veracity of data remains dubious if the way they were originally collected or entered into the system is potentially flawed or biased (Rogerson and Parry 2020). Data entry issues highlight the need for governance standards, but also opportunities for automation, reducing workload on farmers and ensuring that data originating with them are trustworthy.

Second, once data is entered into systems, there is a *centralizing tendency* of which to beware in order to guard against inefficiencies and abuses. With so many different proprietary systems, multiple different centralized databases can operate in parallel. As shown in the farm stories above, this means that data may get stuck in silos with a specific service provider as data controller or that data is not able to flow and be reused in combination. This creates rework for the farmer and reduces the promised efficiency. Data can be copied into different databases, but that is effectively a bridge between Industry 4.0 sensors and Web2.0 centralized databases. Even within decentralized blockchain systems, centralization of data mining can occur, which creates or maintains power imbalances. What is required is a truly decentralized Web3.0 system where sensor data is stored locally, with the farmer as the data controller. Instead of individual services "taking" their data and processing it centrally, they could locally host their algorithms such that processing is done on the farm, creating a system of decentralized edge computing. Freeing the data from centralization would allow greater value to be realized through the combination of datasets to provide novel insights envisioned and required by the agri-food sector.

Third, not only is centralization a risk to the sharing of value, but it also limits *farmers' ability to adopt new technologies independently* for fear of both up-front and ongoing costs. The case examples in Sect. 3 show that much of the collection and management of data does not improve on-farm efficiency or benefit the farmer.

In fact, it often absorbs time and money with little benefit. More work is required to identify what is truly needed and what is "digital waste," such that only minimum-viable datasets are collected. Those key data then need to be available to be combined, transformed, and employed to create insights. However, technology service providers perceive the data their services collect as valuable, and they are reluctant to make available what they perceive as their source of competitive advantage. Changing this business model to gain access may require legislative changes and, if these were possible, data format standardization for transferring data between systems via open application programming interface (AKA APIs). For the data to then be useful, farmers would need to develop skills to manipulate data and extract value. This is now within reach as software developers have created "low-code" systems that require little to no coding skills yet allow users to create applications and analytic processes. Rather than programming languages, low-code systems use visual interfaces with drag-and-drop tools that enable users to undertake complex data analytics and visualizations. The engagement of end users as innovators, and perhaps coders and data analysts, makes agricultural digitalization processes more inclusive and reinforces the value of diverse actor groups within the socio-technical futures digital technologies help to create.

Finally, the potential usefulness of the smart contract technology must be treated with certain caution (Zheng et al. 2020). Despite the name, *smart contracts are not "smart*." They are often simple "if this then that" functions that execute automatically. Malicious actors may trigger them artificially to gain benefit. Smart contracts are difficult to read, as they are written as code, and are also difficult to change. As digital systems are developed over time, multiple smart contracts may end up chaining together, making processes difficult to predict. Neither are they directly legal contracts. Rather, smart contracts are representations of the contractual terms, and their relevance or legality may be open to interpretation by different partners or courts (Dixit et al. 2022).

4.3 Decentralization Paradox

Smart contracts are agreements between two or more parties that can be automatically enforced without the need for an intermediary and thus are a form of decentralized control in a blockchain network. Contracts use agreed objectives as triggers for actions such as payment, for example. The original design philosophy of blockchain as a disintermediating technology with decentralized control sought to remove centralized power. However, care is needed when "decentralization" is used in discussions of blockchain. Control can be decentralized through democratizing governance structures; through decentralizing data by distributing them across a network; and through offering availability of data, and therefore visibility, to different stakeholders. The type and level of decentralization will depend on the design of the system. A (de)centralization paradox exists within blockchain design, as in practice centralizing tendencies compete with the decentralizing philosophy as seen in public blockchain mining cooperatives and groups of core developers holding design authority over changes to the code. Pragmatically, the benefits of decentralization should be coupled with necessary centralization to develop optimized systems.

5 Conclusion

In this chapter, we discussed the increasingly ubiquitous presence that digital technologies have on many modern farms and the promises they hold. But we also presented New Zealand case studies which demonstrate that the adoption of these technologies, and their offered business value, varies significantly between farm systems. This is partly due to farmers' ability to navigate relevant implementation obstacles but also wider socio-technical challenges associated with realizing the full value of digital technologies. Web3.0 approaches, and the vision of openness and decentralization they portray, offer some avenues to resolve these challenges. However, our discussion demonstrates that there is no single magical solution to address the diverse array of intersecting technical, farm system, economic, and sociocultural factors outlined in this chapter. Effectively overcoming associated challenges and focusing on the practical implementation of digital technologies will be crucial for realizing the envisioned beneficial futures of agricultural digitalization. This includes attending to wider considerations for the ways in which digital technologies can support long-term environmental, social, and economic sustainability.

References

Afy-Shararah, M., Rich, N. 2018. Operations flow effectiveness: a systems approach to measuring flow performance. International Journal of Operations & Production Management.

AgriTechNZ. 2022. Baseline of Digital Adoption in Primary Industries. Accessible at https://agritechnz.org.nz/wp-content/uploads/sites/34/2022/08/AgriTechNZ-Digital-Adoption-in-Primary-Industries-Report-2022.pdf

Barrett, H. and Rose, D.C. 2022. Perceptions of the fourth agricultural revolution: What's in, what's out, and what consequences are anticipated? Sociologia Ruralis, 62(2), pp.162–189.

Bronson, K. 2019. Looking through a responsible innovation lens at uneven engagements with digital farming. NJAS-Wageningen Journal of Life Sciences, 90, p.100294.

Chakravorti, B., and Chaturvedi, R.S. 2017. Digital planet 2017: How competitiveness and trust in digital economies vary across the world. Retrieved from https://sites.tufts.edu/digitalplanet/files/2020/03/Digital_Planet_2017_FINAL.pdf

Cook, S., Jackson, E.L., Fisher, M.J., Baker, D. and Diepeveen, D. 2022. Embedding digital agriculture into sustainable Australian food systems: pathways and pitfalls to value creation. International Journal of Agricultural Sustainability, 20(3), pp.346–367.

da Silveira, F., Lermen, F.H. and Amaral, F.G. 2021. An overview of agriculture 4.0 development: Systematic review of descriptions, technologies, barriers, advantages, and disadvantages. Computers and Electronics in Agriculture, 189, p.106405.

Dixit, Deval, V., Dwivedi, V., Norta, A., & Draheim, D. 2022. Towards user-centered and legally relevant smart-contract development: A systematic literature review. Journal of Industrial Information Integration, 26, 100314–. https://doi.org/10.1016/j.jii.2021.100314

Eastwood, C., Klerkx, L., Ayre, M. and Dela Rue, B. 2019. Managing socio-ethical challenges in the development of smart farming: from a fragmented to a comprehensive approach for responsible research and innovation. Journal of Agricultural and Environmental Ethics, 32(5), pp.741–768.

Eastwood, C., Turner, J.A., Romera, A.J., Selbie, D., Henwood, R., Espig, M. and Wever, M. 2023. Transitioning from digital agriculture to a digital bioeconomy: a review of multi-scale barriers to digitalisation of the bio-based economy. CABI Reviews, https://doi.org/10.1079/cabireviews.2023.0002.

Espig, M., Fielke, S., Finlay-Smits, S.C., Jakku, E., Turner, J.A., Robinson, C., Hunter, C. and Lacey, J. 2022. Responsible digital agri-food innovation in Australian and New Zealand public research organisations. Sociologia Ruralis, 62(2), pp. 389–409.

Fernández-Caramés, T.M. and Fraga-Lamas, P. 2018. A Review on the Use of Blockchain for the Internet of Things. IEEE Access, 6, pp. 32979–33001.

Fielke, S., Bronson, K., Carolan, M., Eastwood, C., Higgins, V., Jakku, E., Klerkx, L., Nettle, R., Regan, Á., Rose, D.C. and Townsend, L.C. 2022. A call to expand disciplinary boundaries so that social scientific imagination and practice are central to quests for 'responsible' digital agri-food innovation. Sociologia Ruralis, 62(2), pp.151–161.

Fleming, A., Jakku, E., Lim-Camacho, L., Taylor, B. and Thorburn, P. 2018. Is big data for big farming or for everyone? Perceptions in the Australian grains industry. Agronomy for sustainable development, 38(3), pp.1–10.

Fleming A, Jakku E, Fielke S, Taylor BM, Lacey J, Terhorst A, and Stitzlein C. 2021. Foresighting Australian digital agricultural futures: Applying responsible innovation thinking to anticipate research and development impact under different scenarios. Agricultural Systems, 190, 103120. https://doi.org/10.1016/j.agsy.2021.103120

Ho, C.-F. 1996. Information technology implementation strategies for manufacturing organizations: A strategic alignment approach. International Journal of Operations & Production Management 16, 77–100.

Jakku, E., Fleming, A., Espig, M., Fielke, S., Finlay-Smits, S.C. and Turner, J.A. 2023. Disruption disrupted? Reflecting on the relationship between responsible innovation and digital agriculture research and development at multiple levels in Australia and Aotearoa New Zealand. Agricultural Systems, 204, p.103555.

Jakku, E., Taylor, B., Fleming, A., Mason, C., Fielke, S., Sounness, C. and Thorburn, P. 2019. "If they don't tell us what they do with it, why would we trust them?" Trust, transparency and benefit-sharing in Smart Farming. NJAS-Wageningen Journal of Life Sciences, 90, p.100285.

Abou Jaoude, J. and Saade, R.G. 2019. Blockchain applications–usage in different domains. IEEE Access, 7, pp.45360–45381.

Kache, F., Seuring, S. 2017. Challenges and opportunities of digital information at the intersection of Big Data Analytics and supply chain management. International Journal of Operations & Production Management 37, 10–36.

Klerkx, L. and Rose, D. 2020. Dealing with the game-changing technologies of Agriculture 4.0: How do we manage diversity and responsibility in food system transition pathways?. Global Food Security, 24, p.100347.

Lei, D., Goldhar, J.D. 1991. Computer-integrated manufacturing (CIM): redefining the manufacturing firm into a global service business. International Journal of Operations & Production Management 11, 5–18.

Lioutas, E.D., Charatsari, C. and De Rosa, M. 2021. Digitalization of agriculture: A way to solve the food problem or a trolley dilemma?. Technology in Society, 67, p.101744.

Lajoie-O'Malley, A., Bronson, K., van der Burg, S., and Klerkx, L. 2020. "The future(s) of digital agriculture and sustainable food systems: An analysis of high-level policy documents." Ecosystem Services 45:101183. https://doi.org/10.1016/j.ecoser.2020.101183.

Lu, Y. 2017. Industry 4.0: A survey on technologies, applications and open research issues. Journal of Industrial Information Integration 6, 1–10. https://doi.org/10.1016/j.jii.2017.04.005

Ministry of Business, Innovation and Employment. 2020. Agritech Industry Transformation Plan. Retrieved from https://www.mbie.govt.nz/dmsdocument/11572-growing-innovative-industries-in-new-zealand-agritech-industry-transformation-plan-july-2020-pdf

Mondejar, M.E., Avtar, R., Diaz, H.L.B., Dubey, R.K., Esteban, J., Gómez-Morales, A., Hallam, B., Mbungu, N.T., Okolo, C.C., Prasad, K.A. and She, Q. 2021. Digitalization to achieve sustainable development goals: Steps towards a Smart Green Planet. Science of The Total Environment, 794, p.148539.

Moringiello, Juliet M. and Odinet, Christopher K., The Property Law of Tokens (November 1, 2021). 74 Florida Law Review 607 (2022), U Iowa Legal Studies Research Paper No. 2021-44, Widener Law Commonwealth Research Paper, Available at SSRN: https://ssrn.com/abstract=3928901 or https://doi.org/10.2139/ssrn.3928901

Parry, G., Brax, S.A., Maull, R., Ng, I. 2016. Visibility of consumer context: improving reverse supply with internet of things data. Supply Chain Management: An International Journal, 21, 228–244.

Rogerson, M., & Parry, G.C. 2020. Blockchain: case studies in food supply chain visibility. Supply Chain Management: An International Journal, 25(5), 601–614.

Romer, P. 1990. Endogenous technological change. Journal of Political Economy, 98(5), Part 2, S71–S102.

Samuelson, Paul A. 1954. The Pure Theory of Public Expenditures. The Review of Economics and Statistics, 36, 387–389.

Shepherd, M., Turner, J.A., Small, B., and Wheeler, D. 2020. Priorities for science to overcome hurdles thwarting the full promise of the 'digital agriculture' revolution. Journal of the Science of Food and Agriculture 100 (14):5083–92. https://doi.org/10.1002/jsfa.9346.

van der Burg, S., Bogaardt, M.J., and Wolfert, S. 2019. "Ethics of smart farming: Current questions and directions for responsible innovation towards the future." NJAS - Wageningen Journal of Life Sciences 90–91:100289. https://doi.org/10.1016/j.njas.2019.01.001.

Wolfert, S., Ge, L., Verdouw, C. and Bogaardt, M.J. 2017. Big data in smart farming–a review. Agricultural Systems, 153, pp. 69–80.

Wolfert, S., Verdouw, C., van Wassenaer, L., Dolfsma, W., and Klerkx, L. 2023. "Digital innovation ecosystems in agri-food: design principles and organizational framework." Agricultural Systems, 204:103558. https://doi.org/10.1016/j.agsy.2022.103558.

Zheng, Xie, S., Dai, H.-N., Chen, W., Chen, X., Weng, J., & Imran, M. 2020. An overview on smart contracts: Challenges, advances and platforms. Future Generation Computer Systems, 105, 475–491. https://doi.org/10.1016/j.future.2019.12.019

Zhou, K., Liu, T., Zhou, L. 2015. Industry 4.0: Towards future industrial opportunities and challenges, in: 2015 12th International Conference on Fuzzy Systems and Knowledge Discovery (FSKD). Presented at the 2015 12th International Conference on Fuzzy Systems and Knowledge Discovery (FSKD), pp. 2147–2152. https://doi.org/10.1109/FSKD.2015.7382284

A Review of Innovative Technologies for Sustaining Water Catchment Areas: Toward Sustainability Development

Mariam I. Adeoba and Opeyemi C. Fatayo

1 Introduction

Water is one of the most crucial components of sustainable agriculture in terms of the industry and its effects on the natural environment and other water users. The catchment area remains the primary component of any water supply system. The area of land where all surface water joins a body of water, such as a river, is known as a catchment area. This area of land serves as the source of water and sediment flow through the river (Pant et al. 2018). Water catchments provide humanity with a variety of unique services, such as freshwater supply and partial purification, habitat for flora and fauna, flood control, flow regulation, aesthetic pleasure, and cultural, religious, and inspirational benefits (NHMRC 2017). The quality of water supplied by a catchment is highly affected by the natural characteristics of the catchment, land use, the efficiency of the water quality protection system, and the general state of the environment (Almaarofi et al. 2017). The state of a catchment determines the healthiness of all components that depend on it (Charrière and Aumond 2016).

Also, there is a huge challenge ahead of us to generate nearly 50% more food by 2030 and double output by 2050. As a result, it will be critical in the future for

M. I. Adeoba (✉)
Unisa Biomechanics Research Group, Department of Mechanical Engineering, College of Science Engineering and Technology (CSET) University of South Africa, Pretoria, South Africa

Department of Environmental Management, Pan African University Life and Earth Sciences Institute (Including Health and Agriculture) PAULESI, University of Ibadan, Ibadan, Nigeria

O. C. Fatayo
Department of Environmental Management, Pan African University Life and Earth Sciences Institute (Including Health and Agriculture) PAULESI, University of Ibadan, Ibadan, Nigeria

I. S. Dunmade et al. (eds.), *Sustainable Engineering*, Green Energy and Technology, https://doi.org/10.1007/978-3-031-47215-2_2

farmers to receive the appropriate signals in order to improve water-use efficiency and management, especially since agriculture is the world's largest water user, accounting for over 40% of total water withdrawals (OECD 1998). Because of the demands of rising urbanization, industry, and climate change, this will almost certainly have to be implemented with less water coupled with the need to adopt innovative technologies toward sustaining catchment areas and building resilience in case of anomalies that require systematic and structural readjustment. Sustaining catchment areas involves the use of technologies that help to actualize the "measure it to manage it" goal through the use of water meters (Tom et al. 2011; Liu and Mukheibir 2018). It also incorporates the use of sustainable building materials and the use of treatment technologies as well as smart devices capable of monitoring catchment areas with minimal human supervision (Gould 1995; Malaeb and Ayoub 2011; Abioye et al. 2022). The goal of this chapter is to explore some of the ongoing advanced technologies that are currently applicable to sustaining catchment areas toward Sustainable Development Goal 2 which seeks to achieve food for all (zero hunger) and SDG 6 of sustainable water use (clean water and sanitation).

This chapter, although built on existing scientific knowledge, adequately aggregated the disjointed pool of literature that illuminates the potentials of innovative technologies for the sustenance of water catchment areas. We explored the sectorial approach which gave an insight into the key players in and around water catchment areas, bringing water catchment sustainability efforts to the least stakeholders. This will ease further reviews to build on the coherent structure already established as new technologies emerge and existing ones improved on.

2 Methodology

A literature review was conducted for this study, featuring 29 existing literature pulled from key sectors that play important roles in the utilization and management of water catchment areas and the associated resources. These include geospatial and remote sensing technology, membrane technology, construction, artificial intelligence, and machine learning.

3 Sustaining Catchment Areas with Innovative Technology

A series of developed and emerging technologies have been adopted for catchment sustenance globally. Some of these technologies are discussed in subsequent sections below.

3.1 Geographic Information Systems and Remote Sensing Technology

Geographic information systems (GIS) and remote sensing (RS) technologies are potent drivers of solutions to the earth's system, including water catchment areas (Pullar and Springer 2000; Van Manen et al. 2009). GIS and RS technologies can be employed in several ways in catchment areas. These include climate change prediction, hazard prediction, pollution tracking, and soil erosion estimation.

Lashford et al. (2022) posited that the development of modeling approaches, such as GIS and RS technologies, can imitate fluid movement both temporally and spatially, thereby assisting major stakeholders including community administrators, disaster managers, and municipal authorities in assessing catchment sustainability. These technologies are essential for assessing and comprehending the catchment's effects outside of the immediate vicinity of the water structure. This captures the sustainable aspect of water catchment areas as far as flood hazard is concerned by developing early warning systems and building resilience against catastrophic flood events. GIS and RS technologies help prepare catchment caretakers against the adverse effects of floods through effective simulation and modeling of ever-evolving landscape (Pullar and Springer 2000).

GIS and RS technologies also help to identify pollution sources around a water catchment and provide the choice of a management option, depending on whether the pollution source is point source or a nonpoint source (Pullar and Springer 2000). Land use types around the catchment can be derived from remotely sensed data (Rozenstein and Karnieli 2011), and this can further be utilized to effectively manage the catchment to prevent intrusions of pollutants from different sources. For instance, contaminants can be leeching from a cattle dip, nutrient transport of fertilizer chemicals from agricultural lands into waterways (which can cause eutrophication), salt intrusion from marine water bodies, untreated sewage from households, and industrial effluents. GIS and RS provide a bird's-eye view of the various pollution sources, combining land use and topographic factors to inform prompt remedial actions.

Climate change has developed from being a problem to being a threat to humanity, to lives flourishing on earth, and, of course, to water catchment areas (Pham-Duc et al. 2020). Against this background, GIS and RS help in the prediction of climate change and also monitor changes associated with climate change within a catchment over time. An outstanding example that showcases the impacts of the changing climate on water catchments is the Lake Chad in Nigeria's Sahelian region. Although the size of the lake stands fairly constant in the past 20 years (facilitated by more permanent vegetation and groundwater recharge), the shrinking effects of climate change on the lake have been studied over the past years as the lake presented larger coverage during the mid-twentieth century (Pham-Duc et al. 2020). Trend analysis using these tools illuminates a downturn in annual precipitation within the Chad basin and stream flow of the lake's tributaries. Such observed

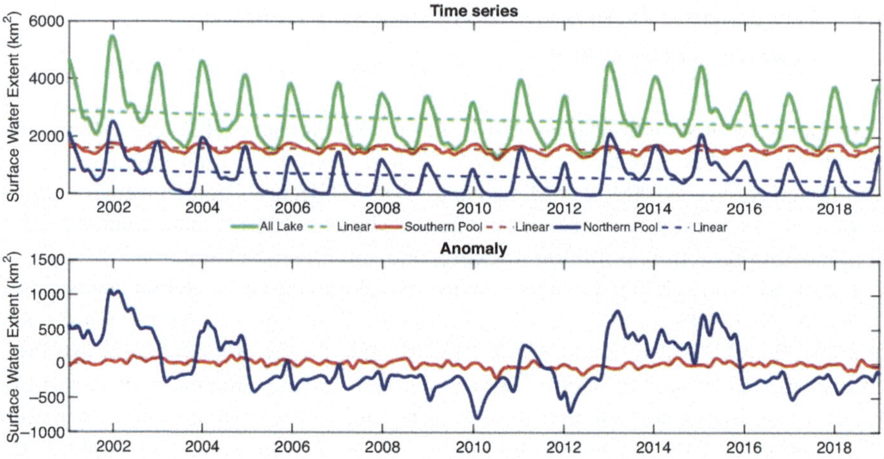

Fig. 1 Temporal (top) and irregularity (bottom) of the size of the northern plash (blue), southern plash (red), and Lake Chad (green) over a 9-year period. (Adapted from Pham-Duc et al. 2020) is an open-access article distributed under the terms of the Creative Commons Attribution License, which permits unrestricted use, distribution, and reproduction in any medium, provided the original author and source are credited

changes in the trend analysis charts (see Fig. 1) can inform decisions for water conservation.

Soil erosion modeling is also possible with GIS and RS (Mitasova et al. 2013; Kabantu et al. 2018). Soil erosion estimation can be used to determine a landscape's sediment yield and load and how they may affect water catchments through siltation, as illustrated in Fig. 2.

3.2 Drip Irrigation Technology

Drip irrigation system has been demonstrated as one of the most effective techniques for sustaining water catchment areas. In agroecosystems, drip irrigation is used for dispensing water to plants with a very low waste stream based on the specific water requirements of specific crops (Agrawal and Singhal 2015). It is now used on less than 2% of the world's irrigated land, and it has been shown to cut water use by 30–70%.

It is considered an effective technology because it uses less water, consumes less energy, and generates less waste materials. To properly conserve water in water catchment that combined agricultural practices such as animal husbandry and plant cropping, drip technology will be essential. Despite the high initial cost, the benefits of drip irrigation include time savings, the elimination of mistaken assumptions in the regulation of soil water content, boosting revenue per hectare, minimizing evaporation and water supplied to noxious plants, and crop watering to an accurate depth of root (Kavianand et al. 2016).

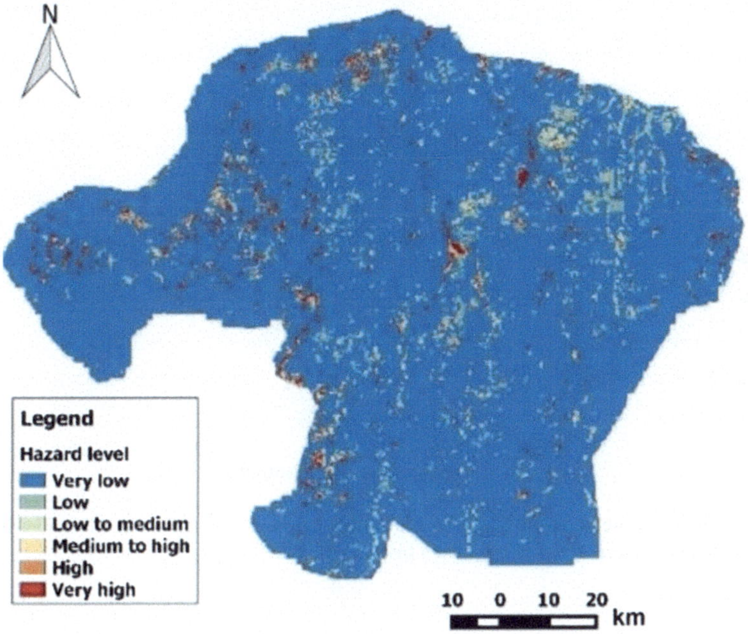

Fig. 2 GIS-based soil erosion modeling in Kinshasa, DR Congo. (Adapted from Kabantu et al. 2018) is an open-access article distributed under the terms of the Creative Commons Attribution License, which permits unrestricted use, distribution, and reproduction in any medium, provided the original author and source are credited

3.3 Artificial Intelligence for Metering, Measuring, and Managing

Digital and intelligent meters have been integral parts of water catchments in the relatively recent past. Globally, they play important roles in improving, managing, and sustaining water catchment areas (Liu and Mukheibir 2018). Water meters have the potential to send water consumption information electronically frequently, which are subsequently used to generate insight into water consumption and recharge levels over a period of time (Figs. 3 and 4). They help in the quantification of reserve and consumption for effective catchment management.

According to Tom et al. (2011), water meters have been able to save about 24.1% of water that would have gone to waste weekly in Sydney, Australia. In their review paper on metered catchments, Liu and Mukheibir (2018) found that mean water savings across reviewed studies is about 5.5% (disregarding the oddities at the extremes).

Gamification is an additional strategy employed by some digital water metering innovations to retrieve usage data across user profiles. This is evident in the Dubuque portal of Iowa, where elements of gamification have presented success in water saving as studied by Erickson et al. (2012). The game feature includes matching a

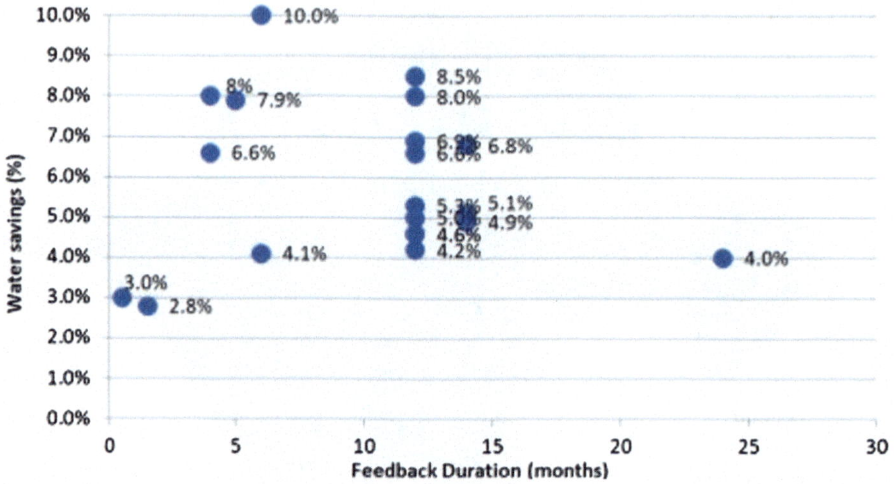

Fig. 3 Percentage water conservation vs. reporting period. (Adapted from Liu and Mukheibir 2018 with the permission given to be reused by the author)

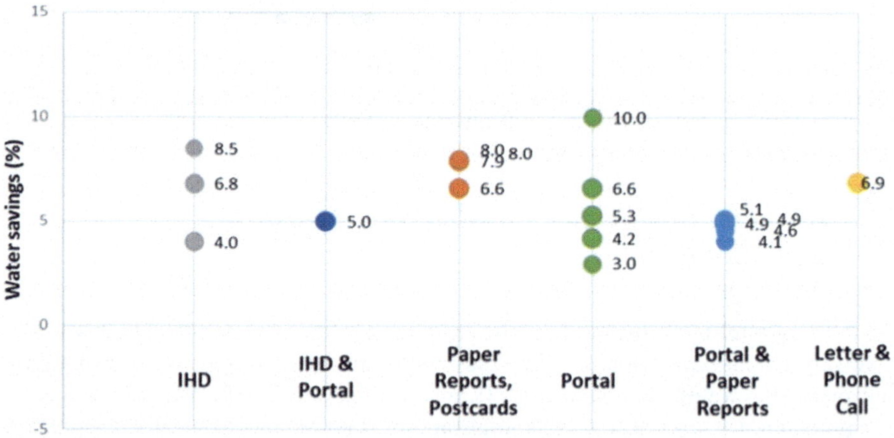

Fig. 4 Percentage water conservation vs. reporting channel. (Adapted from Liu and Mukheibir 2018 with the permission given to be reused by the author). Note. IHD means in-home display

number of water users in a catchment with the goal of using less water than the opposing teams on a weekly basis. The observed water-saving rate among users was about 6.6%, and studies have further shown that this strategy, if implemented for an extended period of time, can guarantee considerable water conservation.

The recent advancements in and incorporation of *artificial intelligence (AI)*, most especially *machine learning (ML) and data analytic (DA) technologies*, into water catchments have increased the sustainability of catchment areas (Rahim et al. 2020). ML and DA effectively leverage on the bulk data generated by water meters to sustain water catchments by giving a critical analysis and forecast of water demand, socioeconomic analysis, behavior analysis (of both the catchment in terms of fluctuations and irregularities and the users based on profile), and water-use feedback. ML and DA have been proven to facilitate water-saving strategies and more efficient demand management, such as water-conscious behaviors, through user profiling (farmers, pastoralists, hunters, households, religious groups) to promote water-conscious behaviors. Artificial intelligence also aids plant breeders to uncover complications such as leakages within the irrigation system with a view to sustaining water catchment areas.

3.4 Nanotechnology and Catchment Sustenance

Silver nanoparticle (AgNp)-coated filters made with nanotechnology are economical and are used to purify water. *E. coli* in water is eliminated by the highly efficient antibacterial property of silver nanoparticles. Owing to the mechanical, catalytic, and antimicrobial properties of nanoscale materials, effective techniques have been built around them to contribute significantly to decontamination, filtration, and desalination processes in catchment areas in Fig. 5 (Praun-Pereira and Backx 2021). Initially, nano-membrane technology was employed for desalination processes, but further advances have shown that they are instrumental in the decontamination of effluents by pesticides. In nanobiotechnology, the flexible filtration membranes installed in bioreactors conveniently retain inorganic particles and organic matter,

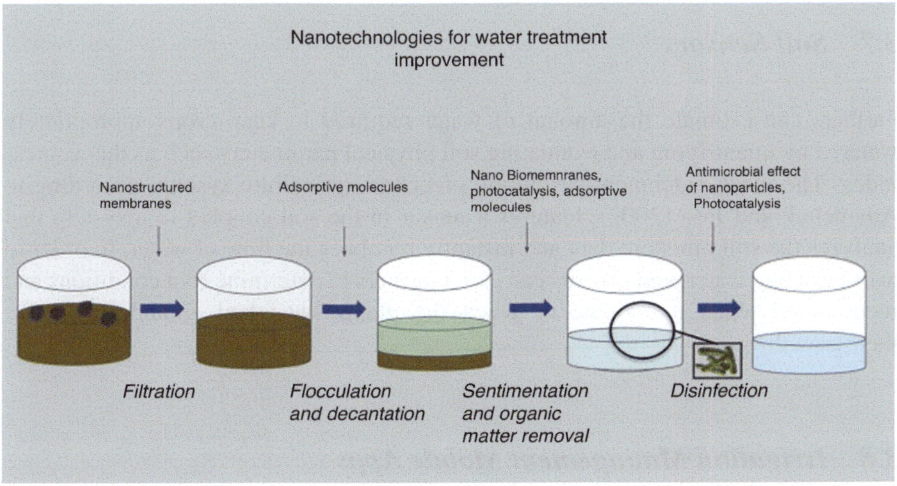

Fig. 5 Nanotechnology for water treatment in catchment areas. (Adapted from Praun-Pereira and Backx 2021 © 2021 Praun Pereira BM, et al. is an open-access article distributed under the terms of the Creative Commons Attribution License, which permits unrestricted use, distribution, and reproduction in any medium, provided the original author and source are credited)

which are utilized more often in sewage treatment. Nanotechnology has an enhanced capability for biomass retention and thus can help with water disinfection before being fed into catchment areas.

3.5 *Reverse Osmosis Technology*

Reverse osmosis (RO) technology is employed mainly for desalination in catchment areas. More than 95% of dissolved salts are left off in the waste stream after a high-pressure pump raises the pressure on the salt side of the RO system and drives the water across the semipermeable RO membrane (Malaeb and Ayoub 2011). In simple terms, when pressure is applied to unfiltered water, a semipermeable membrane in reverse osmosis eliminates pollutants.

3.6 *Greening of Water Infrastructure*

Heck (2021) highlighted the importance of greening water infrastructures for the purpose of storm retention, even in the face of racial tussle. Green catchment infrastructures act as a storm buffer, reduce excess evaporation, provide stability for catchment banks, and help to filter water (Pham-Duc et al. 2020; Brears 2021). They also help to maintain carbon neutrality within the catchment.

3.7 *Soil Sensors*

Farmers can estimate the amount of water required to keep crops appropriately watered by quantifying and evaluating soil physical parameters such as the wetness index. The ideal, catchment-sustaining precision agriculture system, according to Adamchuk and Jasa (2002), features a sensor in the soil coupled to a system that analyzes the soil moisture data and instantly regulates the flow of water. In order to avoid wasting water, soil sensors can assist farmers to determine root conditions and recommend when to irrigate or when the plants have had enough water for their growth.

3.8 *Irrigation Management Mobile Apps*

The most widely used surface irrigation technique uses the largest quantity of water in agriculture sector (Pramanik et al. 2022). The quality and quantity of mobile apps that help farmers monitor their irrigation systems is constantly rising (Abioye et al.

2022). For instance, FieldNET Mobile enables subscribers to operate and closely track irrigation equipment from their iPhone or Android devices, allowing farmers to instantly alter their irrigation in response to changing conditions. Complemented by data from regional weather stations, Smart Irrigation Apps designed by agricultural researchers at the University of Georgia assist southeast farmers in planning irrigation based on specific water requirements of crops. According to Coolong et al. (2018), the smart irrigation software assisted farmers in maintaining crop yields while conserving water and nutrients.

3.9 Systems Design and Construction Techniques

The revamped regard for water catchment technologies has been successful in terms of innovative technology in both the design of the systems and construction techniques. The use of rainwater catchment systems can be traced to ancient times. Roof catchment tanks can provide good quality rainwater clean enough for anyone to drink, farm, and feed herds, so long as nontoxic materials are incorporated into the design (Khoury-Nolde 2010). It is crucial to have a hygienic, impermeable roof composed of nontoxic materials. Lead and asbestos roofs, as well as roofs coated with lead-based paint, should be avoided. According to Ngigi (2003), upscaling rain water harvesting technologies in river basins has the potential of enhancing agricultural production and water availability in regions of eastern Africa characterized with shortage of rainfall.

4 Appraisal of the Innovative Technologies for Water Catchment Sustenance

One major precedence of the technologically managed water catchment areas over those not managed technologically is the very minimal human supervision involved. Although some of these technologies such as the reverse osmosis and nanotechnology may have high capital and maintenance cost, they have shown great potentials for maintaining and improving water quality and quantity where they have been adopted. This has ensured water security in such catchments and improving water availability for irrigation, which is in tandem with the fulfillment of Sustainable Development Goal 6 (clean water and sanitation) and Sustainable Development Goal 2 (zero hunger). Further, owing to their explained inherent abilities to, to an extent, remove physical, chemical, and bacteriological contaminants from water before being fed into catchment areas, these technologies indirectly help to ensure public health and safety, aligning them with Sustainable Development Goal 3 (good health and well-being). Gamification of water meters and the associated competition has also encouraged responsible consumption of water and reduced superfluous uses among the various stakeholders interested in water catchments.

On the contrary, water catchment areas not managed with the ever-advancing pool of innovative technologies tend to have lesser potentials for maintaining or improving water quality and may in turn jeopardize public health and safety. Irresponsible consumption of water will also be encouraged, especially in areas where water is generally available for the most part of the year. Howbeit, these technologies cannot totally downplay the need for humans to continue to get involved in the management of water catchment areas.

5 Conclusion

In summary, ensuring the sustainability of catchment areas cannot be done in isolation from innovative technologies. These technologies provide platforms for prospecting, forecasting, and managing catchments with minimal human supervision. Thus, catchment areas under innovative technologies are most likely to be more sustainably managed and resilient than others with little or no technological base. Therefore, it is recommended that water catchment areas be furnished with technological outfits capable of water filtering, hazard prediction, and effective water management systems to avert crises associated with water shortage and pollution.

References

Abioye, E. A., Hensel, O., Esau, T. J., Elijah, O., Abidin, M. S. Z., Ayobami, A. S., Yerima, O., & Nasirahmadi, A. (2022) Precision Irrigation Management Using Machine Learning and Digital Farming Solutions. ***AgriEngineering, 4(1), 70–103.*** https://doi.org/10.3390/agriengineering4010006

Adamchuk, V.I. and Jasa, P.J.(2002) EC02-178 Precision Agriculture: On-the-Go Vehicle-Based Soil Sensors. ***Historical Materials from University of Nebraska-Lincoln Extension*, p.706.** http://precisionagriculture.unl.edu/

Agrawal, N., & Singhal, S. (2015) Smart drip irrigation system using raspberry pi and arduino. ***International Conference on Computing, Communication & Automation.*** https://doi.org/10.1109/ccaa.2015.7148526

Almaarofi, H., Etemad-Shahidi, A. and Stewart, R.A., (2017) Strategic evaluation tool for surface water quality management remedies in drinking water catchments. ***Water*, *9* (10), p.738.** https://doi.org/10.3390/w9100738

Brears, R.C. (2021) Chapter 5 Greening of grey water infrastructure. ***Water Resources Management, 63–84.*** https://doi.org/10.1515/9783110685640-005

Charrière, S., & Aumond, C. (2016) Managing the drinking water catchment areas: the French agricultural cooperatives feedback. ***Environmental Science and Pollution Research, 23(11), 11379–11385.*** https://doi.org/10.1007/s11356-016-6639-8

Coolong, T., Miller, L. and Vellidis, G., (2018) Using Smartphone Technologies to Manage Irrigation. ***Irrigation in Agroecosystems. IntechOpen.*** https://doi.org/10.5772/intechopen.77304

Erickson, T., Podlaseck, M.E., Sahu, S., Dai, J.D., Chao, T., Naphade, M., (2012) The Dubuque water portal: evaluation of the uptake, use and impact of residential research water consumption feedback, Austin Texas, USA, ACM. ***CHI 2012, May 5–10, 2012 675–684***

Gould, J. E. (1995) Rainwater Catchment Systems Technology: Recent Developments in Africa and Asia. ***Technology and Developing Countries, First Edition. Routledge, 13, 9781315037516***

Heck, S. (2021) Greening the color line: historicizing water infrastructure redevelopment and environmental justice in the St. Louis metropolitan region. ***Journal of Environmental Policy & Planning, 23(5), 565–580.*** https://doi.org/10.1080/1523908x.2021.1888702

Kabantu, T.M., Tshimanga, M.R., Kileshye, O.J.M., Gumindoga, W., & Beya, T.J. (2018) A GIS-based estimation of soil erosion parameters for soil loss potential and erosion hazard in the city of Kinshasa, the Democratic Republic of Congo. ***Proceedings of the International Association of Hydrological Sciences, 378, 51–57***

Kavianand, G., Nivas, V. M., Kiruthika, R., & Lalitha, S. (2016) Smart drip irrigation system for sustainable agriculture. ***2016 IEEE Technological Innovations in ICT for Agriculture and Rural Development (TIAR).*** https://doi.org/10.1109/tiar.2016.7801206

Khoury-Nolde, N. (2010) ***Rainwater harvesting. Zero-M organization, Germany.***

Lashford, C., Lavers, T., Reaney, S., Charlesworth, S., Burgess-Gamble, L., & Dale, J. (2022) Sustainable Catchment-Wide Flood Management: A Review of the Terminology and Application of Sustainable Catchment Flood Management Techniques in the UK. ***Water, 14(8), 1204.*** https://doi.org/10.3390/w14081204

Liu, A., & Mukheibir, P. (2018) Digital metering feedback and changes in water consumption – A review. ***Resources, Conservation and Recycling, 134, 136–148.*** https://doi.org/10.1016/j.resconrec.2018.03.010

Malaeb, L., & Ayoub, G. M. (2011) Reverse osmosis technology for water treatment: State of the art review. ***Desalination, 267(1), 1–8.*** https://doi.org/10.1016/j.desal.2010.09.001

Mitasova, H., Barton, M., Ullah, I., Hofierka, J., & Harmon, R. S. (2013) 3.9 GIS-Based Soil Erosion Modeling. ***Treatise on Geomorphology, 228–258.*** https://doi.org/10.1016/b978-0-12-374 739-6.00052-x

Ngigi, S. N. (2003) What is the limit of up-scaling rainwater harvesting in a river basin?. ***Physics and Chemistry of the Earth, Parts A/B/C, 28(20-27), 943–956***

NHMRC (2017) *Australian drinking water guidelines Paper 6: National Water Quality Management Strategy, version 3.4 Updated October 2017* (Issue October). Canberra ACT 2601

OECD (1998) Sustainable Management of Water in Agriculture. ***In Sustainable Management of Water in Agriculture.*** https://doi.org/10.1787/9789264163225-en

Pant, B., Kaur, R., Soranganba, N., Nazir, I., Lohani, V., & Ram. R.N. (2018) Role of catchment area on water quality and production pattern in two different riverine Ecosystems. ***Journal of Entomology and Zoology Studies, 5(2), 1545–1549***

Pham-Duc, B., Sylvestre, F., Papa, F., Frappart, F., Bouchez, C., & Crétaux, J.-F. (2020) The Lake Chad hydrology under current climate change. ***Scientific Reports, 10(1).*** https://doi.org/10.1038/s41598-020-62417-w

Pramanik, M., Khanna, M., Singh, M., Singh, D.K., Sudhishri, S., Bhatia, A. and Ranjan, R., (2022) Automation of soil moisture sensor-based basin irrigation system. ***Smart Agricultural Technology, 2, p.100032.*** https://doi.org/10.1016/j.atech.2021.100032

Praun-Pereira BM, Backx BP. (2021) Nanotechnology in Water Treatment: An Optimistic Perspective for the Near Future. ***J Nanotechnol Nanomaterials, 2(1): 51–56***

Pullar, D., & Springer, D. (2000) Towards integrating GIS and catchment models. ***Environmental Modelling & Software, 15(5), 451–459.*** https://doi.org/10.1016/s1364-8152(00)00023-2

Rahim, M. S., Nguyen, K. A., Stewart, R. A., Giurco, D., & Blumenstein, M. (2020) Machine Learning and Data Analytic Techniques in Digital Water Metering: A Review. ***Water, 12(1), 294.*** https://doi.org/10.3390/w12010294

Rozenstein, O., & Karnieli, A. (2011) Comparison of methods for land-use classification incorporating remote sensing and GIS inputs. ***Applied Geography, 31(2), 533–544.*** https://doi.org/10.1016/j.apgeog.2010.11.006

Tom, G., Tauchus, G., Williams, J., Tong, S., (2011) The role of communicative feedback in successful water conservation programs. ***Appl. Environ. Educ. Commun. 10 (2), 80–90***

Van Manen, N., Scholten, H. J., & van de Velde, R. (2009) Geospatial Technology and the Role of Location in Science. ***GeoJournal Library, 1–13.*** https://doi.org/10.1007/978-90-481-2620-0_1

A Nano-Based Approach Toward the Sustainable Recovery of Nutrients from the Food Industry Generated Wastewater

Sivaranjani Palanisamy Ravikumar, Rahul Islam Barbhuiya, Charles Wroblewski, Saipriya Ramalingam, S. Sudheer Khan, Winny Routray, Abdallah Elsayed, Gopu Raveendran Nair, and Ashutosh Singh

1 Introduction

The role of water in sustaining life is critical, but its availability is often threatened through increasing demand caused by drastic growth of global populations, urbanization, and industrialization. To make matters worse, contamination of many water sources has put further strain by decreasing availability. According to UNESCO, the global water usage is predicted to continue increasing at an annual rate of 1% resulting in an excess usage of 20–30% by 2050 (UNESCO World Water Assessment Programme, 2019, 2021). This forecast is made worse with the suggestion that by 2030 the world will face a 40% global water deficit (UNESCO World Water Assessment Programme, 2021). These alarming predictions highlight the

S. P. Ravikumar
School of Engineering, University of Guelph, Guelph, ON, Canada

Department of Biotechnology, Bannari Amman Institute of Technology, Sathyamangalam, Tamil Nadu, India

R. I. Barbhuiya · C. Wroblewski · S. Ramalingam · A. Elsayed · A. Singh (✉)
School of Engineering, University of Guelph, Guelph, ON, Canada
e-mail: asingh47@uoguelph.ca

S. S. Khan
Department of Oral Medicine and Radiology, Saveetha Dental College and Hospitals, Saveetha Institute of Medical and Technical Sciences (SIMATS), Chennai, Tamil Nadu, India

W. Routray
Department of Food Process Engineering, National Institute of Technology, Rourkela, India

G. R. Nair
Department of Agriculture and Biological Engineering, University of Illinois at Urbane, Champaign, IL, USA

I. S. Dunmade et al. (eds.), *Sustainable Engineering*, Green Energy and Technology, https://doi.org/10.1007/978-3-031-47215-2_3

importance and necessity of freshwater preservation and wastewater treatment for sustainability. Meanwhile, in many ecosystems, phosphates and nitrates represent essential nutrients crucial for plant growth but can cause potential concerns for wastewater as both excess and deficiency in their concentrations beyond optimum levels can be detrimental to the environment and water quality. These nutrients can be categorized as point which refers to pollution from identified isolated source such as municipal, domestic, and livestock wastewater, anaerobic digestion effluent, landfill leachate, and industrial wastewater with major contributors including dairy and bakery industries. In contrast, nonpoint sources refer to indirect pollution from agricultural farms, forests, and runoffs from open barren lands (Kamilya et al., 2022).

Nutrient excess in aquatic environments through improper wastewater treatment can have detrimental effects on the biodiversity and economy leading to eutrophication, release of toxins, and parasitic infections (Nayak et al., 2021). Increase in nutrient levels promote growth of detrimental algal blooms which reduce levels of dissolved oxygen leading to anoxic conditions and released toxins. In the United States, algal blooms have led to water bans for Lake Erie, affecting over 400,000 people (Morgan M. Steffen et al., 2017). Some of the toxins such as microcystic can result in gastrointestinal disorders, while lipopolysaccharide, a neurotoxin produced by cyanobacteria, can complicate the proper functioning of drinking water treatment plants (Kamilya et al., 2022). Similarly, growth of red algae on seashores has resulted in millions of dollars in economic losses for the fish farming industry and impeded power generation by blocking water pipes and turbines (Almanassra et al., 2021).

The demand for synthetic alternatives of essential nutrients in agriculture has worsened the problem of nutrient contamination using nonrenewable alternatives. Since 1960, the use of nonrenewable phosphate fertilizer has tripled contributing to the collapse of the natural phosphorous cycle (Ownby et al., 2021). Approximately 80–90% of the mined phosphorous rocks are utilized to produce phosphate fertilizers contributing to environmental and ecological destruction through landscape modification, ground erosion, and generation of dust and air pollutants along with spreading radioactive elements. Recent studies suggest phosphate fertilizers with current phosphorus sources can only meet demand for a limited number of years before depletion (Almanassra et al., 2021). In Canada, the discharge limits have also become more stringent setting total maximum discharge for phosphorous of 0.1 mg PO_4-P/L according to pan-Canadian effluent limit (Ownby et al., 2021). As a result, nutrient contaminants can be considered as a valuable resource (Wu & Vaneeckhaute, 2022). These pressures have shifted the focus from removal toward recovery encouraging sustainability, reducing the threat of food insecurity, assisting in meeting the demand of fertilizers, minimizing the impact of mining, reducing generation of sludge, and lowering disposal costs (Ye et al., 2020; Sniatala et al., 2023).

The recovery of nutrients from waste has heavily relied on a verity of approaches both chemical and biological based. These methods have included precipitation, ammonia stripping, ion exchange, nitrification-denitrification, coagulation, sedimentation, polyphosphate accumulating organisms (PAOs) and adsorbents such as clay minerals, zeolite, and porous silica. However, meeting stringent limits set for nutrient discharge using existing methods has necessitated the development of advanced approaches. These innovated techniques encompass a range of processes including partial nitritation, nitrification-denitrification, membrane osmosis, autotrophic nitrogen removal, anaerobic ammonium oxidation, oxygen-limited autotropic nitrification and denitrification, moving bed biofilm reactor, along with microalgae and electrochemical processes (Hasan et al., 2021; Xie et al., 2016). However, a comprehensive evaluation of the scalability and effectiveness of these methods remains pending.

In the past decade, nanotechnology has experienced significant growth with applications in various scientific fields including diagnostics, pharmaceuticals, therapeutics, agriculture, nutraceuticals, environmental remediation (Arshad et al., 2021), and removal of pollutants from the environment (Poornima et al., 2022). Nanotechnology has enabled development of efficacious separation and adsorption systems by incorporating nanoparticles (NPs) such as iron oxides into ion-exchange resins to enhance selectivity (Sánchez & Martins, 2021). The usage of NPs containing micronutrients such as ZnO NPs is drawing attention (Bellani et al., 2020) with studies already exploring nutrients adsorbed by nanomaterials as fertilizers (Li et al., 2021). This area of research is expected to continue growing as demand of micronutrient fertilizers for food production is projected to hit 263 MT by 2050 (Jakhar et al., 2022). This chapter provides an overview of widely employed techniques for nutrient removal, the application of nanotechnology in nutrient recovery, and the state and scope of nutrient recovery in the food industry.

2 Current Methods of Nutrient Recovery

2.1 Chemical Precipitation

One of the conventional methods employed for recovery of nutrients through chemical processes is chemical precipitation. Two commonly targeted precipitates are struvite ($MgNH_4PO_4.6H_2O$) and hydroxyapatite ($Ca_5(OH)(PO_4)_3$). Struvite is a premium slow-release fertilizer consisting equal molar ratios of magnesium, ammonium, and phosphorous. Hydroxyapatite is used in fertilizer production and water purification and as a raw material in the phosphate industry. The synthesis of these precipitates is provided by their reaction (Eqs. 1 and 2).

Struvite precipitation

$$Mg^{2+} + PO_4^{3-} + NH_4^{+} + 6H_2O \rightarrow MgNH_4PO_4 \cdot 6H_2O(\text{precipitate}) \quad (1)$$

Hydroxyapatite precipitation

$$5Ca^{2+} + 3PO_4^{2-} + OH^{-} \rightarrow Ca_5(OH)(PO_4)_3(\text{precipitate}) \quad (2)$$

The efficiency of the precipitation process can be influenced by several factors including pH, molar ratio of constitutive ions, temperature, and presence of foreign substances. In considering the application for fertilizer, it is also important to examine quality of the recovered precipitate for contaminants such as heavy metals, coexisting ions, and emerging pollutants such as pharmaceuticals and hormones which can all have adverse effect on human and environmental health.

The precipitation of struvite occurs when Mg^{2+}, NH_4^+, and PO_4^{3+} ions exceed the solubility product constants of struvite. Several factors need to be considered for struvite crystallization such as pH (8 to 10.5), temperature (25–35 °C), magnesium source, molar N/P ratio, and the concentration of Mg^{2+}, NH_4^+, and PO_4^{3+} ions. The reaction kinetics of struvite precipitation depends on the phosphorous concentration, which should be at least 50 mg/L and above 100 mg/L for effective precipitation while inadequate for NH_4^+ slow crystallization. The high cost of struvite crystallization is due to the requirement of Mg and heavy use of alkali since most wastewater is slightly acidic (Ye et al., 2020; Chen et al., 2022). In cases where wastewater is deficient in nitrogen, calcium phosphate precipitate is more suitable than struvite crystallization. Hydroxyapatite is the most stable bond between calcium and orthophosphate ions with precipitation occurring at a molar ratio of Ca/P of 1.667. Initially calcium phosphate is precipitated as octa-calcium phosphate (OCP) and follows a transformation to amorphous noncrystalline phosphates (ACP), to deficient hydroxyapatite (DAP), and finally to hydroxyapatite.

The precursors of hydroxyapatite include $CaHPO_4 \cdot 2H_2O$, $Ca_4H(PO_4)_3 \cdot 5H_2O$ and amorphous $Ca_3(PO_4)_2$. The main phase recovered is pH dependent producing brushite when the pH is below 10 and hydroxyapatite when pH is above 10. Crystalline precipitates are often more economical than amorphous ones as they have higher recovery efficiency and consume energy in water drainage. The addition of flocculants speeds up the thickening but hampers the crystallization process producing larger particles with little to no crystallinity, while higher alkaline pH (9–11), longer sedimentation, and resistance time in the reactor favor crystallization. When these variables were applied to a continuous precipitation process with intensive mixing, a favorable reproducibility and efficiency of recovery were observed with efficiency of more than 80% even at low Ca/P molar ratio of 1.4 (Cichy et al., 2019).

In a study conducted at Autonoma de Occident University in Colombia, the pH and molar concentration of the constituent ions were found to have a major influence on the precipitation of hydroxyapatite and struvite. The solubility of particles was found to be influenced by the pH with the highest reduction found to be at the pH of 9.51. The addition of natural seeds such as previously formed struvite also promoted growth of crystals. One notable result from this study found raw wastewater had more viability to obtain a larger quantity of sludge in comparison with the treated wastewater, but treated wastewater was preferred due to the presence of fewer contaminants such as oils, fats, and inert materials that significantly affect the formation, growth, and precipitation of salts and the quality of the precipitate (Sarria et al., 2022).

In another study, waste concrete- and cement-based material were found to be capable of precipitating phosphorous as amorphous calcium phosphate and adsorb nitrate through anionic exchange. The study used cement paste to remove fluoride, phosphorous, nitrate, and sulfate from a heavily mixed fluoric acid wastewater obtained from a semiconductor fabrication plant (Park et al., 2008).

2.2 *Adsorption*

Absorption is one of the most preferred methods for wastewater treatment and nutrient recovery due to its simplicity, flexibility, low cost, effectiveness, avoidance of chemical sludge, and regenerative property (Wang et al., 2016). Three major mechanisms are involved in adsorption of nutrients including ion exchange, electrostatic interaction, and surface precipitation (Ye et al., 2020). Commonly used materials for phosphate and nitrate absorption include rocks, clay minerals, and organoclays such as dolomite, bentonite, kaolinite, halloysite, laterite, sandstone, and zeolites with less commonly used materials including aluminum oxides and polymeric ion exchangers (Karaca et al., 2006; Xi et al., 2010; Coulibaly et al., 2016; Zhang et al., 2020). Recently research has focused on developing absorbents from waste materials for nutrient recovery. Babatunde et al. (2010) used waste alum sludge from drinking wastewater treatment plants to adsorb phosphate. Wang et al. (2016) utilized wheat straw to prepare an amphoteric adsorbent for the recovery of $H_2PO_4^-$ and NH_4^+ which was also used as a novel multifunctional slow-release compound fertilizer with excellent water retention and slow-release properties.

Biochar which is a carbon rich solid obtained from biomass through thermal decomposition has garnered much attention in the recent decades as attractive adsorbents (Demiral & Gündüzoğlu, 2010) due to its favorable properties such as availability, low cost, well-developed pore structures, and high specific area with presence of functional groups containing oxygen (Zhang et al., 2020). Adsorption of nutrients (NH_4^+, NO_3^-, PO_4^{3-}) occurs through various mechanisms including electrostatic interaction, surface precipitation, ion exchange, interaction with surface functional groups, ligand exchange, and hydrogen bonding performance (Almanassra et al., 2021; Zhang et al., 2020). Unmodified biochar demonstrates low adsorption capacity, but engineering of biochar with doping of elements such as lanthanum, aluminum, calcium, iron, and magnesium will enhance the adsorption of phosphate. Modification with cationic polymer, namely, poly(diallyldimethylammonium chloride) (pDADMAC), has been shown to significantly increase anionic exchange capacity amplifying phosphorous adsorption by 1000 times in comparison with the unmodified biochar (Wang et al., 2020). Additionally, modifications of with iron offer an additional benefit of magnetic recovery.

The efficiency and properties of the biochar are significantly influenced by the feedstock, doping agent, temperature, pH, and pyrolysis temperature. Properties of biochar prepared from two different plants were studied and showed varied physicochemical properties and adsorption capacity for nitrate and phosphate (Cui et al., 2016). Influence of ions such as Cl^-, HCO_3^-, and HCO_3^- has been shown to negatively influence removal of PO_4^{3-} and/or NO_3^-, whereas cations such as Mg^{2+}, K^+, Na^+, and Ca^{2+} compete with NH_4^+.

2.3 Membrane-Based Process

Membrane technologies employed in the nutrient recovery include electrodialysis, forward osmosis, membrane distillation, and membrane separation process. Membrane separation process can be further separated to include microfiltration, ultrafiltration, nanofiltration, and reverse osmosis. In forward osmosis, a semipermeable membrane is used to recover the nutrients where the nutrients diffuse across the membrane with respect to its osmotic pressure gradient (Ye et al., 2020). This process exhibits low fouling propensity enabling the enrichment of the nutrients prior to struvite precipitation through the supplementation of magnesium ion via reverse salt flux.

Electrodialysis separation involves the selective mitigation of ions through membrane upon exposure to an electric field. Electrodialysis could split the proton into hydroxide and proton via bipolar membrane which in turn assists in the production of ammonia and phosphoric acid (Xie et al., 2016). For instance, phosphorus was recovered from food waste biogas digestate ash in the form of phosphoric acid. The simple leaching method exhibited only 65% recovery, whereas electrodialysis (10 mA) exhibited 90% recovery in comparison (Huang et al., 2022). In addition, electrodialysis can be selectively partitioned with an anionic-selective membrane. Further, the nutrient ions could be concentrated and recovered from the micropollutant contaminated urine using electrodialysis (Xie et al., 2016).

Separation of dissolved constituents using microfiltration, ultrafiltration, nanofiltration, and reverse osmosis is based on the mean pore size, molecular size, or molecular weight cutoff. Microfiltration is used as a pretreatment step to remove relatively larger compounds such as fat particles and suspended casein. Ultrafiltration is to separate proteins, whereas nanofiltration is used to separate lactose and polyphenols from the food processing wastewater. Integration of these membrane processes and their various combinations such as microfiltration-ultrafiltration-nanofiltration reduces the fouling of the succeeding membranes and improves the efficiency of separation (Lee & Stuckey, 2022). Similarly, studies predict that integration of membrane-based process with other nutrient recovery methods especially with precipitation methods could significantly improvise nutrient recovery in turn enhancing recovery efficiency and diversify the products (Xie et al., 2016).

2.4 Biological Recovery

Biological recovery of phosphorous employs PAOs through an enhanced biological removal process (EBPR). Organisms are exposed to alternating aerobic and anaerobic conditions resulting in the release of phosphorus into the surrounding solution, under anaerobic conditions. PAOs enrich phosphate along with metal ions Mg^{2+} and K^{+} in the wastewater. This process also generates energy which is further used to consume the carbon sources such as the volatile fatty acids and is stored as

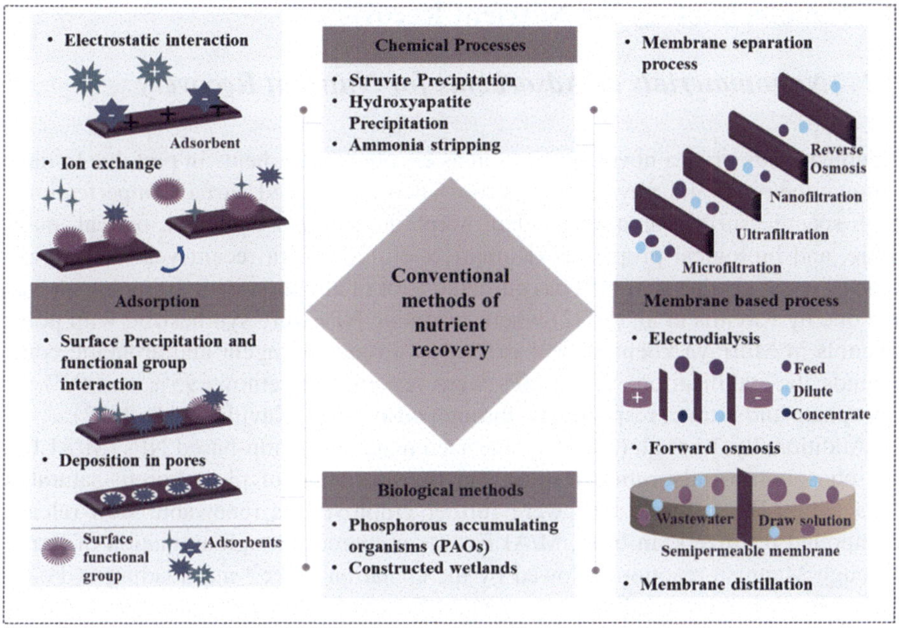

Fig. 1 Conventional methods of nutrient recovery from wastewater

poly-β-hydroxyalkanoates (PHAs). This cycle continues as phosphorous is accumulated by the PAOs on being exposed to the aerobic conditions, and energy for the accumulation of phosphorous in the aerobic conditions is provided by the stored PHAs. Additionally metal ions are potentially able to be adsorbed onto the biomass (Yuan et al., 2012).

Many of these conventional methods face difficulties in treating wastewater with varying loads of pollutants and are further limited by the high costs; in contrast, constructed wetlands are a cost-effective, robust, and effective method in treating nutrient-rich wastewater of varying loads of nutrients with a high degree of nutrient removal. These are eco-friendly, artificially engineered sustainable treatment frameworks which utilize natural purification mechanisms such as normal cycles, wetland vegetation, soil, and microbes to remediate the contaminated destinations (Muthukumaran, 2022). It has been observed that constructed wetlands with vertical flow are more suitable for the removal of phosphorous while horizontal flow is more suitable for nitrogen. Nitrogen recovery and removal can be further enhanced through hybrid constructed wetlands. However, phosphorous recovery is not notably improvised by the hybrid constructed wetlands. Further, the performance of the constructed wetlands could be amplified through bioaugmentation, variation of plant species and plantation, external carbon sources, and the introduction of artificial aeration (Kamilya et al., 2022). A short summary of the conventional methods of nutrient recovery is provided in Fig. 1.

3 Nanotechnology for Nutrient Recovery

3.1 Nanomaterials as Adsorbents for Nutrient Recovery

Nanomaterials offer a novel application as excellent adsorbents in part due to their unique chemical and physical properties such as reduced surface imperfections, high specific surface area, controlled morphologies, and electrical, optical, magnetic, and biological properties (Kumari et al., 2019). In recent years, the focus toward green synthesis of NPs has garnered a lot of attention with examples such as the one by Kavitha et al. (2022) where magnetic NPs were synthesized with polyphenols in Murraya koenigiii leaf extract as a reducing agent and aromatic compounds for stabilization. These NPs were reported to remove 95% and 90% of phosphate and nitrate, respectively, through adsorption (Kavitha et al., 2022).

Additionally, Li et al. (2021) synthesized magnetic lignin-based NPs (M/ALFe) for phosphorous adsorption resulting in the formation of phosphorous-saturated NPs (M/ALFeP). These NPs were further employed as renewable slow-release compound fertilizers. In brief, M/ALFe was prepared through amination of lignin through Mannish reaction followed by the chelation of Fe^{3+} and loading of Fe_3O_4 NPs. The phosphorous adsorption on M/ALFe was facilitated by electrostatic interaction, and the M/ALFeP NPs were recovered through magnetic separation. The cumulative release of Fe and P was found to be 67.2% and 69.1%, respectively, after 30 days (Li et al., 2021). A novel biocompatible nanocomposite was fabricated for phosphate and nitrate recovery through green methods. The polysaccharide obtained from mushroom was used to prepare the polysaccharide induced biogenic magnetite NPs (P@MNP) which was further immobilized on the biochar obtained from jackfruit peel (BC). The fixed bed column studies revealed that higher phosphate removal was observed on employing higher bed depth and lower flow rate. Further, these conditions ensured lower unused column bed as well (Nayak et al., 2021).

The binding interaction between the adsorbate and adsorbent can be determined with the help of the mean free energy of the adsorption process (E_D). The ΔG of the process is negative for spontaneous adsorption to occur and can be influenced by several observable parameters such as contact time, dosage of substrate and the NPs, temperature, bed depth, flow rate, and pH. The general trend portrayed by adsorbate, irrespective of its nature, is that the higher concentration of adsorbate results in lower removal rate. This could be ascribed to the aggregation of the adsorbate at higher concentrations which resists the effective diffusion of adsorbate toward the adsorbent. Diffusion drives the adsorbate molecules toward the surface of the adsorbent. The diffusion could be either through film diffusion or pore diffusion or via both. Mesopores provide transportation routes while micropores provide binding sites. On the other hand, the removal rate increases with the dosage of NPs as more active sites are available for adsorption. Similarly, increase in bed depth implies higher availability of active sites and, thus, results in enhanced breakthrough time and higher volume of treated water. In case of physisorption, the adsorption decreases with increase in temperature which could be ascribed to the weakened interaction between the adsorbate and adsorbent (Kavitha et al., 2022; Li et al., 2021; Nayak et al., 2021).

One of the most reported mechanisms behind nutrient absorption by NPs is electrostatic interactions. The zeta potential of these materials and existing ionic forms of the nutrients or adsorbate is greatly dependent on the pH which in turn is seen as a significant factor in determining the rate of nutrient removal by the NPs (Li et al., 2021). The NPs are protonated below the pH of the point zero charge (pH_{pzc}) and are deprotonated above the pH of pH_{pzc}. For instance, the pH_{pzc} of BC-P@MNP was reported to be 5.8 implying that the NPs possess neutral surface charge at this pH. Therefore, the phosphate ($H_2PO_4^-$) and nitrate (NO_3^-) species are adsorbed by BC-P@MNP at $pH < pH_{pzc}$ and, likewise, repelled the adsorbates ($H_2PO_4^-$, HPO_4^{2-}, NO_3^-) at $pH > pH_{pzc}$. However, below pH 4 lower adsorption was observed which could be ascribed to the limited number of binding sites due to increased proton adsorption. It could be seen that the change in pH also assists in desorption of the adsorbate. Some of the probable reactions explaining the abovementioned phenomenon are provided in Eqs. 3, 4, 5, and 6:

At pH < 5.8

$$-C-(OH)+H_3O^+ \rightleftharpoons -C-(OH_2)^+ + H_2O \text{ (protonation)} \tag{3}$$

$$-C-(OH_2)^+ + H_2PO_4^- \rightleftharpoons -C-(OH_2)^+ H_2PO_4^- \text{ (binding)} \tag{4}$$

At pH < 5.8

$$-C-(OH)+OH^- \rightleftharpoons -C-O^- + H_2O \text{ (deprotonation)} \tag{5}$$

$$-C-O^- + H_2PO_4^- \rightleftharpoons -C-O^- H_2PO_4^- \text{ (repulsion)} \tag{6}$$

The consideration of coexisting anions in wastewater is also an important factor with a crucial influence on the process of adsorption. For instance, the phosphate removal efficiency of BC-P@MNP was not strongly influenced by the coexistence of sulfates, nitrates, chlorides, and carbonates, whereas the removal efficiency of nitrates was notably hindered by the coexistence of the abovementioned ions. This could be due to the relatively higher charge density of sulfates and phosphates in comparison with nitrate, while alternatively carbonates may cause alkalinity of the aqueous medium reducing the removal efficiency of both phosphates and nitrates (Nayak et al., 2021).

3.2 *Nanomaterial-Microalgae Conjugated Systems*

Another approach adopted to recover nutrients using nanomaterials is their integration with algal systems to produce lipids. These systems in literature have been shown to treat wastewater effectively while potentially promoting the growth of microalgae. In the study by Vasistha et al. (2021), ZnO NPs were incorporated into the growth medium of *Chlorosarcinopis* sp. MAS04 where primary and secondary treatment wastewater was employed as the growth medium. The removal efficiencies of total nitrogen (TN), total phosphorous (TP), and total organic carbon (TOC) exhibited by the *Chlorosarcinopis* sp. MAS04-ZnO NP system were 87.20, 82.21, and 97.5%, respectively, which were significantly higher than the controls. In addition, high lipid and biomass content of 36% and 3.43 g L^{-1}, respectively, were observed. The lipids produced by this system were considered acceptable for the production of biodiesel according to the international standards which reflects the potential of primary and secondary wastewater treatment to be employed as an economical replacement of BG-11 medium. One of the remarkable observations of this study was that the ZnO NPs did not pose a threat of secondary pollution during the nutrient removal process which is attributed to the electrostatic attraction between the negatively charged microalgal cells and the positively charged ZnO NPs resulting in the adsorption of ZnO NPs onto the algae. Interestingly, it was noticed that the ZnO NPs promoted the photosynthetic rate of *Chlorosarcinopis* sp. MAS04 as the Rubisco activity increases in the presence of NPs. Further, ZnO imparted less cytotoxicity to the algae at lower concentrations as the reactive oxygen species were effectively captured by the defensive enzymes of algae (Vasistha et al., 2021).

The sensitivity of microalgae to NPs is often species dependent, while other parameters such as surface area, size, crystal structure, periodic vacancies, and opaqueness of the NPs play a significant role in controlling toxicity. For instance, MgO NPs are reported as nontoxic to the green algae *Pseudokirchneriella subcapitata* (EC_{50} = 100 mg L^{-1}) while toxic for green algae *Scenedesmus obliquus* even at low concentrations (EC_{50} = 36.70 mg L^{-1}). Similarly, the influence of Ti NPs on lipid productivity and cell recovery for three mixotrophically cultivated microalgae strains, namely, *Scenedesmus* sp., *Chlorococcum* sp., and *Euglena* sp., was observed to enhance lipid productivity in all three strains with the highest being 1.23-fold in *Chlorococcum* sp. Additionally, cell harvesting efficiency of the latter was found to be 82.46% in 45 min (sedimentation time) by 15 ppm Ti NPs (Khanra et al., 2020). He et al. (2017) investigated lipid production of *Scenedesmus obliquus* exposed to three different NPs, namely, carbon nanotubes (CNTs), α-Fe_2O_3, and MgO, with α-Fe_2O_3 reported the most promising one in comparison. These differences could be attributed to the dissociation of Fe^{3+} from α-Fe_2O_3 at lower concentrations imparting constructive effects on cell growth and the accumulation of biochemical compounds. Low doses of CNT and MgO NPs did promote the lipid productivity, but the high dose limited the productivity due to the reduction in biomass accumulation. Another phenomenon which hinders the growth of the microalgae at higher concentration apart from the oxidative stress is the shading effect. It refers to the reduction of light availability to the algae due to the aggregation of NPs on the cell's surface. Apart from the dosage of NPs, factors including the growth medium, age of culture, light illumination, incubation time, pH, and temperature play an important role in the lipid production by the microalgae-NP system.

One of the major reasons for the co-employment of NPs along with microalgae systems for lipid production is that the microalgae accumulate neutral lipids when prone to photooxidative stress or other adverse environmental conditions. The neutral lipids are accumulated as storage energy source which is around 20–50% of the dry cell weight of the microalgae. The boosted biomass assimilation and lipid production by algal-NP system in the wastewater are ascribed to a couple of possible reasons. Firstly, the photocatalytic activity of NPs facilitates the accessibility to the nutrients by breaking down complex compounds to simpler ones and triggers the activity of the digestive enzymes. Secondly, presence of higher amount of organic carbon and lower total nitrogen and phosphorous supports higher lipid assimilation. In brief, the mixotrophic medium prompts carbohydrate accumulation which leads to the activation of acetyl-CoA. Further, the acetyl-CoA results in the formation of triglyceride and eventually to lipid granules. Similarly, low N and P content results in the upraised production of fatty acids and hence the lipid granules. Studies predict that still many strains of microalgae are yet to be explored for applications associated with wastewater treatment integrated with lipid production (Vasistha et al., 2021; He et al., 2017). Apart from lipids and fatty acids, the microalgae are a source of several high value products including proteins, pigments, vitamins, carbohydrates, and antioxidants which have significant applications in the animal feed, food, and pharmaceutical industries (Lee & Stuckey, 2022).

3.3 *Other Approaches*

Microbes such as bacteria, yeast, fungi, and microalgae own several clusters of metal resistance genes that enable the detoxification of the cell through a set of mechanism such as efflux, complexation, and reduction ending with precipitation. The latter mechanism is used to synthesize metal NPs from the wastewater through the reduction of the metal ions present in them. It is important to recover these metal ions from the wastewater as they are nonbiodegradable and impart risk to the environment when they present beyond their optimum levels. In addition, microbial synthesis of NPs is relatively a safer method of NP synthesis. These reduced metal NPs include micronutrients such as Zn, Se, and Cu. Cu NPs were reported to be synthesized from wild strain of *Chlamydomonas reinhardtii*, a green microalga (Žvab et al., 2021). Similarly, a haloalkaliphilic bacterium called NDSe-7, a strain belonging to the genus *Lysinibacillus*, reduced selenite into Se nanospheres. In addition, it exhibits excellent potential to recover selenite from industrial saline wastewaters with high alkalinity (Won et al., 2021). Currently, these NPs are utilized in a variety of applications. For instance, the Cu NPs could be used in food packaging to preserve fresh-cut fruits (Patel et al., 2021). These NPs which tend to be micronutrients could be potential nano-fertilizers as the usage of NPs containing the micronutrients is gaining interest (Bellani et al., 2020).

Other budding approaches to recover nutrients through the incorporation of nanotechnology include membrane separation and electrochemical reduction reaction. The recovery of aquaculture wastewater by the thin-film composite forward osmosis (TFC-FO) membrane was enhanced by multiple folds through the modification

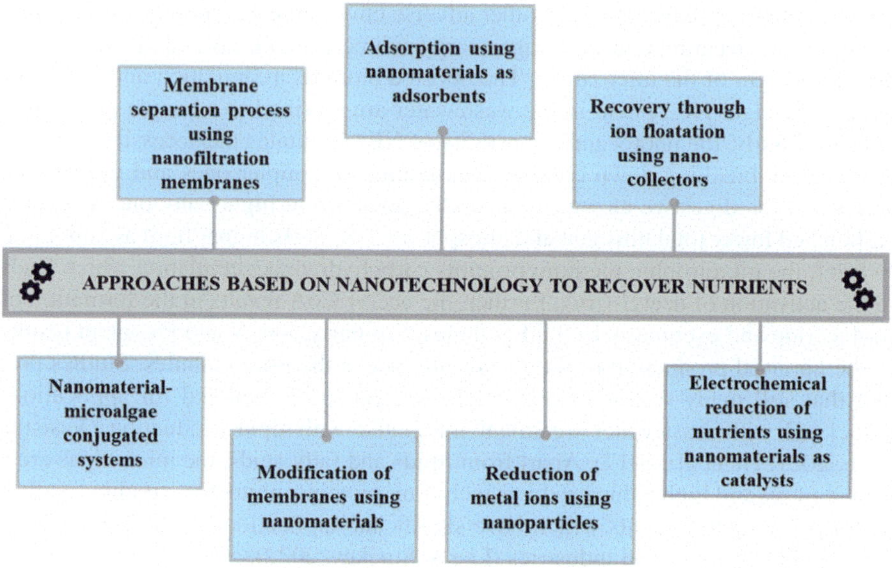

Fig. 2 Nanotechnology based approaches to recover nutrients from wastewater

of the membrane using SiO_2 NPs. The 3-aminopropyltriethoxysilane (APTES)-SiO_2 modified membrane exhibited a recovery rate of around 47% within 84 h, whereas the virgin FO membrane exhibited only 26% in 108 h, and the modified membrane had low reverse salt flux as well. This was attributed to the enhanced membrane hydrophilicity by SiO_2 which in turn increased the water permeate flux and thus the high osmotic pressure difference and water recovery rate (Nguyen et al., 2021). Electrochemical nitrate reduction reaction in the nitrate rich wastewater was achieved through Pd10Cu/BCN, and the reaction was driven by sustainable energy. Pd10Cu/BCN refers to single Cu atoms distributed in the boron-carbon-nitrogen matrix with Pd NPs. Pd10Cu/BCN exhibited excellent nitrate removal of nearly 100%. In addition, it enables the green synthesis of ammonia with higher yield rate of 102,153 μg h^{-1} $mg_{cat}{}^{-1}$ (Zhao et al., 2022). Figure 2 summarizes the nano-based approaches to recover nutrients from the wastewater.

4 State of Food Industry Processing Water

Wastewater from food processing industries is one of the largest sources of high-strength wastewaters in the world. High-strength wastewater refers to the wastewater possessing a 30-day average concentration of biochemical oxygen demand (BOD) >300 mg/L, total suspended solid (TSS) >330 mg/L or fat, oil and grease (FOG) >100 mg/L prior to the septic tank or other onsite wastewater treatment

Table 1 Recovery of valuable products from different food industry effluents

S. no.	Food industry	Valuable product
1	Soy whey	Proteins, isoflavones
2	Cheese whey	Whey proteins, lactose, lactoferrin, lipids
3	Palm oil mill effluent, olive oil mill wastewater	Carotenes, polyphenols, polysaccharides
4	Brewery wastewater, winery wastewater	Phenolic compounds, polysaccharides
5	Fish fillet wastewater, shrimp wastewater	Astaxanthin, proteins, fats
6	Starch wastewater	Proteins, polyphenols oxidase, β-amylase, sporamin
7	Meat processing wastewater	Proteins
8	Dairy whey	Proteins
9	Dairy wastewater	Proteins, lipids
10	Tofu whey	Proteins
11	Potato starch wastewater	Proteins
12	Mustard protein isolate wastewater	Phenolic compounds
13	Fermented starch wastewater	Sporamin

systems (California State Water Resources Control Board). In context, production of 1 kg of cheese demands 10 kg of milk, leaving 8–9 kg of turbid yellowish liquid by-product with high saline and organic content. Similarly, processing of approximately 3.5 tons of potato juice to produce 1 ton of potato starch leaves behind a residual thin liquid by-product waste. It could be observed that for every single unit of processed food, multiple units of liquid waste are generated (Lee & Stuckey, 2022). There are numerous technologies involved in treating the wastewater from food industries which include but are not limited to dissolved air floatation, coated mesh treatment, macroalgae bioremediation, electrocoagulation and catalytic sono-peroxone, combined physiochemical and biological treatment, reverse osmosis, membrane filtration, electrochemical systems, microbial fuel cells, and photocatalytic degradation (Shrivastava et al., 2022).

Generally, food processing industry wastewater possesses significant concentration of biomolecules including but not limited to pigments, proteins, polyphenols, lactose, isoflavones, and polysaccharides which can be recovered and reused by the same or other food industries. Currently, the food industry effluents, namely, fish fillet wastewater, shrimp wastewater, brewery wastewater, winery wastewater, palm oil mill effluent, olive oil mill wastewater, starch wastewater, cheese whey, and soy whey, have been studied for the recovery of valuable products including nutrients. A short summary of value-added products obtained from effluent generated by various food industries is summarized in Table 1.

Some of the physicochemical separation techniques employed to recover the valuable products from food industry wastewater include adsorption, precipitation, aqueous two-phase system, solvent extraction, foam fractionation, and membrane technology. One of the most common approaches with respect to nanotechnology to recover nutrients from the food industrial wastewater is the nanofiltration method. Nanofiltration membranes possess smaller molecular weight cutoff which enables the recovery of small size compounds such as polyphenols and lactose with the overall process usually preceded by microfiltration or ultrafiltration to prevent the fouling of the membrane as well as to enhance to efficiency of separation. In a study by Atra et al. (2005), ultrafiltration membrane with molecular weight cutoffs of 6 to 8 kDa, operated at 50 °C and 3 bar, exhibited 98% rejection of whey proteins. Further, subsequent integration of the nanofiltration membrane (0.4 kDa, 20 bar) leads to the recovery of lactose by 90% from the ultrafiltration permeate. Several studies have focused on the separation of ingredients such as lactose and proteins using membrane filtration from food processing wastewater such as cheese and dairy whey. However, only a few studies have been performed on the recovery of proteins from starch processing wastewater and soy whey through membrane technology. Apart from the recovery of useful ingredients from the food processing wastewater, the development of new food products out of them is also gaining attention. Benedetti et al. (2016) prepared a functional drink, a lactic beverage, through the fermentation of concentrated tofu whey wherein the tofu whey was concentrated using nanofiltration.

Similarly, single cell proteins can be produced with the microbial biomass (algal-bacterial consortium) from food industrial wastewater such as the wastewater from potato processing plant. These single cell proteins tend to be better alternative to conventional protein sources at a lower cost with lower footprints on land, water, and energy associated with production of conventional protein such as soy and meat, contributing to environmental sustainability goals (Rasouli et al., 2018). One of the major limitations in nutrient recovery from the food processing industries has been the vast difference in wastewater chemical composition which appears to be sector dependent and impart complexity to the recovery options and demands of specific systems and methods in recovering targeted compounds. Currently, there is an absence of a universal method to recover a variety of valuable ingredients from different industrial effluents (Lee & Stuckey, 2022).

5 Conclusion and Future Perspectives

To date, most studies dealing with nutrient recovery have focused on the recovery of phosphate, ammonium, and nitrates. Conventionally employed methods such as struvite and hydroxyapatite precipitation, adsorption, and membrane separation still dominate the nutrient recovery operations at the larger scale and can vary depending on the composition of wastewater. This clearly reflects research gaps in the emerging technologies with respect to the nutrient recovery. Studies on the incorporation of nanotechnology to well-established existing processes or the sole employment of nanotechnology and nanomaterials to recover nutrients are relatively under reported in comparison with those employed in the removal of pollutants. Nanomaterials which have been explored have largely been in the application as adsorbents or in the integration with microalgae systems as concerned to nutrient recovery.

With respect to adsorption studies, many have been carried out in synthetic wastewater with spiked concentration of the nutrients; however, this fails to adequately represent the complexity of real-world wastewater samples which can be contaminated with the presence of other ions and pollutants. Hindering of nutrient absorption is a complex matter attributed to many factors such as competition sorption of target nutrients and other contaminants on the vacant sites of the adsorbents and the blockage of pores by the organic and biological contaminants. To properly understand and compete in commercial markets, these methods require research and scaling of operation to develop real time recovery systems and for the recovered NH_4^+, NO_3^-, and PO_4^{3-}. Future studies could include the performance of the adsorbents in multicomponent solutions, column study investigation, selectivity, stability, reusability, regeneration, and lifetime of the adsorbents. Additionally, a lack of studies on the potential use of the adsorbents as fertilizers offers new research opportunities examining nutrient release from the adsorbent, uptake of the nutrients by the plants, and the shortcomings associated with them.

Nutrient recovery has also been heavily focused on selected sources of wastewater and nutrients. The scope should be extended to provide a more complete view of nutrients apart from NH_4^+, NO_3^-, and PO_4^{3-} in addition to broadening the source of wastewater from municipal wastewater, swine wastewater, and the treated water (primary) from the sewage treatment plants. Other sources of nutrient-rich wastewater that could be explored in-depth include that of the food. Effluent from food processing industries possess fewer toxic contaminants and a variety of valuable resources such as lipids and proteins which possess economical value. Additionally, wash water from food industries also possess an untapped source of nutrients which may be exported. This is made more financially appealing as the generation of food industry effluents is expected to soar up as the food production is ramping up to prevent the supply disruption.

Given these circumstances, it could be suggested that the recovery of nutrients is an urgent requirement. The application of nanotechnology in the recovery of nutrients could be used in order to meet food security and environmental goals through sustainability. To date, several nanomaterials have been identified with capabilities in recovering nutrients; unfortunately, they are not considered fully environmentally friendly. They do however provide innovative and effective solution, avoid the bulk usage of synthetic chemicals, enable recyclability, and aid in achieving the stringent discharge limits. Overall nutrient recovery benefits the economy, supports circular economy, makes the wastewater treatment more sustainable, and aids in the conservation of the resources as well as the environment. Despite its pressing need and potential benefits, there are numerous research gaps and opportunities that require exploration to establish them as sustainable solution.

References

Almanassra, I. W., Mckay, G., Kochkodan, V., Ali Atieh, M., & Al-Ansari, T. (2021). A state of the art review on phosphate removal from water by biochars. *Chemical Engineering Journal, 409*, 128211. https://doi.org/10.1016/J.CEJ.2020.128211

Arshad, R., Gulshad, L., Haq, I. U., Farooq, M. A., Al-Farga, A., Siddique, R., Manzoor, M. F., & Karrar, E. (2021). Nanotechnology: A novel tool to enhance the bioavailability of micronutrients. *Food Science & Nutrition, 9*(6), 3354–3361. https://doi.org/10.1002/FSN3.2311

Atra, R., Vatai, G., Bekassy-Molnar, E., & Balint, A. (2005). Investigation of ultra- and nanofiltration for utilization of whey protein and lactose. *Journal of Food Engineering, 67*(3), 325–332. https://doi.org/10.1016/J.JFOODENG.2004.04.035

Babatunde, A. O., & Zhao, Y. Q. (2010). Equilibrium and kinetic analysis of phosphorus adsorption from aqueous solution using waste alum sludge. *Journal of Hazardous Materials, 184*(1–3), 746–752. https://doi.org/10.1016/J.JHAZMAT.2010.08.102

Bellani, L., Siracusa, G., Giorgetti, L., di Gregorio, S., Ruffini Castiglione, M., Spanò, C., Muccifora, S., Bottega, S., Pini, R., & Tassi, E. (2020). TiO2 nanoparticles in a biosolid-amended soil and their implication in soil nutrients, microorganisms and Pisum sativum nutrition. *Ecotoxicology and Environmental Safety, 190*, 110095. https://doi.org/10.1016/J.ECOENV.2019.110095

Benedetti, S., Prudencio, E. S., Müller, C. M. O., Verruck, S., Mandarino, J. M. G., Leite, R. S., & Petrus, J. C. C. (2016). Utilization of tofu whey concentrate by nanofiltration process aimed at obtaining a functional fermented lactic beverage. *Journal of Food Engineering, 171*, 222–229. https://doi.org/10.1016/J.JFOODENG.2015.10.034

Chen, X., Jin, Y., Zhou, Z., Huang, P., Chen, X., Ding, R., & Chen, R. (2022). Spontaneous nutrient recovery and disinfection of aquaculture wastewater via Mg–coconut shell carbon composites. *Journal of Hazardous Materials, 426*, 128119. https://doi.org/10.1016/J.JHAZMAT.2021.128119

Cichy, B., Kużdżał, E., & Krztoń, H. (2019). Phosphorus recovery from acidic wastewater by hydroxyapatite precipitation. *Journal of Environmental Management, 232*, 421–427. https://doi.org/10.1016/J.JENVMAN.2018.11.072

Coulibaly, L. S., Akpo, S. K., Yvon, J., & Coulibaly, L. (2016). Fourier transform infra-red (FTIR) spectroscopy investigation, dose effect, kinetics and adsorption capacity of phosphate from aqueous solution onto laterite and sandstone. *Journal of Environmental Management, 183*, 1032–1040. https://doi.org/10.1016/J.JENVMAN.2016.09.061

Cui, X., Hao, H., He, Z., Stoffella, P. J., & Yang, X. (2016). Pyrolysis of wetland biomass waste: Potential for carbon sequestration and water remediation. *Journal of Environmental Management, 173*, 95–104. https://doi.org/10.1016/J.JENVMAN.2016.02.049

Demiral, H., & Gündüzoğlu, G. (2010). Removal of nitrate from aqueous solutions by activated carbon prepared from sugar beet bagasse. *Bioresource Technology, 101*(6), 1675–1680. https://doi.org/10.1016/J.BIORTECH.2009.09.087

Hasan, M. N., Altaf, M. M., Khan, N. A., Khan, A. H., Khan, A. A., Ahmed, S., Kumar, P. S., Naushad, M., Rajapaksha, A. U., Iqbal, J., Tirth, V., & Islam, S. (2021). Recent technologies for nutrient removal and recovery from wastewaters: A review. *Chemosphere, 277*, 130328. https://doi.org/10.1016/J.CHEMOSPHERE.2021.130328

He, M., Yan, Y., Pei, F., Wu, M., Gebreluel, T., Zou, S., & Wang, C. (2017). Improvement on lipid production by Scenedesmus obliquus triggered by low dose exposure to nanoparticles. *Scientific Reports 2017 7:1, 7*(1), 1–12. https://doi.org/10.1038/s41598-017-15667-0

Huang, M., Zhai, Y., Liu, X., Liu, X., Wang, Z., Zhou, Y., & Xu, M. (2022). Efficient extraction of phosphorus from food waste biogas digestate ash through two-compartment electrodialysis cell. *Journal of Environmental Chemical Engineering, 10*(6), 108701. https://doi.org/10.1016/J.JECE.2022.108701

Jakhar, A. M., Aziz, I., Kaleri, A. R., Hasnain, M., Haider, G., Ma, J., & Abideen, Z. (2022). Nano-fertilizers: A sustainable technology for improving crop nutrition and food security. *NanoImpact, 27*, 100411. https://doi.org/10.1016/J.IMPACT.2022.100411

Kamilya, T., Majumder, A., Yadav, M. K., Ayoob, S., Tripathy, S., & Gupta, A. K. (2022). Nutrient pollution and its remediation using constructed wetlands: Insights into removal and recovery mechanisms, modifications and sustainable aspects. *Journal of Environmental Chemical Engineering*, *10*(3), 107444. https://doi.org/10.1016/J.JECE.2022.107444

Karaca, S., Gürses, A., Ejder, M., & Açikyildiz, M. (2006). Adsorptive removal of phosphate from aqueous solutions using raw and calcinated dolomite. *Journal of Hazardous Materials*, *128*(2–3), 273–279. https://doi.org/10.1016/J.JHAZMAT.2005.08.003

Kavitha, S., Surendran, G., Karunakaran, S., & Kumari, N. (2022). Removal of contaminants from waste water by using Murrayakoenigii nanoparticles. *Materials Today: Proceedings*, *57*, 1906–1912. https://doi.org/10.1016/J.MATPR.2022.01.465

Khanra, A., Vasistha, S., Kumar, P., & Rai, M. P. (2020). Role of C/N ratio on microalgae growth in mixotrophy and incorporation of titanium nanoparticles for cell flocculation and lipid enhancement in economical biodiesel application. *3 Biotech*, *10*(8), 1–12. https://doi.org/10.1007/S13205-020-02323-0/FIGURES/10

Kumari, P., Alam, M., & Siddiqi, W. A. (2019). Usage of nanoparticles as adsorbents for waste water treatment: An emerging trend. *Sustainable Materials and Technologies*, *22*, e00128. https://doi.org/10.1016/J.SUSMAT.2019.E00128

Lee, S. Y., & Stuckey, D. C. (2022). Separation and biosynthesis of value-added compounds from food-processing wastewater: Towards sustainable wastewater resource recovery. *Journal of Cleaner Production*, *357*, 131975. https://doi.org/10.1016/J.JCLEPRO.2022.131975

Li, T., Lü, S., Wang, Z., Huang, M., Yan, J., & Liu, M. (2021). Lignin-based nanoparticles for recovery and separation of phosphate and reused as renewable magnetic fertilizers. *Science of The Total Environment*, *765*, 142745. https://doi.org/10.1016/J.SCITOTENV.2020.142745

Muthukumaran, M. (2022). Advances in bioremediation of nonaqueous phase liquid pollution in soil and water. *Biological Approaches to Controlling Pollutants*, 191–231. https://doi.org/10.1016/B978-0-12-824316-9.00006-9

Nayak, A., Bhushan, B., Gupta, V., & Kotnala, S. (2021). Fabrication of microwave assisted biogenic magnetite-biochar nanocomposite: A green adsorbent from jackfruit peel for removal and recovery of nutrients in water sample. *Journal of Industrial and Engineering Chemistry*, *100*, 134–148. https://doi.org/10.1016/J.JIEC.2021.05.028

Nguyen, T. Q., Tung, K. L., Lin, Y. L., di Dong, C., Chen, C. W., & Wu, C. H. (2021). Modifying thin-film composite forward osmosis membranes using various SiO2 nanoparticles for aquaculture wastewater recovery. *Chemosphere*, *281*, 130796. https://doi.org/10.1016/J.CHEMOSPHERE.2021.130796

Ownby, M., Desrosiers, D. A., & Vaneeckhaute, C. (2021). Phosphorus removal and recovery from wastewater via hybrid ion exchange nanotechnology: a study on sustainable regeneration chemistries. *Npj Clean Water 2021 4:1*, *4*(1), 1–8. https://doi.org/10.1038/s41545-020-00097-9

Park, J. Y., Byun, H. J., Choi, W. H., & Kang, W. H. (2008). Cement paste column for simultaneous removal of fluoride, phosphate, and nitrate in acidic wastewater. *Chemosphere*, *70*(8), 1429–1437. https://doi.org/10.1016/J.CHEMOSPHERE.2007.09.012

Patel, A., Enman, J., Gulkova, A., Guntoro, P. I., Dutkiewicz, A., Ghorbani, Y., Rova, U., Christakopoulos, P., & Matsakas, L. (2021). Integrating biometallurgical recovery of metals with biogenic synthesis of nanoparticles. *Chemosphere*, *263*, 128306. https://doi.org/10.1016/J.CHEMOSPHERE.2020.128306

Poornima, S., Manikandan, S., Karthik, V., Balachandar, R., Subbaiya, R., Saravanan, M., Lan Chi, N. T., & Pugazhendhi, A. (2022). Emerging nanotechnology based advanced techniques for wastewater treatment. *Chemosphere*, *303*, 135050. https://doi.org/10.1016/J.CHEMOSPHERE.2022.135050

Rasouli, Z., Valverde-Pérez, B., D'Este, M., de Francisci, D., & Angelidaki, I. (2018). Nutrient recovery from industrial wastewater as single cell protein by a co-culture of green microalgae and methanotrophs. *Biochemical Engineering Journal*, *134*, 129–135. https://doi.org/10.1016/J.BEJ.2018.03.010

Sánchez, A. S., & Martins, G. (2021). Nutrient recovery in wastewater treatment plants: Comparative assessment of different technological options for the metropolitan region of

Buenos Aires. *Journal of Water Process Engineering*, *41*, 102076. https://doi.org/10.1016/J.JWPE.2021.102076

Sarria, N. V., Rivera Velasco, D. M., Larrahondo Chávez, D. A., Mazuera Ríos, H. D., Gandini Ayerbe, M. A., Goyes López, C. E., & Mejía Villareal, I. M. (2022). Struvite and hydroxyapatite recovery from wastewater treatment plant at Autónoma de Occidente University, Colombia. *Case Studies in Chemical and Environmental Engineering*, *6*, 100213. https://doi.org/10.1016/J.CSCEE.2022.100213

Shrivastava, V., Ali, I., Marjub, M. M., Rene, E. R., & Soto, A. M. F. (2022). Wastewater in the food industry: Treatment technologies and reuse potential. *Chemosphere*, *293*, 133553. https://doi.org/10.1016/J.CHEMOSPHERE.2022.133553

Sniatala, B., Kurniawan, T. A., Sobotka, D., Makinia, J., & Othman, M. H. D. (2023). Macronutrients recovery from liquid waste as a sustainable resource for production of recovered mineral fertilizer: Uncovering alternative options to sustain global food security cost-effectively. *Science of The Total Environment*, *856*, 159283. https://doi.org/10.1016/J.SCITOTENV.2022.159283

Steffen, M. M., Davis, T. W., McKay, R. M. L., Bullerjahn, G. S., Krausfeldt, L. E., Stough, J. M. A., Neitzey, M. L., Gilbert, N. E., Boyer, G. L., Johengen, T. H., Gossiaux, D. C., Burtner, A. M., Palladino, D., Rowe, M. D., Dick, G. J., Meyer, K. A., Levy, S., Boone, B. E., Stumpf, R. P., … Wilhelm, S. W. (2017). Ecophysiological Examination of the Lake Erie *Microcystis* Bloom in 2014: Linkages between Biology and the Water Supply Shutdown of Toledo, OH. *Environmental Science & Technology*, *51*(12), 6745–6755. https://doi.org/10.1021/acs.est.7b00856

UNESCO World Water Assessment Programme. (2019). *The United Nations World Water Development Report 2019: Leaving no one behind.* Paris, UNESCO.

UNESCO World Water Assessment Programme. (2021). *The United Nations World Water Development Report 2021: Valuing water.* Paris, UNESCO.

Vasistha, S., Khanra, A., & Rai, M. P. (2021). Influence of microalgae-ZnO nanoparticle association on sewage wastewater towards efficient nutrient removal and improved biodiesel application: An integrated approach. *Journal of Water Process Engineering*, *39*, 101711. https://doi.org/10.1016/J.JWPE.2020.101711

Wang, X., Lü, S., Gao, C., Feng, C., Xu, X., Bai, X., Gao, N., Yang, J., Liu, M., & Wu, L. (2016). Recovery of Ammonium and Phosphate from Wastewater by Wheat Straw-based Amphoteric Adsorbent and Reusing as a Multifunctional Slow-Release Compound Fertilizer. *ACS Sustainable Chemistry and Engineering*, *4*(4), 2068–2079. https://doi.org/10.1021/ACSSUSCHEMENG.5B01494/SUPPL_FILE/SC5B01494_SI_001.PDF

Wang, Z., Bakshi, S., Li, C., Parikh, S. J., Hsieh, H.-S., & Pignatello, J. J. (2020). Modification of pyrogenic carbons for phosphate sorption through binding of a cationic polymer. *Journal of Colloid and Interface Science*, *579*, 258–268. https://doi.org/10.1016/j.jcis.2020.06.054

Won, S., Ha, M. G., Nguyen, D. D., & Kang, H. Y. (2021). Biological selenite removal and recovery of selenium nanoparticles by haloalkaliphilic bacteria isolated from the Nakdong River. *Environmental Pollution*, *280*, 117001. https://doi.org/10.1016/J.ENVPOL.2021.117001

Wu, H., & Vaneeckhaute, C. (2022). Nutrient recovery from wastewater: A review on the integrated Physicochemical technologies of ammonia stripping, adsorption and struvite precipitation. *Chemical Engineering Journal*, *433*, 133664. https://doi.org/10.1016/J.CEJ.2021.133664

Xi, Y., Mallavarapu, M., & Naidu, R. (2010). Preparation, characterization of surfactants modified clay minerals and nitrate adsorption. *Applied Clay Science*, *48*(1–2), 92–96. https://doi.org/10.1016/J.CLAY.2009.11.047

Xie, M., Shon, H. K., Gray, S. R., & Elimelech, M. (2016). Membrane-based processes for wastewater nutrient recovery: Technology, challenges, and future direction. *Water Research*, *89*, 210–221. https://doi.org/10.1016/J.WATRES.2015.11.045

Ye, Y., Ngo, H. H., Guo, W., Chang, S. W., Nguyen, D. D., Zhang, X., Zhang, J., & Liang, S. (2020). Nutrient recovery from wastewater: From technology to economy. *Bioresource Technology Reports*, *11*, 100425. https://doi.org/10.1016/J.BITEB.2020.100425

Yuan, Z., Pratt, S., & Batstone, D. J. (2012). Phosphorus recovery from wastewater through microbial processes. *Current Opinion in Biotechnology, 23*(6), 878–883. https://doi.org/10.1016/J.COPBIO.2012.08.001

Zhang, M., Song, G., Gelardi, D. L., Huang, L., Khan, E., Mašek, O., Parikh, S. J., & Ok, Y. S. (2020). Evaluating biochar and its modifications for the removal of ammonium, nitrate, and phosphate in water. *Water Research, 186*, 116303. https://doi.org/10.1016/J.WATRES.2020.116303

Zhao, X., Jia, X., Zhang, H., Zhou, X., Chen, X., Wang, H., Hu, X., Xu, J., Zhou, Y., Zhang, H., & Hu, G. (2022). Atom-dispersed copper and nano-palladium in the boron-carbon-nitrogen matric cooperate to realize the efficient purification of nitrate wastewater and the electrochemical synthesis of ammonia. *Journal of Hazardous Materials, 434*, 128909. https://doi.org/10.1016/J.JHAZMAT.2022.128909

Žvab, U., Kukulin, D. S., Fanetti, M., & Valant, M. (2021). Bioremediation of copper polluted wastewater-like nutrient media and simultaneous synthesis of stable copper nanoparticles by a viable green alga. *Journal of Water Process Engineering, 42*, 102123. https://doi.org/10.1016/J.JWPE.2021.102123

Technical Assessment of Agricultural Equipment Condition for Sustainability

Alex Folami Adisa

1 Introduction

Zero hunger/no hunger is Sustainable Development Goal 2 (SDG2) of the United Nations General Assembly agenda to be achieved by 2030 (UN-GA 2015). This study captured aspects of the SDG2 target of agricultural productivity for food production systems by monitoring and controlling the operating conditions of agricultural equipment. In general term, agricultural equipment productivity can be determined by ratio of outputs to inputs in the system (Dharmasiri 2013). This can also be measured by total factor productivity (TFP), whose changes are usually attributed to improvements in technology, such as the equipment management system.

Having a large fleet of farm equipment that are purchased and brought to use at various times (years) for various farm operations, they don't wear out at the same rate even though being maintained under the same workshop management system. Hence, there is a need from time to time to carry out technical assessment of the equipment field operations performance to determine their overall effectiveness levels for the purpose of their optimum usage. Equipment utilization, availability, reliability, and maintenance are all factors that serve as performance indicators of sustainability for agricultural equipment for a functional and sustained agricultural mechanization system for sustained food production. The goal of sustainable agricultural business is to maximize output and minimize production downtime (MaxGrip 2022) by tracking equipment successes and failures from which the organization can develop informed recommendations on when equipment should be replaced, upgraded, or brought in for maintenance (Severino 2019).

A. F. Adisa (✉)
Agricultural and Bio-resources Engineering Department, Federal University of Agriculture, Abeokuta, Nigeria

I. S. Dunmade et al. (eds.), *Sustainable Engineering*, Green Energy and Technology, https://doi.org/10.1007/978-3-031-47215-2_4

Farm machines are used during a relative short period of annual use, and as farms grow, more machines are required (in numbers and sizes) to meet task demands in these short periods' operation. This results in higher ownership and operation costs which are being spread over hundreds of hours every year; hence, there is a need to shorten the time from planting to harvesting (Srivastava et al. 2006).

Equipment breakdowns exist in different degrees like partial or total breakdown in which the system, component, or device can no longer produce/perform desired operations in terms of quantity and quality even when the equipment is still running and producing (Limblecmms 2018). Machinery operational scheduling is determining the time periods during the year when each operation can be performed from the point where the required capacity can be calculated. An obsolete machine is one that is out of production, and repair parts are no longer available, and it can be replaced by another machine or method that will produce at a higher profit (Srivastava et al. 2006).

Reliability method is described as a scientific method of predicting, analyzing, preventing, and reducing failures over a defined period (De Carlo 2013). This is accomplished by evaluation of operation management efficiency and calculation of overall equipment effectiveness (OEE), which is an important parameter for availability estimation (Nakajima 1988). Of significant importance and effect is level of availability which is determined from equipment reliability and maintainability (De Carlo 2013). Equipment reliability and maintenance (maintainability) include machines, tools, and fixtures, but also the technical, operational, and management activities, ranging from equipment specifications to daily operation and maintenance to sustain the performance of equipment throughout its useful life (National Research Council 1991). Equipment reliability and maintenance development are from breakdown maintenance and repair to preventive maintenance to predictive maintenance. Breakdown maintenance and repair is the restoration of failed equipment, while preventive maintenance is the systematic servicing of equipment to reduce the possibility of failure (National Research Council 1991). Function ability/utilization of any asset/equipment depends on the reliability of maintenance services on the equipment (Nkakini et al. 2008).

In equipment day-to-day operation, there are two important basic measures of equipment performance to monitor: actual and schedule output which is the output expected from an operation for a given allocation of time. This is usually based on a published output rate, while actual output reflects the true performance of an operation, including scrap and both scheduled and unscheduled downtime (National Research Council 1991). High equipment availability is delivered by effective maintenance, while information on equipment availability and reliability are important performance measures. It is possible for equipment reliability to be low even when the availability of the equipment is relatively high (Dunn 2002). From this exercise, it will be possible to determine which of the equipment to be retained, replaced, and traded off which is the basic reason for this study.

2 Materials and Methodology

This study was carried out at Sunti Golden Sugar Company, Niger State, Nigeria, a 1300 ha farm, and about 15,000 ha is for future expansion (a major source of the nation's sugar production/supply). It treated 120 equipment on individual basis to obtain information to monitor and control data needed to reduce equipment maintenance costs, optimize expenditure on maintenance, and minimize the total cost of the system (Dunn 2002) for 2017 and 2018 cropping seasons. Figures 1 and 2 are broken equipment awaiting repairs in the open field. Figure 3 is a field fuel depot to ease field fuel supply, while Fig. 4 is a prepared field awaiting sugarcane planting.

Fig. 1 Broken down equipment waiting to be repaired at field site workshop

Fig. 2 A crawler equipment with three tine rippers at field site workshop

Fig. 3 Fuel depot at field workshop to ease fuel supply to machines right on the field

Fig. 4 A field fully prepared for sugarcane planting

Fig. 5 Central workshop where several engines are being overhauled

Fig. 6 Some earthmoving equipment in central workshop awaiting repair

2.1 Workshop Operations

Figures 5 and 6 show the central workshop where major repair work is conducted. Workshop was headed by workshop manager which was divided into the following sections as of the time of this study:

- Heavy duty equipment for earth work (civil), road, dyke, drainage, and irrigation equipment.
- Heavy duty equipment for agriculture land clearing, land development, and land preparation; among them are crawler and wheel tractors, cane loaders, and cane harvesters.

- Implements maintenance and repair sections for plow, harrow, sprayer, and irrigation/drainage pumps just to name a few.

The observed limitations that will not allow the equipment to meet targeted operations are as follows:

- Operator not available to operate the machine during the study cropping seasons
- Machine breakdown and the nature of breakdowns during the studied cropping seasons
- Time spent sourcing and delivering spare parts during the study cropping seasons
- Fuel supply shortage/delay during the study cropping seasons

2.2 Machine Operation Data Capturing for 2017 and 2018 Cropping Seasons

Collection of the following data that was available on stated equipment

- Total obtainable time of machine operation for the study cropping seasons
- Theoretical time per hectare (h) within the study seasons
- Actual time of operation within the study seasons
- Lost operation time due to (not proportional to area) rests, breakdowns, and idle travelling to field
- Lost time per hectare (proportional to area) – filling tank of sprayers, fertilizer box, high yield, etc.
- Number of machine failures/unit time
- Time spent in restoring failures
- Machine rated speed, km/h
- Machine working speed, km/h
- Machine rated output
- Machine working output
- Total operating time
- Number of frequency of breakdowns
- Total failure time
- Number of failures

Equipment detail performance was evaluated for each of 120 equipment during the study period based on total equipment time, effective operation time, idle time, breakdown time, frequency of breakdowns, utilization of equipment, mean time between failures, mean time to repair, failure rate, restoration rate per time, mission time, utilization, availability, reliability, and maintainability of equipment.

2.3 *Machinery Performance and Reliability Operational Determination for Cropping Seasons*

It is possible for equipment fail/breakdown in different degrees such as partial or total. To effectively monitor and control these failures/breakdowns, there are failure metrics to guide the process of determining equipment failures which are MTBF (mean time between failure), MTTR (mean time to repair), and MTTF (mean time to failure) (Limblecmms 2018). Data required for the monitoring and control of factor assessment are labor hours spent on maintenance (time spent), number of breakdowns and repairs (downtime), operational time, equipment uptime, and total operational time which is expected equipment operating hours minus equipment downtime to calculate MTTR and MTBF (Limblecmms 2018).

MTTR is especially useful in determining frequent or regular failures and long downtime periods that affect productivity, such as in agricultural operations. It has an even bigger impact on production results that are sensitive to failure which causes missed production/operation deadlines. It increases labor costs, loss of revenue, and various operational issues (Limblecmms 2018). MTBF measures the time it takes for equipment to fail until another failure occurs. This is an indication of how long equipment can stay in operation over a period between unplanned breakdowns. To prepare for the unexpected, this is useful. A combination of equipment availability and MTBF yields a more accurate measure of equipment performance. A practical indicator of availability performance is reliability; however, availability is a significant "value adder" (Eti et al. 2006). Equipment utilization, availability, reliability, and maintainability calculations were based on labor hours spent on maintenance (time spent), number of breakdowns and repairs (downtime), operational time, equipment uptime data recorded for the period of field operation in question (2017 and 2018 cropping seasons).

2.3.1 Equipment Utilization

Equipment utilization is referred to as the measurement of the use and performance of site machinery. This assists businesses to improve jobsite productivity and reduce the cost of equipment rental and project delays. Equipment utilization was calculated using Eq. 1:

$$E_u,(\%) = \frac{\text{equipment effective}(\text{operation})\text{time}}{\text{total obtainable time}} \times 100 \tag{1}$$

2.3.1.1 Failure Rate

Failure rate was calculated using Eq. 2 (ASAE EP456 DEC 1986 2015):

$$\gamma = \frac{\text{number of failures}}{\text{unit cumulative time}} = \frac{\text{total failures}}{\text{total time}} \tag{2}$$

2.3.1.2 Mean Time Between Failures (MTBF)

Mean time between failures (MTBF) was calculated using Eqs. 3 and 4 (Abernethy 1996):

$$\text{MTBF} = \frac{\text{units of cumulative time}}{\text{number of failures}} \tag{3}$$

$$= \frac{\text{total operating time}}{\text{number of frequency}} = \frac{1}{\gamma} \tag{4}$$

where γ– failure rate.

2.3.1.3 Mean Time to Repair (MTTR)

Mean time to repair (MTTR) was calculated using Eq. 5 (Abernethy 1996):

$$\text{MTTR} = \frac{\text{total failure time}}{\text{number of failures}} = \frac{1}{\tau} \tag{5}$$

2.3.1.4 Repair Rate

Repair rate was calculated by using Eq. 6 (Nkakini et al. 2008):

$$\tau = \frac{\text{restoration time}}{\text{duration of failures}} = \frac{1}{\text{MTTR}} \tag{6}$$

where τ – restoration time to duration of failures, i.e., repair rate.

2.3.2 Reliability of Equipment Operation, *R*

An operational reliability value is equal to 1 minus the probability of downtime when both probabilities are expressed as decimals. Equipment reliability is the probability that the equipment will perform its intended function satisfactorily for a specified period under specified operating conditions (De Carlo 2013). Two reliability figures of merit are the failure rate, and mean time between failures (MTBF), which is the inverse of the failure rate. Reliability of the equipment was calculated using Eq. 7 (ASAE EP456 DEC 1986 2015):

$$R(t) = e^{-\gamma t} \tag{7}$$

where e – base of the natural logarithm (2.7182)

γ – failure rate
t– mission time, h (equipment operational time plus failure time)

2.3.3 Maintainability of Equipment, *M*

Maintainability is the ease with which maintenance activities can be performed on an asset or equipment (Ryan 2019). Its purpose is to measure the probability that a piece of equipment in a failed state can be restored to normal operating condition after undergoing maintenance. Maintainability of equipment was calculated using Eq. 8 (Abernethy 1996):

$$M(t) = 1 - \exp.^{-\tau t} \tag{8}$$

2.3.4 Equipment Availability, *A*

The availability of equipment and machinery is defined as the percentage of time it is operable and committable when needed. It is the actual time that the machine or system is capable of production as a percent of the total planned production time. Equipment availability is the percentage of time during which a process (or equipment) is available to run. Uptime of equipment is when it performed. Equipment availability was calculated using Eq. 9 (Nkakini et al. 2008):

$$A = \frac{1}{1 + \gamma\tau} \tag{9}$$

2.3.5 Worked Example (Table 1 Row 3 Calculation)

A John Deer tractor model 3050 with 67.60 kW power rating was operated for 455 h and idled for 881 h. It broke down three times in 24 h during the season. Calculate the tractor mean time between failures, mean time to repair, utilization, reliability, maintainability, and availability of the equipment within this cropping year, assuming e to be 2.7182.

Solution
From the captured operation performance data, with Eq. 2, failure rate (γ) was calculated to be 0.0066; with Eq. 6, repair rate (τ) is 0.125; mission time (t) is operation plus failure time which is 479 h. With Eq. 4, the calculated value of MTBF is 151.52 h, and MTTR is 8 h with Eq. 5. Based on Eq. 1, equipment utilization is

Table 1 Crop cultivation equipment performance evaluation for the 2017 and 2018 cropping seasons

S/ no.	Equipment type/model	Power rating, kw	MTBF, h	MTTR, h	Equipment utilization, %	Equipment reliability, %	Equip. maintainability	Equipment availability, %
1	John Deer-01-3050	67.6	92.59	8.00	28.2520	3.8427	1	92.0530
2	John Deer-02-2850S	64.0	0.73	8.00	8.2446	0	1	8.35580
3	John Deer-03-3050	67.6	151.52	8.00	33.4558	4.2500	1	94.9896
4	John Deer-04-4960	149.0	38.17	8.00	27.9220	0.0019	1	82.6923
5	John Deer-05-3050	67.6	48.31	8.00	27.7960	0.0286	1	85.7868
6	John Deer-06-3650	85.3	0.00	8.00	0	0	1	0
7	John Deer-08-2650	58.2	36.36	8.00	18.3467	0.2245	1	81.9820
8	John Deer-09-3350	74.6	232.56	8.00	31.8681	35.5410	1	96.6667
9	John Deer-10-3650A	85.3	0.32	8.00	1.6447	0	1	3.8760
10	John Deer-11-4960	149.0	1.83	8.00	8.4183	1.63E−124	1	18.6441
11	John Deer-13-8960	275.0	0.00	8.00	0	0	1	0
12	John Deer-14-8960	275.0	7.51	8.13	36.0537	6.415E−83	1	47.9725
13	John Deer-29-8970	298.4	6.68	6.62	38.7640	4.84E−106	1	50.1818
14	John Deer-30-8970	298.4	59.17	9.43	67.1296	8.511E−10	1	86.3095
15	John Deer-31-8970	298.4	12.33	8.00	18.2086	1.82E−09	1	60.6557
16	John Deer-36-8970	298.4	92.59	10.64	62.0989	0.0015	1	89.7585
17	John Deer-44-7225J	168.0	212.77	9.17	73.7599	0.0675	1	95.8736
18	John Deer-45-7225J	168.0	7.94	8.06	38.3928	3.962E−79	1	49.6941
19	John Deer-46-7225J	168.0	7.80	8.00	32.6675	6.575E−56	1	49.3671
20	Bell Tract- 47-1734AF	224.0	1.00	0.00	61.8783	100.00	1	100.00000
21	Valtra Tract-15-T171	134.0	3.49	8.00	15.1098	9.097E−89	1	30.3867
22	Valtra Tract-16-T171	134.0	42.19	9.01	67.5925	9.833E−19	1	82.4859
23	Vatra Tract-17-T171	134.0	0.00	8.00	0	0	0	0
24	Valtra Tract-18-T171	134.0	151.52	8.00	52.6724	1.4854	1	95.0233

(continued)

Table 1 (continued)

S/ no.	Equipment type/model	Power rating, kw	MTBF, h	MTTR, h	Equipment utilization, %	Equipment reliability, %	Equip. maintainability	Equipment availability, %
25	Valtra TracT-19-T191	141.0	15.72	8.00	39.9088	2.766E−24	1	66.2703
26	Valtra Tract-20-T191	141.0	0.00	0.00	41.3065	100.0000	0	100.00
27	Vatra Tract-21-T191	141.0	3.02	8.00	12.7840	7.64E−105	1	27.4205
28	Valtra Tract-22-T191	141.0	0.00	0.00	38.3445	100.0000	1	100.000
29	Valtra Tract-23-T191	141.0	6.41	8.00	13.9880	6.012E−31	1	44.4795
30	Valtra Tract-24-T191	141.0	0.00	0.00	36.4297	100.0000	0	100.00
31	Valtra Tract-25-A950	71.0	0.00	8.00	0	0	1	0
32	Valtra Tract-26-A950	71.0	526.32	8.00	38.4943	36.2489	1	98.5455
33	Valtra Tract-27-A950	71.0	0.00	0.00	40.0390	100.000	0	100.000
34	Valtra Tract-28-A950	71.0	454.55	8.00	29.7872	36.1369	1	98.2456
35	Valtra Tract-31-T191	141.0	0.00	0.00	0	0	0	0
36	Valtra Tract-32-T191	141.0	769.23	5.99	46.7289	36.5131	1	99.2556
37	Valtra Tract-33-T191	141.0	33.11	8.00	40.3571	8.191E−07	1	80.5574
38	Valtra Tract-34-BM125i	98.4	2.58	8.00	22.8346	4.08E−159	1	24.3698
39	Valtra Tract-35-BM125i	98.4	270.27	8.00	39.2241	12.7632	1	97.1530
40	Valtra Tract-37-T191	141.0	208.33	9.26	44.0148	1.5328	1	95.7373
41	Valtra Tract-38-T191	141.0	0.00	0.00	62.8780	100.0000	0	100.0000
42	Valtra Tract-39-T191	141.0	0.00	0.00	39.375	100.0000	1	100.000
43	Valtra Tract-40-T191	141.0	0.00	0.00	75.5	100.0000	1	100.000
44	Valtra Tract-41-T191	141.0	833.33	9.00	50.7138	36.3951	1	98.9378
45	Valtra Tract-42-T191	141.0	714.29	8.00	45.4189	36.3663	1	98.8604
46	Valtra Tract-43-T191	141.0	0.00	0.00	48.5665	100.00	0	100.000
47	Valtra Tract48- T194H	298.0	0.00	0.00	72.8078	100.00	0	100.000
48	Bulldozer-01-D7G	164.0	10.78	8.00	23.6816	8.921E−33	1	57.3965
49	Bulldozer-02-D7G	164.0	12.69	8.00	44.7704	1.546E−57	1	61.2922

50	Bulldozer-03-D7G	164.0	81.97	8.06	45.3125	0.0002	1	91.0434
51	Bulldozer-04-D7G	164.0	87.72	8.20	70.05	2.521E−06	1	91.4491
52	Bulldozer-05-D7G	161.0	84.75	8.00	65.4494	0.0006	1	91.3726
53	Bulldozer-06-D8R	245.0	13.11	8.06	32.0652	1.219E−17	1	61.9965
54	Bulldozer-07-D8R	245.0	0.00	8.00	0	0	0	0
55	Bulldozer-08-D8R	245.0	5.83	8.00	17.6005	5.55E−42	1	42.1687
56	Bulldozer-09-D8R	245.0	0.67	8.70	6.70872	0	1	7.1691
57	Bulldozer-10-D8R	245.0	0.12	8.47	1.3816	0	1	1.4132
58	JD Cane loader-01-BELL	49.0	238.10	8.00	24.3853	35.5719	1	96.7480
59	JD Cane loader-02-BM100	74.6	0.89	8.00	9.2634	0	1	10.0363
60	JD Cane loader-03-BM100	74.6	92.59	11.76	36.3281	0.3573	1	88.7405
61	JD Cane loader-04-BM100	74.6	22.08	8.00	8.125	1.6807	1	73.4219
62	JD Cane loader-05-1850	78.0	0.00	0.00	47.2994	100.00	0	100.00
63	JD Cane loader-06-1850	78.0	0.00	0.00	67.5336	100.00	0	100.00
64	JD Cane loader-07-EU6	164.0`	0.00	0.00	76.6542	100.00	0	100.00
65	Crane truck-03-JIAN HUAN		0.00	10.31	0	0	1	0
66	Cane truck-04-BEDFORD		0.00	10.31	0	0	1	0
67	Cane truck-05-IVECO	307.0	0.00	8.00	0	0	1	0
68	Cane truck-06-HOWO	251.0	0.15	8.00	0.5769	0	1	1.8405
69	Crane liebher-60T-LIEBHER	270.0	33.56	8.00	73.5825	7.22E−08	1	80.7638
70	Crane liebher-70T	330.0	103.09	8.00	81.4474	0.1557	1	92.8036
71	Motor grader-01-140H	138.0	0.00	8.00	0	0	1	0
72	Motor grader-02-140H	138.0	10.25	8.47	42.0982	1.205E−71	1	54.7937
73	Motor grader-03-140H	138.0	0.00	8.00	0	0	1	0
74	Motor grader-04-140H	138.0	78.13	10.20	66.6382	1.513E−08	1	88.4485
75	DP-03	45.0	0.00	8.00	0	0	1	0
76	Tipper lorry-01-MP380E38H EURTRAKKER	283.0	35.84	8.00	59.4406	8.016E−09	1	81.7308

(continued)

Table 1 (continued)

S/ no.	Equipment type/model	Power rating, kw	MTBF, h	MTTR, h	Equipment utilization, %	Equipment reliability, %	Equip. maintainability	Equipment availability, %
77	Tipper lorry-02- MP380E38H EUROTRAKKER	283.0	344.83	8.00	62.3188	12.918476	1	97.7273
78	MP380E38H Tipper lorry-03	283.0	0.00	8.00	0	0	0.8	0
79	Tipper lorry-04- MP380E38H EUROTRAKKER	283.0	2.93	8.00	8.3970	2.741E−47	1	26.8293
80	Tipper lorry-05- MP380E38H EUROTRAKKER	283.0	6.30	8.00	34.8993	9.038E−64	1	44.0678
81	Tipper lorry-06- MP380E38H EUROTRAKKER	283.0	2.34	8.00	19.5489	1.09E−169	1	22.6087
82	Tipper lorry-07- MP380E38H EUROTRAKKER	283.0	3.95	8.00	11.1281	2.255E−47	1	33.0317
83	Roller compac. -01-CP563E	108.0	0.00	8.00	0	0	1	0
84	Roller compac. -02- CP563E	108.0	0.00	8.00	0	0	1	0
85	Roller compac. -04-CS563E	108.0	2.57	8,26	21.6634	2.44E−308	1	23.7057
86	Roller compac. -05-CS563E	108.0	0.02	8.00	0.2577	0	1	0.2755
87	Roller compac. -06-CS563E	108.0	8.91	9.80	27.7237	3.052E−57	1	47.5
88	Excavator-01-325C	141.0	34.60	8.00	41.5240	3.28E−06	1	81.2395
89	Excavator-02-320DL	97.0	68.97	8.55	63.4868	1.458E−05	1	88.9401
90	Excavator-03-320DL	97.0	0.32	8.00	3.8961	0	1	3.8961
91	Excavator-04-325C	141.0	8.62	8.00	27.5	1.049E−29	1	51.8699
92	Excavator-05-325C	141.0	62.11	8.62	58.5481	1.299E−09	1	87.7656
93	Excavator-06-325C	141.0	0.00	8.00	0	0	0	0
94	Excavator-07-325D	141.0	0.00	8.00	0		1	0
95	JCB Backhoe-01-3CX	55.0	5.18	7.14	18.1429	2.49E−49	1	42.0530
96	JCB Backhoe-02-3DX	68.0	23.26	9.09	36.3281	4.448E−18	1	71.8147

97	JCB Backhoe-03-3DX	68.0	1000	8.00	66.9837	36.4907	1	99.1952
98	JCB Backhoe-04-4CX	74.6	14.37	8.20	46.3294	4.917E−43	1	63.7108
99	Ditch witch-01	45.0	0.00	8.00	0	0	1	0
100	Ditch witch-02	45.0	6.00	8.00	9.375	7.13E−10	1	42.8571
101	Ditch witch-03	45.0	5.00	8.00	7.8125	3.795E−11	1	38.4615
102	DP-02	45.0	0.00	8.00	0	0	1	0
103	TelehandlerTH-01-TH460bB	55.0	8.94	8.00	42.4051	4.047E−48	1	52.7559
104	ST-01	65.0	0.00	0.00	42.6073	0	0	0
105	Hydra Crane-01-15XW	80.9	666.67	8.00	56.3758	36.3526	1	98.8235
106	Hydra Crane-02-15XW	80.9	2.29	8.00	19.5565	1.88E−246	1	22.2647
107	Forklift-02	40.0	0.00	0.00	100.00	100.00	0	100.00
108	Forklift-03-7FDU30, 3 tons	40.0	19.16	8.00	70.5263	5.726E−16	1	70.5263
109	Forklift-04-DP35N,3.5 tons	40.0	9.60	8.00	30.6122	1.244E−18	1	54.5455
110	Forklift-05-DP30N, 3 tons	40.0	3.78	8.00	26.5777	2.769E−77	1	32.0644
111	Forklift-06-DP30N, 3 tons	40.0	31.75	8.00	68.0556	5.677E−08	1	79.8519
112	Forklift-32T-SVE TRUCK	40.0	21.28	8.00	54.4034	1.752E−09	1	72.6755
113	Wheel loader-01-938G	134.0	0.07	8.00	0.7143	0	1	0.9174
114	Wheel loader-02-938G	134.0	0.02	8.00	0.2737	0	1	0.2750
115	Wheel loader-03-950H	162.0	14.35	8.00	44.8171	1.825E−26	1	64.1921
116	Wheel loader-04-950H	162.0	68.97	8.00	62.7016	0.0044	1	89.6254
117	Wheel loader-05-928G	116.0	9.97	8.00	50.3989	1.817E−28	1	55.4905
118	Wheel loader-06-928G	116.0	37.45	8.00	77.9167	0.0005	1	82.3789
119	Fuel Bowser-05	52.2	0.00	0.00	0	100.00	0	100.00
120	Fuel Bowser-06	52.2	0.00	0.00	0	100.00	0	100.00

33.46%; using Eq. 7, equipment reliability is 4.25%; using Eq. 8, equipment maintainability is 1, while using Eq. 9, equipment availability is 94.99%.

3 Result and Discussion

3.1 *Equipment Operational Condition Level Indicator Determination for the Cropping Seasons*

A total of 120 fleets of agricultural production equipment were analyzed over the 2017 and 2018 cropping seasons of 1300 ha for the failure rate, repair rate, mission time, mean time between failure (MTBF), and mean time to repair (MTTR). The results of this were used to obtain Table 1, which provides clear indicators of each equipment operational state condition indicator – utilization, reliability, maintainability, and availability – so that the management and ownership decisions of the system can assess their technical soundness and operational cost-effectiveness.

3.1.1 Equipment Utilization

The data in Table 1 column 6 shows the calculated values of fleet utilization, which ranged from 0% to 81%. This gives the knowledge of how equipment utilization helps to

(i) Minimize equipment hoarding and get it to the job site as needed.
(ii) Prevent machines sitting unused on sites, eating up maintenance costs while contributing nothing to the project.
(iii) Enhance uptime with an effective maintenance schedule, and identify machine underutilization for manager's management planning.

The evaluation of mixed fleets of equipment was found to be useful for evaluating the operational level of individual equipment and machine fleets.

3.1.2 Equipment Reliability

Table 1 column 7 shows the results from the reliability calculation for equipment which were as low as 36.50% and even in some cases as low as 0. This is the percentage of what was okay for the period in question. The inverse of this is the percentage of its failure at the given time. Equipment with a zero-reliability level is not dependable for operational work in the field. Differentiating availability from reliability, reliability differs from availability by quantifying the probability that equipment will work as intended without any failures. This is in contrast with just measuring the time it is in operation.

3.1.3 Equipment Maintainability

The equipment being evaluated here are repairable types; hence, there was a need to determine the state of the technical soundness condition of each equipment or maintainability level of each of them at the time in question which varied between being in functioning (1) or at failure state (0) as shown in Table 1 column 8 (INFRALERT 2016). Most of the equipment maintainability values were 1, indicating that the equipment was functioning at the time of evaluation. Very few equipment maintainability value was 0, an indication that some equipment was in failure state at the time of evaluation. By interpretation, a lower mean time to repair would correspond to a higher level of maintainability, and conversely, maintainable equipment takes less time to repair. Maintainability picks up where reliability might fall short. While reliability characterizes how long an asset can operate without issues, maintainability describes the likelihood that the same asset can be restored once a failure occurs.

3.1.4 Factors That Affect Maintainability (Thompson 2018)

(i) Better training for the team to improve work/job performance
(ii) Maintenance drawings and checklists for standardization and repeatability
(iii) Standardizing equipment and inventory
(iv) Increasing planned and unplanned preventive maintenance reduces amount of unscheduled downtime of equipment.

3.1.5 Equipment Availability

Table 1 column 9 shows the calculated equipment availability during the 2017 and 2018 cropping seasons' operation. It showed good values of 100% except for a few cases where equipment was down throughout the seasons. As a result, it helps monitor and control equipment availability, as well as reduce downtime to a minimum. However, planned downtime must be used to take initiative-taking measures to keep equipment in efficient working order.

Generally, it can be said of the fleets of these 120 equipment assessed that maintainability was easier to improve because of proper documentation of repair procedures (historic repair records) and availability of repair tools and materials which could significantly reduce the time it takes to restore broken down equipment. All these factors will improve availability and reliability and decrease downtime of equipment and help determine which ones to be retained, replaced, or traded off. In Table 1, the combined values of reliability and maintainability shows a clear level of availability results for each equipment which is in accordance with the findings of De Carlos (2013).

3.2 Equipment Operational Planning and Control

For crop cultivation, establishment, and weed control up to harvesting and transporting to the postharvest processing store, tractor power and machinery size must match in terms of row spacing and tramline width to reduce soil compaction over time (Adisa 2012). The headland or trace's width must be determined by the length of the tractor plus the attached implement or its multiples to reduce turning time at the end of each field block operation.

The use of a cropping field operations calendar is essential for crop and operational sequencing and should be put in place since crop and operational periods overlap. Resources like fuel and oil, equipment spare parts, fertilizer, insecticides, and herbicides and their availability are limiting factors to be considered as well (Kepner et al. 2005).

During planned equipment downtime, major repairs and maintenance are to be conducted before critical field operation periods to improve pre-field and in-field efficiency (Deere 1984). Proper breakdown communication reporting and daily record of every equipment operation will improve machinery performance efficiency (Adisa 2012).

3.3 Equipment/Plant Control Unit

This unit is overly critical for releasing and allocating equipment periodically to various sections. A range of operations will be allocated based on the equipment budget, machinery costs/hour, and the size of the field section's jobs. Now it will be left for sectional manager to effectively use, monitor to minimize equipment time lost to the barest minimum, and properly report and record the equipment uptime and downtime.

3.4 Communication and Information Dissemination in the System

This was made possible through the following:

- Radio and telephone were the means of communication between the fields, workshops, and other service units.
- Management meetings of project update reporting by sections were done every Monday of the week, while sections hold meetings every day for effective running of the system.

3.5 Field Technology Unit

A central laboratory was needed for the following reasons. The soil is the major component of crop cultivation, and because there is always a long-term effect of mechanical tillage, chemical application, and equipment continuous traffic on soil texture and structure, adequate technical precautions must be put in place from the start. Hence, a soil map and periodic soil survey and analysis are needed to determine soil texture and structure condition, soil nutrient level, soil tilth condition, and chemical component level. The outcome of this information includes, aside from normal soil preparation practices, special soil preparation, the amounts and types of additional nutrient supplementary and soil treatment required, disease control, and a few others. Also crop breeding for improved, adaptable, higher yielding, and drought- and disease-resistant varieties must be developed for commercial production.

4 Conclusion

The calculated values of equipment utilization, availability, reliability, and maintainability of the 120-equipment fleet for the 2017 and 2018 cropping seasons of 1300 ha show variation from values of almost 100% to 0%. Thus, the majority was well utilized, readily available, reliable to use, and of good maintainability; only a few items were unutilized, unavailable, and unreliable to use, which were of low maintainability. This gives clear picture of their technical soundness and operational cost overall effectiveness which is extremely useful in system management and ownership decisions.

In general, it can be said that these assessment indicators, equipment utilization, availability, reliability, and maintenance of 120 fleets of equipment operation data, were easier to monitor and control because proper documentation of repair procedures (historic repair records) and availability of repair tools and materials resulted in a significant reduction in the time taken to restore broken down equipment.

This knowledge of equipment utilization helps to minimize hoarding and get it to worksites where needed and prevents machines sitting unused on sites, eating up maintenance costs while contributing nothing toward the project. It also improves uptime through an effective maintenance schedule and understanding which machine is underutilized and which is useful for planning management actions. Determination of equipment maintainability provided information for better team training to improve business performance, providing maintenance drawings and checklists that are standardized and repeatable, standardizing equipment and inventory, and increasing planned and unplanned preventive maintenance to reduce the amount of unscheduled downtime of equipment. As a result, equipment availability can be checked and controlled, and downtime can be reduced to a minimum. These factors all contributed to improve availability and reliability and decreasing downtime of equipment and clearly determined specifically which equipment to be maintained/retained, replaced, or traded off.

These equipment technical performance assessment indicators, operational system monitoring, and control are all factors determined for the purpose of SDG2 to sustain continuous food production/supply on large scale basis for the ever-increasing human population. Sustainability of agricultural equipment system, where agricultural business is to maximize output and minimize production downtime for functional and sustained agricultural mechanization system, will help to achieve the SDG2 goal of zero hunger/no hunger.

Acknowledgments I hereby acknowledge the World Bank Sponsored Project of African Centre of Excellence in Agricultural Development and Sustainable Environment (CAEDESE); the Federal University of Agriculture, Abeokuta, Nigeria, for sponsoring this Faculty Outreach (2018) study at Flour Mills Nigeria Plc, Sunti Golden Sugar Company, Niger State, Nigeria; and, also, the Sunti Golden Sugar Company staff who contributed immensely to the success of this study.

Conflict of Interest There is no conflict of interest on this manuscript; the work is purely original author's work.

References

Abernethy, R. B. (1996). *Reliability (Eingineering).* The New Weibull Handbook, 3rd Ed. Gulf Publishing Company. Hunston TV.

Adisa, A. F. (2012). *Appropriate Management and Operational Planning Technique for Mechanized Soybean and Maize Production in Nigeria.* Journal of Science, Technology, Mathematics and Education. Published by Department of Science Education, Federal University of Technology, Minna, Nigeria, **8 (2).**

ASAE EP456 DEC 1986 (R2015). *Test and Reliability Guidelines.* American Society of Agricultural and Biological Engineers, ASABE Standard.

De Carlo, F. 2013. *Reliability and Maintainability in Operations Management.* Published by IntechOpen edited by Massimilliano M. Schiradi. Chapter 4: p81–111.

Deere, J. 1984. *Machinery Management, Fundamentals of Machine Operation,* Deere, and Company Service Training. Illinois, USA.

Dharmasiri, M. 2013. *Measuring Agricultural Productivity using the Average Productivity* Index (APL). Archived 2013 at the Wayback machine.

Dunn, S. (2002). *The Reliability Revolution.* Webmaster, Plant Maintenance Resource Centre, info@maintenanceresources.com.

Eti, M. C., Nkakini, S. O., and Nkoi, B. 2006. *Application of Availability and Reliability Analysis Techniques to Improve Bayelsa Water Board Productivity.* Nigeria Journal of Engineering Research and Development.

INFRALERT, 2016. *D5.1 – RAMS Data Collection and Failure Rate Analysis at Component Level, Revision 4.* Linear Infrastructure Efficiency Improvement by Automated Learning and Optimized Predictive Maintenance Techniques: P14.

Kepner, R. A., Bainer, R., and Barger, E. L. 2005. *Principles of Farm Machinery.* 3rd Edition, CBS Publishers & Distributors, India.

Limblecmms, 2018. *MTTR vs MTBF vs MTTF: A visual simple guide to failure metrics,* Streamline work with Limble.

MaxGrip, 2022. *Reliability Metrics 101: Mean Time Between Failure (MTBF)*-MaxGrip.

Nakajima, S. 1988. *Introduction to TPM: Total Productive Maintenance.* Productivity Press, Inc. p:129.

National Research Council. (1991). *The Competitive Edge: Research Priorities for U.S. Manufacturing: Equipment Reliability and Maintenance.* Washington, DC: The National Academies Press. https://doi.org/10.17226/1618; p54–55.

Nkakini, S. O., Eti, M. E. and Barinyima, N. 2008. *Tractor Functionability Assessment using Reliability and Availability Technique.* Journal of Agricultural Engineering and Technology *(JAET),* Published by The Nigeria Institution of Agricultural Engineers **16** **(2)**: p4–13.

Ryan, C. 2019. *Maintainability Definition and Calculation. Maintenance Question and Answer Preventive Maintenance.* Upkeep Maintenance Management.

Severino, J. 2019. *Incident Management Handbook.* Atlassian, Lex Neva, SRE weekly.

Srivastava, A. K., Goering, C. E., Rohrbach, R. P. and Buckmaster, D. R. 2006. *Engineering Principles of Agricultural Machines,* America Society of Agricultural and Biological Engineers, USA. ASAE publication 801M0206, second edition: pp525–549.

Thompson, P. 2018. *Manufacturing Maintenance Tips: 5 factors that affect maintainability.* SAGE Group Automation Company.

UN-GA, 2015. *Sustainable Development Goals.* United Nations General Assembly. Wikipedia.

Sustainable Food Packaging Solutions: Polysaccharide-Based Films and Coatings

Ozgun Guzdemir

1 Introduction

Over the past several decades, excessive usage of petroleum-based polymers has become a serious environmental problem due to their nonbiodegradability, non-renewability, and recycling difficulties (Yang et al. 2021). Polymers, plastics, are broken down into small pieces after many years of existence in marine or soil environments (Jang et al. 2020; Andrady 2011; Yang et al. 2021); however, they are not degraded in the nature. These tiny pieces called microparticles having the size between 0.1 μm and 5000 μm have already entered the food chain and been causing various health damages to all living creatures. One of the known damages in the human body was reported as gastrointestinal problems, in the literature (Ateia et al. 2022; Lehel and Murphy 2021; Waring et al. 2018).

Among the global plastic waste, 50% of all plastic consumption belongs to the packaging industry (Ncube et al. 2021). In recent years, food consumers have become more conscious of sustainability problems and been looking for biobased and biodegradable alternatives. Therefore, many packaging researchers and material scientists have focused on the studies to develop biodegradable, biobased, renewable, and environmentally-friendly packaging materials as alternatives to synthetic plastic packages. Biodegradable polymers are classified into two categories according to their resources: biobased (renewable) and non-biobased (non-renewable). Polymers that are both biodegradable and biobased are obtained from agro-biopolymers, biomonomer chemical synthesis, or microbial polymers. Agro-biopolymers are natural sources like plants, animals, marine life, and microbial products which are mainly categorized as proteins, lipids, polysaccharides, and their derivatives (Nešić et al. 2020; Wankhade 2020; Ferreira et al. 2016).

O. Guzdemir (✉)
Department of Food Engineering, Aydın Adnan Menderes University, Aydın, Turkey
e-mail: ozgun.guzdemir@adu.edu.tr

I. S. Dunmade et al. (eds.), *Sustainable Engineering*, Green Energy and Technology, https://doi.org/10.1007/978-3-031-47215-2_5

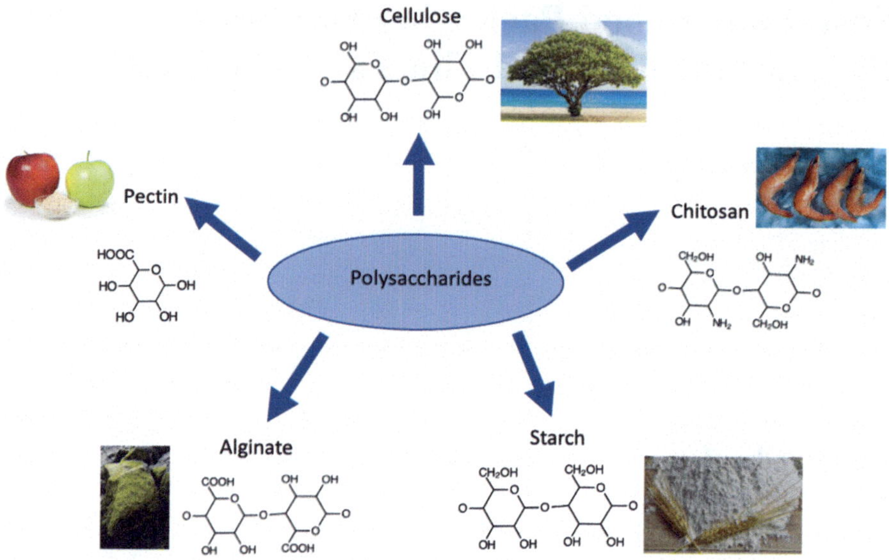

Fig. 1 Main types of polysaccharides with their repeating monomer units

Polysaccharides are the most abundant biopolymers, carbohydrates, and organic materials on Earth (Theng 2012; Song et al. 2012). More than 90% of the carbohydrates are found in nature as polysaccharides. Polysaccharides have a polymeric structure consisting of a long chain of monosaccharide units linked by glycosidic bonds with linear and branched parts. Their degree of polymerization (number of monosaccharides) varies between 100 and 15,000. According to the monosaccharide building units, polysaccharides are classified in two categories: homopolysaccharides and heteropolysaccharides. Homopolysaccharides (homoglycan) are the ones having the same type of monosaccharides, such as cellulose and starch, while heteropolysaccharides (heteroglycans) have at least two types of monosaccharides, such as pectin and alginate (Naseri-Nosar and Ziora 2018; Nie et al. 2017; Vasile et al. 2020). Polysaccharides can also be categorized depending on their sources as plant-based (e.g., cellulose, starch) and animal-based (chitosan, alginate), as shown in Fig. 1 (Nešić et al. 2020).

In the industry, a food packaging material is produced mainly by continuous methods such as melt extrusion and blown-film extrusion. During research studies, various batch film processing methods like solution casting and compression molding are used to test the film-forming ability of the material. Recently, electrospinning has become attractive in food packaging processing studies, which results in packaging in the form of nonwovens. However, these batch methods are not practical in the industry due to the low processing rate (Güzdemir et al. 2021).

Polymers can be converted into food packaging film via its film formability and biocompatibility characteristics, so it should have enough extensional and appropriate shear viscosity. Polysaccharides are polymeric substances possessing a low degree of film formability, but they are biocompatible materials, which makes them to convert into food packaging material in the form of films and coatings. Processing aids like compatibilizers, plasticizers, and emulsifiers are used to improve the film processability. Additionally, the composites of the polysaccharides with other polymers may have improved viscosity and better processability compared to those of its pure form (Nešić et al. 2020; Cazón et al. 2017).

A food packaging material should not permeate aroma and gases (water vapor, oxygen, nitrogen, and carbon dioxide) and should have appropriate mechanical properties like ductility, tensile strength, and increased shelf life (Senturk et al. 2018). Polysaccharides have been investigated in many food packaging researches, especially as the matrix polymer in biodegradable active packaging and as bioactive and sensor materials in smart packaging. Due to polysaccharide films' low mechanical properties and poor water vapor barrier characteristics, they need modifications during processing by nano/micro modifiers. Also, the other desired properties can be gained to polysaccharide films by incorporating functional additives like antioxidant, antimicrobial, or other biologically active components. Therefore, currently, R&D focus is to improve the properties of these biodegradable food packaging materials besides their processability so that they can replace or reduce the number of synthetic polymers used in the packaging industry (Nešić et al. 2020).

The main objective of this chapter was to review recently published literature on polysaccharide-based films. The review was done mainly on the processing techniques and packaging properties of cellulose and derivatives, chitosan, starch, alginate, and pectin-based food packaging materials.

2 Polysaccharides for Biodegradable Films

2.1 Cellulose and Derivatives

Cellulose is the most abundant polysaccharide in nature. It has a linear structure bonded with covalent linkage between two anhydroglucose rings and an oxygen in a β1-4 glycosidic bond. Cellulose is a plant-based polysaccharide available in almost all plants and in agricultural residues. Although its main sources are wood, cotton, and bacteria, it is found in agricultural waste like shell, husk and peel, and sugarcane bagasse, which can be valorized as a low-cost material. It exists in the form of a composite containing hemicellulose, cellulose, and lignin, while its pure state is synthesized by microorganisms (Nešić et al. 2020; Ummartyotin and Manuspiya 2015). Cellulose has many derivatives, some of which are cellulose sulfate, acetate, and nitrate, carboxymethyl cellulose, ethyl and methyl cellulose, and micro and nanocelluloses (Cao et al. 2016; Liu et al. 2021).

Owing to cellulose's inexpensive, biobased, biodegradable, chemically stable, and nontoxic nature, there is a significant amount of research on cellulose-based films (Cazón et al. 2017). Besides its advantages, pure cellulose doesn't have enough

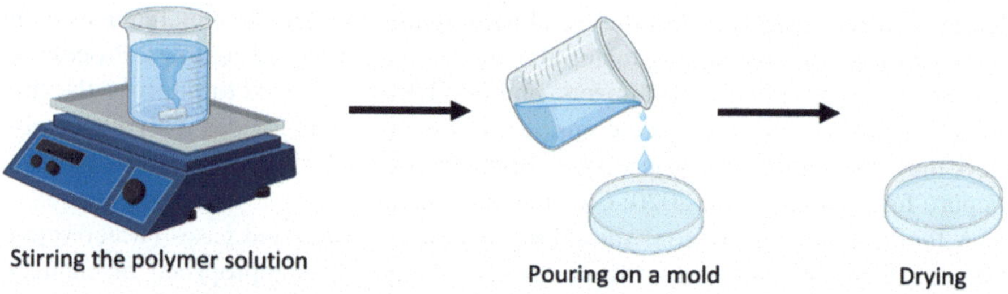

Fig. 2 Solution film casting method

extensional viscosity to be able to form films (Moberg et al. 2014). Therefore, a majority of cellulose films are polymer composites filled with cellulose particles or its derivatives. Melt extrusion, electrospinning, and solution casting are the methods for production of cellulose-filled composites.

Solution casting is one of the easiest methods to produce films in laboratories for testing film-forming ability of a polymer-polymer composite and then for investigating the film properties, but not suitable for large numbers of production. In solution casting method, polymer is solved in a suitable solution, and then polymer solution is poured in a mold. After drying, polymer film is formed, as shown in Fig. 2 (Wang et al. 2016a). Cellulose is not soluble in water and common organics because of its long and closely packed molecular chains and complex patterns of its hydrogen bonds (Wang et al. 2016b; Cazón et al. 2017). Therefore, a great effort has been made on the improvement of cellulose solubility. It is soluble in some carboxylic acids, dimethylsulfoxide (DMSO), *N*-methylmorpholine *N*-oxide, ionic liquids, LiCl/*N*,*N*-dimethylacetamide, NaOH aqueous solution, alkali/urea, NaOH/thiourea aqueous solution, tetrabutylammonium fluoride/dimethyl sulfoxide system, and molten inorganic salt hydrates (Cazón et al. 2017; Jedvert and Heinze 2017).

Methyl cellulose has the highest hydrophilicity among cellulose derivatives so that it is more easily solved in the organic solvents. Methyl cellulose is also known for its low permeability properties to oxygen, carbon dioxide, and lipid, but it permeates moisture (Cazón et al. 2017; Bedane et al. 2016). De Dicastillo et al. (2016) produced methyl cellulose films cross-linked with glutaraldehyde and incorporated with maqui berry extract by solution casting method using ethanol solution. Incorporation of glutaraldehyde and maqui berry extract resulted in better water vapor barrier properties.

Cellulose and cellulose-based composites have been blended with other additives to improve some of the properties. Dashipour et al. (2015) worked on the improvement of antioxidant and antimicrobial properties of cellulose. They casted methyl cellulose films by using water as solvent, glycerol as plasticizer, and *Zataria multiflora* (ZM) essential oil as an additive. ZM essential oil led the films with higher antioxidant and antimicrobial activities but resulted in lower water vapor barrier properties. De Dicastillo et al. (2016b) used ethanol to solve methyl cellulose for the solution casting process. Glutaraldehyde, as cross-linking agent, improved mechanical strength of methyl cellulose. Also, murta fruit extract was incorporated into the solution which contributed antioxidant and antimicrobial properties to the films. In a recently published work, carboxymethyl cellulose-polyvinyl alcohol-aloe vera blends were cross-linked with citric acid, and packaging films were produced by solution casting method (Kanatt and Makwana 2020). The films were observed to have improved water vapor barrier properties owing to the presence of citric acid. Citric acid also strengthened interfacial bonding resulting in higher mechanical properties. The films applied on minced chicken meat led to delay in lipid peroxidation and inhibition of bacteria growth due to aloe vera. In another study, Cazon et al. (2018) casted cellulose-polyvinyl alcohol (PVA) composites by using sodium hydroxide-urea aqueous solution and plasticizing with glycerol. PVA contributed toughness enhancement of the films; in contrast, glycerol addition resulted in a reverse effect on toughness. These films demonstrated UV-protection properties with optimal transparency. Furthermore, in another work, cellulose nanofiber based films were functionalized with carbon dots (Ezati et al. 2022). The films showed high antioxidant activity and increased opaqueness with a good level of transparency. On the other hand, they had a worse water vapor barrier property. When tangerine and strawberries were coated with the film solution, shelf live is prolonged by more than 10 and 2 days, respectively, due to the films' antimicrobial activity.

Melt extrusion, another method to produce polymeric films, is an industrial continuous processing method. Cellulose and its derivatives have been extruded into films, especially when it is blended with other polymers. Cellulose nanofillers (CNF) were incorporated in PLA for food packaging applications by Claro et al. (2016). CNF improved the mechanical, thermomechanical, and barrier properties of PLA. The challenge here was to increase the content of cellulose in composite due to the increased shear viscosity and decreased film-drawing property. Similarly, nanocellulose and cellulose nanocrystals were extruded with PLA into packaging films (Ariffin et al. 2018; Dhar et al. 2017). Improved tensile strength, crystallinity, wettability, and reduced respiration were observed for nanocellulose-filled PLA films (Liu et al. 2021; Ariffin et al. 2018). Thermal and barrier properties were enhanced when cellulose nanocrystals were incorporated into PLA (Dhar et al. 2017; Liu et al. 2021). In another recently conducted study, spray-dried carboxymethyl cellulose (CMC) extracted from bleached bagasse was filled into PLA, and then the composite films were melt-extruded (Kamthai and Magaraphan 2019). The films were applied on mangoes as an active packaging. Increasing CMC content resulted in longer shelf life due to the water vapor absorption property of CMC. In fact, water vapor held by CMC retarded the ripening by slowing down the respiration rate.

Cellulose fibers can be produced by electrospinning, and then these fibers are either converted into nonwovens used for food packaging or incorporated into polymer matrices to be converted into composite films. Several examples of the polymer matrices used for cellulosic nanofibers were soybean isolate, chitosan, and polyvinyl alcohol. Electrospun cellulose nanofibers have high light transmission capacities and acceptable mechanical properties so they can be used as edible films. In a study, cellulose nanofibers were filled into mango puree-based edible films, resulting in enhanced mechanical properties and also water vapor barrier properties (Rezaei et al. 2015). Furthermore, electrospinning is a suitable method to produce active and intelligent packaging from cellulose and its derivatives. In a recent study, zein-ethyl cellulose fibers were electrospun (Niu et al. 2020). Mushroom samples were packaged with this composite packaging material that extended the shelf life. The electrospun fibers presented enhanced water resistance due to the reduced content of free hydrophilic groups via hydrogen bonding between the hydroxyl groups of ethyl cellulose and the amino groups of zein. Aydogdu et al. (2019) processed hydroxypropyl methyl cellulose (HPMC) fibers by electrospinning and collected the fibers on PLA films resulting in bilayer packaging films. The bilayer films represented increased opacity. In another study, HPMC-polyethylene oxide fibers were spun as packaging material. Polyethylene oxide helped the films to get lower water resistance compared to pure HPMC films (Aydogdu et al. 2018). When HPMC-polyethylene oxide blend was functionalized with gallic acid, the processed fibrous package delayed the oxidation during storage (Aydogdu et al. 2019b).

Cellulose and its derivatives have been also used as materials of intelligent food packaging in several works. In a recent study, bacterial cellulose films were produced by polypyrrole-zinc oxide nanocomposite as intelligent packaging material (Pirsa and Shamusi 2019). Chicken thigh samples were packaged with these films. As a result, pH and microbial growth in the chicken thigh was controlled by the films. Sensory properties of the chicken thigh samples were enhanced while its shelf life prolonged. The electrical resistance of the film affected microbial activity, sensory properties, and acidity of the chicken samples. The changes in electrical resistance of the film were sign for the expiration date determination, which made the package an intelligent packaging material.

2.2 *Chitosan*

Chitosan is a natural linear polymer consisting of (1,4)-linked 2-amino-deoxy-b-D--glucan (Dutta et al. 2009). It is the second most abundant polysaccharide and the most abundant animal-originated biopolymer, which is obtained by partial deacetylation of chitin. Chitin is found in terrestrial arthropods like spiders and scorpions; in marine crustaceans like crabs, lobsters, and krill; mollusk like squids; and microorganisms like fungi (Haghighi et al. 2020; Salari et al. 2018). Chitin is the main by-product of seafood industry that is extracted from crustacean shell waste (Priyadarshi and Rhim 2020). Chitosan has become a widely studied material, especially for edible films and coatings, by packaging material researchers due to its film-forming and antimicrobial properties (against bacteria, yeast, molds, and fungi), biodegradability, biocompatibility, chemical stability, and nontoxicity (Cazón and Vázquez 2020).

Neat chitosan, insoluble in water, can be converted into film only via solvent casting method by acidified water. Formic acid, lactic acid, acetic acid, and propionic acid are used for solvent (water) acidification. Type of the acid, acetylation degree, and solution viscosity directly affect the film properties. Films obtained from acetic and propionic acid showed lower water vapor permeabilities compared to those obtained from formic and lactic acid (Mujtaba et al. 2019; Van den Broek et al. 2015).

In the solvent casting process, plasticizer agents are also used to improve film formability, mechanical properties, and water vapor permeability. Neat chitosan films and coating are generally edible films, so their plasticizers are chosen from edible materials. Glycerol, polyethylene glycol, polypropylene glycol, olive oil, and corn oil are the plasticizers used in chitosan solutions (Suyatma et al. 2005; Giannakas et al. 2017; Srinivasa et al. 2007; Leceta et al. 2013; Monte et al. 2018; Frick et al. 2018; Pereda et al. 2012). Glycerol showed higher plasticizing efficiency compared to the other studied plasticizers (Mujtaba et al. 2019).

The important mechanical properties for the food packaging, especially tensile strength and elongation at break, increase by plasticizer incorporation, and the films become more plastic/ductile and less brittle. In the literature, tensile strength and strain to failure of neat chitosan films were reported at 23–66 MPa and 46–66%, respectively, which seem comparable to the values of synthetic polymer films like low- and high-density polyethylene. Nevertheless, in fact, due to their low stability in humid environment, the mechanical properties of chitosan films deteriorate quickly by either it being swollen or solved in moisture (Mujtaba et al. 2019; Van den Broek et al. 2015). Additionally, literature studies showed that the properties of neat chitosan films were affected by storage conditions (time/temperature) and molecular weight of chitosan. The mechanical properties (tensile strength and elongation at break) improved with increased molecular weight and storage time but decreased with storage temperature (Van den Broek et al. 2015). Property enhancement is done by incorporation of essential oil or other additives into chitosan. In a recent study, lemongrass essential oil helped to reduce water vapor permeability of the chitosan matrix. It elongated the shelf life of chicken patties by reducing lipid oxidation (Diao et al. 2020).

Polymer blending is another way to process chitosan films (Sirinivasa et al. 2007). By polymer blending with chitosan, the target can be to increase the biodegradation rate and antimicrobial properties or to reduce the synthetic polymer content in the packaging film, if the blended polymer is synthetic and nonbiodegradable. Chitosan was blended with various polymers such as polyethylene oxide (PEO), poly (lactic acid) (PLA), poly(vinyl alcohol) (PVA), silica, polyethylene, and polyethylene terephthalate; and various cross-linking agents such as fatty acids, borate, tripolyphosphate, and glutaraldehyde were used in the solvents to make edible antimicrobial films (Nešić et al. 2020; Aydogdu et al. 2018; Pirsa and Shamusi 2019; Dutta et al. 2009).

Polymer blending is conducted by either solution casting or extrusion. During blending, various additives have been used in the blends to improve desired properties. Lei et al. (2021) prepared chitosan-PVA films functionalized with a catechol-containing compound by using solution casting method at 10 wt% of chitosan content. Functionalized chitosan-PVA films with an average thickness of 80 μm had improved opacity and showed higher tensile strength and strain to failure than those of neat PVA films. The films exhibited excellent antibacterial properties against *Escherichia coli* (*E. coli*). In another study, *Syzygium cumini* leaf extract (*S. cumini*) was incorporated into chitosan-PVA blend (Kasai et al. 2018). *S. cumini* contributed to improvement of antibacterial property of the films and also mechanical properties at low concentrations due to its attribution to H-bonding interactions between chitosan and PVA.

Extrusion is a continuous film processing method, which is mostly preferred in the industry, especially for the traditional synthetic plastics. One of the most widely used plastics in the packaging industry is linear density polyethylene (LDPE). There have been a great number of studies about chitosan-synthetic polymer composites. In one of the recent ones, Giannakas et al. (2021) blended LDPE with chitosan and used basil oil as a modifier in the blend. Extruded films with basil oil showed improved moisture barrier properties besides higher mechanical and antioxidant properties. Chicken filets packed with this active film had a longer shelf life than those packed with neat LDPE and LDPE-chitosan. Furthermore, various synthetic films were extruded with chitosan as composite films. Chitosan contributed antibacterial activity and hydrophilicity to the films. Some examples are polypropylene and polystyrene (Wang et al. 2018).

Poly(lactic) acid (PLA) is a biodegradable polymer possessing the disadvantages of a high cost and low degradation rate (Güzdemir et al. 2020). Using chitosan with PLA may provide to the films a lower cost and faster degradation rate, while it provides a better processability and higher mechanical properties to chitosan. There have been a good number of studies about chitosan-PLA composites. Latest studies have been concentrated on the advancement of the interfacial adhesion between chitosan and PLA. Râpă et al. (2016) plasticized PLA with tributyl o-acetyl citrate (ATBC) and extruded chitosan-PLA composite films. ATBC contributed in a better transparency, mechanical properties, and thermal properties than those of without ATBC. The migration test results offered these composite films to be used for the packaging of nonfatty foods at refrigeration temperature. In another recent study, Salmas et al. (2021) produced chitosan-PLA film by melt extrusion at 30 wt% chitosan concentration. Basil oil was used as an antimicrobial and compatibilizer agent in chitosan-PLA blends. The composite films with basil oil showed improved thermal, mechanical, and barrier properties compared to those of the films without basil oil.

There have been few studies on the film extrusion of chitosan blended with natural polymers. Lately, Llanos et al. (2021) extruded chitosan-starch blends filled with bamboo fibers and nanoclay into films at maximum 25 wt% chitosan content. It was reported that higher concentrations of chitosan caused insufficient extensional viscosity and reduced film-forming ability of the blend. The fillers enhanced the mechanical properties and water vapor permeability of the starch-chitosan films.

Other than solution casting and extrusion, processing methods of in situ polymerization and electrospinning have been used to produce chitosan and chitosan-based food packaging materials. One of the latest studies demonstrated that chitosan-PVA films were produced by in situ development of silica. Silica led to the building of hydrogen bonds resulting in higher mechanical properties and improved oxygen and moisture barrier properties. The packaging films were applied to cherries and observed an elongated shelf life by three times (Yu et al. 2018). Moreover, in several studies, nonwovens of chitosan were electrospun for food packaging purposes. Similar to chitosan films, nanofibers of chitosan electrospun with poly(ethylene) oxide/poly(vinyl alcohol)/gliadin protein isolate prolonged the shelf life of food products by exhibiting antibacterial properties, especially against Gram-negative *Escherichia coli* and *Salmonella Typhimurium* and Gram-positive *Staphylococcus aureus* and *Listeria innocua*. In a study, chicken samples were packaged with ε-polylysine/chitosan nanofibers; later, it was observed that the color and flavor of the chicken stayed unchanged while its shelf life extended (Kumar et al. 2019). Also, another work pointed out that chitosan nanofibers were used to wrap unprocessed meat, and the results were the prolonged shelf life to 1 week with the conservation of its quality and freshness (Arkoun et al. 2018).

Chitosan has also been used as a component of intelligent packaging besides active packaging. Koosha and Hamedi (2019) fabricated smart packages by black carrot anthocyanins incorporated into chitosan-PVA-bentonite nanocomposites. While bentonite caused a reduction in water vapor permeability, anthocyanin worked as an antibacterial agent. In another recent study, electrospun chitosan-based nanofibers with poly(ethylene) oxide and curcumin were tested on chicken breast samples. During 5-day storage, the color and pH change of the nanofibers were the responses of the alteration in the total volatile basic nitrogen of chicken. Thus, the package was used to monitor the shelf life of the chicken samples (Yildiz et al. 2021).

Chitosan or chitosan composites have also been used as a component of edible coating. Chitosan-based coatings lead to longer shelf life of various whole and fresh-cut fruits because it postpones the alteration on the fruit due to reactive oxygen species by fortifying enzymatic and nonenzymatic antioxidant activities (Adiletta et al. 2021). One example is that the banana fruit exhibited longer shelf life when coated with chitosan filled starch coating (Lustriane et al. 2018). Another edible coating prepared with water chestnut starch, glycerol, fruit extracts, essential oils, and chitosan led to a better cheese quality and longer shelf life when applied onto bod ljong cheese (Mei et al. 2015). Furthermore, chitosan-based coatings demonstrated an inhibitory effect on fruit browning, which occurs through an enzymatic oxidation reaction in which phenolics turn into quinones via polyphenol oxidase. The literature studies showed that enzymatic browning of strawberry, sweet cherry, cherimoya, loquat, longan, guava, grape, fig, apple, mango, pistachio, lemon, and pineapple was inhibited when coated by chitosan-based coatings (Adiletta et al. 2021).

2.3 Starch

Starch is the most abundant polysaccharide and the major carbohydrate found in plants, produced mainly in plant tubers and endosperm. It can be extracted from rice, maize, potato, wheat, barley, soya, cassava, tapioca, etc. Starch consists of two main components, amylopectin and amylose. Amylopectin is a highly branched, high molecular weight glucan with glycosidic linkages, constituting approximately 70% of starch structure. On the other hand, amylose, a linear glucan, forms 20–30% of the structure (Nešić et al. 2020; Niranjana and Prashantha 2018; Kumar 2019).

Starch is an inexpensive flexible polymer with the charming properties of biodegradability, nontoxicity, oil and oxygen impermeability, and heat stability for food packaging applications. It is soluble in warm water and has a gelation characteristic (Khan et al. 2017). It has an odorless, colorless, and tasteless chemistry. Amylose part of starch provides film-forming properties to it due to its linear nature (Kumar 2019). These characteristics have drawn attention of packaging researchers onto starch. Especially, thermoplastic starch (TPS), which is known as plasticized starch with plasticizers like water, glycerol, glycol, and sorbitol, made a difference from the other types of polysaccharides owing to its easy extrudability. Therefore, TPS is easily processable into films via the methods of melt extrusion, compression molding, and blow molding (Nešić et al. 2020).

Starch has three main drawbacks when used in the neat form as food packaging material. It has low mechanical properties, poor water resistance, and retrogradation (increasing brittleness over time). In order to improve the properties of starch, it has blended either various reinforcements such as carbon nanotubes, nanoclays, silver, zinc oxide, etc. or various polymers such as polyethylene, polyvinyl alcohol, polycaprolactone, PLA, PHB, and PBS (Nešić et al. 2020). In addition, starch-based plastics have already been commercialized. The Eco-Go company produces food packaging materials from cassava and corn starch, while Plantic Technologies Limited sells biodegradable packaging films made of corn starch and polyethylene/polypropylene that have superior barrier properties (Ferreira et al. 2016).

Essential oils, plant extracts, and other natural products have been mixed with starch to provide the film's antimicrobial and antioxidant properties for active packaging purposes (Theng 2012). In a recent study, Ferreira et al. (2021) prepared TPS films functionalized with chitosan nanocapsules and Ho wood-cinnamon essential oil. Incorporation of Ho wood or cinnamon essential oil at 3 wt% into TPS-chitosan resulted in better water barrier properties and antibacterial characteristics against *E. coli* or *Bacillus subtilis*. The films with cinnamon essential oil preserved the freshness and quality of strawberries by preventing fungi contamination and reducing fruit's weight loss.

Solvent casting method has been used to produce starch films, especially for R&D purposes. In a recent work, solvent-casted cassava starch-based biodegradable films were blended with rosemary leave polyphenols and plasticized with glycerol (Piñeros et al. 2017). Rosemary extract improved UV-blocking properties of the films and provided antioxidant activity to the films. Another work was on the solvent-casted starch-based nanocomposite films reinforced with silver, copper oxide, and zinc oxide nanoparticles (Peighambardoust et al. 2019). The metallic nanoparticles improved the mechanical performance, UV-blocking, water barrier, and antibacterial properties of the composite films compared to neat starch films. However, metallic reinforcements made the packaging film more expensive and favorable to toxicity against eukaryotic cells. Wang et al. (2019) found that corn starch-chitosan films, processed via solvent casting, having cinnamaldehyde additive represented increased moisture barrier and antibacterial properties by virtue of cinnamaldehyde. The films were packed on strawberry samples and retarded the senescence of strawberries while preserving the nutritional value. Xu et al. (2018) used tapioca starch to prepare packaging material with cellulose nanocrystal and grape pomace extracts via solvent casting. Cellulose nanocrystals contributed significantly to boost mechanical strength and water barrier, while grape pomace extract supported antibacterial activity of the films. The nanocomposite films extended shelf life of ready-to-eat deli meat products by inhibiting *L. monocytogenes*. Gao et al. (2020) incorporated soybean oil into corn/octenylsuccinate starch and casted the starch films. Soybean oil decreased water permeability of the films by increasing surface hydrophobicity. Soybean oil also improved film surface uniformity and mechanical properties. Junlapong et al. (2019) produced corn starch-polyvinyl alcohol composite with a compatibilizer, triethylamine. Triethylamine improved the mechanical properties by building intermolecular bonding between starch and PVA. Water resistance of the films increased with increasing PVA content, while their moisture absorption susceptibility decreased. Liu et al. (2018) casted sweet potato starch films to test the effect of ultrasonic treatment on amylose-lipid complex formation. Tensile strength and UV transmissivity improved at low power density ultrasound technique, while tensile modulus and water vapor barrier properties went down.

Compression molding is another method of batch processing that is mostly used for R&D purposes. In this processing method, polymers or premixed composite materials are shaped in a mold by compression at a temperature above the composite's softening point. Starch-based composite films have also been processed by compression molding. A recent study conducted by Menzel (2020) showed that compression-molded starch films filled with sunflower hull extract with its phenols and cellulose fibers had better mechanical properties compared to unfilled starch films due to cellulose fibers and better antioxidant properties due to the phenolic compounds valorized from sunflower waste. When citric acid was also added to the blend as cross-linking agent, these edible films showed a higher tensile strength than that without citric acid. The films were offered as edible coating preventing lipid oxidation. Lopez et al. (2015) fabricated starch-based packaging bags filled with talc nanoparticles by compression molding. Talc at a concentration of higher than 3 w% improved yield stress and modulus of elasticity. The bags, later, were sealed thermally. It was concluded that talc improved water-oxygen barrier properties, tightness of the bags, and seal strength. Castillo et al. (2017) obtained thermoplastic corn starch films with a chitosan oligomer that was extracted from shrimp shell waste. Sachet type packages with antimicrobial activity against molds and yeasts were molded by thermal compression. In another recent study, Moreno et al. (2018) fabricated starch-gelatin composites functionalized with N-α-lauroyl-l-arginine ethyl ester monohydrochloride and coated the films inside of vacuum polyamide-polyethylene pouches containing chicken breast. The films were both produced by compression molding and solvent casting. The solvent-casted films activated oxidation due the compounds formed during Maillard reaction in the films resulting in a change of pH and color of the chicken samples. On the other hand, compression-molded films having better quality than the solvent-casted ones prolonged the shelf life of chicken breast filets without any color or pH change.

The other method to produce starch films is extrusion. Both blown extrusion and melt extrusion have been used to draw starch-based films in the literature. In a study, fructose, sucrose, glucose, and glycerol were used as plasticizers to starch nanocomposite (do Val Siqueira et al. 2021). Although incorporation of sugar into starch led to more flexible film, it made the properties of the film such as tensile strength and water resistance to deteriorate. When the blend of acetylated cassava starch and pea protein isolate was obtained into films via blown extrusion, the films become more hydrophobic, thermally stable, and less permeable to water vapor and oxygen compared to starch films without pea protein isolate; however, they represented less flexibility and light transmissivity. Furthermore, melt extrusion was used to fabricate composite starch-based films having enhanced characteristics. Zarski et al. extruded potato starch esters into films that possessed more hydrophobic surface and so less moisture absorption susceptibility compared to neat starch films (Zarski et al. 2020). Besides, Fitch et al. acetylated starch films filled with acetylated sugarcane fiber via melt extrusion (Fitch-Vargas et al. 2019). While acetylation brought in higher moisture barrier to the films, it also led to mechanical interlocking of the reinforcement to the matrix meaning higher tensile strength. Ghanbari et al. (2018) extruded thermoplastic corn starch filled with cellulose nanofibers, having lower water affinity and better mechanical properties. Calderon et al. (2018) optimized the extrusion conditions for starch plasticized with glycerol. They applied the optimized films on mango samples and observed the fruit freshness for 16 days.

Starch-based films have also been used as intelligent packaging material. The functionalization of anthocyanins and limonene in starch-PVA films generated colorimetric differences when pH was changed. These functionalized films revealed milk acidification by color change and postponed milk spoilage (Liu et al. 2017). Separately, nutmeg oil, zinc oxide, and jamun extract in starch-PVA films and anthocyanin in cassava starch were also developed intelligent films reacting as pH change indicators (Jayakumar et al. 2019; Vedove et al. 2021).

2.4 Alginate

Alginates are indigestible polysaccharides commonly produced from the cell walls of brown algae (called brown seaweed, Phaeophyceae). It is extracted from algal alginates such as *Laminaria hyperborean*, *Macrocystis pyrifera*, *Ascophyllum nodosum*, and *Ecklonia species* at higher amounts and from bacterial alginates such as *Azotobacter* (as exopolysaccharides) at lesser concentrations. Alginates, the salts of alginic acids, are composed of (1→4)-linked α-l-guluronic acid (G) and β-d-mannuronic acid (M). The monomers of alginate polymer, guluronic and mannuronic acids, are alternately arranged in the polymer chains as GG, MM, and MG blocks, providing a linear structure to the polymer (Nešić et al. 2020; Cazón et al. 2017; Senturk et al. 2018; Gheorghita et al. 2020). The physical and chemical characteristics of alginate depend on M/G ratio that relates on the source of the alginate (Senturk et al. 2018).

Alginate is widely used for many purposes including as stabilizing, thickening, emulsifying, swelling, gelling, or chelating agent in various industries (Senturk et al. 2018). It has a high level gelation property; in fact, it forms gels without heating or cooling unlike other polysaccharides. Gel formation of alginates is occurred by two methods: ionic cross-linking with cations and acid precipitation. Alginates have the capacity to react with di- and trivalent cations such as calcium, manganese, magnesium, and aluminum ions to form hydrogels having 3D network structure called "eggbox" model. The most effective gelling agent of alginates is calcium ions (Nešić et al. 2020; Cazón et al. 2017).

Sodium alginate is the most common salt of alginate among commercially available alginates (Gheorghita et al. 2020; Ching et al. 2017). It has become attractive in the bio-packaging industry because of its low cost, nontoxicity, and biodegradability besides its film-forming ability. However, it has a high moisture absorption susceptibility, hygroscopicity, transparency, uniformity, and low stability. Due to these disadvantages, the properties of the alginate films have been modified with various fillers (Nešić et al. 2020; Cazón et al. 2017; Gheorghita et al. 2020).

Alginate films are produced via various methods such as solvent casting, extrusion, and electrospinning. Hilbig et al. processed alginate casings via coextrusion by aiming to get rid of white efflorescence formation during sausage storage (Hilbig et al. 2020). Alginate was treated with calcium and magnesium. It was observed that calcium alginate casing was more effective than magnesium alginate casing to reduce the amount of the white efflorescence on sausage.

Gutierrez et al. (2020) obtained sodium alginate with polyethylene oxide filled with curcumin into nanofibers via electrospinning. Trifluoric acid was used as a cross-linking agent in the blend. The nanofiber composite had higher tensile strength and tensile modulus owing to curcumin and trifluoric acid.

Alginate films have been commonly researched as edible films and coatings. Nair et al. (2018) formulated the edible coating compositions for alginate-chitosan-pomegranate peel extract blend to preserve the quality and extend the shelf life of guava fruit. The coated guava exhibited preserved nutritional values and enhanced appearance. On the other hand, Yin et al. (2019) prepared sodium alginate-based coating and chitosan-cinnamon essential oil coating separately and applied them on mangoes as multilayers. The multilayered coating on mangoes resulted in extended shelf life by retained nutritional content, reduced weight loss rate, and improved visual attributes compared to uncoated mangoes. It was also reported that the adhesion of the edible coatings on the mango surface was strong. Guerreiro et al. (2015)

developed alginate-based coating functionalized with citral and eugenol essential oil additives. Raspberry samples were coated with this edible coating and presented good sensory attributes for 7 days; however, after 7 days, the sensorial properties worsened. The coating showed antimicrobial properties due to the essential oils. Puscaselu et al. (2019) designed sodium alginate-agar coating plasticized by glycerol to package powdered products such as coffee grounds, powdered milk, teas, and dehydrated vegetables. The coating films were enriched with *Stevia rebaudiana*, which is a sugar substitute. In the presence of stevia, the films became more ductile, shiner, and less viscous. The films made of alginate and stevia without agar were formed with high solubility, moderate roughness, and good mechanical properties. The coating films were suggested to be used on powdered products, dried and fresh vegetables and fruits, cheese slices, and sausages. Matiacevich et al. (2015) produced alginate-based edible coating with antimicrobial agents of propionic acid and thyme essential oils to extend shelf life of fresh chicken breast fillets. Although the coating did not affect sensorial attributes of chicken, it extended shelf life by about 33% by decreasing dehydration and adding antimicrobial activities.

2.5 Pectin

Pectin is a heteropolysaccharide component composed of α-(1,4)-linked D-galacturonic acid residues, consisting of more than 17 monosaccharides (Nešić et al. 2018). The main monosaccharide of pectin is galacturonic acid (~70% of total sugars in pectin), and the rest are mostly rhamnose, arabinose, galactose, glucose, and mannose (Mu et al. 2017). Amidation of the galacturonic acid gelates pectin while forming an eggbox structure. The viscosity and the solubility of pectin is determined by the methylated or amidated percentage of its galacturonic acid. Pectin is found in the structure of the plant cell wall, usually available with cellulose, lignin, and polyphenols (Cazón et al. 2017; Mellinas et al. 2020). Since it exists especially in fruit and vegetable peels, pectin production is a good way of waste valorization. While more common sources are apple peel, with a pectin content of 10–15 w%, and citrus peel, with a pectin content of 20–30 w%, some of the other sources include the peels of banana, papaya, tomato, and grapefruit, kiwifruit pomace, green husk of walnut, and sugar beet (Cazón et al. 2017; Naqash et al. 2017; La Torre et al. 2021).

Pectin is used in various applications in the food industry as a thickening agent, emulsifier, texturizer, and gelling agent. It has been also used as packaging material, coating, and microencapsulation agents (Mellinas et al. 2020). It has become attractive in the packaging industry due to its low cost, abundance, and biodegradability (Cabello et al. 2015).

Likewise for the other polysaccharide films, the most widely used technique to process pectin-based films is the solvent casting method (Mellinas et al. 2020). Pectin needs to be mixed with an appropriate plasticizer to turn into film state. Commonly used plasticizer for pectin is glycerol that increases the amorphous region of pectin by weakening intermolecular bonding. Polyethylene glycol is another plasticizer that can be used in pectin-based composites. Cabello et al. (2015) showed that polyethylene glycol acted as an external plasticizer forming a separate phase in the continuous pectin phase. In contrast, glycerol behaved as an internal plasticizer forming glycerol ester during pectin cross-linking. Lorevice et al. (2016) prepared pectin-based films without any plasticizer, but they mixed chitosan nanoparticles into pectin films, resulting in higher tensile strength but lower thermal stability compared to neat pectin films. The pectin films filled with chitosan nanoparticles showed pure strain to failure values.

Similarly to the other polysaccharides, antioxidants and antimicrobial agents have been added into pectin solution to get active films. In a recent study, Nisar et al. (2018) incorporated clove essential oil into pectin films and obtained not only antimicrobial activities against Gram-positive *S. aureus* and *L. monocytogenes* but also improved barrier, mechanical, and thermal properties. Almasi et al. (2020) used marjoram essential oil and Pickering emulsion blended into pectin for film formation. Marjoram essential oil brought antioxidant activity to the films where Pickering emulsion contributed to the tensile strength and water barrier improvement. Chaiwarit et al. (2018) extracted pectin from mango peel, plasticized it with glycerol, and used orange oil as antibacterial agent in the casted films. However, orange oil did not affect the antibacterial activity of the films. Azedero et al. (2016) used pomegranate juice and citric acid in pectin films. Pomegranate juice as a plasticizer enhanced ductility while weakening the other tensile properties and water vapor permeability. On the other hand, citric acid empowered tensile and barrier properties.

Most of the pectin-based edible coatings have been prepared as a solution. Recently, Naqash et al. (2022) 9 coated fresh-cut apples with pectin-cinnamon oil solution resulting in an antimicrobial protection and improved visual properties of the fruit. Priyadarshi et al. (2022) coated peanuts by pectin-pullulan blend functionalized with grape-seed oil. The grape-seed oil made a contribution to the coatings with antibacterial activities against *E. coli* and *L. monocytogenes*, a high UV barrier. It also extended the shelf life of peanuts by delaying the lipid oxidation. Grape-seed oil was suggested as an additive in pectin-based coatings for high fat content foods. In another recent study, Gedikoglu (2022) coated bologna slices with pectin-based solution added with *Thymus vulgaris* and *Thymbra spicata* essential oils. The bologna slices showed a delayed spoilage due the decreased lactic acid bacteria, yeast, mold, and *Salmonella typhimurium* growth by the addition of the essential oils.

Finally, there have also been studies on the pectin-based films processed by different processing techniques than solvent casting such as compression molding, extrusion, and electrospinning. Recently, Gouveia et al. (2019) used compression molding at 110–130 °C to prepare pectin films plasticized with glycerol-choline chloride blend. De Oliveira et al. (2021) successfully extruded high methoxyl pectin-based films plasticized with citric acid and water through a twin screw extruder at 100 °C. The extruded films showed darker color compared to the films compression-molded at 85 °C due to the higher temperature exposure.

3 Summary

In recent years, it has become more crucial to package food products and prevent them from contamination and spoilage, especially understood during the COVID-19 pandemic situation, where the fabrication and transportation shut down. Since the commonly used synthetic plastics for packaging have hauled the world to transform into a relentlessly growing plastic garbage, there is a rising demand for biobased, biodegradable packaging materials to replace the synthetic plastics.

Polysaccharides are inexpensive and sustainable alternatives to petroleum-based packages with their polymeric structure. Their processability requires to be increased with appropriate plasticizers and additives. Although neat polysaccharide films don't possess enough barrier and mechanical characteristics to pack foods for shelf time, they have shown improved properties when combined with other polymers or fillers. Moreover, they can be used in active packaging applications when combined with functional materials. With the formulations for an appropriate processing and properties, the commercialization of polysaccharide films has already been started. In the near future, we will see more polysaccharide-based food packages in the market.

References

Adiletta G, Di Matteo M, Petriccione M. Multifunctional role of chitosan edible coatings on antioxidant systems in fruit crops: A review. *International journal of molecular sciences*. 2021;22(5):2633. https://doi.org/10.3390/ijms22052633

Almasi H, Azizi S, Amjadi S. Development and characterization of pectin films activated by nanoemulsion and pickering emulsion stabilized marjoram (origanum majorana L.) essential oil. *Food Hydrocoll*. 2020;99:105338. https://doi.org/10.1016/j.foodhyd.2019.105338

Andrady AL. (2011) Microplastics in the marine environment. *Mar Pollut Bull*. 62(8):1596–1605. https://doi.org/10.1016/j.marpolbul.2011.05.030

Ariffin H, Norrrahim MNF, Yasim-Anuar TAT, et al. Oil palm biomass cellulose-fabricated polylactic acid composites for packaging applications. In: *Bionanocomposites for packaging applications*. Springer; 2018:95–105. https://doi.org/10.1007/978-3-319-67319-6_5

Arkoun M, Daigle F, Holley RA, Heuzey MC, Ajji A. Chitosan-based nanofibers as bioactive meat packaging materials. *Packaging Technology and Science*. 2018;31(4):185–195.https://doi.org/10.1002/pts.2366

Ateia, M., Ersan, G., Alalm, M. G., Boffito, D. C., & Karanfil, T. (2022). Emerging investigator series: microplastic sources, fate, toxicity, detection, and interactions with micropollutants in aquatic ecosystems–a review of reviews. Environmental Science: Processes & Impacts, 24(2), 172–195. https://doi.org/10.1039/D1EM00443C

Aydogdu A, Sumnu G, Sahin S. A novel electrospun hydroxypropyl methylcellulose/polyethylene oxide blend nanofibers: Morphology and physicochemical properties. *Carbohydr Polym*. 2018;181:234–246. https://doi.org/10.1016/j.carbpol.2017.10.071

Aydogdu A, Sumnu G, Sahin S. Fabrication of gallic acid loaded hydroxypropyl methylcellulose nanofibers by electrospinning technique as active packaging material. *Carbohydr Polym*. 2019b;208:241–250. https://doi.org/10.1016/j.carbpol.2018.12.065

Aydogdu A, Yildiz E, Ayhan Z, Aydogdu Y, Sumnu G, Sahin S. Nanostructured poly (lactic acid)/soy protein/HPMC films by electrospinning for potential applications in food industry. *European Polymer Journal*. 2019;112:477–486. https://doi.org/10.1016/j.eurpolymj.2019.01.006

Azeredo HM, Morrugares-Carmona R, Wellner N, Cross K, Bajka B, Waldron KW. Development of pectin films with pomegranate juice and citric acid. *Food Chem*. 2016;198:101–106. https://doi.org/10.1016/j.foodchem.2015.10.117

Bedane AH, Eić M, Farmahini-Farahani M, Xiao H. Theoretical modeling of water vapor transport in cellulose-based materials. *Cellulose*. 2016;23(3):1537–1552. https://doi.org/10.1007/s10570-016-0917-y

Cabello SP, Takara EA, Marchese J, Ochoa NA. Influence of plasticizers in pectin films: Microstructural changes. *Mater Chem Phys*. 2015;162:491–497. https://doi.org/10.1016/j.matchemphys.2015.06.019

Calderón-Castro A, Vega-García MO, de Jesús Zazueta-Morales J, et al. Effect of extrusion process on the functional properties of high amylose corn starch edible films and its application in mango (mangifera indica L.) cv. tommy atkins. *Journal of food science and technology*. 2018;55(3):905–914. https://doi.org/10.1007/s13197-017-2997-6

Cao J, Sun X, Lu C, Zhou Z, Zhang X, Yuan G. (2016) Water-soluble cellulose acetate from waste cotton fabrics and the aqueous processing of all-cellulose composites. *Carbohydr Polym*. 149:60–67. https://doi.org/10.1016/j.carbpol.2016.04.086

Castillo LA, Farenzena S, Pintos E, et al. Active films based on thermoplastic corn starch and chitosan oligomer for food packaging applications. *Food Packaging and Shelf Life*. 2017;14:128–136. https://doi.org/10.1016/j.fpsl.2017.10.004

Cazón P, Vázquez M, Velazquez G. Cellulose-glycerol-polyvinyl alcohol composite films for food packaging: Evaluation of water adsorption, mechanical properties, light-barrier properties and transparency. *Carbohydr Polym*. 2018;195:432–443. https://doi.org/10.1016/j.carbpol.2018.04.120.

Cazón P, Vázquez M. Mechanical and barrier properties of chitosan combined with other components as food packaging film. *Environmental Chemistry Letters*. 2020;18(2):257–267. https://doi.org/10.1007/s10311-019-00936-3

Cazón P, Velazquez G, Ramírez JA, Vázquez M. Polysaccharide-based films and coatings for food packaging: A review. *Food Hydrocoll*. 2017;68:136–148. https://doi.org/10.1016/j.foodhyd.2016.09.009

Chaiwarit T, Ruksiriwanich W, Jantanasakulwong K, Jantrawut P. Use of orange oil loaded pectin films as antibacterial material for food packaging. *Polymers*. 2018;10(10):1144. https://doi.org/10.3390/polym10101144

Ching SH, Bansal N, Bhandari B. Alginate gel particles–A review of production techniques and physical properties. *Crit Rev Food Sci Nutr*. 2017;57(6):1133–1152. https://doi.org/10.1080/10408398.2014.965773

Claro P, Neto A, Bibbo A, Mattoso L, Bastos M, Marconcini JM. Biodegradable blends with potential use in packaging: A comparison of PLA/chitosan and PLA/cellulose acetate films. *Journal of Polymers and the Environment*. 2016;24(4):363–371. https://doi.org/10.1007/s10924-016-0785-4

Dashipour A, Razavilar V, Hosseini H, et al. Antioxidant and antimicrobial carboxymethyl cellulose films containing zataria multiflora essential oil. *Int J Biol Macromol*. 2015;72:606–613. https://doi.org/10.1016/j.ijbiomac.2014.09.006

De Dicastillo CL, Rodríguez F, Guarda A, Galotto MJ. Antioxidant films based on cross-linked methyl cellulose and native chilean berry for food packaging applications. *Carbohydr Polym*. 2016;136:1052–1060. https://doi.org/10.1016/j.carbpol.2015.10.013

De Dicastillo CL, Bustos F, Guarda A, Galotto MJ. Cross-linked methyl cellulose films with murta fruit extract for antioxidant and antimicrobial active food packaging. *Food Hydrocoll*. 2016b;60:335–344. https://doi.org/10.1016/j.foodhyd.2016.03.020

de Oliveira, Ana Carolina Salgado, Ferreira LF, de Oliveira Begali D, et al. Thermoplasticized pectin by extrusion/thermo-compression for film industrial application. *Journal of Polymers and the Environment*. 2021;29(8):2546–2556. https://doi.org/10.1007/s10924-021-02054-0

Dhar P, Gaur SS, Soundararajan N, et al. Reactive extrusion of polylactic acid/cellulose nanocrystal films for food packaging applications: Influence of filler type on thermomechanical, rheological, and barrier properties. *Ind Eng Chem Res*. 2017;56(16):4718–4735. https://doi.org/10.1021/acs.iecr.6b04699

Diao X, Huan Y, Chitrakar B. Extending the shelf life of ready-to-eat spiced chicken meat: Garlic aqueous extracts-carboxymethyl chitosan ultrasonicated coating solution. *Food and Bioprocess Technology*. 2020;13(5):786–796. https://doi.org/10.1007/s11947-020-02428-7

do Val Siqueira L, Arias, Carla Ivonne La Fuente, Maniglia BC, Tadini CC. Starch-based biodegradable plastics: Methods of production, challenges and future perspectives. *Current Opinion in Food Science*. 2021;38:122–130. https://doi.org/10.1016/j.cofs.2020.10.020

Dutta PK, Tripathi S, Mehrotra GK, Dutta J. Perspectives for chitosan based antimicrobial films in food applications. *Food Chem*. 2009;114(4):1173–1182. https://doi.org/10.1016/j.foodchem.2008.11.047

Ezati, P., Rhim, J. W., Molaei, R., Priyadarshi, R., & Han, S. (2022). Cellulose nanofiber-based coating film integrated with nitrogen-functionalized carbon dots for active packaging applications of fresh fruit. Postharvest Biology and Technology, 186, 111845. https://doi.org/10.1016/j.postharvbio.2022.111845

Ferreira, A. R., Alves, V. D., & Coelhoso, I. M. (2016). Polysaccharide-based membranes in food packaging applications. *Membranes, 6*(2), 22.

Ferreira, R. R., Souza, A. G., Quispe, Y. M., & Rosa, D. S. (2021). Essential oils loaded-chitosan nanocapsules incorporation in biodegradable starch films: A strategy to improve fruits shelf life. International Journal of Biological Macromolecules, 188, 628–638. https://doi.org/10.1016/j.ijbiomac.2021.08.046

Fitch-Vargas PR, Camacho-Hernández IL, Martínez-Bustos F, et al. Mechanical, physical and microstructural properties of acetylated starch-based biocomposites reinforced with acetylated sugarcane fiber. *Carbohydr Polym*. 2019;219:378–386. https://doi.org/10.1016/j.carbpol.2019.05.043

Frick, J. M., Ambrosi, A., Pollo, L. D., & Tessaro, I. C. (2018). Influence of glutaraldehyde crosslinking and alkaline post-treatment on the properties of chitosan-based films. Journal of Polymers and the Environment, 26, 2748–2757. https://doi.org/10.1007/s10924-017-1166-3

Gao W, Wu W, Liu P, Hou H, Li X, Cui B. Preparation and evaluation of hydrophobic biodegradable films made from corn/octenylsuccinated starch incorporated with different concentrations of soybean oil. *Int J Biol Macromol*. 2020;142:376–383. https://doi.org/10.1016/j.ijbiomac.2019.09.108

Gedikoğlu A. The effect of thymus vulgaris and thymbra spicata essential oils and/or extracts in pectin edible coating on the preservation of sliced bolognas. *Meat Sci*. 2022;184:108697. https://doi.org/10.1016/j.meatsci.2021.108697

Ghanbari A, Tabarsa T, Ashori A, Shakeri A, Mashkour M. Preparation and characterization of thermoplastic starch and cellulose nanofibers as green nanocomposites: Extrusion processing. *Int J Biol Macromol*. 2018;112:442–447. https://doi.org/10.1016/j.ijbiomac.2018.02.007

Gheorghita Puscaselu R, Lobiuc A, Dimian M, Covasa M. Alginate: From food industry to biomedical applications and management of metabolic disorders. *Polymers*. 2020;12(10):2417. https://doi.org/10.3390/polym12102417

Giannakas AE, Salmas CE, Leontiou A, et al. Synthesis of a novel chitosan/basil oil blend and development of novel low density polyethylene/chitosan/basil oil active packaging films following a melt-extrusion process for enhancing chicken breast fillets shelf-life. *Molecules*. 2021;26(6):1585. https://doi.org/10.3390/molecules26061585

Gouveia TI, Biernacki K, Castro MC, Gonçalves MP, Souza HK. A new approach to develop biodegradable films based on thermoplastic pectin. *Food Hydrocoll*. 2019;97:105175. https://doi.org/10.1016/j.foodhyd.2019.105175

Guerreiro AC, Gago CM, Faleiro ML, Miguel MG, Antunes MD. Raspberry fresh fruit quality as affected by pectin-and alginate-based edible coatings enriched with essential oils. *Scientia Horticulturae*. 2015;194:138–146. https://doi.org/10.1016/j.scienta.2015.08.004

Gutierrez-Gonzalez J, Garcia-Cela E, Magan N, Rahatekar SS. Electrospinning alginate/polyethylene oxide and curcumin composite nanofibers. *Mater Lett*. 2020;270:127662. https://doi.org/10.1016/j.matlet.2020.127662

Güzdemir Ö, Bermudez V, Kanhere S, Ogale AA. Melt-spun poly (lactic acid) fibers modified with soy fillers: Toward environment-friendly disposable nonwovens. *Polymer Engineering & Science*. 2020;60(6):1158–1168. https://doi.org/10.1002/pen.25369

Güzdemir Ö, Kanhere S, Bermudez V, Ogale AA. (2021) Boron nitride-filled linear low-density polyethylene for enhanced thermal transport: Continuous extrusion of micro-textured films. *Polymers*.13(19):3393. https://doi.org/10.3390/polym13193393

Haghighi H, Licciardello F, Fava P, Siesler HW, Pulvirenti A. Recent advances on chitosan-based films for sustainable food packaging applications. *Food Packaging and Shelf Life*. 2020;26:100551. https://doi.org/10.1016/j.fpsl.2020.100551

Hilbig J, Hartlieb K, Herrmann K, Weiss J, Gibis M. Influence of calcium on white efflorescence formation on dry fermented sausages with co-extruded alginate casings. *Food Res Int*. 2020;131:109012. https://doi.org/10.1016/j.foodres.2020.109012

Jang M, Shim WJ, Cho Y, Han GM, Song YK, Hong SH. (2020) A close relationship between microplastic contamination and coastal area use pattern. *Water Res*. 171:115400. https://doi.org/10.1016/j.watres.2019.115400

Jayakumar A, Heera KV, Sumi TS, et al. Starch-PVA composite films with zinc-oxide nanoparticles and phytochemicals as intelligent pH sensing wraps for food packaging application. *Int J Biol Macromol*. 2019;136:395–403. https://doi.org/10.1016/j.ijbiomac.2019.06.018

Jedvert K, Heinze T. Cellulose modification and shaping–a review. *Journal of Polymer Engineering*. 2017;37(9):845–860. https://doi.org/10.1515/polyeng-2016-0272

Junlapong K, Boonsuk P, Chaibundit C, Chantarak S. Highly water resistant cassava starch/poly (vinyl alcohol) films. *Int J Biol Macromol*. 2019;137:521–527. https://doi.org/10.1016/j.ijbiomac.2019.06.223

Kamthai S, Magaraphan R. Development of an active polylactic acid (PLA) packaging film by adding bleached bagasse carboxymethyl cellulose (CMCB) for mango storage life extension. *Packaging Technology and Science*. 2019;32(2):103–116. https://doi.org/10.1002/pts.2420

Kanatt SR, Makwana SH. Development of active, water-resistant carboxymethyl cellulose-poly vinyl alcohol-aloe vera packaging film. *Carbohydr Polym*. 2020;227:115303. https://doi.org/10.1016/j.carbpol.2019.115303

Kasai D, Chougale R, Masti S, Chalannavar R, Malabadi RB, Gani R. Influence of syzygium cumini leaves extract on morphological, thermal, mechanical, and antimicrobial properties of PVA and PVA/chitosan blend films. *J Appl Polym Sci*. 2018;135(17):46188. https://doi.org/10.1002/app.46188

Khan B, Bilal Khan Niazi M, Samin G, Jahan Z. Thermoplastic starch: A possible biodegradable food packaging material—A review. *J Food Process Eng*. 2017;40(3):e12447. https://doi.org/10.1111/jfpe.12447

Koosha, M., & Hamedi, S. (2019). Intelligent Chitosan/PVA nanocomposite films containing black carrot anthocyanin and bentonite nanoclays with improved mechanical, thermal and antibacterial properties. Progress in Organic Coatings, 127, 338–347. https://doi.org/10.1016/j.porgcoat.2018.11.028

Kumar N. Polysaccharide-based component and their relevance in edible film/coating: A review. *Nutr Food Sci*. 2019. https://doi.org/10.1108/NFS-10-2018-0294

Kumar TSM, Kumar KS, Rajini N, Siengchin S, Ayrilmis N, Rajulu AV. A comprehensive review of electrospun nanofibers: Food and packaging perspective. *Composites Part B: Engineering*. 2019;175:107074. https://doi.org/10.1016/j.compositesb.2019.107074

La Torre C, Caputo P, Plastina P, Cione E, Fazio A. Green husk of walnuts (juglans regia L.) from southern Italy as a valuable source for the recovery of glucans and pectins. *Fermentation*. 2021;7(4):305. https://doi.org/10.3390/fermentation7040305

Leceta, I., Guerrero, P., & De La Caba, K. (2013). Functional properties of chitosan-based films. Carbohydrate polymers, 93(1), 339–346. https://doi.org/10.1016/j.carbpol.2012.04.031

Lehel, J., & Murphy, S. (2021). Microplastics in the food chain: food safety and environmental aspects. Reviews of Environmental Contamination and Toxicology Volume 259, 1–49. https://doi.org/10.1007/398_2021_77

Lei Y, Mao L, Yao J, Zhu H. Improved mechanical, antibacterial and UV barrier properties of catechol-functionalized chitosan/polyvinyl alcohol biodegradable composites for active food packaging. *Carbohydr Polym*. 2021;264:117997. https://doi.org/10.1016/j.carbpol.2021.117997

Liu B, Xu H, Zhao H, Liu W, Zhao L, Li Y. Preparation and characterization of intelligent starch/PVA films for simultaneous colorimetric indication and antimicrobial activity for food packaging applications. *Carbohydr Polym*. 2017;157:842–849. https://doi.org/10.1016/j.carbpol.2016.10.067

Liu P, Wang R, Kang X, Cui B, Yu B. Effects of ultrasonic treatment on amylose-lipid complex formation and properties of sweet potato starch-based films. *Ultrason Sonochem*. 2018;44:215–222. https://doi.org/10.1016/j.ultsonch.2018.02.029

Liu Y, Ahmed S, Sameen DE, et al. (2021) A review of cellulose and its derivatives in biopolymer-based for food packaging application. *Trends Food Sci Technol*. 112:532–546. https://doi.org/10.1016/j.tifs.2021.04.016

Llanos JHR, Tadini CC, Gastaldi E. New strategies to fabricate starch/chitosan-based composites by extrusion. *J Food Eng*. 2021;290:110224. https://doi.org/10.1016/j.jfoodeng.2020.110224

López OV, Castillo LA, Garcia MA, Villar MA, Barbosa SE. Food packaging bags based on thermoplastic corn starch reinforced with talc nanoparticles. *Food Hydrocoll*. 2015;43:18–24. https://doi.org/10.1016/j.foodhyd.2014.04.021

Lorevice MV, Otoni CG, de Moura MR, Mattoso LHC. Chitosan nanoparticles on the improvement of thermal, barrier, and mechanical properties of high-and low-methyl pectin films. *Food Hydrocoll*. 2016;52:732–740. https://doi.org/10.1016/j.foodhyd.2015.08.003

Lustriane C, Dwivany FM, Suendo V, Reza M. Effect of chitosan and chitosan-nanoparticles on post harvest quality of banana fruits. *Journal of Plant Biotechnology*. 2018;45(1):36–44. https://doi.org/10.5010/JPB.2018.45.1.036

Matiacevich S, Acevedo N, López D. Characterization of edible active coating based on alginate–thyme oil–propionic acid for the preservation of fresh chicken breast fillets. *J Food Process Preserv*. 2015;39(6):2792–2801. https://doi.org/10.1111/jfpp.12530

Mei J, Guo Q, Wu Y, Li Y. Evaluation of chitosan-starch–based edible coating to improve the shelf life of bod ljong cheese. *J Food Prot*. 2015;78(7):1327–1334. https://doi.org/10.4315/0362-028X.JFP-14-402

Mellinas C, Ramos M, Jiménez A, Garrigós MC. Recent trends in the use of pectin from agro-waste residues as a natural-based biopolymer for food packaging applications. *Materials*. 2020;13(3):673. https://doi.org/10.3390/ma13030673

Menzel C. Improvement of starch films for food packaging through a three-principle approach: Antioxidants, cross-linking and reinforcement. *Carbohydr Polym*. 2020;250:116828. https://doi.org/10.1016/j.carbpol.2020.116828

Moberg T, Rigdahl M, Stading M, Bragd EL. Extensional viscosity of microfibrillated cellulose suspensions. *Carbohydr Polym*. 2014;102:409–412. https://doi.org/10.1016/j.carbpol.2013.11.041

Monte, M. L., Moreno, M. L., Senna, J., Arrieche, L. S., & Pinto, L. A. (2018). Moisture sorption isotherms of chitosan-glycerol films: Thermodynamic properties and microstructure. Food Bioscience, 22, 170–177. https://doi.org/10.1016/j.fbio.2018.02.004

Moreno O, Atarés L, Chiralt A, Cruz-Romero MC, Kerry J. Starch-gelatin antimicrobial packaging materials to extend the shelf life of chicken breast fillets. *Lwt*. 2018;97:483–490. https://doi.org/10.1016/j.lwt.2018.07.005

Mu T, Sun H, Zhang M, Wang C. *Sweet potato processing technology*. Academic Press; 2017.
Mujtaba M, Morsi RE, Kerch G, et al. Current advancements in chitosan-based film production for food technology; A review. *Int J Biol Macromol*. 2019;121:889–904. https://doi.org/10.1016/j.ijbiomac.2018.10.109
Nair MS, Saxena A, Kaur C. Effect of chitosan and alginate based coatings enriched with pomegranate peel extract to extend the postharvest quality of guava (psidium guajava L.). *Food Chem*. 2018;240:245–252. https://doi.org/10.1016/j.foodchem.2017.07.122
Naqash F, Masoodi FA, Ayob O, Parvez S. Effect of active pectin edible coatings on the safety and quality of fresh-cut apple. *Int J Food Sci Tech*. 2022;57(1):57–66. https://doi.org/10.1111/ijfs.15059
Naqash F, Masoodi FA, Rather SA, Wani SM, Gani A. Emerging concepts in the nutraceutical and functional properties of pectin—A review. *Carbohydr Polym*. 2017;168:227–239. https://doi.org/10.1016/j.carbpol.2017.03.058
Naseri-Nosar, M., & Ziora, Z. M. (2018). Wound dressings from naturally-occurring polymers: A review on homopolysaccharide-based composites. Carbohydrate polymers, 189, 379–398. https://doi.org/10.1016/j.carbpol.2018.02.003
Ncube LK, Ude AU, Ogunmuyiwa EN, Zulkifli R, Beas IN.(2021) An overview of plastic waste generation and management in food packaging industries. *Recycling*; 6(1):12. https://doi.org/10.3390/recycling6010012
Nešić A, Cabrera-Barjas G, Dimitrijević-Branković S, Davidović S, Radovanović N, Delattre C. (2020) Prospect of polysaccharide-based materials as advanced food packaging. *Molecules*. 25(1):135. https://doi.org/10.3390/molecules25010135
Nešić A, Gordić M, Davidović S, et al. Pectin-based nanocomposite aerogels for potential insulated food packaging application. *Carbohydr Polym*. 2018;195:128–135. https://doi.org/10.1016/j.carbpol.2018.04.076
Nie, S., Cui, S., & Xie, M. (2017). Bioactive polysaccharides.
Niranjana Prabhu T, Prashantha K. A review on present status and future challenges of starch based polymer films and their composites in food packaging applications. *Polymer Composites*. 2018;39(7):2499–2522. https://doi.org/10.1002/pc.24236
Nisar T, Wang Z, Yang X, Tian Y, Iqbal M, Guo Y. Characterization of citrus pectin films integrated with clove bud essential oil: Physical, thermal, barrier, antioxidant and antibacterial properties. *Int J Biol Macromol*. 2018;106:670–680. https://doi.org/10.1016/j.ijbiomac.2017.08.068
Niu B, Zhan L, Shao P, et al. Electrospinning of zein-ethyl cellulose hybrid nanofibers with improved water resistance for food preservation. *Int J Biol Macromol*. 2020;142:592–599. https://doi.org/10.1016/j.ijbiomac.2019.09.134
Peighambardoust SJ, Peighambardoust SH, Pournasir N, Pakdel PM. Properties of active starch-based films incorporating a combination of ag, ZnO and CuO nanoparticles for potential use in food packaging applications. *Food Packaging and Shelf Life*. 2019;22:100420. https://doi.org/10.1016/j.fpsl.2019.100420
Pereda, M., Amica, G., & Marcovich, N. E. (2012). Development and characterization of edible chitosan/olive oil emulsion films. *Carbohydrate polymers*, *87*(2), 1318–1325. https://doi.org/10.1016/j.carbpol.2011.09.019
Piñeros-Hernandez D, Medina-Jaramillo C, López-Córdoba A, Goyanes S. Edible cassava starch films carrying rosemary antioxidant extracts for potential use as active food packaging. *Food Hydrocoll*. 2017;63:488–495. https://doi.org/10.1016/j.foodhyd.2016.09.034
Pirsa S, Shamusi T. Intelligent and active packaging of chicken thigh meat by conducting nano structure cellulose-polypyrrole-ZnO film. *Materials Science and Engineering: C*. 2019;102:798–809. https://doi.org/10.1016/j.msec.2019.02.021
Priyadarshi R, Rhim J. Chitosan-based biodegradable functional films for food packaging applications. *Innovative Food Science & Emerging Technologies*. 2020;62:102346. https://doi.org/10.1016/j.ifset.2020.102346

Priyadarshi R, Riahi Z, Rhim J. Antioxidant pectin/pullulan edible coating incorporated with vitis vinifera grape seed extract for extending the shelf life of peanuts. *Postharvest Biol Technol.* 2022;183:111740. https://doi.org/10.1016/j.postharvbio.2021.111740

Puscaselu R, Gutt G, Amariei S. Biopolymer-based films enriched with stevia rebaudiana used for the development of edible and soluble packaging. *Coatings.* 2019;9(6):360. https://doi.org/10.3390/coatings9060360

Râpă M, Miteluţ AC, Tănase EE, et al. Influence of chitosan on mechanical, thermal, barrier and antimicrobial properties of PLA-biocomposites for food packaging. *Composites Part B: Engineering.* 2016;102:112–121. https://doi.org/10.1016/j.compositesb.2016.07.016

Rezaei A, Nasirpour A, Fathi M. Application of cellulosic nanofibers in food science using electrospinning and its potential risk. *Comprehensive Reviews in Food Science and Food Safety.* 2015;14(3):269–284. https://doi.org/10.1111/1541-4337.12128

Salari M, Khiabani MS, Mokarram RR, Ghanbarzadeh B, Kafil HS. Development and evaluation of chitosan based active nanocomposite films containing bacterial cellulose nanocrystals and silver nanoparticles. *Food Hydrocoll.* 2018;84:414–423. https://doi.org/10.1016/j.foodhyd.2018.05.037

Salmas CE, Giannakas AE, Baikousi M, Leontiou A, Siasou Z, Karakassides MA. Development of poly (L-lactic acid)/chitosan/basil oil active packaging films via a melt-extrusion process using novel chitosan/basil oil blends. *Processes.* 2021;9(1):88. https://doi.org/10.3390/pr9010088

Senturk Parreidt T, Müller K, Schmid M. (2018)Alginate-based edible films and coatings for food packaging applications. *Foods.* 7(10):170. https://doi.org/10.3390/foods7100170

Song E-, Shang J, Ratner DM. 9.08 - Polysaccharides. In: Matyjaszewski K, Möller M, eds. *Polymer science: A comprehensive reference.* Amsterdam: Elsevier; 2012:137–155. https://doi.org/10.1016/B978-0-444-53349-4.00246-6.

Srinivasa PC, Ramesh MN, Tharanathan RN. Effect of plasticizers and fatty acids on mechanical and permeability characteristics of chitosan films. *Food Hydrocoll.* 2007;21(7):1113–1122. https://doi.org/10.1016/j.foodhyd.2006.08.005.

Suyatma NE, Tighzert L, Copinet A, Coma V. Effects of hydrophilic plasticizers on mechanical, thermal, and surface properties of chitosan films. *J Agric Food Chem.* 2005;53(10):3950–3957. https://doi.org/10.1021/jf048790+

Theng BKG. (2012) *Formation and properties of clay-polymer complexes.* Elsevier.

Ummartyotin S, Manuspiya H. (2015) A critical review on cellulose: From fundamental to an approach on sensor technology. *Renewable and Sustainable Energy Reviews.* 41:402–412. https://doi.org/10.1016/j.rser.2014.08.050.

van den Broek, Lambertus A. M., Knoop RJI, Kappen FHJ, Boeriu CG. Chitosan films and blends for packaging material. *Carbohydr Polym.* 2015;116:237–242. https://doi.org/10.1016/j.carbpol.2014.07.039.

Vasile, C., Pamfil, D., Stoleru, E., & Baican, M. (2020). New developments in medical applications of hybrid hydrogels containing natural polymers. Molecules, 25(7), 1539. https://doi.org/10.3390/molecules25071539

Vedove TM, Maniglia BC, Tadini CC. Production of sustainable smart packaging based on cassava starch and anthocyanin by an extrusion process. *J Food Eng.* 2021;289:110274. https://doi.org/10.1016/j.jfoodeng.2020.110274

Wang H, Qian J, Ding F. Emerging chitosan-based films for food packaging applications. *J Agric Food Chem.* 2018;66(2):395–413. https://doi.org/10.1021/acs.jafc.7b04528

Wang, L. F., Rhim, J. W., & Hong, S. I. (2016a). Preparation of poly (lactide)/poly (butylene adipate-co-terephthalate) blend films using a solvent casting method and their food packaging application. LWT-Food Science and Technology, 68, 454–461. https://doi.org/10.1016/j.lwt.2015.12.062

Wang S, Lu A, Zhang L. Recent advances in regenerated cellulose materials. *Progress in Polymer Science.* 2016b;53:169–206. https://doi.org/10.1016/j.progpolymsci.2015.07.003

Wang Y, Li R, Lu R, Xu J, Hu K, Liu Y. Preparation of chitosan/corn starch/cinnamaldehyde films for strawberry preservation. *Foods.* 2019;8(9):423. https://doi.org/10.3390/foods8090423

Wankhade V. (2020) Animal-derived biopolymers in food and biomedical technology. In: *Biopolymer-based formulations*. Elsevier; 139–152. https://doi.org/10.1016/B978-0-12-816897-4.00006-0

Waring, R. H., Harris, R. M., & Mitchell, S. C. (2018). Plastic contamination of the food chain: A threat to human health?. Maturitas, 115, 64–68. https://doi.org/10.1016/j.maturitas.2018.06.010

Xu Y, Rehmani N, Alsubaie L, Kim C, Sismour E, Scales A. Tapioca starch active nanocomposite films and their antimicrobial effectiveness on ready-to-eat chicken meat. *Food packaging and shelf life*. 2018;16:86–91. https://doi.org/10.1016/j.fpsl.2018.02.006

Yang L, Zhang Y, Kang S, Wang Z, Wu C. (2021) Microplastics in soil: A review on methods, occurrence, sources, and potential risk. *Sci Total Environ*. 780:146546. https://doi.org/10.1016/j.scitotenv.2021.146546

Yildiz E, Sumnu G, Kahyaoglu LN. Monitoring freshness of chicken breast by using natural halochromic curcumin loaded chitosan/PEO nanofibers as an intelligent package. *Int J Biol Macromol*. 2021;170:437–446. https://doi.org/10.1016/j.ijbiomac.2020.12.160

Yin C, Huang C, Wang J, Liu Y, Lu P, Huang L. Effect of chitosan-and alginate-based coatings enriched with cinnamon essential oil microcapsules to improve the postharvest quality of mangoes. *Materials*. 2019;12(13):2039. https://doi.org/10.3390/ma12132039

Yu Z, Li B, Chu J, Zhang P. Silica in situ enhanced PVA/chitosan biodegradable films for food packages. *Carbohydr Polym*. 2018;184:214–220. https://doi.org/10.1016/j.carbpol.2017.12.043

Zarski A, Bajer K, Raszkowska-Kaczor A, Rogacz D, Zarska S, Kapusniak J. From high oleic vegetable oils to hydrophobic starch derivatives: II. physicochemical, processing and environmental properties. *Carbohydr Polym*. 2020;243:116499. https://doi.org/10.1016/j.carbpol.2020.116499

Food Poisoning: Strategic Implementation of Hazards and Quality Analyses of Critical Control Points for Cassava Processing

Taiwo M. Samuel, Babatunde O. Adetifa, Festus O. Uzzi, Nurudeen S. Lawal, and Ayoola A. Babalola

1 Introduction

That man shall live by food is never in contention. This is even corroborated by the Book. Since the world began, there has never been an alternative to food for man's survival and never will be, regardless of advancements in technology. Food preparations and methods may differ, and taste may change with quantity and population vis-à-vis civilization, but production and consumption must continue for human survival without borders, boundaries or barriers. Whether locally-made or imported, the source matters less. But the substance must be fit for human consumption. Otherwise, the purpose of eating – to live – may inadvertently change to serious harm and possible sudden death. It could translate to eating to die!

In animal life and general living, feeding is key. It is the major aspect of raising animals of any type. Even in the jungles, feeding constitutes a major activity daily, and animal life revolves around this for true and sustainable living. The same goes for man and their sustainability. Feeding goes beyond satisfying appetite – it is the means of deriving energy for physical activity or metabolism and well-being. In a way, food is medicine in the nutraceuticals (foods that are designed to improve health and lower the risk of disease) as against the pharmaceuticals (a medicinal drug). A population adequately taken care of in a consistent/regular balanced diet is

T. M. Samuel (✉) · B. O. Adetifa · N. S. Lawal · A. A. Babalola
Department of Agricultural and Biosystems Engineering, Faculty of Engineering and Technology, Ibogun Campus, Olabisi Onabanjo University, Ago-Iwoye, Ogun State, Nigeria
e-mail: taiwosamuel@oouagoiwoye.edu.ng; olusola.adetifa@oouagoiwoye.edu.ng; nslawal@oouagoiwoye.edu.ng; ayobabalola@oouagoiwoye.edu.ng

F. O. Uzzi
Department of Fine and Applied Arts, Faculty of Environmental Studies, Olabisi Onabanjo University, Ibogun Campus, Ago-Iwoye, Ogun State, Nigeria
e-mail: uzzi.festus@oouagoiwoye.edu.ng

I. S. Dunmade et al. (eds.), *Sustainable Engineering*, Green Energy and Technology, https://doi.org/10.1007/978-3-031-47215-2_6

healthy and would most likely have much fewer health issues throughout life. It is observed life expectancy was higher in those days of food sufficiency when the population was agrarian than in our technological and information age when we are lacking in sufficiency, have more pollution and thus, less food safety.

In ancient times, the natural man generally may not need medical attention or rather need much less of it with good supply of nutritious food, and a guided, active lifestyle in their natural environment, just as it is with the animals in their natural state in the jungle, where they intuitively select their food and are always well through life without the necessity for medication or treatments, once their environmental equilibrium is unaltered. In the animal kingdom, they may not suffer food poisoning unless the equilibrium is interrupted by human activities in or around their natural ecology. Should humans be able to maintain the status quo at inception or control their activities, the same scenario could probably have played out without the incidence of food poisoning, its fears and/or susceptibility.

Food poisoning is a condition brought on by consuming food or water that has been contaminated with bacteria, poisons, parasites, viruses, or chemicals. Food poisoning is a catastrophe in any population, and it could be a foe of epidemic magnitude, if not guarded against or at least controlled to a high degree. It kills and perpetrates its harms in a moment before help could be received when it is not premeditated or suspected. Therefore, prevention is the key, rather than cure. The effect on the affected population depends on the spread or prevalence of the poison, the food (soluble or insoluble) in question, and the concentration of the poison in relation to the susceptibility of the affected human organs involved. This determines the extent of the damage that may result therefrom. If it occurs in a family, the entire generation might be wiped out suddenly without prior indication.

Food poisoning is a critical illness that is triggered by the consumption of dirty, sour, or harmful foods which is due to several factors including but not limited to poor food preparation, unhygienic food packaging, and pure ignorance of the consumer. The major sources of food poisoning are viruses, bacteria and parasites. Food poisoning can have various sources depending on the country and region. In Nigeria, the leading causes of food poisoning include contaminated water, poor hygiene practices, and improperly cooked food. Bacterial infections such as *Salmonella*, *Escherichia coli* (*E. coli*), and *Campylobacter* are frequently responsible for food poisoning in Nigeria. The use of expired or adulterated food products can also lead to food poisoning in the country (Ogbolu et al., 2017). In Africa, food poisoning is often caused by bacterial infections such as *Salmonella*, *Listeria*, and *E. coli*, as well as toxins produced by bacteria like *Staphylococcus aureus* and *Clostridium botulinum*. These toxins can contaminate food products such as meat, fish, and dairy. Poor food handling and storage practices also contribute to food poisoning in Africa (WHO, 2015). In Asia, food poisoning is commonly caused by bacterial infections from contaminated food, poor food handling practices, and unsanitary cooking conditions. Some of the most common bacteria responsible for food poisoning in Asia include *Salmonella*, *E. coli*, and *Vibrio cholerae*. Food products such as shellfish, raw eggs, and uncooked meat can also be sources of foodborne illnesses (Akhtar et al., 2020). Pathogens like *Staphylococcus aureus*, *Toxoplasma* spp., *Streptobacillus moniliformis*, *Trichinella* spp., *Yersinia pestis*, *Francisella tularensis*, and *Hantaviruses* can be deposited by rodents (Oladayo et al., 2022).

It, is therefore, critical to determine factors that precipitate food poisoning. Food poisoning can be caused by a variety of factors, both within and outside of the food itself.

Within-factors include the presence of harmful bacteria, viruses, parasites, and toxins that contaminate the food during production, processing, handling, and storage. These microorganisms and toxins can cause illness when ingested by humans, leading to symptoms such as nausea, vomiting, diarrhea, and abdominal pain. Common bacterial culprits of food poisoning include *Salmonella*, *E. coli*, *Listeria*, and *Campylobacter*, while viruses such as norovirus and *hepatitis A* can also cause foodborne illnesses (CDC, 2021).

Without-factors refer to external circumstances that can contribute to food poisoning, such as poor sanitation, inadequate refrigeration or heating, and contaminated water or soil used in food production. Inadequate handwashing, cross-contamination between raw and cooked foods, and improper cooking temperatures are also common without-factors that can lead to food poisoning (WHO, 2015).

Another-without factor is the increasing globalization of the food supply chain, which makes it difficult to trace the origin of foodborne illnesses and prevent their spread. For example, contaminated produce from one country can be distributed and consumed in another country, leading to widespread outbreaks of food poisoning (FDA, 2021).

Nausea, vomiting, and diarrhea are among the signs of food poisoning that people typically experience. The following are the top 5 risk factors for contracting a foodborne illnesses, according to the Centres for Disease Control (Kralj, 2020):

1. Unsafe source of food
2. Misuse of time and temperature
3. Incorrect meal preparation
4. Using contaminated equipment
5. Engaging in poor personal hygiene

2 Food Safety Standards, Organizations, and Enforcement

Food safety standards, organizations, and enforcement differ across countries, but they all aim to prevent or reduce the incidence of food poisoning.

In the United States, the Food and Drug Administration (FDA) and the US Department of Agriculture (USDA) are the primary regulatory agencies responsible for ensuring the safety of food products. The FDA oversees most foods, including imported foods, while the USDA regulates meat, poultry, and egg products. Both agencies have set stringent food safety standards, such as the Food Safety

Modernization Act (FSMA) and Hazard Analysis and Critical Control Points (HACCP) system, to minimize the risk of food poisoning (FDA, 2021).

In the European Union (EU), food safety is regulated by the European Food Safety Authority (EFSA) and the European Commission. The EFSA conducts scientific risk assessments on food safety issues, while the European Commission develops and implements food safety regulations. The EU's food safety standards are generally regarded as among the strictest in the world, with regulations covering all stages of the food supply chain (European Commission, 2021).

In China, the China Food and Drug Administration (CFDA) is responsible for food safety regulation and enforcement. The agency sets standards for food production, processing, and distribution and conducts inspections and testing, to ensure compliance with these standards. However, food safety in China has been a major public health concern, with numerous high-profile incidents of foodborne illness and contamination in recent years (Kan et al., 2014).

In Nigeria, the National Agency for Food and Drug Administration and Control (NAFDAC) is the primary agency responsible for regulating food safety. The agency sets standards for food production, processing, and labeling and enforces these standards through inspections and product testing. However, enforcement of food safety regulations in Nigeria has been challenging due to limited resources and a large informal food sector (Ogbolu et al., 2017).

Effective food safety regulation requires a combination of strong standards, effective enforcement, and ongoing monitoring and testing, to identify and address emerging food safety risks.

Here are some examples of food safety regulations:

1. Hazard Analysis and Critical Control Points (HACCP). This is a preventive approach to food safety that identifies potential hazards and puts in place controls to prevent or reduce the risk of contamination. The HACCP system is required by law for certain food businesses in many countries, including the United States, Canada, and the European Union (FAO/WHO, 2019).
2. Food Safety Modernization Act (FSMA). This is a US law that aims to ensure the safety of the US food supply by shifting the focus from responding to contamination to preventing it. The FSMA requires food businesses to implement preventive controls, including the HACCP system, and to have a plan in place to address any potential contamination (FDA, 2021).
3. Good manufacturing practices (GMP). These are guidelines that outline the basic requirements for the production and processing of food products, including hygiene, sanitation, and quality control. GMP is a mandatory requirement for all food businesses in many countries, including the European Union and Australia (WHO, 2020).
4. Codex Alimentarius. This is a set of internationally recognized food safety standards developed by the World Health Organization (WHO) and the Food and Agriculture Organization (FAO) of the United Nations. The Codex standards cover all aspects of food production, processing, and distribution and are used as a reference by many countries in setting their food safety regulations (FAO/WHO, 2020).

The following are the important national laws in Nigeria that address food safety (Food Safety Standards, 2021):

A. Public Health Ordinance Cap 164 of 1958, which replaced the Public Health Laws of 1917
B. 1974's Food and Drugs Decree No. 35
C. Decree No. 56 of 1971 of the Standards Organizations of Nigeria (SON)
D. No. 10 of 1988 Animal Disease Control Decree
E. Decree No. 41 of 1990 on the Marketing of Breast Milk Substitutes
F. Decree No. 15 of 1993, Laws of the Federation, National Agency for Food and Drug Administration and Control (NAFDAC)

In achieving zero hunger and good health and well-being of the Sustainable Development Goals, serious attention must be given to this subject, to prevent a dire situation of what this evil portends. Food poisoning reduces the zero-hunger strategy SDG in that food in this category is a minus to the available food for consumption when such food with such a potential for poisoning is fully identified. Besides, food poisoning is a clarion call to hunger as it creates fear in the populace, making the populace to be highly selective in their food intake. Where this is prevalent, the achievement of zero hunger becomes a mirage and unattainable.

Achievement of a healthy population and well-being is another goal that touches on food poisoning and food safety. A hungry population is malnourished, weak, and unhealthy. An interesting part of medical treatment is that medication must be taken after meals most of the time, while dieting is also part of treatments. The goal of good health and well-being, therefore, depends very much on food, while food poisoning is counterproductive toward the achievement of this incredibly important goal.

3 Critical Control Points in the Case Study of Cassava Processing into *Gari* - a Case Study

3.1 Unit Operations in Gari *Production*

Cassava handling (processing) into *gari* is the case study under consideration. Assuming pure cassava production and harvesting, then the postharvest begins with peeling. Other unit operations include grating, pressing/dewatering/fermentation, pulverizing and frying, and packaging.

The grating operation could be susceptible to contamination, only where the machine parts where the material in contact with are contaminated or the water used is contaminated. The same goes for pressing. During pressing, fermentation takes place where cyanide content within the cassava mash is neutralized.

The next operation is frying in which case the moisture content is lowered/reduced under heat which raises the temperature. Ordinarily, this process can remove/neutralize any contamination of poisoning agents which could not stand the heat. The only susceptibility remains contact with cookware, especially after the cassava mash attains room temperature. This has to do with the cookware material, since it is in direct contact with the finished food product, particularly when directly consumed without further processing – heating, boiling, or "washing"/ 'decanting" unless the active poisoning agent is active even over the heating/boiling temperature.

Various materials abound which cookware is made up of, with their attendant composition of varying contaminations.

In their investigation of the traditional processing of *gari*, *iru*, *ogiri*, and *soy-ogi* delicacies, Asagbra et al. (2019) used the decision tree technique to identify systemic hazards and critical regulatory phases in the processing of these products. Prerequisite programs (environmental cleanliness and personal hygiene) are recommended as options for increasing the safety of these traditional fermented foods and sauces (Fig. 1).

Rots (soft, dry, or dark) in cassava roots indicate that tubers have already been contaminated with fungus, some of which may be pathogenic (Vilas et al., 2017). If these are not completely killed out during the roasting process, the food will deteriorate or pose a health risk. As a result, the cassava roots must remain strong and free of softening or rotting. The fermentation stage is not only required for product digestibility and scent creation, but it is also an effective procedure for removing the toxin, the cyanogenic glucosides found in bitter cassava roots.

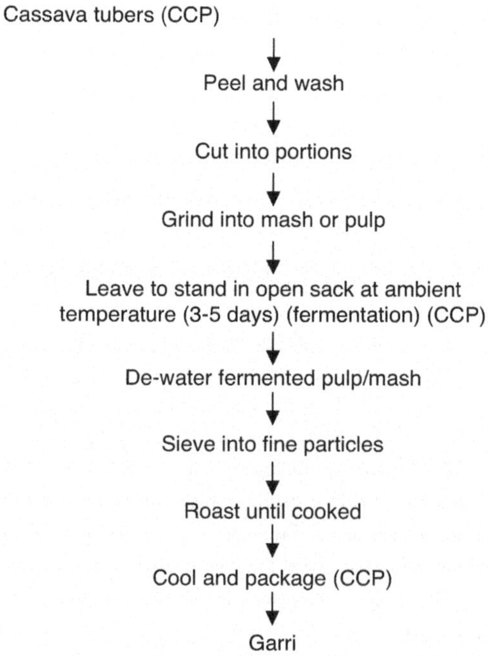

Fig. 1 Unit operations of *gari* production with identified CCPs (Asagbra et al. 2019)

3.2 Risks of Food Poisoning in Processing Cassava into Gari

Cassava is a starchy root vegetable popular worldwide, notably in Africa, Asia, and South America. While cassava is a healthy and flavorful meal, it can also offer significant hazards if not properly processed and cooked. Following are some potential sources of food poisoning to be careful of when preparing cassava:

1. Cassava has a chemical called linamarin, which when eaten, crushed, or otherwise broken can emit hydrogen cyanide. Cyanide poisoning can produce headaches, dizziness, nausea, and vomiting and, in extreme cases, can result in death. Cassava should be soaked and fermented before cooking to minimize cyanide poisoning, or it can be processed using other ways that lower cyanide levels.
2. Contamination. During harvesting, processing, and storage, cassava can get contaminated with hazardous bacteria, viruses, and other diseases. These pollutants can induce foodborne sickness, resulting in symptoms including diarrhea, stomach cramps, and vomiting. Cassava should be collected and processed under sanitary circumstances, and kept at the optimum temperature and humidity levels, to avoid infection.
3. Inadequate cooking/frying. Cassava should be fully cooked to kill any possible germs and make it safe for ingestion. Failure to perfectly cook cassava might result in food poisoning.
4. Mold growth. Cassava can also be infected with mold, which can create chemicals that are harmful to humans. If the circumstances are not favorable, mould development might occur during harvesting, processing, and storage. Cassava should be stored in a dry and cold environment to avoid mold formation, and it should be examined for symptoms of mold frequently.

By being aware of these food poisoning points and responding appropriately to avoid them, cassava can be a safe and healthy food to consume. In this regard, it is important to investigate critical unit operations in processing cassava into *gari* where food poisoning can occur.

3.2.1 Potential Food Poisoning During Cassava Peeling

Cassava peeling is done manually in most underdeveloped/developing nations, particularly Nigeria, using blades made of either mild steel or stainless steel (Owolarafe et al., 2021). Knives, if not thoroughly cleaned and sterilized before use, can be a source of infection and lead to food poisoning. Maertens et al. (2014) discovered that knives infected with listeria monocytogenes, a pathogenic bacterium that may cause severe food poisoning, had a significant role in listeriosis outbreaks. To lessen the risk of contamination and later food poisoning, the research suggests washing and sanitizing knives and other food preparation equipment. While contaminated knives can contribute to food poisoning, basic food handling and hygiene measures can significantly lower the risk. It is advised that knives and other food preparation equipment be properly washed with hot water and soap before and after use and that they be sanitized with a bleach-water solution or other authorized sanitizing agents. To further limit the risk of food poisoning, it is also necessary to observe proper food handling techniques such as washing hands often, keeping raw and cooked foods separate, and boiling food to the optimum temperature.

Kolawole et al. (2021) reported peeling methods under consideration to include chemical means whereby roots are soaked for a short while in a non-harmful liquid compound and are later removed and washed thoroughly to remove both the peel and the chemical compound, as well as application of mechanical means whereby the periphery of the root and tuber crop is rubbed against hard objects without human handling the metal directly as noticed in manual peeling. The latter seemed to gain acceptance particularly in the rural areas and among the small-scale processors for obvious reasons. In the first instance, chemical application is alien to many communities and apart from the fact that it may be beyond their technical awareness, they may not be able to exercise the necessary precaution required to avoid food poisoning in the process of food preparation. Hence, mechanical method remains relevant. Aside from the peeling process, another key aspect of peeling that might lead to food poisoning is the peeling delay. According to Owolarafe et al. (2018), 65% of cassava processors begin hand peeling a day after harvesting, 18% after 2 days, and 2% after 4–5 days. In contrast, 9% of cassava processors begin mechanical peeling a day after harvesting, 3% after 2 days, 1.5% after 3 days, and 1% after 4–5 days. Cassava has significant quantities of cyanide compounds, which can convert to hazardous levels if not processed within a set timeframe, increasing the risk of food poisoning. Onabolu et al. (2011) studied the effect of delayed processing on cassava cyanide levels and protein quality. The researchers discovered that delayed cassava processing increased the cyanide concentration of the goods, potentially leading to food poisoning. Furthermore, it was shown that delayed processing reduced the protein content of cassava products, which might have significant health consequences. To avoid the risk of food poisoning caused by delayed cassava processing, cassava should be processed as soon as possible after harvest. This entails peeling, soaking, and fermenting the cassava before cooking and eating it to minimize its cyanide concentration. Furthermore, storing picked cassava in a cool, dry area can help avoid rotting and reduce the risk of food poisoning.

3.2.2 Potential Food Poisoning While Washing Cassava Tubers

Cassava processors utilize water from a variety of sources to wash peeled tubers in processing centers. According to Owolarafe et al. (2021), the largest percentage of enterprises utilizing well as their source of water for washing was 72.2%, while only 5% of industries had access to pipe-borne water. Boreholes, streams/rivers, and ponds were the remaining sources, accounting for 10%, 11.7%, and 1.1%, respectively. If the water used to wash cassava is contaminated with pathogenic bacteria or other dangerous elements, washing it may contribute to food poisoning. FAO (2013) outlines the possible hazards of washing cassava. Cassava, according to the paper, may absorb pollutants from the soil, and washing cassava in polluted water can result in pathogen translocation to the cassava.

Oyeyiola et al. (2008) also evaluated the microbiological quality of cassava tubers, peeled tubers, and fufu (a cassava-based food product) from Lagos, Nigeria, marketplaces. The researchers discovered that the water used to wash cassava and the equipment used to prepare cassava might be contaminated with dangerous microorganisms such as *E. coli*, *Salmonella*, and *Staphylococcus aureus*. To limit the danger of contamination and consequent food poisoning, the study emphasizes the significance of good sanitation methods while processing cassava. Similarly, Oluwafemi et al. (2012) looked at the microbiological quality of cassava tubers sold in Nigerian marketplaces. The researchers discovered that the water used to wash the cassava may be a source of contamination since it had excessive amounts of bacteria, including *Escherichia coli*, *Salmonella*, and *Staphylococcus aureus*.

Okolie et al. (2007) assessed the microbiological and sensory quality of *gari*, a cassava-based food widely available in Nigeria. The water used to wash the cassava was discovered to have significant amounts of bacteria, including *E. coli*, *Salmonella*, and *Staphylococcus aureus*, making it a possible source of contamination, according to the researchers. To limit the danger of food poisoning from washing cassava, it is critical to utilize clean water and fully sterilize any processing equipment. To further limit the risk of food poisoning, it is also necessary to observe proper food handling techniques such as washing hands often, keeping raw and cooked foods separate, and boiling food to the optimum temperature. Owolarafe et al. (2021) verified that the usage of well water in cassava processing centers in southern Nigeria has increased while the use of stream water has decreased. This might be due to the broad knowledge in the research region of drilling wells as an alternate source of clean water for both home and industrial needs.

3.2.3 Potential Food Poisoning While Grating Cassava

According to Owolarafe et al. (2021), the abrasive disc grater and hammer mill are commonly used for cassava grating. Although this has boosted the rate of *gari* manufacturing, most of the grating discs were constructed of mild steel, which corrodes and rusts readily. Food poisoning can develop during cassava grating if the grater or other processing equipment is infected with dangerous microorganisms.

Oyewole and Adejumo (2010) investigated the microbiological quality of cassava mash and *gari*, a popular cassava-based food offered in Nigerian supermarkets. The researchers discovered that the cassava processing equipment, such as graters and fermenting containers, might be a source of infection. According to Owolarafe et al. (2021), the abrasive disc grater and hammer mill are commonly used for cassava grating. Although this has boosted the rate of *gari* manufacturing, many of the grating discs were constructed of mild steel, which corrodes and rusts readily. Food poisoning can develop during cassava grating if the grater or other processing equipment is infected with dangerous microorganisms.

Oyewole and Adejumo (2010) investigated the microbiological quality of cassava mash and *gari*, a popular cassava-based food offered in Nigerian supermarkets. The researchers discovered that the cassava processing equipment, such as graters and fermenting containers, might be a source of infection. Elevated levels of bacteria, including *Escherichia coli* and *Staphylococcus aureus*, were found in some samples.

Ezeama et al. (2018) studied the microbiological quality and mycotoxin contamination of cassava tuber-derived *gari*, fufu, and starch in Nigeria. The researchers discovered that the cassava grating equipment was a possible source of contamination since it contained significant quantities of germs such as *E. coli*, *Salmonella*, and *Staphylococcus aureus*. The study also discovered that certain cassava samples tested positive for mycotoxins, which are harmful substances generated by some types of fungi.

Ijabadeniyi and Shodehinde (2012), on the other hand, studied the sanitary procedures and microbiological quality of cassava flakes and flour in Nigeria. The researchers discovered that the cassava grating machines were a possible source of infection since they were frequently not cleaned or sterilized sufficiently between usages.

Aside from the contamination from the graters, Sanni et al. (2003) reported that microbial and physicochemical changes can occur during cassava fermentation. The researchers discovered that incorrect grating and storage procedures might cause cassava contamination, resulting in deterioration and the proliferation of dangerous bacteria.

To reduce the risk of food poisoning from cassava grating, use clean equipment and adhere to safe food handling practices such as frequent handwashing, keeping raw and cooked foods separate, and cooking food to the proper temperature. Furthermore, it is advised that cassava be purchased from trusted sources and processed as soon as possible after harvest to decrease the danger of rotting and contamination. Food poisoning can occur during cassava grating if the grating equipment is not adequately cleaned and sterilized or if the cassava itself is contaminated with harmful bacteria or toxins.

3.2.4 Potential Food Poisoning During Cassava Mash Fermentation

Cassava fermentation is an important stage in the manufacturing of cassava-based goods like *gari* and fufu. Inadequate fermentation procedures, on the other hand, might result in food poisoning. Here are several concerns associated with cassava fermentation that may result in food poisoning:

1. Contamination. Cassava contamination during fermentation can result in the growth of pathogenic microorganisms that can cause food poisoning. According to Owolarafe et al. (2021), the bagging technique of fermentation is more sanitary since it protects the cassava mash from bacterial contact.
2. pH and acidity. Maintaining proper pH and acidity levels during fermentation is critical for preventing the formation of dangerous microorganisms. There is a

risk of spoiling and infection by harmful bacteria if the pH or acidity levels are too low (Sanni et al. 2003).

3. Temperature. Temperature management is crucial for ensuring the final product's safety during fermentation. Over-fermentation can occur if the temperature is too high, resulting in spoiling and the development of harmful bacteria. If the temperature is too low, fermentation may not occur, resulting in an incomplete fermentation process that might lead to food poisoning.

Oluwole and Oluwamukomi (2013), Njoku and Obeta (2013), Amoa-Awua et al. (2005), and Oboh and Akindahunsi (2004) researched various soaking and fermenting procedures and their efficiency in lowering cassava cyanide concentration. Oluwole and Oluwamukomi (2013) used response surface methods to improve the soaking and fermentation conditions for cassava flour and discovered that soaking for 48 hours and fermenting for 72 hours at 30 °C resulted in the lowest cyanide concentration. Njoku and Obeta (2013) discovered that soaking cassava in water for 48 hours followed by 72 hours of fermentation lowered the cyanide concentration by 87.5%. Amoa-Awua et al. (2005) and Oboh and Akindahunsi (2004) investigated the effects of fermentation on cassava and discovered that fermentation can lower cyanide concentration while improving nutritional quality.

It is critical to utilize clean water and adequately sterilize processing equipment to decrease the danger of food poisoning during cassava fermentation. It is also advised to check pH and acidity levels as well as maintain correct temperature management during the fermentation process.

3.2.5 Potential Food Poisoning During *Gari* Frying

Gari frying is the process of simultaneously cooking and dehydrating/heating dewatered cassava mash that has been crushed into grains to create gelatinized and dried grains known as *gari*. The method of *gari* frying used influences the product's quality and, of course, its market worth (Samuel et al., 2010). Several problems have been identified with various *gari* frying techniques (Adetifa and Samuel 2012, 2018; Samuel et al., 2015; Samuel and Adetifa, 2012, 2013). To address some of these issues, several solutions have been proposed (Samuel et al., 2016a, b, c). One of the concerns linked with this procedure is food poisoning.

The formation of acrylamide is one of the leading causes of food poisoning during the frying of *gari*. Acrylamide is a chemical molecule that forms when starchy foods reach elevated temperatures. Acrylamide has been found in studies to be a possible carcinogen and to have harmful effects on the neurological system (Tareke et al., 2002). It is important to use correct food handling and preparation practices while frying *gari* to reduce the danger of food illness. This involves ensuring that the cassava used to create *gari* is mold-free and that it is processed and kept appropriately to avoid contamination with aflatoxins. It is also advised to fry *gari* at lower temperatures to reduce the generation of acrylamide (World Health Organization, 2016). While frying *gari* is a common method of preparing this West African delicacy, it is crucial to be aware of the hazards of food illness connected with this method. To prevent these dangers, adequate food handling and preparation practices should be used, and *gari* should be fried at lower temperatures to reduce acrylamide generation.

Gari contamination while frying is a major problem since it might result in foodborne diseases. Contamination during *gari* frying can occur from a variety of causes, including microbiological contamination, heavy metal and other harmful material contamination, and aflatoxin contamination. According to one study conducted in Nigeria, the frying process of *gari* significantly increased the microbial load, with a higher proportion of *Staphylococcus aureus* and *Escherichia coli* bacteria (Oluwajoba et al., 2017). These bacterial species have been linked to foodborne illnesses, causing symptoms such as nausea, vomiting, and diarrhea. Another research in Ghana discovered that *gari* samples acquired from retail establishments had significant amounts of lead, a heavy element that can cause lead poisoning (Gbedema et al., 2011). Lead may build up in the body and cause major health problems, including neurological impairment.

Aflatoxins are another major worry when it comes to *gari* pollution. Findings in research done in Nigeria revealed that over 90% of the *gari* samples tested positive for aflatoxins (Apeh et al., 2016). Aflatoxins are poisonous chemicals generated by fungi that grow on crops like cassava. Aflatoxin-contaminated meals can cause liver damage and raise the chance of developing liver cancer.

It is critical to use correct food handling and preparation practices while *gari* frying to reduce the danger of contamination. This involves ensuring that the cassava used to create *gari* is mold-free and that it is processed and kept appropriately to avoid contamination with aflatoxins. Furthermore, excellent hygiene standards during *gari* frying, including handwashing and using clean equipment, are critical to preventing microbiological infection.

Owolarafe et al. (2021) recognized *gari* frying containers composed of mild steel, stainless steel, or cast aluminum. Cast aluminum had the greatest rate of 37.2%, followed by black mild steel tray, which had a percentage of 36.1%. The percentage of people who used stainless trays was the lowest, at 26.7%. *Gari* contamination during frying might also arise because of the *gari* frying kettle or equipment utilized. If the *gari* frying vessel or equipment is not thoroughly cleaned and sterilized, it can become a source of infection.

A Nigerian investigation discovered that *gari* frying containers were a significant source of contamination during *gari* processing (Oluwajoba et al., 2017). The frying vessels showed significant microbiological loads, including *Staphylococcus aureus* and *Escherichia coli*, both of which are known to cause foodborne diseases, according to the research. To reduce the possibility of infection from the frying vessel or equipment, thoroughly clean and sterilize the equipment before use. This includes cleaning the equipment with hot water and soap and disinfecting it with sanitizer. To avoid contamination, it is also critical to use clean equipment and utensils while *gari* frying. Furthermore, it is critical to utilize proper materials for *gari* frying containers. The use of improper materials, such as copper and brass, might contaminate the *gari* with harmful compounds such as lead and zinc (Gbedema et al., 2011).

Food poisoning could be avoidable if all food safety strategies are dedicatedly implemented and enforced. Primarily, it is everybody's business. Identification of critical control points through hazard analysis, as in the case study of processing cassava into *gari*, could help, in addition to doing due diligence in food hygiene, among others. The Sustainability Development Goal of zero hunger is attainable through this preventive mechanism.

4 Quality Management for *Gari* Production

4.1 Total Quality Management

The key to arresting food poisoning incidence is quality assurance in the production of the food product. The quality assurance strategy preempts good quality management in the production cycle of the product. With this in place, quality can be assured while the incidence of food poisoning is remote. Quality is the fitness of a product for use by consumers. While reduced quality or lack of it in the production of nonfood products results in deficient products that are less useful or usable to the customer, the consequence in the production of food items has a rather grave implication, that is, food poisoning with its attendant serious crises/danger to the consumer's health and even life-threatening! Thus, the importance of quality management in food production to prevent food poisoning cannot be overemphasized. On a large scale, the concept of total quality management is apt and could be applied for long-term success in this venture, which continuous production on a large organization demands.

Total quality management (TQM) is a strategic tool that is based on the belief that an organization can achieve long-term success by focusing all of its members, from entry-level employees to top executives, on improving quality and, as a result, delivering customer (consumer) satisfaction and meeting regulatory standards. But since most agricultural enterprises at the rural level or semi-urban level may not demand as much, the scope could be minimized to small-scale, rural settings involving manual production/operation upon which bulk production in developing countries depends.

4.2 Development of HACCP in Gari Processing

HACCP is a systematic method to food safety management that identifies possible risks and applies control measures to avoid or decrease their occurrence. HACCP has now gained widespread acceptance in the food sector as a preventative approach to food safety management. The hazard analysis critical control point HACCP system, as a scientific and systematic technique, identifies hazards and proposes options for their management. The HACCP system is used to assure food safety when manufacturing gari. HACCP promotes food safety by monitoring and controlling the components and procedures that may threaten the safety of final food products. The Quality Analysis Critical Control Point (QACCP) system was developed to control both quality and product safety, ensuring that food quality is maintained throughout the manufacturing process (Dziedzoave et al., 2006).

The implementation of HACCP in gari processing entails identifying possible risks in the manufacturing process and adopting control measures to avoid or decrease their occurrence. The seven HACCP principles outlined by the Codex Alimentarius Commission can be utilized in the gari processing business as follows:

1. Conduct a risk assessment. Identify possible dangers in the gari processing process, such as microbiological contamination, chemical risks, and physical risks.
2. Identify critical control points (CCPs). Identify the essential control points in the gari processing process where control measures may be implemented to prevent or decrease hazard occurrence.
3. Set important boundaries. Establish critical limits for each CCP, which are the limits that must be fulfilled in order to prevent or decrease the danger.
4. Establish processes to monitor each CCP to verify that the critical limits are fulfilled.
5. Develop corrective steps to be implemented when a critical limit is not met, such as determining the cause of the problem and applying corrective measures.
6. Create verification procedures. Create methods for assessing the efficacy of control measures and remedial actions.
7. Establish processes for recordkeeping and documentation. Create procedures for recordkeeping and documentation to ensure that the HACCP system is implemented and maintained.

The application of HACCP in gari processing can help to prevent or reduce the occurrence of hazards, improve product quality, and ensure compliance with food safety regulations. Adebayo et al. (2012) highlighted the unit operations in *gari* production according to their potential hazard (Table 1).

Adebayo et al. (2012) identified seven processes in the installation of a HACCP/QACCP management system for gari manufacturing:

1. Evaluation of possible dangers and quality difficulties in gari manufacture
2. Establishment of CCPs
3. Establishment of critical limits for individual CCPs
4. Establishment of a monitoring system
5. Definition of corrective actions to be taken when operations are all out of line
6. Creating a mechanism for testing the system's efficacy
7. Documenting and maintaining track of all system implementation operations

Table 1 Critical control points for *gari* processing (Adebayo et al., 2012)

Critical control point	Unit operation	Potential risk
Critical control point 1	Fresh cassava root delivery	Roots that have become old or spoiled or have a high cyanogen content were provided
Critical control point 2	Washing	Pathogenic organisms from the water or the soil on cleaned roots
Critical control point 3	Grating	Pathogenic thermophiles from dirty or inadequately cleaned graters
Critical control point 4	Fermentation	Pathogenic organisms from contaminated troughs in contaminated environments or inadequate or excessive acidity due to wrong fermentation time
Critical control point 5	Dewatering/ pressing	In squeezed cassava mash, there is cyanogen residue
Critical control point 6	Cake splintering	Pathogenic organisms originating from filthy hands
Critical control point 7	Roasting	Rust contamination of food items in roasting pans
Critical control point 8	Packaging	Underweight packages from leaks caused by package breakage or faulty packaging closure and moisture absorption by items due to leaks
Critical control point 9	Storage	Infestation of packed goods by weevils

4.3 Design of Gari *Processing Facilities and Implementation of Quality Management*

In many West African nations, gari processing is a major part of using cassava, so it's important to make sure the processing facility is built to satisfy high standards to avoid food poisoning. To reduce the danger of contamination and guarantee that the finished product is safe for consumption, good hygiene practices (GHP) should be used throughout the whole processing process.

Here are some design requirements and factors for a superior gari processing facility that makes use of GHP:

(a) Location. The processing plant should be situated in a place where there are no open sewage systems, landfills, or other industrial facilities that could serve as possible sources of contamination. Additionally, it is crucial to guarantee that the area has access to electricity and clean water (WHO, 2018).
(b) Layout. To reduce the risk of cross contamination, the processing facility should be constructed with a layout that enables a smooth movement of materials and workers. Separate areas of the plant should be set aside for various processing steps, such as peeling, washing, grating, pressing, and frying (FAO, 2016).
(c) Equipment. The processing equipment should be composed of materials that are safe for food, including stainless steel, and it should be simple to clean and

maintain. Additionally, it should be regularly examined for wear and tear and replaced, as necessary (WHO, 2018).

(d) Ventilation. The processing facility needs to have adequate ventilation to get rid of smoke and fumes produced during the frying process, which might be bad for the health of the workers (FAO, 2016).

(e) PPE (personal protective equipment). To reduce the chance of contamination, the employees should put on the proper PPE, such as gloves, aprons, and face masks (WHO, 2018).

(f) Cleaning and sanitation. The processing facility must have a thorough washing and sanitation program that includes routine cleaning of the machinery, the floors, the walls, and the ceilings. You can utilize sanitizing treatments like chlorine.

(g) Pest control. To avoid infestations by rodents, insects, and other pests that can contaminate the product, the processing facility should have a pest control program in place (WHO, 2018).

(h) Quality control. To make sure that the finished product fulfills the necessary quality standards, the processing facility should have a quality control program in place. At various stages of processing, samples should be gathered and evaluated for moisture content, pH, and microbiological contamination (FAO, 2016).

Designing a high-quality gari processing plant that employs GHP is essential for producing safe and healthy products. The plant's design should prioritize location, layout, equipment, ventilation, PPE, cleaning and sanitation, pest control, and quality control. By implementing these criteria and considerations, food poisoning can be minimized, and the quality of the final product can be assured.

5 Plant Layout and Materials Handling for Quality Management

5.1 Plant Layout

At the base of any meaningful total quality management in any organization are plant layout and materials handling upon which the organization could be safely established. While the operations/activities of the organization are very central, on which quality management revolves, this must be built/established upon a solid structure of plant layout and materials handling in the industry. Otherwise, quality management which is out to stamp out all acts inimical to the safe production and running of the business would be a mirage.

Plant layout can be defined as the physical location or configuration of departments, workstations, and equipment or physical resources used to create the product. This is sometimes referred to as facility layout. Plant layout begins with the selection of the location of the plant itself based on known criteria to achieve total quality management at a minimum cost of production, among others. This includes the decision to locate the industry in a rural, semi-urban, or urban area with its attendant implications on the running of the industry (Sharma, 2007).

The most significant challenge in achieving the lowest overall cost of materials handling is the layout of an industry. Since once the building is made and the plant is commissioned, it is difficult to change it. To have an efficient layout that would achieve the set purpose, a comprehensive knowledge of the intended industry in terms of operations, type of building needed, needed equipment, workers and workstations, space utilization, material flow, etc. must be well known. By implication, materials handling is always designed into the plant layout.

5.2 *Materials Handling*

Materials handling involves the movement of materials from one point to the other. The goal is to assure and improve production quality at minimal costs with efficient use of the facility in the best way possible to maximize the profitability of the enterprise. A suitably designed handling system helps in planning a layout.

Materials handling equipment is operated mechanically, electrically, hydraulically, pneumatically, or a combination of these methods. Materials handling equipment is not production machinery. These are auxiliary equipment which improve the flow of materials which in turn reduces stoppages in production machines and thus increase their production and product quality. These are conveying equipment (conveyors) that follow horizontal, vertical, or compound motions through the air, over fixed routes by gravity or by power; transportation and transferring equipment (hoists); and containers and supports, auxiliary equipment, etc.

5.3 Gari *Production and Plant Layout-cum-Materials Handling*

For the cassava processing into *gari*, the material flow from the point of receiving the raw cassava mash to the finished quality *gari* product must be designed by providing a suitable plant layout and materials handling. This assignment takes into consideration the plant and equipment of interest, the cassava tuber (the material to be handled) which enters the system for processing and must move from one point to the other during processing from one piece of equipment to the other, together with necessary manipulations by workers at various levels, till the product leaves the system.

In this case, the specification of the *gari* product is of fundamental importance in determining the type of plant, because the nature and volume of production determine the method to be employed. This production process then determines the equipment, machinery used, and the plant layout. Therefore, for a good layout, product design and its analysis are prerequisites. This design begins when the product enters the system till it leaves for storage and/or to the market for consumption.

It is no gainsaying that the success of quality management to produce quality *gari* devoid of poison depends more on the correct plant layout from the conception of the project. This section reviews how this is done in the *gari* processing industry, to achieve the lofty aim of eradicating food poisoning. If things are wrong from this level, most probably subsequent efforts would never yield desired maximum output. At best, the result would be less than optimum. In the end, food poisoning would not be averted, thereby subjecting the life of numerous consumers of the eventual *gari* product to danger as initially reeled out in the introductory section of the chapter.

6 Certification and Conformity to Standards for Quality Production

Foods are contaminated when at least one unwanted/unsafe substance is found in them, which can happen during production, sales, cooking, packaging, transportation, and storage, as well before and during harvest. Food contamination can be chemical, physical, or biological (FDA, 2017). There is a global concern over foodborne diseases and their associated illness and death. To combat this problem, hazard analysis and critical points (HACCP) and food safety/quality management are employed, using existing and emerging food safety/quality management technologies with traditional and modern/novel techniques to improve the traditional HACCP, food safety, and quality management in food and agricultural systems (Chinaza, 2023).

In developing countries, incorrect application of good agricultural practices (GAP) and the abuse or misuse of agrochemicals by farmers and in storage have grave health implications for the populace. Sometimes farmers use dangerous chemicals as food ripening agents to hasten the ripening process and use pesticides inappropriately for fishing which could result in foodborne illness and death in some cases. Unsafe packaging materials are not left out among the causes of contamination of food in the public.

Given the serious negative public health and economic impact of unsafe food, which is more pronounced in sub-Saharan Africa, many regulatory agencies in several countries require the mandatory application of specific HACCP programs for different foods.

The complexities of regulating foodborne contaminants and toxicants can be traced to variations in the quality of supplied raw materials supplied by food processors (because many of these raw materials are supplied from diverse sources before they are pooled together). Furthermore, having many food handlers, intermediaries, and vendors along the value chain poses more risks to food quality and significantly exposes the food to contamination. Other mitigating factors such as inadequate storage and handling infrastructure, poor quality assurance during checks, insufficient manpower and technical know-how in laboratory procedures of food quality, low number of regulatory personnel, inadequate coordination of food safety activities along the value chain, and insufficient knowledge by food handlers are responsible for the low productivity indices recorded in food quality and safety, respectively. Other inhibiting factors include poor postharvest handling, insufficient storage facilities, poor transport infrastructure such as good road networks, absence or shortage of regular electricity supply and clean water, as well as a general lack of requisite knowledge and expertise in good agricultural practices at the farm level.

7 Conclusion

Arresting food poisoning malaise to enhance food security for sustainable development has been addressed in this piece by the strategic implementation of the hazard and quality analyses of critical control points by putting in place veritable plant layout and materials handling system specifically designed for *gari* production, our case study. It cannot be overemphasized that implementation of proper hazard and quality analyses would only depend on good plant layout and materials handling which mirrors the production processes, upon which identification of critical control points depends, on which subsequent hazard analysis and later quality analysis also depend. This underscores the fact that conformity to standards to eradicate food poisoning begins at the conception stage of the food industry, albeit at the inception. The implication is that for seriousness in dealing with ensuring food safety and eliminating food poisoning, the stakeholders must be constituted and commence their works at an earlier stage than is being done in practice. In other words, the food industry is facing a significant issue in addressing this prevalent global menace than is often realized or attended to, since human sustainability depends much on it. With increasing challenges of technological development and the attendant side effects, including pollution and other health challenges, both the stakeholders in the food industry and standard regulatory and certification agencies have great roles to play and at a much earlier time in the food production business as often realized.

References

Adebayo BA, Nanam TD, Bamidele EA, Braima DJ (2012) Quality management manual for the production of *gari* IITA, Ibadan, Nigeria

Adetifa BO, Samuel TM (2012) Characterisation of *Gari* Frying Stations in Ifo Local Government, Ogun State. Nigeria Journal of Research in National Development 10 (3):34–46

Adetifa BO, Samuel TM (2018) Prevalence of the Risk of Work-Related Musculoskeletal Disorder in the Stirring Task of *Gari* Frying as a Result of Different Sitting Postures. International Journal of Engineering Research in Africa 37:43–51, https://doi.org/10.4028/wwwscientificnet/JERA3743

Akhtar, S., Sarker, M. R., Hossain, A., Rahman, M., & Khan, M. A. (2020). Food safety in Asia: a systematic review of policy and research priorities. International journal of environmental research and public health, 17(19), 7038. https://doi.org/10.3390/ijerph17197038

Amoa-Awua WK, Sampson E, Tano-Debrah K, Alemawor F (2005) Microbial studies of cassava fermentation for the production of "Agbelima". International Journal of Food Microbiology 99(1):63–72, https://doi.org/10.1016/jijfoodmicro200407002

Apeh DO, Ojukwu UD, Odu NN, Orji M U (2016) Assessment of the levels of aflatoxin contamination in *gari*, a fermented and roasted cassava product. Journal of Food and Nutrition Research 4(5):328–332

Asagbra AE, Onawola OO, Okeagu MO (2019) Identification of Critical Control Points in Some Nigerian Fermented Foods and Seasonings World Journal of Innovative Research (WJIR) 6(4):55–60

CDC, Centers for Disease Control and Prevention , National Center for Emerging and Zoonotic Infectious Diseases (NCEZID) (2021): Foodborne Disease (Food Poisoning). https://www.cdc.gov/ncezid/what-we-do/our-topics/foodborne-disease.html#:-:text=According%20to%20CDC%20estimates%2C%20the,%2C%20Campylobacter%2C%20and%20Staphylococcus%20aureus. Accessed 10th November 2023

Chinaza GA (2023) HACCP, quality, and food safety management in food and agricultural systems. Cogent Food & Agriculture 9:1

Dziedzoave NT, Abass AB, Amoa-Awua WKA, Sablah M (2006) Quality management Manual for the production of high-quality cassava flour, edited by GO Adegoke and L Brimer International Institute of Tropical Agriculture, Ibadan 68

European Commission. (2021). Food safety. https://ec.europa.eu/food/safety_en Accessed 03 May 2023

Ezeama C F, Nwakaudu AA, Ezeama NN (2018) Microbial quality and mycotoxin contamination of *gari*, fufu and starch processed from cassava (Manihot esculenta Crantz) tubers. Journal of Food Safety and Hygiene 4(1): 1–6

FAO. (2016). Gari processing in Africa: A review of processing methods, past and present. http://www.fao.org/3/a-i6589e.pdf Accessed 03 May 2023

FAO/WHO. (2019). Hazard analysis and critical control point (HACCP). https://www.who.int/foodsafety/areas_work/food-hygiene/haccp/en/ Accessed 03 May 2023

FAO/WHO. (2020). Codex Alimentarius. https://www.who.int/foodsafety/areas_work/food-standards/codex/en/ Accessed 03 May 2023

FDA. (2017). Foodborne illness: What you need to know. https://www.fda.gov/food/buy-store-serve-safe-food/foodborne-illness-what-you-need-know Accessed 03 May 2023

FDA. (2021). Food safety modernization act (FSMA). U.S. Food and Drug Administration. https://www.fda.gov/food/food-safety-modernization-act-fsma Accessed 03 May 2023

Food and Agriculture Organization of the United Nations (2013) Cassava processing and utilization Rome. http://www.fao.org/3/a-i3208epdf Accessed 24 April 2023

Food Safety Standards (2021) Legislations (Laws) and Enforcement Agencies in Nigeria (How to Get Approvals).

Gbedema SY, Glover-Amengor M, Osei J, Darfour, B (2011) Lead concentrations of *gari*, a cassava-based food product, in Ghana. Journal of Environmental Science and Health Part B 46(5): 441–448

Ijabadeniyi AO, Shodehinde SA (2012) Hygienic practices and microbial quality of cassava flakes and flour in Nigeria. Journal of Food Research 1(3):204–214, https://doi.org/10.5539/jfrv1n3p204

Kan, C., Shi, Y., Wu, S., & Wang, Q. (2014). Food safety regulation in China: a retrospective and comparative analysis. Food Control, 46, 111–117. https://doi.org/10.1016/j.foodcont.2014.05.026

Kolawole, O. P., Abass, A. B. Kulakow, P., Diallo, T, Samuel, T.M. (2021): Comparative Evaluation of Four Different Cassava Peeling Devic es. Agricultural Mechanization in Asia, Africa and Latin America AMA. 2021 Vol.52 No.2, 74–81 pp.

Kralj RA (2020) Five Common Risk Factors for a Foodborne Illnesses Outbreak. Food Safety and Quality Extension Educator, Penn State Extension

Maertens de Noordhout, Devleesschauwer C, Angulo B, Verbeke F J, et al (2014) The global burden of listeria monocytogenes: a systematic review and meta-analysis. The Lancet Infectious Diseases 14(11):1073–1082, https://doi.org/10.1016/S1473-3099(14)70870-9

Njoku PC, Obeta, JA (2013) Effect of soaking, boiling and fermentation on the cyanide content of cassava (Manihot esculenta, Crantz) tubers. International Journal of Current Microbiology and Applied Sciences 2(8):254–258

Oboh G, Akindahunsi AA (2004) Change in the chemical composition of cassava products (flour and *gari*) subjected to traditional fermentation during production. Journal of Food Composition and Analysis 17(3–4): 451–457, https://doi.org/10.1016/jjfca200309003

Ogbolu, D. O., Alli, J. A., & Ogunleye, V. F. (2017). Food hygiene practices among food vendors in secondary schools in Ogun state, Nigeria. Journal of public health in Africa, 8(2), 675. https://doi.org/10.4081/jphia.2017.675

Okolie NJC, Olayemi AB, Fagade OE (2007) Microbiological and sensory evaluation of *gari* sold in Ota, Ogun state. Nigeria African Journal of Biotechnology 6(1): 55–58

Oladayo T, Miteu G, Adeh I, Folayan E, Olayinka T, et al (2022) Most prominent factors of food poisoning in Africa – Nigeria based perspective. IPS Journal of Nutrition and Food Science1(1):11–17, https://doi.org/10.54117/ijnfsv1i11

Oluwafemi F, Awosika SA, Osundahunsi OF (2012) Microbiological quality of cassava tubers sold in some markets in Ogun State, Nigeria. African Journal of Microbiology Research 6(5): 956–960

Oluwajoba SO, Igbeneghu OA, Adeniji JA, Ayeni FA (2017) Microbial contamination of locally processed cassava (*gari*) in Ogun State, Nigeria. Sky Journal of Microbiology Research 5(1):1–6

Oluwole OB, Oluwamukomi MO (2013) Optimization of soaking and fermentation conditions for improving the quality of cassava flour using response surface methodology. Journal of Food Processing and Preservation 37(1):66–74, https://doi.org/10.1111/j1745-4549201100645x

Onabolu AO, Sadiq IO, Aboaba, OO (2011) Effect of delayed processing on the cyanide content and protein quality of cassava (Manihot esculenta Crantz) products. Journal of Applied Sciences Research 7(12): 2122–2127

Owolarafe OK, Adetifa BO, Oyekanmi AC, Lemikan TG, Samuel TM (2018) Assessment of Fufu Production Technologies in Ogun State, Nigeria. Arid Zone Journal of Engineering, Technology and Environment 14(4):547–558

Owolarafe OK, Ayelegun TA, Samuel TM, Adetifa BO (2021) An Assessment of Technologies for Processing Cassava into *Gari* in Southwestern Nigeria. Innovative Solutions in Engineering (ISIE): A Journal of the Nigerian Academy of Engineering 3(1):47–67

Oyewole OB, and Adejumo IO (2010) Microbiological quality of cassava mash and *gari* at retail outlets in Ibadan. Nigeria African Journal of Food Science 4(2):38–42

Oyeyiola GP, Ogundana SK, Sanni LO (2008) Microbial quality of cassava tubers, peeled tubers and fufu from markets in Lagos, Nigeria. African Journal of Biotechnology 7(14): 2316–2320, https://doi.org/10.5897/AJB08114

Samuel T M,, Igbeka J.C., and Kolawole O. P. (2010) A Survey of Improved *Gari* Frying Methods. International Journal of Food Engineering 6(2), https://doi.org/10.2202/1556-37581473

Samuel TM and Adetifa BO (2013) Assessing Musculoskeletal Risks in *Gari*-Frying Workers. Leonardo Journal of Sciences 12(23):61–76

Samuel TM, Adetifa BO (2012) Investigating Heat Stress in *Gari* Frying Agricultural Engineering International: CIGR Journal 14(3):183–189.

Samuel T.M., Aremu O.O., Adetifa B.O., Ismaila S.O., Onu L.I., Adegbite S.A., and Olokoshe AA (2015). Variation of Anthropometric and B ody Composition Parameters Within Different Age Groups of Gari-frying Workers in Southwest Nigeria. Annals of Faculty of Engineerin g Hunedoara — International Journal of Engineering. Vol. 3, pp. 99–104, Universitatea Politechnica Timisoara, Romania, ISSN:1584-2673, (Indexed in DOAJ).

Samuel TM, Adetifa BO, Aremu OO (2016a) Design of Ergonomic Stirring Tool for *Gari* frying Workers. Journal of Applied Sciences and Engineering Technology 16(1):37–43

Samuel TM, Aremu OO, Salami IO, Adetifa BO et al. (2016b) Anthropometric studies for designing to fit *gari*-frying workers. Agricultural Engineering International: CIGR Journal 18(1):180–191

Samuel TM, Aremu OO, Salami IO, Adetifa BO et al. (2016c) Variation of Anthropometric and Body Composition Parameters within Different Age Groups of *Gari* Frying Workers in Southwest Nigeria. Annals of Faculty Engineering Hunedoara – International Journal of Engineering 3:99–104

Sanni AI, Oyewole OB, Ogunlana, OE (2003) Microbial and physicochemical changes during cassava fermentation for fufu flour production using mixed cultures of lactic acid bacteria. World Journal of Microbiology and Biotechnology 19(5):499–506, https://doi.org/10.1023/A:1025120828954

Sharma SG (2007) Plant layout and Materials Handling. Second Edition, Khana Publishers.

Tareke E, Rydberg P, Karlsson P, Eriksson S et al. (2002) Analysis of acrylamide, a carcinogen formed in heated foodstuffs. Journal of agricultural and food chemistry 50(17): 4998–5006

Vilas SA, Boas SA, Santos de Oliveira CA, Braganca D et al (2017) Survey of fungi associated with cassava root rot from different producing regions in Brazil. Sci Agric 74(1):60–67, https://doi.org/10.1590/0103-9016-2015-0336

WHO. (2015). Food safety in Africa: challenges and opportunities. World Health Organization. https://www.who.int/foodsafety/publications/food-safety-in-africa/en/ Accessed 03 May 2023

WHO. (2018). Guidelines on good manufacturing practices for the production of traditional medicines. World Health Organization. https://www.who.int/medicines/areas/quality_safety/quality_assurance/GMPtraditionalmedicinesTRS1003Annex7.pdf Accessed 03 May 2023

WHO. (2020). Good manufacturing practices for pharmaceuticals. World Health Organization. https://www.who.int/medicines/areas/quality_safety/quality_assurance/gmp/en/ Accessed 03 May 2023

World Health Organization (2016) Acrylamide in Food: WHO Food Additives, Series 64. https://appswhoint/iris/bitstream/handle/10665/246197/9789241660649-engpdf Accessed 22 April 2023

Part II
Sustainability in Materials Recovery and Processing

Sustainable Development Goal: An Engineering Approach to Health and Well-Being Implication of Wet Milling

B. G. Jahun, L. G. Abubakar, A Mohammed, and J. Y. Dibal

1 Introduction

The world has made absurd progress since 2000 against several of the leading causes of illness and death. Life expectancy has increased (World Health Statistics (2016)), infant and maternal mortality have declined (Global Health Observatory (GHO) data 2018a, b, c), malaria deaths have more than halved, and significant progress has been made against the human immunodeficiency virus epidemic (World Health Organization 2018a, b).

However, this progress has been fragile and uneven both between and within countries. There remains a 31-year discrepancy between the countries with the shortest and longest life expectancies (Global Health Observatory (GHO) data 2018a, b, c). While some countries have made impressive health gains, national averages can hide in-country disparities in health outcomes (World Health Organization 2018a, b), for example, in marginalized populations.

The incidence of chronic respiratory diseases, diabetes, various types of cancer, and road traffic injuries is on the rise. Noncommunicable diseases account for some 70% of premature deaths, predominantly in low- and middle-income countries (WHO 2017).

Engineering solutions can improve the quality of the healthcare systems through new technologies and research findings where there is lack of healthcare facilities in the remote areas. Engineers contribute to good health and ensure healthy lives and promote well-being at all ages. Sustainable Development Goal (SDG) 3 sets out nine key targets and four additional targets for resource mobilization and policy to be achieved by 2030. One of the key targets by 2030 is to substantially reduce the

B. G. Jahun · L. G. Abubakar · A Mohammed · J. Y. Dibal
Department of Agricultural and Bioresource Engineering, Abubakar Tafawa Balewa University, Bauchi, Nigeria
e-mail: gjbala@atbu.edu.ng; lgabubakar@atbu.edu.ng; aminu@atbu.edu.ng

I. S. Dunmade et al. (eds.), *Sustainable Engineering*, Green Energy and Technology, https://doi.org/10.1007/978-3-031-47215-2_7

number of deaths and illnesses from hazardous chemicals and air, water, and soil pollution and contamination (Diaz-Sarachaga et al. 2018).

The achievement of SDG 3 is interlinked to every other SDG in some way. For example, SDG 6 – clean water and sanitation – can largely reduce the mortality and morbidity from the environmental hazards, the poor performance in nutrition (SDG 2 – no hunger) has negative impact on children's physical and mental health and well-being, and education and awareness on disease prevention is a necessity to ensure a sustainably healthy population (SDG 4 – quality education) (Copestake 2008).

The current status of achieving SDG 3 in Nigeria was the release of grant to investigate the effects of depositional level of wear elements in wet milling which require immediate attention in the health sector under SDG 3 targets with the help of other crosscutting sectors (Yamey et al. 2014), and communicable disease such as malaria needs support from environmental sector, while noncommunicable diseases such as cardiovascular diseases and chronic kidney diseases of unknown etiology (CKDu) require attention from sectors such as environment and water.

Particle size reduction of wet corn-milling by mechanical means, also known as milling or comminution, is undoubtedly a very important unit operation in agricultural and food industries. The reduction in size of agricultural products is brought about by mechanical means without change in chemical properties of the materials (Ortega-Rivas 2012). The size of agricultural products may be reduced in several ways. The main methods used are crushing, impact, shearing, and cutting (Fellows 2009).

Removal of metal from the surfaces of corn-grinding plates occurs in both dry and wet corn-grinding. The factors affecting wear are governed by the distribution and characteristics of microconstituents in the metal or alloy. Additional corrosive wear is apparent during wet corn-grinding in which less resistant corrosion product films are abraded away, contaminating the processed food. The degree of contamination depends on pH of the environment, chemical composition, microstructure, and hardness of the corn-mill plate materials. The metallurgy of food processing equipment such as corn-grinding mill machine usually uses cast iron plates because of its good mechanical properties and is also relatively cost-effective. Nevertheless, cast iron is strongly susceptible to corrosion attack in food product environments due to their acidic nature. A quantitative understanding of the corrosion rate of cast iron plates under wet corn-grinding would be a key factor in providing an accurate risk assessment of the attack of the equipment part made from these materials (Kwofie and Chandler 2006).

These grinding plates produced from the casting method are not regulated by any standard procedure for defects (standard chemical composition) and health impacts before being introduced to the market; hence, they may pose high risk to end users. Although wearing of these plates into the food during the milling process is inevitable, the rate of these wear and the levels of contamination they introduce into our food are not known. A defective plate (in chemical structure) will wear faster (Adeti 2015). The health impacts of using a defective grinding plate are enormous.

Health has been recognized as central to international development for more than 20 years, and major efforts have been made to reduce morbidity and mortality either universally or through a focus on specific population subgroups (e.g., "the poor," "women and children") (United Nations 2015). Meanwhile, increased prosperity around the world, changes in diets and lifestyles, and rapid and unplanned urbanization have brought new health threats. The incidence of chronic respiratory diseases, diabetes, various types of cancer, and road traffic injuries are all on the rise. Noncommunicable diseases account for some 70% of premature deaths, predominantly in low- and middle-income countries (Amina and Tedros 2018). Food toxicity could be caused by trace elements because of local food processing. It could reach a significant stage, where it will be fatal with few exceptions, as nearly all foods consumed locally are milled before they are consumed. These metals are potential environmental contaminants with the capability of causing human health problems if present in excess in the food we eat (Salama and Radwan 2005).

1.1 Adverse Health Effects of Wear Elements

Some metals which are known to be toxic are not required by the body in any amount (Mieiro et al. 2012). Wear elements with adverse health effects in human metabolism (including lead, cadmium, and mercury) present obvious concerns due to their persistence in the environment (Kabir et al. 2015). Acute wear element (heavy metal) intoxications may damage the central nervous function, cardiovascular and gastrointestinal systems, lungs, kidneys, liver, endocrine glands, and bones (Yakubu et al. 2015). Chronic heavy metal exposure has been implicated in several degenerative diseases of these same systems and increases the risk of some cancers (Lakherwal 2014). Many investigators have considered the toxic effects of exposure to metals such as arsenic (As), cadmium (Cd), lead (Pb), and mercury (Hg), and there had been substantial expenditure on measures to reduce human exposure. Although toxic effects are clearly established for occupationally exposed people and for people in locations where particularly high levels are present, demonstration of harmful effects population-wide has been more difficult (Kwofie et al. 2011). Lung inflammation and damage to the nasal cavity have been observed in animals exposed to nickel compounds. At high concentrations, lung damage is severe enough to affect lung function. Oral exposure of humans to high levels of soluble nickel compounds through the environment is extremely unlikely (Kalasuramath et al. 2015). Chronic exposure to As has been associated with diabetes (Coronado-Gonzalez et al. 2007; Navas-Acien et al. 2006), cardiovascular disease (Wang et al. 2003), intellectual impairment in children (von Ehrenstein et al. 2007; Wasserman et al. 2007), and lung and bladder cancers (Navarro Silvera and Rohan 2007). Cadmium primarily affects the kidneys, with tubular proteinuria providing a useful marker of the renal damage. Studies in Cd-polluted areas have reported an excess mortality risk among people showing tubular or glomerular proteinuria, but the causes of this excess mortality are unclear (Satarug et al. 2003). Other claimed adverse effects of Cd include low bone mineral density and osteoporosis (Alfven et al. 2000, 2004) and lung cancer (Navarro Silvera and Rohan 2007). Hg has been implicated in severe but localized poisonings (Ekino et al. 2007). Widely publicized concerns about the effects of Hg from vaccine preservatives have been refuted (Thompson et al. 2007), but effects on intellectual functioning or behavior have been reported or suggested (Axelrad et al. 2007; Davidson and Swisher 2004). Cardiovascular (Virtanen et al. 2007) and renal (Hodgson et al. 2007) disease have been associated with low-level Hg exposure. For lead, the main concerns have been about impaired intellectual development in infancy and childhood (Canfield et al. 2003; Needleman 2004; Pocock et al. 1994; Schwartz 1994), but adult cognitive function may also be affected (Schwartz and Hu 2007; Shih et al. 2007). There is also evidence that higher Pb values are associated with hypertension (Navas-Acien et al. 2007), peripheral vascular disease (Navas-Acien et al. 2004), increased adult mortality (Lustberg and Silbergeld 2002; Schober et al. 2006), and cognitive decline in older people (Weisskopf et al. 2004). Studies on overall and cause-specific mortality suggest a 40–60% increase in adjusted mortality with increasing Pb concentration in the US population (Jemal et al. 2002; Lustberg and Silbergeld 2002; Menke et al. 2006; Schober et al. 2006).

Processing of grain flour is as vital and important as the food itself. During this processing such as milling with machine and pounding with wooden pestle and mortar, toxic materials or elements from the processing machine are introduced into the matrix of the food material.

In this work, wearability behaviors of four different cast iron corn-mill plates and three corn-grinding machine speed variations have been investigated on three different crops. The relationship between corn-grinding machine and corn-grinding plates on depositional rate has been developed and subsequently linked to its effect on the wear elements of corn-mill plates when used in wet grinding environments. The characterization and effect of depositional level of wear elements on the wearability of cast iron in corn-mill different grindig plates basd on corn-grinding machine speeds variations and compared with the World Health Organization (WHO) standard on daily wear elements intake were performed.

Studies have been reported on the elemental composition of a number of milled products such as cassava flour, corn flour, wheat flour, etc. There are limited data in the literature regarding elemental profile, particularly metal levels of crops such as (beans, millet, and tomatoes) paste. It is therefore important to determine the metal levels, exposure, and risk associated with the consumption of crops (beans, millet and tomatoes) paste processed. This work will therefore help fill that knowledge gap by providing comprehensive information on the heavy metal profile of samples prior to and after milling that are health death traps. It is important for attaining the SDGs and African Union's Agenda 2063 by providing some of the relevant knowledge required in the areas of food safety and associated effect on public health. In this work, information on the entire metal profile of mechanically milled bean, millet, and tomato paste is obtained from a single scan for each sample, and those with serious health threats were targeted as few numbers of heavy metals at a time, hence the novelty of this study.

2 Materials and Methods

The research activities were carried out within Bauchi metropolis, Nigeria, Food Processing Laboratory of the Department of Agricultural and Bioresource Engineering, and the Laboratory of the Department of Civil Engineering of the Faculty of Engineering and Engineering Technology, Abubakar Tafawa Balewa University Bauchi Nigeria were used for analysis.

2.1 Reagents and Chemicals

HNO_3, H_2O_2, and H_2SO_4 are analytical grade. All aqueous solutions and dilutions were prepared with ultrapure water (18.2 MΩ cm) obtained from an SG Ultra Clear UV Plus system (Wasseraufbereitung und Regenerierstation GmbH, Germany). All glassware was cleaned by soaking in 15% (v/v) HNO_3 for 24 h and rinsed with ultrapure water prior to use. High purity argon (99.99%) was used as a plasma sustaining gas.

2.2 Sample Collection

Corn-grinding plates made from India and Nigeria were purchased from the Nigerian open market and used for this research. The mode of selection of these corn-grinding plates was at random sampling. The corn-grinding plates are in a form of a disc having a diameter of about 30 cm and a hollow core diameter of about 11 cm. The corn-grinding plates are not of uniform thickness. The thickness ranges from about 10 mm at the outer circumference and tapers down to 3 mm at the center. In addition, the plates' surfaces have grooves/striations for effective grinding/milling operation. Crop samples (beans, millet, and tomatoes) were purchased from identified farmers not in an open market in order to control over the mix-up in the crops and the soil where such crops were produced.

2.3 Determination of Heavy Metals Prior to Wet Milling

All samples of beans, millet, and tomatoes were washed with deionized water to remove soil particles or dusts adhering to the grain surface. Samples were oven-dried at 45 °C for 24 hours then ground in a stainless-steel pestle and mortar to a fine powder and stored in plastic bags for wear element analysis prior to milling (Plate I). The crops samples were analyzed for the heavy metals As, Cd, and Pb. Then, it is digested with a mixture of nitric acid (HNO_3) and hydrogen peroxide (H_2O_2) (Asere 2022).

Plate I Fine powder ground crops for heavy element determination prior to wet milling

2.3.1 Dry Ashing

2.0 g of dried samples (beans, millet, and tomatoes) was accurately weighed in a porcelain crucible; the porcelain crucible was then heated in muffle furnace at 200 °C for 1.5 h and gradually heated to 350 °C for 4 h in order to dry ash the sample as shown in Plate II. The ash samples were digested with 5 ml HNO_3, 3 ml HCl, and 2 ml H_2O_2 of concentrated mixed acid (HCl: HNO_3 = 3:2:1), and then the digestion solution was heated up on electric hot plate at 150 °C until being evaporated to near dryness. The residue was filtered and transferred into a volumetric flask and made up to 25 ml with 3% HNO_3. The blank digestion experiments were also carried out in the same way as reported by Banks et al. (1997) (Figs. 1 and 2).

Before the elements were determined, the instrument was cleaned thoroughly by spraying clean water into the flame and then set to zero. 10 g of each sample of millet, beans, and tomatoes were weighed and put into crucibles, respectively. The crucibles were arranged accordingly on the gas burner and ignited gently over a low flame until it charred and then moved into already switched on muffled furnace with temperature controlling regulator. The samples were heated gradually to about 600 °C with an instrument called muffle furnace (see Fig. 3). Appropriate temperature for ashing the selected samples is at 6 hours to obtain the ash and to avoid volatilization or interaction between constituents. The ash samples were allowed to cool in desiccators to avoid moisture and contaminants from coming in contact with them (Fig. 4).

The ash sample obtained from muffle furnace after getting cooled from desiccators was moistened with water inside a conical flask, and 10 ml diluted hydrochloric acid (1:1) was added. Each diluted sample was evaporated to dryness on a regulated electrical hot plate and allowed to cool. Then later 20 ml water and 10 ml diluted hydrochloric acid were added, boiled, and filtered into 250 ml volumetric flask. The remaining contents in the filter were thoroughly washed with hot water; the solution is then allowed to cool and diluted to the mark with water. Atomic absorption spectrophotometer instrument was used to determine the trace elements in the aliquots

Fig. 1 Tomato sample used

Fig. 2 Bean and millet samples used

of the digested samples (Sinayobye and Saalia 2011). The heavy metals that were determined include lead, cadmium, and arsenic.

2.4 Wet Milling Digestion Procedures of Crops

The crops were soaked and rinsed repeatedly with tap water and then rinsed three times (each time for 3 min) with ultrapure water in an ultrasonic cleaner and then placed in constant temperature oven drying at 45 °C to constant weight. The milling process was carried out using three different corn-grinding milling machines (GX 160, GX 200, and GX 390) with varying horse power (5.5, 6.5, and 13 hp., respectively). The Adex, Nas, Lotus-Zamfara, and Lotus-India made corn-grinding plates were placed in each of the corn-milling machine; 2 kg weight of the crop (beans, millet, and tomatoes) samples were milled in each of the three corn-milling machine

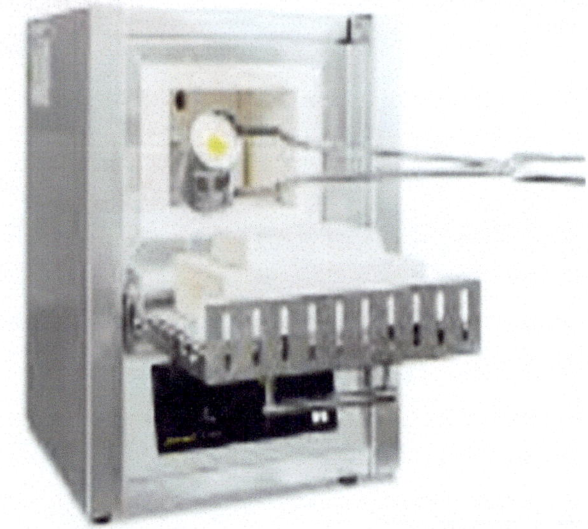

Fig. 3 Muffle furnace (Axelrad and Swisher 2007)

Fig. 4 Desiccator

immediately after the insertion of the corn-grinding plates into the machines. Clearance of the plates was adjusted with the help of adjusting screw to the required level. The process was repeated for three replications to assess the wearing rate of the grinding plate as shown in Plate V. The corn-grinding milling machines were located at the food processing laboratory of Abubakar Tafawa Balewa University, Bauchi, Nigeria (Fig. 5).

The wet samples were homogenized using an agate homogenizer and stored in polyethylene bags until analysis as shown in Plate VI. Bean, millet, and tomato

Fig. 5 One of the grinding machines during wet milling of one of the sample

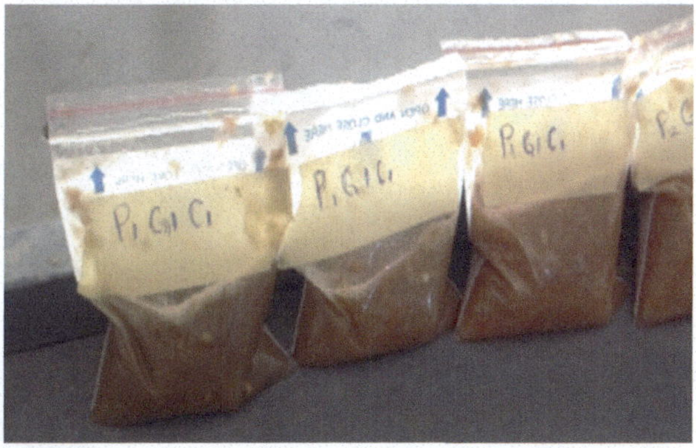

Fig. 6 Ground crops labeled based on replications

pastes were stored in polyethylene (PE) bags. Care was taken at each stage of preparation to avoid contamination. Samples collected were thoroughly mixed to make sure that representative samples were obtained. The ground crops from the corn-grinding milling were labeled based on replications consecutively. A total of 108 samples were then obtained (Fig. 6).

2.4.1 Wet Ashing

Recent studies have shown that mixtures of HNO_3/H_2O_2 are better than HNO_3, HCl, or H_2SO_4 in terms of complete dissolution in a short time period for wet digestion. That was why the mixtures of HNO_3/H_2O_2 were used in the wet digestion

Fig. 7 Digestion block during sample digestion

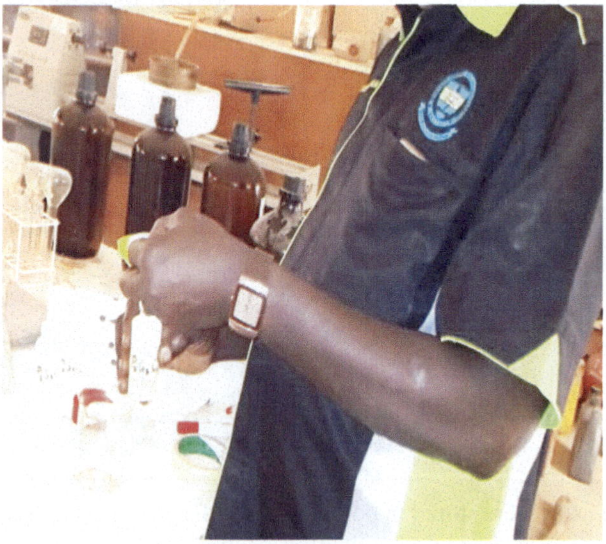

Fig. 8 Digested solutions transferred into volumetric bottles

procedures. 5.0 gm of dried samples (beans, millet, and tomatoes) was accurately weighed into 50 mL beaker and added with 10 mL of concentrated HNO_3 and 2 ml of 30% H_2O_2, and 3 ml of concentrated HCl were mixed and then heated in a digestion block at 100 °C. After the disappearance of brown fumes, the digestion solution evaporated to near dryness. The residue was filtered and transferred into a volumetric flask and made up to 100 ml as shown in Fig. 7; the blank digestion experiments were also carried out in the same way (Figs. 8 and 9).

Fig. 9 Sample solution for determination of presence of wear metallic elements

2.5 Determination of Wear Element in the Samples

The filtered samples were assessed for the presence of wear elements on a WINCOM AAS 320 N Atomic Absorption Spectrophotometer with an air-acetylene burner or nitrous oxide-acetylene burner for flame and a graphite furnace for electrothermal determinations, with appropriate background (nonatomic) correction (Coronado-Gonzalez et al. 2007) with the recommended instrument parameters including detection limits for each metal determined. The determination of structural elucidation of various substances was done by atomic absorption spectroscopy (Plate VIII). The samples in solution state were sprayed over a burner. This leads to evaporation of solvent and leaves fine dry residue behind which is nothing but neutral atoms in ground state. To these atoms, a light of specific wavelength was passed and the unabsorbed light is recorded over a detector. The cathode lamps made of different elements were used and so for all the other metal elements analyzed. Blanks and standard reference reagents were also run concurrently with the metal analyses to ascertain reproducibly and quality assurance as presented by Grace (2009) (Fig. 10).

2.6 Statistical Analysis

Statistical analysis and independent t-tests were performed using Statistical Analysis System (SAS16.0). Non-normal data were normalized through square root transformation to improve normal distribution and to reduce the influence of high analytical data. T-test and Pearson's correlation analysis were conducted, and analysis of variance (ANOVA) was performed on log-transformed wear element analytical data. Correlation analysis based on the correlation matrix was conducted for the wear element data set. The aim of using t-test was to compare the means of two independent groups to determine whether there is statistical evidence that the associated

Fig. 10 Atomic absorptive spectrophotometric machine (ASS) for the determination of structural elucidation of various substances in the solution

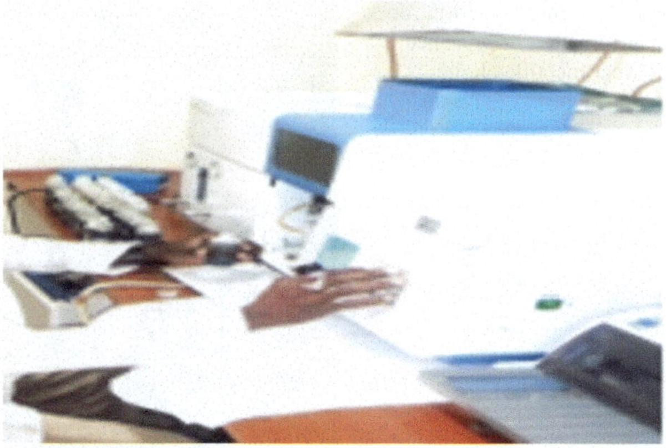

population means are significantly different and ascertain any patterns in the crop samples in relation to these chemical characteristics and then make a preliminary conclusion to the possible relationship between heavy metal concentrations and corn-grinding plate properties. The data from the analysis on wear element depositional levels were subjected to ANOVA and one sample t-test at a 95% confidence level.

After the initial profiles and core samples were obtained, the milling operations were performed according to the experimental design layout. Three factors of influence were chosen for this study, namely, three corn-grinding milling machines (GX 160, GX 200, and GX 390) as G_1, G_2, and G_3, respectively; three different crops (beans, millet, and tomatoes) as C_1, C_2, and C_3, respectively; and four different grinding plates (Adex, Nas, Lotus-Zamfara, and Lotus-India) as P_1, P_2, P_3, and P_4, respectively. This gives a total of $4 \times 3 \times 3 = 36$ treatments. This was fitted into a completely randomized design (CRD). Consequently, the experiment was replicated through randomized complete design, and as such, the variations are minimized and the error due to the variations was removed. The treatments were randomly assigned to experimental units such that each treatment occurs equally. Each test was conducted in three replications, which gave a total of $36 \times 3 = 108$ experimental tests.

2.7 *Results and Discussion*

Normality of the data and equality of variances across comparison groups were maintained. Both analyses are performed on log-transformed data and compare the means of the groups.

Hypotheses.

H_o = there is no significant difference between the depositional level of the wear elements in the studied factor.

H_1 = there is significant difference between the depositional level of wear elements in the studied factor.

$$t(\text{degrees of freedom}) = \text{the } t \text{ statistic}, p = p \text{ value} \tag{1}$$

2.7.1 Independent *t*-Test for Tomatoes on the Studied Factors

From the results of the pretest corn-milling (mean = 0.035; variance = 0.000) and posttest (mean = 0.008; variance = 0.000), there was no significant increase of Pb in the tomato-milled, $t(3)$ =2.24, p = *0.155*, while in the pretest tomato-milling (mean = 0.003; variance = 0.000) and posttest tomato-milling (mean = 0.009; variance = 0.000), there was a significant increase in the depositional level of wear element (Cd) in the corn wet milled, $t(3) = -6.57$ p = *0.02*. Since the p-value is 0.155 and 0.565, we conclude that at the 5% level of significance, we accept the null hypotheses (H_o), that there is not enough evidence to reject the claim that there is no significant difference between the depositional level of wear element in tomatoes for pre- and posttest. We can accept the alternative hypothesis (H_1) that there is enough significant difference at 5% level of significance of depositional level of (Cd) at *p-value is 0.02*. The findings showed that wear element (Pb) depositional level pre- and posttest on corn wet milling does not differ significantly among a variety of tomatoes. However, findings revealed a significant difference between the mean of wear elements (Cd).

2.7.2 Independent *t*-Test for Beans on the Studied Factors

From the pretest (mean = 0.007; variance = 0.003) and posttest (mean = 0.019; variance = 0.000) bean-milling, there was no significant increase of Cd in the beans milled, $t(3) = -3.84$, p = *0.06*, while in the pretest (mean = 0.428; V = 0.000) and posttest (mean = 0.010; variance = 0.000) bean-milling, there was a significant increase in the depositional level of wear element (Pb) in the beans wet milled, $t(3) = 12.44$, p = *0.006*. Since the *p-value* is 0.06, we conclude that at the 5% level of significance, we accept the null hypotheses (H_o), that there is not enough evidence to reject the claim that there is no significant difference between the depositional level of wear element in beans for pre- and posttest. We can accept the alternative hypothesis (H_1) that there is enough significant difference at 5% level of significance of depositional level of (Pb) at *p-value is 0.006*. The findings showed that wear element (Cd) depositional level pre- and posttest on beans wet milled does not differ significantly among a variety of beans.

2.7.3 Independent *t*-Test for Millets on the Studied Factors

From results of the pretest (mean = 0.051; variance = 0.000) and posttest (mean = 0.014; variance = 0.000) corn-milling, there was no significant increase of Pb in the corn milled, $t(3) = 2.66$, $p = 0.117$, while in the pretest (mean = 0.001; variance = 0.000) and posttest (mean = 0.009; variance = 0.000) corn-milling, there was a significant increase in the depositional level of wear element (Pb) in the corn wet milled, $t(3) = -8.31$, $p = 0.014$. Since the *p-value* is 0.177, we conclude that at the 5% level of significance, we accept the null hypothesis (H_o), that there is not enough evidence to reject the claim that there is no significant difference between the depositional level of wear element in millets for pre- and posttest (Pb). We can accept the alternative hypotheses (H_1) that there is enough significant difference at 5% level of significance of depositional level of (Pb) at *p-value is 0.014*. The findings showed that wear element (Pb) depositional level pre- and posttest on corn wet milled does not differ significantly among a variety of millets. However, findings revealed a significant difference between the mean of wear elements (Pb).

In fact, if the actual *p-value* of the first study was 0.048 and that of the second study was 0.052, the two studies are entirely consistent with each other. The conventional value for statistical significance ($p < 0.05$) should always be viewed in context, and a *p-value* close to this arbitrary cutoff point should perhaps lead to the conclusion that further work may be necessary before accepting or rejecting the null hypothesis.

2.7.4 Statistical Analysis for Grinding Models, Grinding Plates, and Different Types of Crops for the Determination of Depositional Level of Wear Elements

Laboratory data for grinding different crops were collected and analyzed using completely randomized design (CRD). The CRD results in smaller error variance and hence better estimates of effect. Ground samples were collected from the laboratory to investigate the effects of four types of grinding plates (P_1, P_2, P_3, and P_4) at three different crops (C_1, C_2, and C_3) and three different grinding models (G_1, G_2, and G_3). All the data generated were subjected to analysis of variance (ANOVA) to evaluate the level of significance ($p > 0.05$). Tukey's mean separation method was performed in each case for factors that are significantly different. Similarly, predictive model on the relationship between the variables studied gives rise to the use of multiple regression analysis.

2.7.4.1 Analysis of Variance (ANOVA) on Effects of Grinding Models, Grinding Plates, and Crops on Depositional Level of Lead

Using the statistical method, an analysis of variance (ANOVA) was conducted to check whether the interactions between grinding models, grinding plates, and different types of crops have any significant effect on depositional level of lead while

Table 1 Effects of grinding models, grinding plates, and crops on depositional level of lead

Source	df	Mean square	*F* value
Model	35	0.00006047	1.16**
Grinding plates	3	0.00007147	1.38**
Grinding models	2	0.00000260	0.05*
Crops	2	0.00013279	2.56**
Grinding plates* grinding models	6	0.00003251	0.63*
Grinding models* crops	4	0.00010880	2.10*
Grinding plates* crops	6	0.00004483	0.86**
Grinding plates* grinding models* crops	12	0.00006102	1.18*
Error	72	0.00005191	
Corrected total	107		

ns Nonsignificant
*Significant; **Highly significant

grinding of crops during wet milling operation as indicated in Table 7.1. From the results of ANOVA, it revealed that the grinding plates had highly significant effect (*F = 1.38, p = 0.0001 (p-value) < 0.05*) on the depositional level of lead. It clearly showed that the variation in the material contents of the grinding plates had significant effects on the depositional level of lead, meaning depositional levels of lead were not the same among grinding plates while milling crops. Grinding models while milling crops likewise showed significant effects on depositional level of lead with $F = 0.05$, $p = 0.005$ (p-value) < 0.05. This indicates that the more the speeds, the higher the depositional level of lead. Conclusively, at least the grinding models were significantly different. The crops that were ground during milling operation showed highly significant different effects on depositional level of lead ($F = 2.56$, $p = 0.0001$ (p-value) < 0.05). This implies that different types of crop had significant effect on depositional level of lead.

The interactions of grinding plates and grinding models and crops of the wet milling of crops all contributed significant effects on the depositional level of lead. The level of significance was $F = 0.63$, $p = 0.044$ (p-value) < 0.05, and $F = 2.10$, $p = 0.014$ (p-value) < 0.05, while grinding plates and crops indicated a highly significant difference ($F = 0.86$, $p = 0.0001$ (p-value) < 0.05). This indicated that interaction of two factors had high significant effect on the average of depositional level of lead on wet milling of crops. This shows interaction effects between grinding models, grinding plates, and crops with lead relying on the grinding models speeds variation, grinding models different types and different types of crops. The interactions of three factors also showed high significant difference effect on depositional level of lead ($F = 1.83$, $p = 0.0045$ (p-value) < 0.05). These entail that at least one of the average depositional level of lead values for the interaction of the grinding models, grinding plates, and crops was significantly different. Concluding that analysis of variance indicating significance in main and interaction treatment would not be adequate to test the level of significance. Further analysis using comparison of mean was carried out to validate where the difference falls in terms of level and significance.

2.7.4.2 Effects of Grinding Plates, Grinding Models, and Different Crops on Lead

Also, Tukey's mean comparison test was conducted, and the mean of available lead indicates that means having same letters are not significantly different. Table 7.2 presents the mean of available depositional level of lead in wet milled crops. The interactions of grinding model (G_1) and crop (C_1) indicate a significant difference between grinding plates P_1 and P_2 with mean values of 0.202 and 0.173 mg/L, while grinding plates P_3 and P_4 did not show significant difference on level of lead having averages of 1.84 and 1.84 mg/L, respectively. Likewise, at grinding model (G_1) and crop (C_2), grinding plates P_1 and P_3 were not significantly different on available lead with averages as 0.194 and 0.190 mg/L, and also, P_2 and P_4 were not significantly different with mean values of 0.182 and 0.185 mg/L accordingly. Similarly, grinding model (G_1) and crop (C_3) interactions and grinding plates P_1 and P_2 did not record significant difference on available lead with averages 0.190 and 0.194 mg/L, but grinding plates P_3 and P_4 show a significant difference on level of lead with mean values of 0.173 and 0.218 mg/L, respectively.

The combination of grinding model (G_2) and crop (C_1) and grinding plates indicates that there is significant difference on depositional level of lead between grinding plates P_2 and P_4 with average values as 0.168 and 0.192 mg/L, while there is no significant difference existing between grinding plates P_1 and P_3 with mean values of 0.175 and 0.177 mg/L accordingly. Also, grinding model (G_2) and crop (C_2) interactions showed a significant difference on depositional level of lead between grinding plates P_2 and P_1, P_3, and P_4 with mean values as 0.174 and 0.215, 0.234, and 0.202 mg/L, respectively. Similarly, interactions of grinding model (G_2) and crop (C_3) and grinding plates P_1 and P_2 indicate nonsignificant difference with an average of 0.180 and 0.186 mg/L, but grinding plates P_3 and P_4 were significantly different at depositional level of lead with mean values of 0.175 and 0.160 mg/L correspondingly. Lastly, grinding model (G_3) and crop (C_1) and grinding plates P_1 and P_4 showed nonsignificant difference with mean values of 0.184 and 0.185 mg/L, while grinding plates P_2 and P_3 were not significantly different with averages 0.174 and 0.177 mg/L accordingly. Interactions between grinding model (G_3) and crop (C_2) indicated that there was a significant difference between grinding plates P_1 and P_4

Table 2 Effects of grinding plates, different crops, and grinding models on lead

	Lead								
	G_1			G_2			G_3		
	C_1	C_2	C_3	C_1	C_2	C_3	C_1	C_2	C_3
P_1	0.202^{c}	0.194^{ab}	0.190^{ab}	0.175^{ab}	0.215^{ab}	0.180^{ab}	0.184^{c}	0.187^{c}	0.224^{a}
P_2	0.173^{b}	0.182^{c}	0.194^{ab}	0.168^{c}	0.174^{c}	0.186^{ab}	0.174^{a}	0.192^{ab}	0.178^{b}
P_3	0.184^{a}	0.190^{ab}	0.173^{c}	0.177^{ab}	0.234^{a}	0.175^{c}	0.177^{ab}	0.196^{b}	0.190^{c}
P_4	0.184^{a}	0.185^{c}	0.218^{d}	0.192^{d}	0.202^{ab}	0.167^{d}	0.185^{c}	0.178^{d}	0.177^{ab}

Means having the same letters are not significantly different within the column at $P < 0.05$ level using Tukey's test with averages of 0.187 and 0.178 mg/L, while grinding plates P_2 and P_4 did not show any significant difference with mean values of 0.192 and 0.196 mg/L, respectively. Grinding model (G_3) and crop (C_3) when interacting showed that grinding plates P_1, P_2, and P_3 were significantly different with depositional level of lead having averages of 0.224, 0.178 and 0.190 mg/L, respectively. There is no significant difference between grinding plates P_2 and P_4 with mean values of 0.178 and 0.177 mg/L, respectively. The overall results show that the highest depositional level of lead was obtained by the grinding plate P_3, grinding model (G_2), and crop (C_2).

From the results pertaining to lead in the grinding plates, the result obtained from the milled crops showed the concentration value in all the milled crops was higher than that obtained in the crops prior to grinding. This shows that the milling process introduced some amount of the lead metal into the milled crops. Considering the rate of decreasing lead contents at the different times of the milling, it can be seen that the rate of wear varied with the four different plates. From Table 7.2, the concentration of the lead ranges between 0.167 mg/L (cowpea) and 0.234 mg/L (millet) for Indian and Nigerian made plates. Generally, the lead concentrations in all the four plates are below WHO standard, and the EU Commission (EC) regulation for lead in food both of which are 0.3 mg/L. Lead was detected but at lower amounts, and these amounts were relatively higher compared to those reported by other studies in Africa. Wyasu et al. (2010) in Nigeria reported 0.013 mg/kg.

The results of the present study are much lower than the results of Lin (1991) and Kwofie and Chandler (2006) who found higher lead concentrations of 0.43 and 0.74 mg/kg, respectively, in cereals. Higher concentration of lead may cause brain complications; coma and death may occur if not treated instantly (Kwofie et al. 2011).

2.7.4.3 Analysis of Variance (ANOVA) on Effects of Grinding Plates, Grinding Models, and Crops on Depositional Level of Arsenic

An analysis of variance (ANOVA) was conducted to determine whether there were significant effects of grinding plates, grinding models, and crops as shown in Table 7.3. The results of the ANOVA pointed that there is a significant effect of the grinding plates ($F = 0.43$, $p = 0.0023$ (p-value) < 0.05) on the depositional level of arsenic, suggesting that the difference in the grinding plates had a significant effect on the depositional level of arsenic generated due to the wet milling, and mean depositional levels were not equal among grinding plates. Grinding model speed levels also show a significant effect on the depositional level of arsenic ($F = 0.78$, $p = 0.0044$ (p-value) < 0.05), showing that the higher the speed level, the more quantity of arsenic. The crops which the milling operation was performed had a significant effect on the depositional level of arsenic ($F = 8.48$, $p = 0.0095$ (p-value) < 0.05).

The interactions between grinding plates and grinding models and crops as well as the combination of grinding plates and crops of the wet milling operation all contributed a significant effect on depositional level of arsenic, with levels of

Table 3 ANOVA on effects of grinding plates, grinding models, and crops on arsenic

Source (arsenic)	df	Mean square	F value
Model	35	0.00004009	1.50*
Grinding plates	3	0.00001155	0.43*
Grinding models	2	0.00002079	0.78*
Crops	2	0.00022603	8.48*
Grinding plates* grinding models	6	0.00003906	1.46*
Grinding models* crops	4	0.00002050	0.77*
Grinding plates* crops	6	0.00004050	1.52*
Grinding plates* grinding model* crops	12	0.00002630	0.99*
Error	72	0.00002667	
Corrected total	107		

ns Nonsignificant
*Significant; **Highly significant

significance $F = 1.46$, $p = 0.0046$ (p-value) < 0.05, $F = 0.77$, $p = 0.0033$ (p-value) < 0.05, and $F = 1.52$, $p = 0.0037$ (p-value) < 0.05. The treatments of two-way combinations had significant effects on the mean depositional level of arsenic. Interaction effect indicates that the relationship between grinding plates, grinding models, and crops with depositional level of arsenic depends on the grinding plates, grinding models, and crop milling speeds. The treatment of three-way interaction of grinding plates, grinding models, and crops also showed significant effects on the depositional level of arsenic ($F = 0.99$, $p = 0.0029$ (p-value) < 0.05). These entail that at least one of the mean depositional levels of arsenic for the combinations of grinding plates, grinding models, and crops was significantly different. We can conclude that there was an interaction of grinding plates, grinding models, and crops in the mean depositional level of arsenic. Analysis of variance indicating significance in main and interaction treatment will not generally be enough for concluding the level of significance. Therefore, there is a need to determine where the difference lies. The level and variation in the significance level can only be resolved using comparison of mean.

2.7.4.4 Effects of Grinding Plates, Grinding Models, and Different Crops on Arsenic

Similarly, mean separation using Tukey's methods stated that means on arsenic having same letters were not significantly different as shown in Table 7.4. The interaction between the crop (C_1) and grinding model (G_1) and grinding plates showed significant difference with mean values 0.179, 0.167, 0.205, and 0.182 mg/L, respectively. Crop (C_2) and grinding model (G_1) show that there was no significant difference with grinding plate P_1 and P_4 having averages of 0.178 and 0.175 mg/L, respectively. Also, grinding plate P_2 and P_3 did not exhibit significant difference with mean values of 0.166 and 0.165 mg/L accordingly. Similarly, at crop (C_3) and

Table 4 Effects of grinding plates, different crops, and grinding models on arsenic

	Arsenic								
	G_1			G_2			G_3		
	C_1	C_2	C_3	C_1	C_2	C_3	C_1	C_2	C_3
P_1	0.179^{a}	0.178^{a}	0.188^{a}	0.200^{a}	0.189^{a}	0.198^{ab}	0.190^{a}	0.178^{a}	0.175^{a}
P_2	0.167^{b}	0.166^{b}	0.183^{a}	0.172^{b}	0.175^{b}	0.204^{d}	0.175^{b}	0.180^{bc}	0.204^{de}
P_3	0.205^{c}	0.165^{b}	0.176^{b}	0.175^{b}	0.177^{bc}	0.178^{c}	0.179^{bc}	0.176^{a}	0.218^{e}
P_4	0.182^{d}	0.175^{a}	0.205^{c}	0.174^{b}	0.172^{bc}	0.192^{ab}	0.170^{b}	0.187^{bc}	0.209^{de}

Means having the same letters are not significantly different within the column at $P < 0.05$ level using Tukey's test.

grinding model (G_1), grinding plates revealed that there was no significant difference with grinding plates P_1 and P_2 having average values of 0.188 and 0.183 mg/L but showed significant difference with grinding plates P_3 and P_4 having mean values of 0.176 and 0.206 mg/L correspondingly.

Also based on Tukey's mean separation, there was significant effect on arsenic on interaction between the crop (C_1) and grinding model (G_2), and grinding plates showed significant difference between grinding plate P_1 with mean value 0.200 mg/L and grinding plates P_2, P_3, and P_4 with averages 0.172, 0.175, and 0.174 mg/L respectively. Crop (C_2) and grinding model (G_2) and grinding plates show that there was a significant difference with grinding plate P_1 with mean value of 0.189 mg/L and grinding plates P_2, P_3, and P_4 having averages of 0.175, 0.177, and 0.172 mg/L accordingly. Similarly, at crop (C_3) and grinding model (G_2), grinding plates revealed that there was no significant difference with grinding plates P_1 and P_4 having mean values of 0.198 and 0.192 mg/L but showed significant difference with grinding plates P_2 and P_3 having mean values of 0.204 and 0.178 mg/L correspondingly.

At the grinding plates P_2, P_3, and P_4, there were no significant difference between interaction of crop (C_1) and grinding model (G_3) with their average values as 0.175, 0.179, and 0.170 mg/L, respectively, but there was a difference between grinding plate P_1 significantly with a mean value of 0.190 mg/L. Similarly, grinding plates P_1 and P_3 were not significantly different with crop (C_2) and grinding model (G_3) interaction given a mean value of 0.178 and 0.176 mg/L, respectively. It also showed that there was no significant difference on the level of arsenic between grinding plates P_2 and P_4 with the mean values as 0.180 and 0.187 mg/L accordingly.

Likewise, grinding plates P_2, P_3, and P_4 were not significantly different when interacting with crop (C_3) and grinding model (G_3), and the mean values were 0.204, 0.218, and 0.209 mg/L correspondingly. Equally, there were significant difference effect with grinding Plate I with the remaining plates with average value as 0.175 mg/L. Conclusively, at grinding plate (P_3), grinding model (G_1) gave the least amount of arsenic in crop (C_2), and the highest depositional level was obtained by grinding plate (P_3) and grinding model (G_3) in crop (C_3). The highest obtained was from tomatoes, when compared with WHO, Food and Agriculture Organization (FAO), and other standards.

The arsenic concentration recorded in all the milled crops exceeded that obtained in the ones prior to grinding. This shows that the milling process introduced some amount of the arsenic metal into the wet milled crops. Crop samples from the first millings with the four plates recorded significantly high contamination of arsenic. The contamination deceased from the other subsequent millings.

From Table 7.4, arsenic concentration ranges between 0.165 and 0.218 mg/L, respectively, for all the plates. Lotus-Zamfara had the least concentration in millet and had the highest concentration in tomatoes.

Generally, the concentrations of arsenic in the crops with all the four plates are below level of trace amount proposed by FAO (2001). Abrefah et al. (2011) recorded arsenic levels below detection limit. Too much arsenic can also damage the liver, kidneys, and nerves, and it may cause irregular heart rhythm.

2.7.4.5 Analysis of Variance (ANOVA) on Effects of Grinding Plates, Grinding Models, and Crops on Depositional Level of Cadmium

The data collected during laboratory analysis of grinding models with different speeds of operation were analyzed using analysis of variance (ANOVA) to determine if there were significant effects of grinding models, grinding plates, and crops on depositional level of cadmium as shown in Table 7.5. The results of the ANOVA indicate that there is a significant difference of the grinding plates ($F = 1.82$, $p = 0.0029$ (p-value) <0.05) on the depositional level of cadmium. This explains clearly that the change in the makes of grinding plates on grinding plates had a significant effect on the depositional level of cadmium, meaning that average depositional level of cadmium is not the same among grinding plates. Different grinding model speeds during milling operation also indicates that there is significant difference on depositional level of cadmium ($F = 1.27$, $p = 0.0405$ (p-value) <0.05). It reveals that the higher the speeds of the milling operation, the more depositional

Table 5 ANOVA on effects of grinding plates, grinding models, and crops on depositional level of cadmium

Source	df	Mean square	*F* value
Model	35	0.00010175	1.57*
Grinding plates	3	0.00011803	1.82*
Grinding models	2	0.00008221	1.27*
Crops	2	0.00067940	10.48**
Grinding plates* grinding models	6	0.00008669	1.34*
Grinding models* crops	4	0.00010782	1.66*
Grinding plates* crops	6	0.00002622	0.40**
Grinding plates* grinding models* crops	12	0.00004793	0.74*
Error	72	0.00006484	
Corrected total	107		

ns Nonsignificant
*Significant; **Highly significant

level of milled cadmium. It can be concluded that at most, one of the mean values of the grinding model level of cadmium deposits was significantly different. The crops showed a highly significant effect on depositional level of cadmium ($F = 10.48$, $p = 0.0001$ (p-value) <0.05). It shows that crops had a significant effect on the depositional level of cadmium.

The combinations between grinding plates and grinding models and crops for wet milling of crops had a significant effect on depositional level of cadmium. Their levels of significant difference are $F = 0.34$, $p = 0.0036$ (p-value) <0.05, and $F = 1.66$, $p = 0.0043$ (p-value) <0.05. Expressively, combination effect shows that the relationship depends on the grinding plates, grinding models, and crops. The factor interactions between grinding plates and crops also indicated a highly significant effect ($F = 0.40$, $p = 0.0001$ (p-value) <0.05) on the depositional level of cadmium. This means that at least one of the mean depositional levels of wet milling for the interaction is significantly different. We cannot rely on the ANOVA to verify the significance of the main and interaction effects. However, to verify where the parity lies, the level and change in the significance level can best be determined using separation of mean.

2.7.4.6 Effects of Grinding Plates, Grinding Models, and Crops on Depositional Level of Cadmium

Analysis of variance depicted that the effects of grinding models, grinding plates, and crops on depositional level of cadmium were significant (<0.05) effect. The mean separation using Turkey's method has shown that mean depositional levels of cadmium with same letters are not significantly different as illustrated in Table 7.6. The combination of grinding model (G_1) and crop (C_1) and grinding plates P_1 and P_4 did not show significant difference, with average mean of depositional level of cadmium of 0.185 and 0.181 mg/L, respectively. There was a significant difference with a mean depositional level of 0.173 and 0.191 mg/L between grinding plates P_2 and P_3. At grinding model (G_1) and crop (C_2) interacting, grinding plates P_1 and P_4 were not significantly different, with the mean depositional level of cadmium of 0.175 and 0.173 mg/L, respectively. Grinding plates P_2 and P_3 were significantly

Table 6 Effects of grinding plates, different crops, and grinding models on cadmium

	Cadmium								
	G_1			G_2			G_3		
	C_1	C_2	C_3	C_1	C_2	C_3	C_1	C_2	C_3
P1	0.185^{bc}	0.175^{de}	0.201^{ab}	0.184^{bc}	0.189^{cd}	0.191^{c}	0.181^{b}	0.173^{cd}	0.192^{c}
P2	0.173^{de}	0.186^{bc}	0.185^{c}	0.179^{d}	0.176^{ab}	0.206^{ab}	0.179^{cd}	0.170^{c}	0.210^{ab}
P3	0.196^{a}	0.195^{a}	0.229^{ab}	0.188^{bc}	0.173^{ab}	0.176^{de}	0.193^{a}	0.186^{b}	0.266^{a}
P4	0.181^{bc}	0.173^{de}	0.202^{ab}	0.196^{a}	0.175^{ab}	0.203^{ab}	0.175^{cd}	0.195^{a}	0.231^{ab}

Means having the same letters are not significantly different within the column at $P < 0.05$ level using Tukey's test different with depositional level of cadmium giving a 0.186 and 0.195 mg/L accordingly. Similarly, with grinding model (G_1) and crop (C_3), grinding plates P_1, P_3, and P_4 were not significantly different with amount of cadmium averages of 0.201, 0.229, and 0.202 mg/L respectively. It is significantly different with grinding plate P_2 with mean of 0.185 mg/L. The interaction between grinding model (G_2) and crop (C_1) indicated that grinding plates P_1 and P_3 were not significantly different, with an average depositional level of cadmium of 0.184 and 0.188 mg/L, respectively. The grinding plates P_1 were significantly different with P_2, P_3, and P_4 with means of 0.189, 0.176, 0.173, and 0.175 mg/L when grinding model (G_2) and crop (C_2) interacted. Grinding model (G_2) interaction with crop (C_3) and grinding plates P_1 and P_3 were significantly different with means of 0.191 and 0.176 mg/L, while differences between grinding plates P_2 and P_4 were not significantly different with averages as 0.206 and 0.203 mg/L, respectively. Lastly, with grinding model (G_3) and crop (C_1), grinding plates P_1 and P_3 were significantly different with averages of 0.181 and 0.193 mg/L, while grinding plates P_2 and P_4 were not significantly different with mean values of 0.179 and 0.175 mg/L, respectively. The grinding model (G_3) and crop (C_2) grinding plates P_1 and P_2 were not significantly different having a mean depositional level of cadmium of 0.173 and 0.170 mg/L, while there is significant difference between grinding plates P_3 and P_4 having mean values of 0.186 and 0.195 mg/L accordingly. Finally, grinding model (G_3) and crop (C_3) were interacting, and grinding plate P_1 was significantly different with P_2, P_3, and P_4 with mean averages of 0.192, 0.210, 0.266, and 0.231 mg/L correspondingly. The highest depositional level of cadmium is obtained by G3, C3, and P3 which is higher than the results obtained by Jorhem (1999).

The cadmium concentration recorded in all the wet milled crops was higher than that obtained prior to grinding. This shows that the milling process introduced some amount of the cadmium metal into the crop pastes during the wet milling process.

The observed trend of lower concentrations recorded with time shows that the rate of wear decreases with time as the plates are being used. The rate of wear can be said to vary for the four different plates.

Generally, the concentrations of cadmium in the wet milled crop samples from three of the four grinding plates (Adex, Nas, and Lotus India) are below the WHO standard and also the EC regulation for cadmium in food (0.2 mg/L) except that of Lotus-Zamfara which has a value of 0.266 mg/L. Algül and Kara (2014) and Frazzoli et al. (2007) had cadmium concentrations in corn flour samples from 0.004 to 0.38 and 0.005 to 0.49 mg/kg, respectively, for the two different works.

The exceeding levels of cadmium can affect human health. Cadmium can disturb kidney functions, and some studies indicate a cancerous effect. Research has proven that in those countries whose main food is cereal, the consumption of these cereals causes an intake of cadmium. Machiwa (2010) had reported that 50% of cadmium intake comes from the cereal consumption, and in Japan, this amount was 40 to 60%.

The Asian governments have established critical maximum levels of heavy metals in cereals to protect the health of their citizens. In Japan, the maximum level of cadmium in unpolished cereal is 1.0 mg/kg, while in China, the maximum permitted level is 0.4 mg/kg of polished cereals (Chen 2000).

The European Union (EU) food legislation provides an overview of maximum permissible level of cadmium concentration in cereal as 0.2 mg/kg (EU 2006).

2.8 Pearson's Correlation Analysis

Quantitative traits like wear elements express themselves in close association with many other traits. Change in the expression of one trait is usually associated with changes in the expression of many other traits. Therefore, the correlations obtained in the present study are useful in the selection of traits having direct and significant correlation in improving wear elements. Study of the values of the wear elements based on Pearson's correlation coefficients indicated in Table 7.7 the diagonal line shows that iron was found to be negative and significant ($p > 0.05$) and correlated with copper ($r = -0.229$*). It was observed that magnesium has a positive significant ($P > 0.05$) correlation with manganese ($r = 0.199$*). Lead gave highest positive significant correlation ($P > 0.01$) with copper ($r = 0.352$**) and manganese ($r = 0.358$**). Similarly, a significant positive correlation ($P > 0.05$) of lead with magnesium ($r = 0.204$*). Cadmium was found to be highly significant with positive correlation ($P > 0.01$) with arsenic, and cadmium also has positive correlation ($P > 0.05$) with copper ($r = 0.224$*) and negative significant correlation ($P > 0.05$) with magnesium($r = -0.2.40$*). Nickel at 1% significant showed highly significant positive correlation with copper ($r = 0.342$**) and cadmium ($r = 0.432$**). A positive and significant correlation (P.0.05) was found with arsenic (0.205*). The trait also showed high significant positive correlation ($P > 0.01$) of zinc with magnesium ($r = 0.626$**) and also positive significant correlation ($P > 0.05$) with copper($r = 0.208$*). Cobalt was found to have highly positive significant correlation ($P > 0.01$) with iron ($r = 0.392$**) and a positive significant correlation ($P > 0.05$) with nickel ($r = 0.216$*). The trait also had highly positive correlation ($p > 0.01$) of chromium with arsenic ($r = 0.344$**), cadmium ($r = 0.379$**), and nickel ($r = 0.389$**) respectively. Lastly, selenium had significant positive correlation ($p > 0.01$) with lead ($r = 0.251$**) and chromium ($r = 0.289$**) and positive significant correlation ($p > 0.05$) with manganese.

2.9 Conclusion

The presence of relatively high concentration of wear elements in comparison with the controls could be attributed to wearing from aging milling machines and milling plates. The levels were however within the permitted limits set by WHO and FAO. Routine monitoring by regulatory agents is therefore recommended to enforce the replacement of overage milling machines in order to protect consumers from potential metal exposure and attain the Sustainable Development Goal (SDG) 3.

Table 7 Pearson's correlation among the wear element traits in the studied crops

	Cu	Fe	Mn	Mg	Pb	As	Cd	Ni	Zn	Co	Cr	Se
Cu	1											
Fe	−0.229*	1										
Mn	0.069	0.015	1									
Mg	0.095	−0.120	0.199*	1								
Pb	0.352**	−0.162	0.358**	0.204*	1							
As	0.079	−0.093	−0.019	−0.098	−0.078	1						
Cd	0.224*	−0.027	−0.017	−0.240*	0.000	0.498**	1					
Ni	0.342**	0.027	−0.092	−0.087	−0.017	0.205*	0.432**	1				
Zn	0.208*	0.156	0.112	0.626**	0.111	0.094	0.105	0.068	1			
Co	−0.059	0.392**	−0.093	−0.165	−0.047	−0.132	−0.030	0.216*	0.007	1		
Cr	0.103	−0.064	0.064	−0.058	−0.007	0.344**	0.379**	0.389**	−0.000	0.004	1	
Se	0.072	−0.048	0.233*	0.108	0.251**	0.178	0.131	0.158	0.076	−0.086	0.289**	1

*Significance at 5%; **Significance at 1%

Acknowledgments The authors sincerely thank all staff of the Department of Agricultural and Bioresource Engineering of the Abubakar Tafawa Balewa University and staff of the Chemistry Laboratory of the Abubakar Tatari Ali Polytechnic, Bauchi, Nigeria, for their support and contributions toward the success of this research work.

Ethical Approval Ethical clearance was not required for this study as no human participants were involved.

Funding The research was supported by the Tertiary Educational Fund (TETFUND) Nigeria for the purchase of instruments, grinding machines, grinding plates, and crops and data acquisition using research grant no. TETTF/DR&D/CE/UNIV/BAUCHI/IBR/2019.

References

Abrefah RG, Sogbadji RMB, Ampomah-Amoako E et al (2011) Comparison of the effects of cadmium-shielded and boron carbide-shielded irradiation channel of the Ghana research reactor-1. Nuclear Engineering and Design 241(8). https://doi.org/10.1016/j.nucengdes.2010.05.005.

Adeti PJ (2015) Assessment of heavy metals introduced into food through milling process: health implications. International Atomic Energy Agency (IAEA)/International Nuclear Information System (INIS). http://inis.iaea.org/search/search.aspx?orig_q=RN:47048409

Asere AM (2022) Milling Method: Level of Heavy Metal Content in Foodstuffs and Soup Ingredients. NIPES Journal of Science and Technology Research 4(2):136 – 143. https://doi.org/10.37933/nipes.e/4.2.2022.14

Chen ZS (2000) Relationship between heavy metal on concentrations health risk due to dietary intake of heavy metals in the industrial area of Huludao City. China. Sci Total Environ 387: 96–104

Copestake J (2008) Wellbeing in International development: what's New? Journal of International Development 20:577–97

Coronado-Gonzalez AH, Sweyn M (2007) Determination of lead in water samples by graphite furnace atomic absorption spectrometry after cloud point extraction. Talanta 67(5): 992–996

Davidson C, Swisher EH (2004) Determination of cadmium, copper, lead and zinc in water samples by flame atomic absorption spectrometry after cloud point extractionJournal of American Oil Chemistries Association, 65(2004):1704–1706

Diaz-Sarachaga JM, Jato-Espino D, Castro-Fresno D (2018) Is the Sustainable Development Goals (SDG) index an adequate framework to measure the progress of the 2030 agenda? Sustainable Development 26:663–671

Ekino M, Almeida EM, Nogueira NN (2007) Household food insecurity in Canada, In research to identify policy options to reduce food insecurity. Toronto, Canada. Journal of Engineering Sciences

Fellows PJ (2009) Food processing technology principles and practice. Woodhead Publishing Series in Food Science Technology and Nutrition, 3rd edn. ISBN 1845696344, 9781845696344

Frazzoli C, Roberto D, Alberto M et al (2007) Functional toxicity and tolerance patterns of bioavailable Pd(II), Pt(II), and Rh(III) on suspended Saccharomyces cerevisiae cells assayed in tandem by a respirometric biosensor. Analytical and Bioanalytical Chemistry 389:2185–2194

Global Health Observatory (GHO) data (2018a) Child health. World Health Organization 2018. http://www.who.int/gho/ child health/en

Global Health Observatory (GHO) data (2018b) Life expectancy. World Health Organization 2018. http:// www.who.int/gho/mortality_burden_disease/life_tables/en

Global Health Observatory (GHO) data (2018c) Maternal mortality. World Health Organization 2018. http://www.who.int/gho/maternal health/mortality/maternal/en
Hodgson G, Basu SS, Batie CJ et al (2007) Material Science and Processes. 2nd edn. New Delhi, cbs, p 170–176
Jemal R, Hosseini MRM, Assadi Y (2002) Combination of dispersive liquid–liquid microextraction with flame atomic absorption spectrometry using microsampleintroduction for determination of lead in water samples. Anal ChimActa 610(1): 135–141
Jorhem L (1999) Determination of metals in foods by atomic absorption spectrometry after dry ashing; NMKL1 Collaborative Study. Journal of Association of Agricultural Chemists (AOAC) International 2000:83(5)
Kabir ER, Rahman MS, Rahman I (2015) A review on endocrine disruptors and their possible impacts on human health. Environmental toxicology and pharmacology. 40(1): 241–258
Kalasuramath S, VinodKumar C, Kumar, MM et al (2015) Impact of anemia, iron deficiency on physical and cardio respiratory fitness among young working women in India. Indian Journal of Basic and Applied Medical Research 4: 119–126
Kwofie S, Andrews A, Mensah E (2011) The quality of locally manufactured cornmill grinding plates. Journal of Science and Technology 31(1)
Kwofie S, Chandler HD (2006) A potential health effects of locally manufactured cornmill grinding plates. Journal of Science and Technology 26 (2): 137–147
Lakherwal D (2014) Adsorption of heavy metals: a review. International Journal of Environmental Research and Development 4(1): 41–48
Lin CS, Boltz RC, Siekierka JJ et al (1991) Cell. Immunol.133: 269–284.
Machiwa JF (2010). Heavy metal levels in paddy soils and rice (Oryza sativa) from wetlands of Lake Victoria basin, Tanzania. Tanzania Journal of Science 36(2010):451–459.
Mieiro CL. Coelho JP, Pacheco M et al (2012). Trace elements in two marine fish species during estuarine residency: Non-essential versus essential. Marine pollution bulletin.64(12):2844–2848.
Navas-Acien, Candir S., Narin I. and Soylak M. (2007). Determination of cadmium, copper, lead and zinc in water samples by flame atomic absorption spectrometry after cloud point extraction. Analytica Chemica Acta 450(1): 215–222.
Needleman (2004) Air quality criteria for lead. Research triangle park (NC): environmental criteria and assessment 600(80):83–028.
Ortega-Rivas, E. (2012) Size reduction. non-thermal food engineering operations. Publisher-Springer Science and Business Media, p71–87 ISBN 1461420385, 9781461420385.
Salama AK, Radwan MA (2005) Heavy metals (Cd, Pb) and trace elements (Cu, Zn) contents in some food stuffs from egyptian market. Emirate Journal of Agricultural Sciences 17(1):34–42. https://doi.org/10.9755/ejfa.v12i1.5046
Sinayobye, E, Saalia, FK (2011) Size reduction of food commodities and food safety in some accra markets. African Journal of Food Agriculture Nutrition and Development 11(6):5282–5299
Thompson JRC, Soliva P, Elez M et al (2007) Innovative food science and emerging technologies" Journal of Food Science 134:261–268
United Nations (2015). Transforming our World: the 2030 agenda for sustainable development. Resolution adopted by the 70th Session of the UN General Assembly. New York: 12/35. Report No. A /RES/70/1.
Wang T, Jia X, Wu J (2003) Direct determination of metals in organics by inductively coupled plasma atomic emission spectrometry in aqueous matrices. Journal of pharmaceutical and biomedical analysis, 33(2003), 639–646. https://doi.org/10.1016/s0731-7085(03)00357-1
World Health Organization (2018a) HIV/AIDS Key facts Geneva. http://www. who.int/newsroom/fact-sheets/detail/HIV-AIDS [cited 2018 August 10].
World Health Organization (2017) Time to deliver: report of the WHO Independent High-Level Commission on Non-communicable diseases. Geneva. World Health Organization; 2017. http://www.who.int/ncds/management/time-to-deliver/ en/ [cited 2018 August 10]
World Health Organization (2018b) WHO health equity monitor Geneva: World Health Organization 2018. http://www.who.int/gho/health_equity/en/ [cited 2018 August 10].

World Health Statistics (2016) Monitoring Health for the SDGs. Geneva: World Health Organization. World Health Statistics 2016. http://www.who.int/gho/ publications/world health statistics/2016/en/ [cited 2018 August 10].

Yakubu Y, Adoum O, Wudil AM, Ladan, Z (2015). Toxicity study of ethanol root extract of Terminalia macroptera Guill. Perr (Combretaceae) and assessment of some heavy metals. African Journal of Pure and Applied Chemistry, 9(9), 193–196. https://doi.org/10.5897/ajpac2015.0647

Yamey G, Shretta R, Binka FN (2014) The 2030 sustainable development goal for health. BMU349(2022):5295. https://doi.org/10.1136/bmj.g5295

Biochar Development in the Urban Environment: A Case Study of Sludge Char Production in Finland

Oana Iliescu and Mikko Jalas

1 Introduction

Historically, our approach to socioeconomic development has been detrimental to the environment and the rest of species. For a very long time, human development has depended heavily on intensive resource use. And indeed, while the Human Development Reports have shown the improvements in education, life expectancy, domestic gross product, etc. in parts of the world, it also acknowledges that our influence shapes the natural environment to an alarming extent, destabilizing the very systems we depend on for our survival (UNDP 2020). And with a renewed understanding of the situation, an extensive scientific community has been coming forward with different proposals for the climate crisis.

The Intergovernmental Panel on Climate Change (IPCC), as one of the most comprehensive publications on the matter out there, has indicated two possible paths, both equally valuable for safeguarding the future. First, reduce greenhouse gas (GHG) emissions on a large scale as soon as possible, and second, deploy a wide range of carbon dioxide removal (CDR) and carbon capture and sequestration (CCS) techniques (Rogelj et al. 2018). There is a considerable number of challenges for both approaches, ranging from financial, economic, and social issues and debates.

One of the most prominent carbon drawdown technologies emerging in the past years, intensely researched and tested, has been biochar. According to the European Biochar Certification, "biochar is a porous, carbonaceous material that is produced

O. Iliescu (✉)
Department of Built Environment, Aalto University, Espoo, Finland
e-mail: oana.iliescu@aalto.fi

M. Jalas
Department of Design, Aalto University, Espoo, Finland
e-mail: mikko.jalas@aalto.fi

I. S. Dunmade et al. (eds.), *Sustainable Engineering*, Green Energy and Technology, https://doi.org/10.1007/978-3-031-47215-2_8

by pyrolysis of biomass and is applied in such a way that the contained carbon remains stored as a long-term C (carbon) sink" (EBC 2022). By no means a new creation, biochar has a rich history as a soil amendment in agriculture across the globe. However, the interest it raises now for its carbon sequestration potential has increased. With the appropriate technology, the feedstock for biochar production can generally be any organic material. Usually, the most common feedstock is woody biomass (garden and forestry waste, demolition wood), food waste, manure, and sewage sludge.

The purpose of this work is to serve as an introduction to the topic of biochar and sewage sludge char, some of their physicochemical characteristics and the related possible applications within the urban environment. As much as possible, it is meant to give an overview of the topic, based on the published literature, sludge-char producers' web sources, discussions, and interviews with experts in this domain. The aim is to raise awareness and interest in the biochar and sludge-char topic and encourage future research, testing, and development as it is still an emerging subject.

2 Biochar and the Sustainable Development Goals (SDGs)

Generally, the applications of biochar have been long researched in agriculture, applied as a raw soil amendment capable of retaining nutrients and water, thus leading to an increase in crop yields and a decrease in fertilizer use. Studies show that by applying fertilizer in combination with biochar, there was a 15% average yield increase compared to the same fertilization technique without the biochar (Schmidt et al. 2021). According to this focus, biochar covers a wide range of SDG targets, especially within SDG 2 (Zero hunger). More recently, efforts have been made to move in a different direction and combine biochar with compost, fertilizer, or even manure to be applied on agricultural land to lower the overall GHG emissions (Kammann et al. 2017), aiding in the efforts to accomplish SDG 13 (Climate action).

However, the benefits of biochar use in the urban environment are a somewhat newer pursuit, generating a lot of research in academic circles, driven as well by the environmental and economic value of the material. In the urban environment, biochar has been established as a carbon sequestration method, as well as a useful tool for climate adaptation objectives (SDG 13). Biochar can be used as an ingredient in manufactured soil, as a contaminated soil remediation method (SDG 15 – life on land), filtering systems (SDG 6 – access to water and sanitation for all), and construction material additive (e.g., concrete, asphalt, or mortar) (SDG 9 – industry innovation and infrastructure) (Azzi 2021).

2.1 Influencing Factors

The two main deciding factors for the properties of biochar are the feedstock options and the pyrolysis conditions – heating rate and temperature of pyrolysis. These properties then in turn determine the stability of the biochar and, thus, its longevity in the soil (Crombie et al. 2013). Pyrolysis occurs at temperatures between 300 °C and 700 °C in limited oxygen conditions (Ippolito et al. 2020). The heating rate in combination with the temperature range affects the amount of biochar produced. Slower heating rates and lower temperatures (300–500 °C) lead to a higher yield of solids (biochar) – approximately 30% of input. On the other hand, higher heating rates and higher temperatures (600–700 °C) favor the production of by-products and only around 10% yield of biochar (Pokharel et al. 2020). Figure 1 illustrates an example of biomass pyrolysis at medium temperature and heating rate.

However, studies show that temperature remains the factor with the strongest effect on biochar composition. By increasing the pyrolysis temperature, the stability, specific surface area, pore volume, C content, and pH also increase. For instance, specific surface area is related to nutrient and contaminant retention, while pore volume affects water availability (Ippolito et al. 2020). Depending on the intended biochar application, these factors must be considered in the production process to ensure the benefits of the final product.

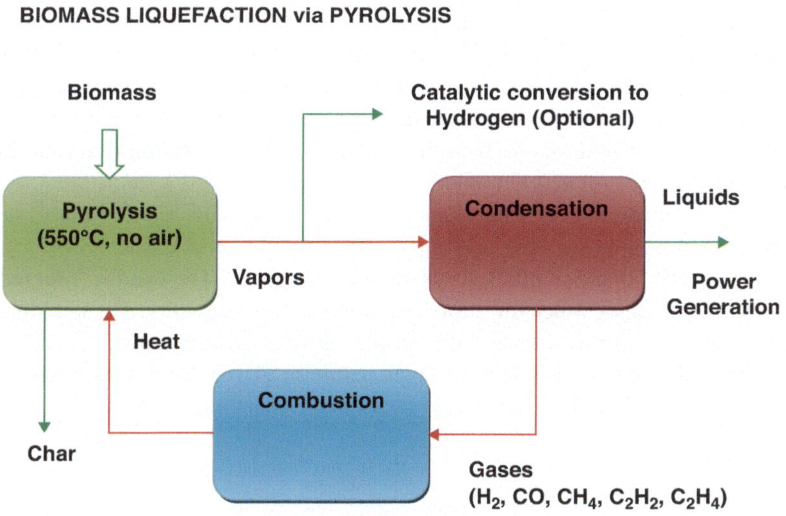

Fig. 1 Example of biomass pyrolysis at medium temperature (550 °C), the resulting by-products, and their interaction for sustained functioning (Hossain et al. 2011)

3 Biochar in Soil Applications

The threat of climate change in the urban environment brings varied and unprecedented consequences on the infrastructure and inhabitants. Often, due to the inherent nature of a city (high density population, intense activity, developed gray infrastructure), the effect of extreme weather conditions is accentuated – be it flooding, drought, storms, etc. On the other hand, cities and urban activities are major contributors to greenhouse gas emissions; an estimated 75% of the global carbon dioxide (CO_2) emissions arise in the cities according to the United Nations Environment Programme (UNEP 2022). Thus, cities constitute a focal point. They have a great potential for action and results, by implementing emission reduction strategies, carbon drawdown programs, or blue-green-gray sustainable infrastructure projects.

3.1 *Biochar in Urban Tree Planting*

Following this logic, the benefits of developing green areas in the urban environment are well-known for the mental and physical health of the inhabitants (Bennett and Jones 2018), the cultural essence of a city, supporting biodiversity (Filazzola et al. 2019), as well as aiding in CO_2 emission reductions. However, as any development project, the process of building and planting trees in the urban environment comes with its own emissions. A study in the Helsinki area estimates that the newly planted trees need approximately 30 years before the initial emissions are offset and they start the net CO_2 sequestration (Riikonen et al. 2017). As a way to offset the initial emissions or to compensate, biochar could be mixed in the soil of urban trees. This has been identified as one benefit of using biochar in urban tree planting, in addition to other soil-related benefits.

The urban environment is rather harsh for plants, due to its higher rates of air and soil pollution, drought exposure, and soil compaction. In some cases, growing conditions for trees and other perennial plants are less than ideal. They suffer from reduced soil volumes, improper water filtration, and unreliable soil quality. Thus, adding biochar to a granular topsoil can prevent excess settling and compaction, as well as providing a nutrient and water retention layer (Embrén 2016). Literature suggests that between 5% and 20% of biochar should be added to the soil mix (ibid.).

3.2 *Biochar in Urban Roof Gardens, Rain Beds, and Green Walls*

Roof gardens and green walls have a long history of usage in the urban environment spanning thousands of years (Abass et al. 2020). In each climate zone, the concept has been adapted accordingly, with varying degrees of popularity over the years.

Green infrastructure is a powerful climate adaptation and mitigation tool in urban resilience plans. Generally, vegetated roofs, walls, and rain beds are included under the umbrella term "blue-green infrastructure" because their purpose is twofold: as part of the stormwater management system, they retain water and help manage urban runoff levels, and as green areas, they help maintain the local biodiversity (Filazzola et al. 2019) and add to human health and well-being benefits.

The benefits of developing green solutions for roofs are best represented by the three sustainability pillars. From the environmental standpoint, roof gardens have a cooling effect on the surroundings, thus lowering the urban heat island effect. From the economic view, they act as insulators for buildings, lowering the heating/cooling needs and saving on energy costs. Lastly, they are aesthetically pleasing and act as a relaxing green area for people, covering the social benefits.

However, green roofs are part of the built environment, and they require careful planning and construction. The structure of a green roof contains several layers of substrate, including a waterproofing layer (most common a bitumen membrane), a drainage layer with a filter, sometimes combined with a rainwater storage tank, and the growing medium which should be lightweight and help with drainage (Setherton 2022). Therefore, there is an added permanent load to any building, and their structural integrity must be assessed in advance. Another impediment for green roofs, especially in the Nordic climate, is represented by the repeated freezing/thawing cycles and the temperature variations. They could be detrimental to the waterproofing layer, increase maintenance needs, as well as harm any nonnative plants (Andenæs et al. 2018).

There are several studies in the past years indicating that biochar use in roof gardens improves soil quality, by increasing the porosity and the soil moisture, regulating the pH values, improving the nutrient and water holding capacity, and reducing the soil's bulk density (Cao et al. 2014) – an important benefit because the same area could be covered by a thicker soil layer without adding weight to the structure. Additionally, there is evidence of increased microbial diversity in the soil due to increased levels of carbon and phosphorus from the added biochar (Chen et al. 2021). There are ongoing studies into the characteristics of biochar suitable for green roofs due to its ability to affect the runoff quality and quantity, especially concerning nutrient leeching (Kuoppamäkia et al. 2015).

Depending on the vegetation type from grass to small trees, the growing medium can vary between 25 and 1000 mm. Most roof gardens in the urban environment should be as self-sufficient as possible, and extensive planting especially in the Nordic climate would not be sustainable in the long term. Literature indicates that 30% biochar mix by volume in the substrate increases the plant available water, while 40% biochar mix significantly increases the water-holding capacity, therefore the optimal choice for stormwater retention (Cao et al. 2014).

3.3 Biochar in Sports and Neighborhood Fields

Sports fields and neighborhood sport sites are high traffic areas, which require certain soil conditions in order to maintain their intended uses over time. Some pitches may also require a more careful planning of the substrate to allow for proper water drainage and limit soil movements. Under the natural grass turf, it is ideal to include a highly permeable rootzone layer (approximately 300 mm) and a coarse subbase (approximately 100 mm) (Estimation QS n.d.). Mixing biochar with the sand in the rootzone layer can lead to faster drainage while improving the water retention and availability for the turf above. Similarly, biochar in the coarse subgrade layer aids in water drainage, lessening the reliability on drainpipes (Major 2010). Finally, there may be a financial and/or aesthetic reason for utilizing biochar in sport pitches (both professional turfs and neighborhood fields) due to prolonged serviceability. The grass turf has access to the necessary nutrients and water to endure longer into the season with lower maintenance/fertilizer needs.

3.4 Biochar in Construction (Road and Railway Embankments, Structural Soils, Bioswales)

The IPCC recognizes biochar as a viable option for carbon sequestration, indicating that the stored carbon can remain in the soil between decades and centuries depending on the soil's type and management and the biochar's production temperature (Masson-Delmotte et al. 2018). Thus, one option to advance any city's climate action plan and develop its carbon storing capacity would be to include biochar into new soils. In this scenario, the benefits of biochar in soil applications take a secondary place, being replaced by the carbon sequestration function.

For instance, one possible storage for the biochar, suggested by the International Biochar Initiative, is in the substrate soil layer under buildings, roads, etc. (Major 2010). Most of the time, before an infrastructure or urban development project can commence, there are extensive earth works on the site. One of the first steps, after the geotechnical investigations of the site, is removing the top layer of soil because of its organic content, poor engineering qualities, contamination, and so forth. A layer of biochar or biochar/soil mix in these conditions could represent a significant use in the urban environment.

However useful as a carbon storing solution, it would be desirable for biochar to serve other purposes in its application to account for the investment. In the construction industry, as an amendment to new soils, biochar could also be useful to improve soil engineering properties. The applications range from road and railway embankments, landfill cover, bioengineered slopes, etc. The geotechnical properties of the soil in these structures require careful design, and the long-term effect of biochar application has not been studied as thoroughly as its agricultural uses (Hussaina et al. 2019).

4 Biochar in Filters

The blue-green-gray infrastructure shows a growth trend in the urban environment, advancing the climate change mitigation and adaptation plans. Blending the built and natural environments is a challenging task, considering the already existing infrastructure and the substantial investments necessary to renew and innovate said infrastructure. Biochar could prove a useful tool in certain sectors such as water and wastewater treatment and stormwater management due to its ability to act as a filter.

4.1 Biochar in Wastewater Treatment

Studies suggest that biochar could be used at different stages of the wastewater treatment process to improve the efficiency of activated sludge treatment and nutrient recovery. The innovation factor for this application is the cascading effect of biochar use. Biochar added during the secondary treatment in the aeration tank improves the efficiency of the process (increasing the settling ability of the sludge by adsorption of inhibitors and toxic compounds). While in the tertiary treatment, biochar adsorbs nutrients (nitrogen and phosphorus in particular). These improvements enhance the quality of the resulting biosolids which, in turn, could be used to produce nutrient-enriched biochar.

4.2 Biochar in Stormwater Retention Basins

Developing stormwater management plans has been on the forefront of cities' priorities, especially in parts of the world regularly dealing with floods or water storage issues. Generally, based on the continuous growth of the city and the evolving legislation, the reasoning of the plan is twofold. It first takes into account the growth of the urban sphere, the expansion of the inner city, and the development of smaller centers, thus leading to a denser city with more impermeable surfaces. Second, it considers the forecasted effects of climate change consisting of more extreme weather events (Nordic climate: more rainfall in the winter, dry periods followed by heavy rainfall during the summer period). Thus, the stormwater flow rates increase, potentially overwhelming the combined sewer system, endangering the health of the residents, and creating the possibility of flash floods.

Special attention should be drawn to utilizing the stormwater as a resource in the urban environment, rather than an inconvenience. However, that implies that the quality of the water becomes a concern. Suspended solids binding nutrients, heavy metals, and hazardous substances from industrial activity that might reach the waterways are especially harmful (Peltosaari et al. 2020). Green solutions for infrastructure are preferable, and indeed, they should be an integral part of a city's urban planning strategy. Thus, stormwater filtration measures could benefit from adding biochar, which, just like activated carbon, has the advantage of retaining harmful substances. It is also a considerably cheaper material to produce and to recycle at the end of its lifecycle compared to other solutions.

5 Biochar in Construction Materials

Integrating biochar in construction materials can have numerous benefits, both from an environmental standpoint and from an engineering perspective. The construction industry seeks to regulate and reduce their CO_2 emissions (Kuittinen and Häkkinen 2020); thus, biochar as a carbon sequestration method can help minimize the emissions. On the other hand, biochar has a low thermal conductivity, and it has high water-holding capacity, meaning that it can be used to insulate buildings and regulate humidity (Schmidt et al. 2014). It can be used both on the inside and outside walls of a building, in a cement mix, additive to bricks or plaster, or with clay.

The effort to develop carbon negative concrete has been growing rapidly in recent years. One concrete example comes from Norway, where Snøhetta (together with Skanska and several other partners) have developed and tested *biocrete* – a mix of concrete and biochar (obtained from construction wood waste) (Snøhetta 2022). There is also research into using biochar in asphalt binder, with some promising results regarding the high-temperature properties (Ma et al. 2022). Their tests and projects are ongoing, but these are example of large investments in biochar and its use in construction materials.

6 Background of Sludge-Derived Biochar

The days of Joseph Bazalgette and the first modern sewer system (De Feo et al. 2014) seem long behind us when we consider the technological advancements in sewer systems and wastewater treatment processes that are available today. Rapid population growth and industrialization are some of the main factors that brought about these advancements, which, in turn, have created great strain on the environment.

Wastewater has been seen for a very long time as an inconvenience rather than an opportunity. And the global standing on wastewater treatment ranges dramatically depending on the location and the socioeconomic background of each country. The United Nations Water estimates that 80% of global wastewater is not adequately treated (UN Water 2017). This represents a huge risk for the environment as well as human health and well-being. Even if properly treated, there is another environmental challenge. The wastewater treatment operations result in large quantities of sewage sludge which need further processing. Sewage sludge represents up to 40% of the total GHG emissions associated with the operations of the wastewater treatment plant (Callegari and Capodaglio 2018). For a while, the most common disposal methods have been landfills, incineration (with or without energy recovery), and agricultural applications/land spreading. However, they come with a host of issues (e.g., leaching, water, air, and land pollution) and challenges (e.g., odor management, transport, energy consumption, high costs) (EC 2001).

More sustainable disposal methods through pyrolysis, hydrothermal carbonization, and anaerobic digestion can add value by recovering nutrients and other compounds, biochar, and bioenergy production. This leads to the interest expressed by the scientific community in determining the possible environmental and financial benefits of producing biochar from sewage sludge (Singh et al. 2020). As discussed, there are generally several drives behind the idea, namely, the increasing quantity of sludge (creating handling issues), the high cost of disposal, and the sustainability factor prompting recovery and reuse rather than disposal.

The next chapters will concentrate mainly on the provenience of sewage sludge, its treatment and processing for pyrolysis, and the resulting characteristics of the char. As mentioned, there are other methods of biochar production than pyrolysis, but they are beyond the scope of this report.

6.1 Feedstock and Production Method

The methods and level of treatment for wastewater depend largely on the size and capabilities of the facilities. Centralized wastewater treatment plants, commonly serving populous urban hubs, have the ability to process larger quantities of wastewater more efficiently. Generally, municipal wastewater and certain industrial streams (limited by national standards on allowable chemicals) are treated together in two or three stages. After screening (to remove large objects from the effluent), primary treatment consists of large sedimentation tanks to separate part of the suspended solids and the liquid, which then passes to secondary treatment. This biological treatment phase may use aerobic or anaerobic digestion to remove as much organic material as possible. Tertiary treatment may include biological or chemical nutrient removal from the effluent, micropollutant removal, and disinfection.

This short introduction is necessary to understand the resulting components in sewage sludge. At the end of the treatment process, raw sewage sludge may contain a combination of valuable and harmful components, organic matter, and other elements (Lasaridi et al. 2018). Besides the useful nutrients (nitrogen, phosphorus, potassium, etc.), there may be other contaminants such as heavy metals (Cr, Ni, Cu, Zn, Cd, and Pb), potentially toxic elements (PTEs), polycyclic aromatic hydrocarbons (PAHs), human bacterial pathogens (HBPs) (Singh et al. 2020), and other microcontaminants (Ratola et al. 2012). The presence and often the interactions between some of these compounds can lead to harmful effects on the environment (Bondarczuk et al. 2016). Additionally, there is the real possibility that with an increase in wastewater treatment efficiency, more pollutants will contaminate the sewage sludge (Collivignarelli et al. 2019) – thus increasing the need for long-lasting sustainable solutions.

The sewage sludge resulting from the treatment process has different components depending on its provenience (primary, secondary, tertiary sludge). Combined in one stream, the mixed sludge needs further treatment to reduce its water content and prepare it for pyrolysis. By weight, sludge may contain over 90% moisture, in different forms (Chen et al. 2020). The dewatering process is essential in reducing the amount of sludge, thus minimizing transportation needs. Current dewatering technologies include chemical preconditioning and electrical mechanical dewatering and subsequent treatments, with plenty of research on the matter (Cao et al. 2021).

Finally, after drying and crushing them to the necessary size, the biosolids and any additional biomass are ready for pyrolysis – defined as a thermal degradation process which results in varying amounts of biochar, bio-oil, and gases. The advantages of pyrolyzing biosolids can be observed in volume reduction, microorganisms' degradation, pathogens' destruction, useful by-products, and ultimately climate change mitigation (Singh et al. 2020).

6.2 *Sludge Char Characteristics*

Similar to biochar produced from other feedstock, the sludge char's characteristics are mainly determined by the components and particle size of the feedstock and the pyrolysis conditions, such as temperature, residence time, heating rate, and method (Chen et al. 2020). The choice of input conditions naturally depends on the available technology and resources but most importantly on the intended biochar application.

Solids Yield Biochar yield decreases as pyrolysis temperature increases, in some examples, decreasing from 72.3% to 52.4% at pyrolyzing temperatures of 300 °C and 700 °C, respectively (Hossain et al. 2011). The percentage is considerably higher compared to biochar obtained from biomass; however, there are notable differences in the characteristics, notably the stable carbon and ash content (Ali et al. 2022). Of note is the conclusion of several studies (Chen et al. 2019) indicating that co-pyrolysis of sewage sludge and a woody biomass yields lower amounts of biochar than sewage sludge alone. However, additional biomass in the process aids in gas generation, which can be harnessed to produce energy (Gopinath et al. 2021) – so a careful balance should be considered.

Physical Characteristics Among these, porosity and specific surface area are the two most important factors that determine the biochar water retention capabilities (Edeh et al. 2020). A recent study conducted in Finland concludes that porosity and specific surface area values of sludge derived biochar were approximately two to four times lower than those obtained for wood-based biochar (Turunen et al. 2021). Unless the pore structure is modified, the study points out that these characteristics greatly affect the usability of the char (ibid.). The electrical conductivity is shown to increase until pyrolysis temperature reaches 500 °C, after which it reduces substantially (Singh et al. 2020).

Chemical Characteristics The ash content in sludge char has been reported as generally higher than in other biochars (different feedstock). It has also been noticed that the ash content increases with increasing pyrolyzing temperature (Gopinath et al. 2021). The pH of sludge char (as other biochars) increases with increasing pyrolysis temperature, with some studies placing it between 8.7 and 11.1 (Zhang et al. 2019).

One of the most important considerations for sludge char is the fate of the harmful contaminants after pyrolysis. Studies suggest that the concentrations of heavy metals (Pb, Cr, Zn, Cu) increase with higher temperatures. However, the heavy metals are in oxidized form; thus, they are more stable and exhibit lower bioavailability (Gopinath et al. 2021). Regarding PAHs, there are studies suggesting that their content reduces in the pyrolysis process; however, the trend is decreasing as the temperature increases (Zielińska and Oleszczuk 2016).

Lastly, depending on the intended use, the characteristics of biochar can be changed or improved by physical activation (using activating agents as carbon dioxide, air, nitrogen, steam, oxygen (Wang et al. 2017)) and chemical activation (using inorganic agents such as zinc chloride, sulfuric acid, potassium hydroxide, sodium hydroxide, and potassium carbonate or organic agents such as citric acid (Devi and Saroha 2017)). Singh et al. reported that the activation process can increase the porosity and the surface area of the sludge char. In addition, it is mentioned that physical activation is more economical and sustainable compared to chemical activation (Singh et al. 2020).

7 Case Study: Sludge Char Production in Finland

Finland has developed its biochar production greatly over the past 10 years, with biochar projects in over ten cities across the country (Nordic Biochar Network 2021). Such great interest and expertise in the area have increased the public awareness and have prompted further research in sludge char production as well. The Helsinki Region Environmental Services (HSY, abbreviation in Finnish) is Finland's largest water and wastewater operator, waste management facility, and environmental services provider covering the capital's metropolitan area.

The Greater Helsinki area is served by two wastewater treatment plants, Viikinmäki and Suomenoja. The former captures the sewerage from all of Helsinki and seven other municipalities around the capital. Around 85% of its flow is due to domestic wastewater, amounting to approximately 860,000 inhabitants, while the rest is accounted by the industrial wastewater and other nondomestic sources (HSY 2018). It is considered the largest facility in the Nordic countries, based on the treated volume, and it has been commissioned in 1994, followed by several updates in capacity (ibid.).

The treatment process is similar in the two locations, using the activated sludge method. In the pretreatment stage, the sewage is screened, the sand removed, and there is a preliminary aeration stage to allow aerobic biodegradation of the organic materials. This is followed by the primary sedimentation stage which allows the clumped bacteria and suspended solids to settle, another aeration phase, secondary sedimentation, and finally, a biological filter. Phosphorus removal is conducted in a two-phase simultaneous precipitation, using ferrous sulfate ($FeSO_4$). The ensuing sediment is bound to the sludge. The nitrogen removal occurs first in the activated sludge process and then in the biological filter (HSY 2018).

The two wastewater treatment plants in Helsinki adhere to strict environmental standards concerning the amount of nitrogen and phosphorus removed from the wastewater. This is due to the effluent being discharged in the Baltic Sea – prone to eutrophication. The high levels of removal from the wastewater in current locations lead to high concentrations of nitrogen and phosphorus in the resulting sludge.

7.1 Sludge Pyrolysis Process

Therefore, since 2016, HSY and their partners have started investigating methods for nutrient recovery, both through further treatment of wastewater as well as sludge pyrolysis (HSY 2021). The pilot plant, a tenth of the full scale needed, is located at the Ämmässuo eco-industrial center. The sludge pyrolysis unit is a TECAM model (TECAM 2022) designed with slight modifications to accommodate HSY's needs and able to provide more accurate data for possible scale-ups. Currently, the pilot plant is able to process approximately 30,000 tons of wet sewage sludge per year. This translates in a rough production of 3000 tons of sludge char annually. Except for some downtime during the year (accounted as 10% of operational hours) due to technical issues, the plant operates continuously.

HSY's process involves mixing approximately 20% wood chips (different amounts have been tested) in the sludge pyrolysis process for energy efficiency. The wet sludge and the wood chips are fed into the thermal dryer, which has a moving belt to convey the material to the drying furnace. The heating energy necessary for the process is obtained by combusting the gas, by-product of pyrolysis. The exhaust air is then treated by an acid scrubber and a biofilter before release to reduce odor.

The pyrolysis unit is a rotary kiln, and it is designed to operate at temperatures ranging from 450 °C to 650 °C at a set time between 30 and 120 min. The sludge char is cooled with indirect water cooling upon exiting the kiln, and it is then transferred by a conveyor to the storage silo. Due to the chosen settings, including the tested temperature ranges, as well as the economic activities in the region, concentrations of heavy metals have been found to be low and not a primary concern. At the moment, the sludge char does not have official status under Finnish law as biochar but rather as a waste product. Therefore, its application is limited to compost mixing. The necessary steps have been taken to change its status.

The pilot project for sludge char has been finalized in 2022, and the results of the research are soon to be published. HSY's plans for a potential full-scale sludge pyrolysis plant are under consideration, depending on the cost efficiency of implementing and operating such a project, as well as the overall environmental benefits.

8 Conclusion

This work intends to offer an overview of biochar and some of the applications suitable for the urban environment, ranging from uses in natural and manufactured soils to more innovative developments in filtration systems and construction materials. Current research has been synthesized into a practical and accessible form, aiming to inform and raise awareness of biochar applications. It is intended to address SDG 13 target 13.3 – to improve education and awareness-rising in climate mitigation, as well as serve as a starting point for local and regional administration and their policy plans according to target 13.2.

Acknowledgments This work has been funded by Carbon Neutral Cities Alliance and by the Strategic Research Council grant number 352478 at the Academy of Finland.

References

UNDP. Human Development Report 2020: The next frontier, Human development and the Anthropocene. 1 UN Plaza, New York: United Nations Development Programme, 2020.

Rogelj, J., et al. Mitigation Pathways Compatible with 1.5°C in the Context of Sustainable Development. Global Warming of 1.5°C. An IPCC Special Report on the impacts of global warming of 1.5°C above pre-industrial levels… s.l. : Cambridge University Press for the Intergovernmental Panel on Climate Change, 2018, p. Ch. 2.

EBC. European Biochar Certificate – Guidelines for a Sustainable Production of Biochar. Arbaz, Switzerland: European Biochar Foundation, 2022.

Biochar in agriculture – A systematic review of 26 global meta-analyses. Schmidt, Hans-Peter, et al. 11, 2021, GCB – Bio, Vol. 13, pp. 1708–1730.

Biochar as a tool to reduce the agricultural greenhouse-gas burden – knowns, unknowns and future research needs. Kammann, Claudia, et al. 2017, Journal of Environmental Engineering and Landscape Management, pp. 114–139.

Azzi, Elias S. Biochar systems across scales in Sweden. Stockholm: KTH – Doctoral Thesis in Industrial Ecology, 2021.

The effect of pyrolysis conditions on biochar stability as determined by three methods. Crombie, Kyle, et al. 2013, Global Change Biology (GCB) Bioenergy, pp. 122–131.

Feedstock choice, pyrolysis temperature and type infuence biochar characteristics: a comprehensive meta-data analysis review. Ippolito, James A., et al. 2020, Biochar, pp. 421–438.

Pokharel et al. Biochar-Assisted Wastewater Treatment and Waste Valorization. Applications of Biochar for Environmental Safety. [Online] 2020. https://www.intechopen.com/chapters/71992.

UNEP. Cities and climate change. UN Environment Programme – Resource Efficiency – Cities. [Online] 2022. https://www.unep.org/explore-topics/resource-efficiency/what-we-do/cities/cities-and-climate-change

The health benefits of the great outdoors: A systematic review and meta-analysis of greenspace exposure and health outcomes. Bennett, Caoimhe Twohig – and Jones, Andy. 2018, Environmental Research, Vol. 166, pp. 628–637.

The contribution of constructed green infrastructure to urban biodiversity: A synthesis and meta-analysis. Filazzola et al. 2019, Journal of Applied Ecology, Vol. 56, pp. 2131–2143.

High carbon losses from established growing sites delay the carbon sequestration benefits of street tree plantings – a case study in Helsinki, Finland. Riikonen et al. 2017, Urban Forestry & Urban Greening, Vol. 26, pp. 85–94.

Planting Urban Trees with Biochar. Embrén, Bjorn. Arbaz, Switzerland: s.n., 2016, Biochar Journal. ISSN 2297-1114.

A Review of Green Roof: Definition, History, Evolution and Functions. Abass, F, et al. Melaka, Malaysia: s.n., 2020. IOP Conference Series: Materials Science and Engineering. Vol. 713.

Setherton, Gregg. Green Roof Construction – How to Guide. Permagard. [Online] 2022. https://www.permagard.co.uk/advice/green-roof-construction.

Performance of Blue-Green Roofs in Cold Climates: A Scoping Review. Andenæs, Erlend, et al. s.l. : Norwegian University of Science and Technology, 2018, buildings.

Biochar makes green roof substrates lighter and improves watersupply to plants. Cao, Cuong T.N., et al. 2014, Ecological Engineering, pp. 368–374.

Biochar Improves Sustainability of Green Roofs via Regulate of Soil Microbial Communities. Chen, Haoming, et al. 2021, agriculture.

Biochar amendment in the green roof substrate affects runoff quality and quantity. Kuoppamäkia, Kirsi, et al. 2015, Ecological Engineering, pp. 1–9.

Estimation QS. How To Build a Football Field for your Professional Club, Community or Backyard – Natural Grass and Artificial Turf Pitch. [Online] n.d. https://estimationqs.com/how-to-make-a-football-field/.

Major, Julie. Guidelines on Practical Aspects of Biochar Application to Field Soil in Various Soil Management Systems. s.l. : International Biochar Initiative, 2010.

IPCC. Global warming of 1.5°C Special Report on the impacts of global warming of 1.5°C above pre-industrial levels and related global greenhouse gas emission pathways, in the context of strengthening the global response to the threat of climate change. s.l. : Edited by: Masson-Delmotte, V., P. Zhai, H.-O. Pörtner, D. Roberts, J. Skea, P.R. Shukla, A. Pirani, W. Moufouma-Okia, C. Péan, R. Pidcock, S. Connors, J.B.R. Matthews, Y. Chen, X. Zhou, M.I. Gomis, E. Lonnoy, T. Maycock, M. Tignor, and T. Waterfield, 2018.

Influence of biochar on the soil water retention characteristics (SWRC): Potential application in geotechnical engineering structures. Hussaina et al. 2019, Soil & Tillage Research.

Peltosaari et al. Storm water management in the city of Helsinki. [Online] 30 6 2020. https://www.bsrwater.eu/news/storm-water-management-helsinki.

Reduced carbon footprints of buildings: new Finnish standards and assessments. Kuittinen, Matti and Häkkinen, Tarja. 2020, Buildings and Cities, pp. 182–197.

The use of biochar as building material. Schmidt et al. 2014, Biochar Journal.

Snøhetta. Biocrete – a carbon negative concrete. Snøhetta. [Online] 4 2022. https://snohetta.com/project/558-biocrete-a-carbon-negative-concrete.

Biochar for asphalt modification: A case of high-temperature properties improvement. Ma, Feng, et al. 2022, Science of The Total Environment, Vol. 804.

The Historical Development of Sewers Worldwide. Feo, Giovanni De, et al. 2014, Sustainability, Vol. 6, pp. 3936–3974.

UN Water. Water Quality and Wastewater. [Online] 2017. https://www.unwater.org/water-facts/quality-and-wastewater/.

Properties and Beneficial Uses of (Bio)Chars, with Special Attention to Products from Sewage Sludge Pyrolysis. Callegari, Arianna and Capodaglio, Andrea Giuseppe. 20, 2018, resources, Vol. 7.
EC. Disposal and recycling routes for sewage sludge – Scientific and Technical Report. Luxembourg: European Commission, 2001.
A sustainable paradigm of sewage sludge biochar: Valorization, opportunities, challenges and future prospects. Singh, Simranjeet, et al. 2020, Journal of Cleaner Production, Vol. 269.
The Evaluation of Hazards to Man and the Environment during the Composting of Sewage Sludge. Lasaridi, Konstantia-Ekaterini, et al. 2018, Sustainability, Vol. 10.
Occurrence of organic microcontaminants in the wastewater treatment process. A mini review. Ratola, Nuno, et al. 2012, Journal of Hazardous Materials.
The urgent need for risk assessment on the antibiotic resistance spread via sewage sludge land application. Bondarczuk, Kinga, Markowicz, Anna and Piotrowska-Seget, Zofia. 2016, Environmental International, Vol. 87, pp. 49–55.
Biosolids: What are the different types of reuse? Collivignarelli, Maria Cristina, et al. 2019, Journal of Cleaner Production, Vol. 238.
Production, properties, and catalytic applications of sludge derived biochar for environmental remediation. Chen, Yi-di, et al. 2020, Water Research, Vol. 187.
Enhanced technology based for sewage sludge deep dewatering: A critical review. Cao, Bingdi, et al. 2021, Water Research, Vol. 189.
Influence of pyrolysis temperature on production and nutrient properties of wastewater sludge biochar. Hossain, M.K., et al. 2011, Journal of Environmental Management, Vol. 92, pp. 223–228.
Characteristics of Biochars Derived from the Pyrolysis and Co-Pyrolysis of Rubberwood Sawdust and Sewage Sludge for Further Applications. Ali, Liaqat, et al. 7, Basel, Switzerland: sustainability, 2022, Vol. 14.
Structure characteristics of bio-char generated from co-pyrolysis of wooden waste and wet municipal sewage sludge. Chen, Q., et al. 2019, Fuel Processing Technology, Vol. 183, pp. 48–54.
Conversion of sewage sludge into biochar: A potential resource in water and wastewater treatment. Gopinath, Ashitha, et al. 2021, Environmental Research, Vol. 194.
A meta-analysis on biochar's effects on soil water properties–New insights and future research challenges. Edeh, I.G., Mašek, O. and Buss, W. 2020, The Science of the Total Environment, Vol. 714.
Pore structure of wastewater sludge chars and their water retention impacts in different soils. Turunen, Mika, et al. 2021, Biosystems Engineering, Vol. 206.
Sludge-based biochar activation to enhance Pb(II) adsorption. Zhang, J., et al. 2019, Fuel, Vol. 252, pp. 101–108.
Effect of pyrolysis temperatures on freely dissolved polycyclic aromatic hydrocarbon (PAH) concentrations in sewage sludge-derived biochars. A. Zielińska, P. Oleszczuk. 2016, Chemosphere, Vol. 153, pp. 68–74.
Characterization and Pb(II) removal potential of corn straw- and municipal sludge-derived biochars. Wang, Shujuan, et al. 9, 2017, Royal Society Open Science, Vol. 4.
Utilization of sludge based adsorbents for the removal of various pollutants: A review. Devi, Parmila and Saroha, Anil K. 2017, Science of The Total Environment, Vol. 578, pp. 16–33.
Nordic Biochar Network. Nordic Biochar Map. [Online] 2021. https://www.nordicbiochar.org/about-us/map/.
HSY. Helsinki Region Environmental Services Authority. Viikinmäki wastewater treatment plant. [Online] 2018. https://niini.fi/wp-content/uploads/2021/01/HSY0014_Viikinmaki_wastewater_treatment_plant.pdf.
HSY. Comprehensive recovery of nutrients and carbon from wastewater (RAHI). [Online] 2021. https://www.hsy.fi/en/environmental-information/projektit-ja-hankkeet/rahi-project/.
TECAM. Tecam-Sludge and slurry waste-Waste-to-product technology. [Online] 2022. https://tecamgroup.com/waste-valorization/sludge-and-slurry-waste/.

Sustainability with Energy Efficiency: A Touch on Power-Hungry Tunnel-Boring Machines

Nazife Erarslan

1 Introduction

The importance of mechanical rock-cutting machines is increasing in parallel with technological developments because the importance of underground structures in the mining and construction sectors is increasing day by day. Structures such as electricity, water, sewerage, telephone, natural gas, mining, and subway tunnels are examples of underground structures where mechanical excavators are most often used. Until the last 10 years, mechanical excavating machines have been very successful in cutting medium-hard rocks (uniaxial compressive strength ≥150 MPa). The increase in costs, along with the decrease in the production rate of mechanical excavators when excavating rocks above 150 MPa, could not ensure economical excavation. The research in this field has gained momentum and has focused on cutting head design with new tools and new technologies in order to make economical and sustainable excavations in hard rock (Hood and Alehossein 2000; Ramazanzadeh and Hood 2010; Shao and Li 2014; Vogt 2016). There are basically two types of machine groups in hard rock-cutting: full-face and partial-face machines. The cutter head of the full-face machines is the same size as the opening to be drilled (e.g., tunnel-boring machines (TBM)), whereas the cutter head of the partial-face machines is smaller and completes the opening in several sweeps (e.g., roadheaders). The developments in hard rock-cutting machines are covered under these two machine groups. TBMs are the most widely used hard rock-cutting machines.

Proposed section where the chapter should be categorized: SECTION D: Sustainability in Process, Materials, Mining and Metallurgical Engineering

N. Erarslan (✉)
Civil Engineering Department, Izmir Democracy University, İzmir, Turkey
e-mail: nazife.dogan@idu.edu.tr

I. S. Dunmade et al. (eds.), *Sustainable Engineering*, Green Energy and Technology, https://doi.org/10.1007/978-3-031-47215-2_9

Fig. 1 Hard rock-cutting technologies developed in last decade (**a**) KR150 roadheader with mini-discs, (**b**) HDRK-Wirth, (**c**) Sandvik Tamrock ARM 1100, (**d**) ODC, (**e**) Komatsu MC51

However, they consume huge amounts of energy due to the need for very high thrust forces on the cutting head.

An important development in hard rock excavation is the use of small disc cutters (Fig. 1a). Reducing the dimensions makes the machine more flexible and less energy consuming. This technology was developed by the Remag company. It is preferred in selective mining, especially in narrow and scattered ore deposits in South Africa. In the last two decades, it has been focused on the tensile strength of the rocks without overcoming the compressive strength in hard rock-cutting machines, and machine designs have been made accordingly. Undercutting technology (cutting parallel to the surface instead of perpendicular to the surface) is a typical example of cutting rocks by considering the tensile strength of rocks. The HDRK-Wirth

Fig. 2 Tunnel-boring machine (TBM)

continuous hard rock-cutting machine was developed based on the undercutting technology in cooperation with Germany and Canada (Fig. 1b). Four freely rotating disc cutters break particles by assisting each other and overcoming the tensile strength of the rock. Another important undercutting machine is the Sandvik Tamrock ARM 1100 (Fig. 1c), developed by Sandvik Tamrock Voest-Alpine. The oscillating disc cutter (ODC) (Fig. 1d) was designed in Australia using an undercutting technique, presented as cutting-edge technology in hard rock-cutting machines (SmartCUT (CSIRO)). Joy Mining Machinery and later Komatsu companies made new machine designs by transferring the ODC technology (Fig. 1e).

In recent years, besides the new technology research in mechanical cutting, the machine designs in which these technologies will be applied have not been easy at all. Tunnel-boring machines (TBMs) have a high investment cost and have been used in this field for many years. They have become the preferred excavation machines today because they prevent undesired underground movements, are soundless and vibration-free, operate quickly, and make the placement of tunnel support systems possible at the same time as rock-cutting. An average of 1000 kW of power is required to operate a TBM. About 70% of this power is used to drive the cutterhead hydromotors. The average weight of a TBM operating in a subway tunnel is around 600 tons. The power requirements of TBMs depend on size, and they range in general from 1.5 MW for a 3 m diameter TBM up to 8 MW for a 12–14-m-diameter TBM. The ventilation and lighting of TBMs need extra power during tunnel construction (Fig. 2).

The energy consumption will be reduced by thousands of kilowatts using the new machine technologies, making a significant contribution to a sustainable future even with a 10% decrease in the energy consumed by machines that consume such a high amount of energy. New technologies have been investigated to reduce rock strength as well as machine dimensions in order to reduce energy consumption in mechanical rock-cutting machines developed in recent years. It is known that many brittle materials, including rock, have a 20–30% decrease in their statically measured strength due to fatigue deformation under cyclic loads (Tao and Mo 1990; Bagde and Petros 2009; Li et al. 2000; Ko and Lee 2020). At this point, rock fatigue and mechanical rock-cutting technologies have been combined, and new technologies have been developed that reduce the energy consumed by 30% by reducing the rock-cutting forces (Hood and Alehossein 2000; Karekal 2000; Erarslan 2011; Grashof et al. 2019; Mohammadi et al. 2020). New technologies have been developed that perform the mechanical cutting process by using the dynamic/periodic loading technique, consuming less force and energy than conventional rock-cutting methods. In recent years, such dynamic rock-cutting methods have been used in new machine designs. The oscillating disc cutting (ODC) machine using this method was developed for the first time in the world by the CRC Mining and Research Institution in Australia in 2005. Assoc. Dr. Nazife Erarslan took part in the ODC project as a doctoral student and researcher between 2005 and 2008. ODC technology was transferred by the international machinery manufacturing giant Joy Mining Machinery and later by Komatsu, and new machine design studies with this technology continue today. The rock-cutting forces and specific energy values were reduced by 35% with the ODC technology by combining the rock fatigue mechanism and the undercutting method, which is parallel to the surface. The 35% energy reduction in the tunnel boring area, which consumed a huge amount of energy, would be a true sustainable technology.

2 Dynamic Cutting Discs (DCD) Technology

The DCD technology is the result of new rock excavation research done for the first time in the field using field-size oscillating discs (Fig. 3). This innovative technology was contemplated to investigate whether integrating rock fatigue into the rock-cutting field leads to any degradation of cutting forces and relative parameters under oscillating loading by DCD.

However, DCD technology does not cut parallel to the surface like ODC technology; it cuts rocks by applying dynamic loads perpendicular to the surface like traditional TBM discs. Because the machines that cut parallel to the surface with the undercutting method, such as ODC, are not full-face machines, they are generally boom-type, partial-face machines. Thus, it is possible to suggest that the application of DCD technology is more convenient for boom-type machines than for TBMs.

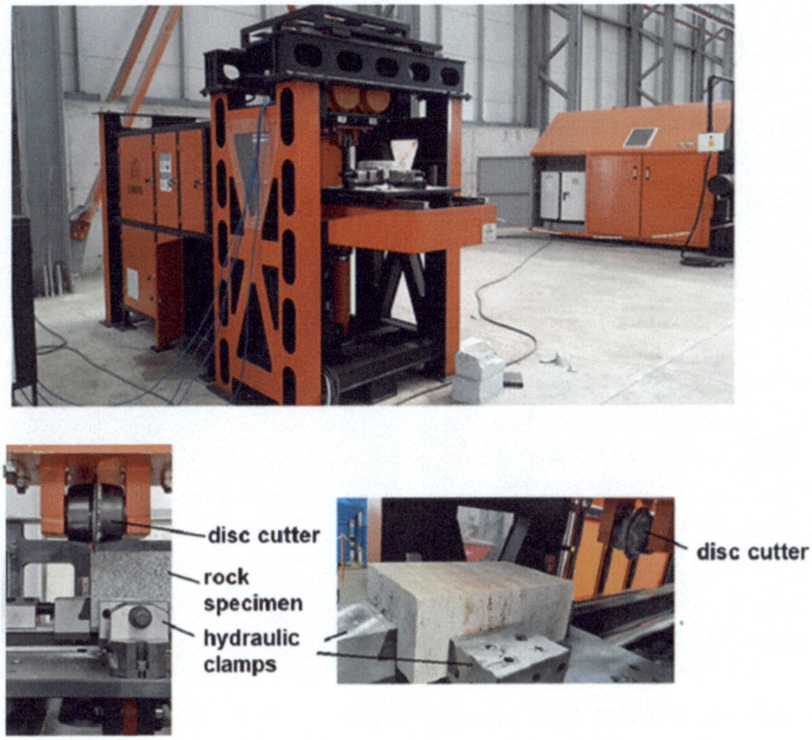

Fig. 3 General view of the DCD technology prototype and an oscillating single disc cutter

2.1 *Rock Fatigue*

The reduction in rock strength under cyclic loading is called "rock fatigue." Although the fatigue of metals, concrete, ceramics, and composite materials is well known, its effect on rocks has only attracted attention in recent years (Bagde and Pedros 2005, 2009; Karekal 2000; Gatelier et al. 2002; Haimson 1978; Tao and Mo 1990; Ritchie and Murakami 2003). However, the rock fatigue research is limited, e.g., studies on the amplitude and frequency effect in rock fatigue, and therefore, there is still no accepted criterion/theory that can explain the rock fatigue and failure mechanism. A typical fatigue behavior in the fatigue tests is to produce a progressive permanent (plastic deformation) strain in the specimen depending on the number of load cycles. Erarslan (2011) explained the rock fatigue damage mechanism that causes the decrease in rock strength. According to the results obtained by Erarslan (2011) using SEM analysis, it was shown that the rock fatigue damage mostly occurs when the plastic deformation accumulation is greater than mineral fracture due to failure in the matrix located between minerals. The main deformation characteristic of rock fatigue is propagation of intergranular fractures at the boundaries of the minerals and the production of very small particles due to the mineral indentation on the weak matrix

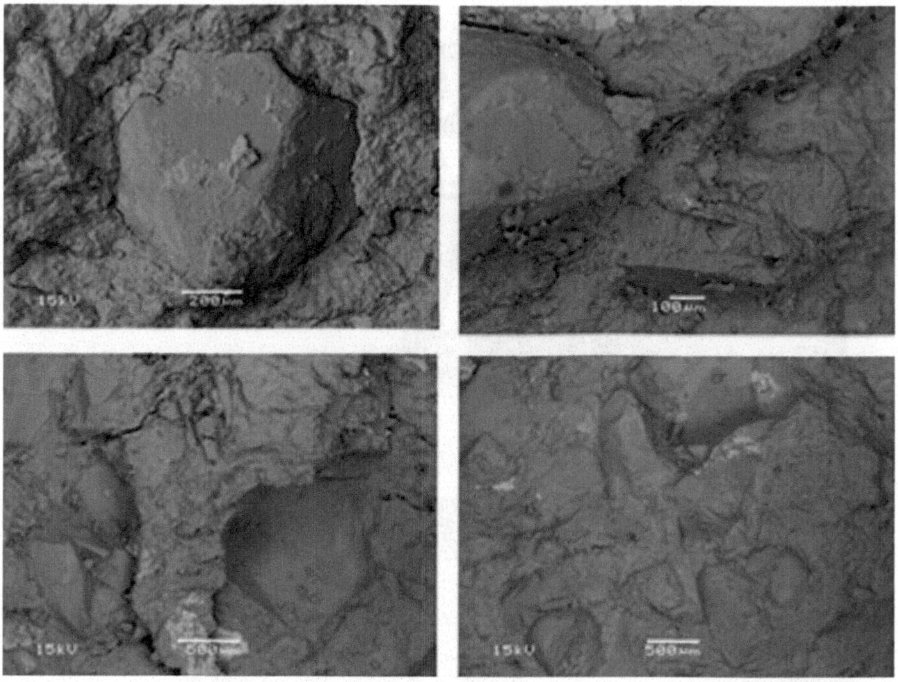

Fig. 4 SEM images of a failed rock under cyclic loading

Table 1 Cutting forces obtained by DCD cutting tests

Cutting depth	Frequency, Hz	Cutting forces	
		FN (kN) (avr.)	FR (kN) (avr.)
4 mm	0	43.2	9.9
6 mm	0	56.8	11.6
4 mm	20	35.3	8.4
6 mm	20	44.1	9.3
4 mm	45	33.8	8.1
6 mm	45	43.2	9.1

surface (Fig. 4). For this reason, this study is a kind of technology research for more efficient mechanical cutting of hard rock or ore using the rock fatigue technique under cyclic loading with lower cutting forces and energy.

3 Cutting Test Results

The lateral force (FS), rolling force (FR), and normal force (FN) acting on the cutter were continuously recorded by the data acquisition device during the rock-cutting process. The results of rock-cutting tests are shown in Table 1. The average FN is

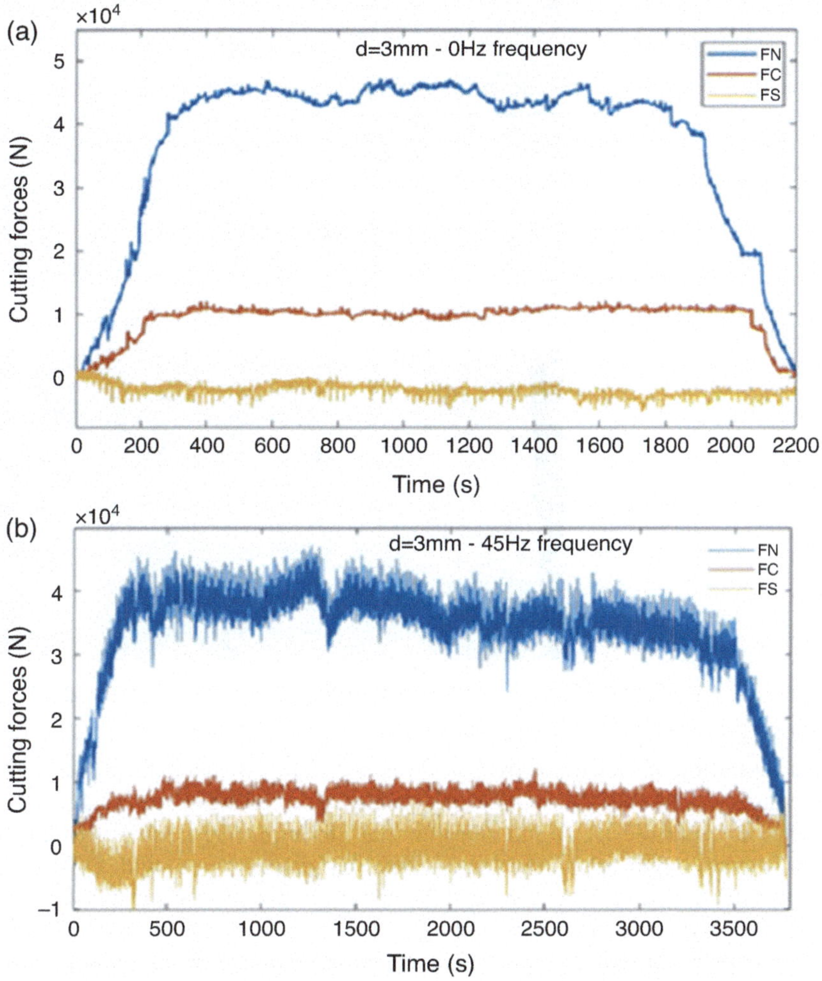

Fig. 5 Comparison between the cutting forces: (**a**) conventional disc cutting without oscillation and (**b**) DCD cutting – 45 Hz

approximately 43.4 kN at 4 mm cutting depth and 56.5 kN at 6 mm cutting depth for conventional (oscillation-free) cutting. These values decreased to 35.3 and 44.03 kN, respectively, when the rock-cutting was conducted with DCD at 20 Hz. The reduction in FN and FR with 45 Hz cutting in general is 25%, while the cutting force reduction at 20 Hz is 18–20%. The cutting force components, especially FN, were observed to be significantly reduced as the oscillation frequency increased.

Figure 5 shows how the cutting forces FN and FR at different cutting depths change with increasing frequency. Figure 5 also shows that both FN and FR cutting forces decrease with the increasing frequency. However, FN, as mentioned before, decreased more than FR with increasing frequency.

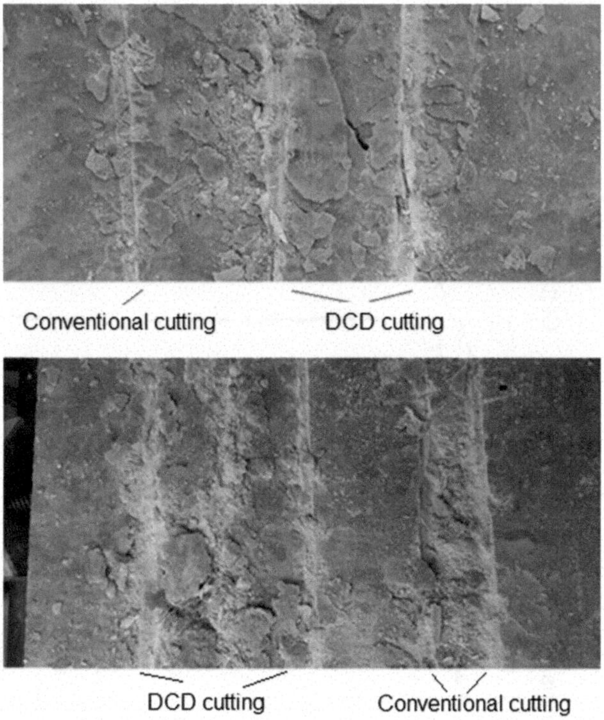

Fig. 6 Comparison of the rock chips and fragments between the DCD and conventional disc cuttings

The formation of rock chips (removed rock particles) in mechanical rock-cutting is the primary subject of the rock-cutting studies. The cutting grooves and cut material obtained with the DCD and conventional vibration-free cutting are shown in Fig. 6. A much larger amount of material and larger size pieces with DCD cutting are shown in Figs. 6 and 7, compared to the vibration-free conventional disc cuttings.

Specific energy (SE) in a rock-cutting area is defined as the energy necessary to excavate a specific amount of rock, and increasing SE means decreased rock-cutting performance. The rock material excavated by the DCD cutting was found to be almost two times more abundant than the material cut with a conventional disc cutter without oscillation. This is a very effective finding because the most effective parameters in evaluating the rock-cutting performance in mechanical rock-cutting research are the FN, FR, FS, and SE data estimated with the amount of rock material removed. It is concluded that an almost 30% decrease in the specific energy values compared to conventional vibration-free cutting was obtained in DCD cutting, with an almost 200% increase in the amount of cut rock and a decrease of approximately 23% in cutting forces. The coarseness index (CI) is a unitless variable that is estimated using sieve analysis. It was found in pick and disc cutter research studies in the literature that the SE was found to be decreasing with the increasing CI at various depths of cuts (Roxborough 1973; Kahraman et al. 2004; Tuncdemir et al. 2008; AbuBakar et al. 2014; Jeong and Jeon 2018). A significant relationship between CI and SE was also found in this research, in accordance with the literature. It is shown

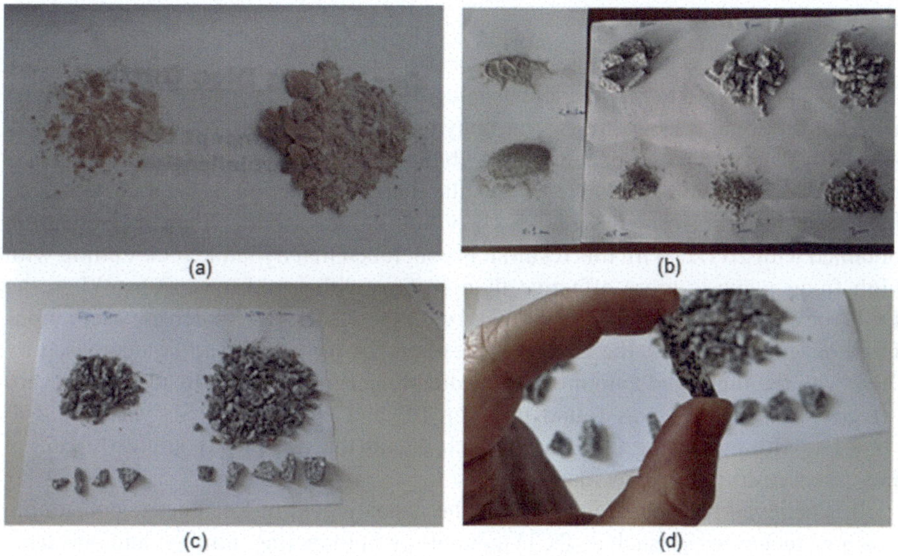

Fig. 7 Comparison of cut material obtained with the conventional and DCD technology: (**a**) conventional cutting (left) and (right) DCD cutting with 45 Hz at 3 mm depth of cut, (**b**) cut material obtained with DCD with 45 Hz at 5 mm depth and classified with sieves, (**c**) traditional cutting (left) and cutting with 45 Hz DCD (right), (**d**) example of a rock chip profile cut with DCD

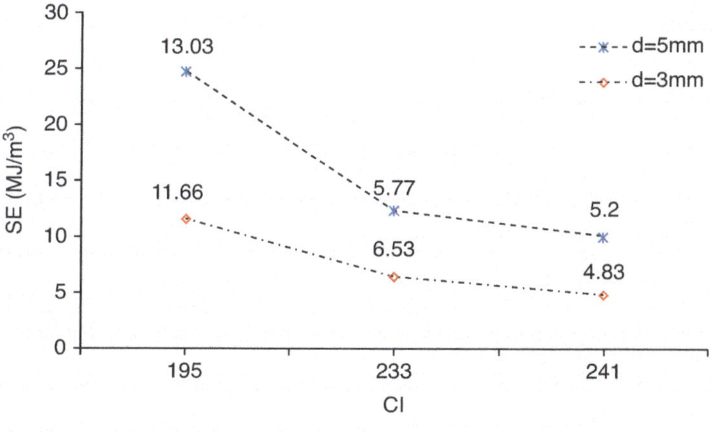

Fig. 8 Decreasing SE with the increasing CI

in Fig. 8 that SE decreases significantly with increasing CI values. An important point to note here is that the decrease in specific energy is due to the increasing amount of excavated rock rather than the decreasing cutting forces. This finding is very important in the field of mechanical rock-cutting because it is evidence that the energy spent to create fractures by cyclic loading is less than the energy spent to obtain more crushed or very fine material (lower CI) obtained with conventional rock-cutting.

4 Conclusions

The great innovation in this research is integrating damage properties of rock fatigue with conventional rock-cutting technology to increase the hard rock-cutting performance by decreasing the spent energy with the increased amount of excavated material. It was found with the DCD test results that the reduction in FR and FN forces in cutting tests performed with 45 Hz vibration was greater than the reduction obtained with 20 Hz. With this result, it is concluded that dynamic rock-cutting with high frequencies is more effective than cutting with low frequencies. When the DCD rock-cutting test results are checked with the TBM discs, which are oscillation-free conventional discs, the cutting forces and SE were found to be decreased by 30% with the increased amount of excavated material by 200%. In addition, a very effective relationship was found between SE and CI, and the SE was found to decrease with increasing CI values, which means the size of the obtained material increased, resulting in a decreasing SE.

It is believed from this research that the development of low-energy consuming machine technologies such as DCD technology in tunneling, mining, and construction areas where hard rock-cutting machines are constantly used will make great contributions to the sustainable economies of countries, the environment, and science.

References

Abu Bakar MZ, Gertsch LS, Rostami J (2014) Evaluation of fragments from disc cutting of dry and saturated sandstone. Rock Mech Rock Eng 47(5): 1891–1903

Bagde MN, Pedros V (2009). Fatigue and dynamic energy behaviour of rock subjected to cyclical loading. Int J Rock Mech Min Sci 46: 200–209

Erarslan N (2011) Static and Cyclic Laboratory Testing of Brisbane Rocks. PhD Thesis. The University of Queensland, Australia

Gatelier, N., Pellet, F. and Loret, B. (2002). Mechanical damage of an anisotropic porous rock in cyclic triaxial tests. Int. J. Rock Mech. Min. Sci., 39, 335–354.

Grashof B, Alvarez G, Schot DL, Helmons RLJ (2019) Oscillating disc technology for rock excavation. DEM8: 8th International Conference on Discrete Element Methods, Enschede, Netherlands, 9-12 July 2019

Haimson, B.C. (1978). Effect of cyclic loading on rock. Dynamic Geotechnical testing ASTM STP, 654, 228–245.

Hood M, Alehossein H (2000) A Development in Rock-cutting Technology. Int J Rock Mech and Min Sci 37(1–2):297–305

Jeong H, Jeon S (2018) Characteristic of size distribution of rock chip produced by rock-cutting with a pick cutter. Geomech and Eng 15(3): 811–822

Kahraman S, Develi K, Yasar E (2004) Predicting the penetration rate of percussive blasthole drills using coarseness index and median particle size. CIM Bull 97 (15):1–4

Karekal S (2000) The effect of variable amplitude cyclic loading on hard rock under triaxial stress condition. PhD thesis, The University of Queensland, Australia.

Ko TY, Lee S (2020) Characteristics of Crack Growth in Rock-Like Materials under Monotonic and Cyclic Loading Conditions. Appl Sci, 10: 719–732. https://doi.org/10.3390/app10020719

Li H, Zhao BJ, Li TJ (2000) Micromechanical modelling of the mechanical properties of a granite under dynamic uniaxial compressive loads, Int J Rock Mech Min Sci 37(6): 923–935

Mohammadi M, Khademi JH, Rostami J, Goshtasbi K (2020) 'A Closer Look into Chip Shape/Size and Efficiency of Rock-cutting with a Simple Chisel Pick: A Laboratory Scale Investigation'. Rock Mech Rock Eng 53: 1375–1392

Ramazanzadeh A, Hood M (2010). A state of the art review of mechanical rock excavation technologies. Int J of Min and Env 1 (1): 29–38

Ritchie R.O. and Y. Murakami: Comprehensive Structural Integrity, Elsevier, Boston, MA, 2003, vol. 4.

Roxborough FF (1973) Cutting rock with picks. The Mining Eng, 32(3): 445–452.

Shao W, Li X (2014) Laboratory comparison of smart*cut picks with WC picks. Adv Mater Res 1017:323–328

SmartCUT (CSIRO). https://www.csiro.au/en/work-with-us/industries/mining-resources/mining/smartcut

Tao Z, Mo H (1990) An experimental study and analysis of the behaviour of rock under cyclic loading. Int J Rock Mech Min Sci Geomech Abs 27: 5–15

Tuncdemir H, Bilgin N, Copur H, Balci C (2008) Control of rock-cutting efficiency by muck size. Int J Rock Mech Mining Sci, 45: 278–288. https://doi.org/10.1016/j.ijrmms.2007.04.010

Vogt D (2016) A review of rock-cutting for underground mining: past, present, and future. J South Afr Ins Mining Metall 26: 381–399

Circular Economy Based Model for End-of-Life Tire Management in Emerging Economies

Felix A. Ishola, Israel Sunday Dunmade, Olufunmilayo O. Joseph, Joshua O. Okeniyi, and Esther T. Akinlabi

1 Introduction

End-of-life tires (EOLTs), also referred to as "waste tires," are tires that have surpassed their service lives for usage in all kinds of vehicles (*Managing End-of-Life Tires*, 2015). This category of tires includes those that have burst, peeled, cupped, wobbled, or other forms of damages that may not be visible, putting usage in vehicles for transportation beyond a safety limit. Deterioration of some specific properties such as eccentricity, flat, stable, and well-treaded radial surface for adequate grip on intended road surfaces (Lebreton and Tuma, 2006) can make use of a tire unsafe for various transportation purposes (Dufton, 2001). Initially, tire consists of a combination of natural rubber and synthetic rubber, to which some definite constituents are added to guarantee performance, safety, and durability (Rosendorfová et al. 1998). These enhancing substances include fillers used as reinforcements (silica and carbon black) as well as vulcanizing agents (sulfur), which supports the vulcanization process (Karagiannidis and Kasampalis 2010). All these combinations make tires, as it were, a nonbiodegradable material classified as a "bulk waste"

F. A. Ishola (✉)
MBA, University of East London, London, UK
e-mail: felix.ishola@jll.com

I. S. Dunmade
Department of Earth and Environmental Sciences, Mount Royal University, Calgary, AB, Canada
e-mail: idunmade@mtroyal.ca

O. O. Joseph · J. O. Okeniyi
Mechanical Engineering Department, Covenant University, Ota, Ogun State, Nigeria
e-mail: funmi.joseph@covenantuniversity.edu.ng; joshua.okeniyi@covenantuniversity.edu.ng

E. T. Akinlabi
Northumbria University, Newcastle, UK

I. S. Dunmade et al. (eds.), *Sustainable Engineering*, Green Energy and Technology, https://doi.org/10.1007/978-3-031-47215-2_10

in waste management and mainly found problematic to the ecosystem (Zedler et al. 2020). All things being equal, a budget tire's average life span should be greater than 8 years (Weissman et al. 2003). However, due to factors such as relatively poor operating conditions and intensity of use or even poor storage conditions before use, the tires show considerable aging damage after about 6 years. Concerns about aging and wears prompt many consumers to replace their tires in less than 5 years, even without any visible damage (Ishola et al. 2018a).

In most developed countries, the objectives of waste management and maximization of energy consumption are usually clearly defined. They put in place guidance to review the use and disposal of used tires to avoid their indiscriminate disposal and maximizing the resources contained in the out of service tires (Winternitz et al. 2019). On the contrary, effective waste management, in developing countries, has been hampered by difficulties such as inadequate substructure, underfunding, an unhealthy culture of the populace, and disruption of civil, legal, and political frameworks (Omole et al. 2016). Many developing and underdeveloped countries are still struggling to advance the management of nonbiodegradable wastes in order to harness the resources therein (Godfrey and Oelofse 2017). Most government policies in developing countries have been unable to explore the socioeconomic advantages in solid wastes (Marshall and Farahbakhsh 2013). The problem is aggravated by the fact that some fairly used tires in the developed world are imported into the developing countries in a long-age thriving used tire international trade (Ahmed and Batool 2015), (Kinobe et al. 2012). This necessitated a need for the design of a new waste management system model that will suit the emerging economies.

Tires undergo the bath-tube curve trail (Wang et al. 2002), which means that the life span of a tire at secondhand is nearer to a culmination of the life cycle (Ishola et al. 2018b). This connotes a higher number of used tires per capita in developing countries. Many other identified personal-based factors may contribute to a higher number of used tires per capital in the developing world. Those personal-based factors include attributed bad maintenance culture (Ishola et al. 2019c), lack of knowledge, and technicality of tire care (in accordance to standards described by Weissman et al. (2003)). Some tire simulations had established how bad roads and difficult terrains could contribute to the deformation of a tire due to greater out-of-plane forces and moments exerted on the tire belt per unit time (Gipser 2000). All these further indicate a need for a specialized tire waste management strategy that will accommodate the peculiarities associated with developing economies.

The recycling of the EOLT is adjudged the most sustainable method of keeping the bulk waste from ending up on landfills, thus creating an environmental nuisance (Bittencourt et al. 2020). It is then necessary to keep researching on moving from the linear model of its life cycle to the circular economy model (Gravagnuolo et al. 2019). In the linear economy, the resources are extracted, combined, processed, consumed, and discarded (Hartley et al. 2020). Nevertheless, circular-based economy (CBE) has been defined as a business model emphasizing on the reduction of the waste generation, reuse, recycling, and recovery of materials at different levels to achieve sustainable development (Vanhamäki et al. 2020). Figure 1 depicts the typical process of transition model from linear to a circular economy, highlighting its components, barriers, and benefits.

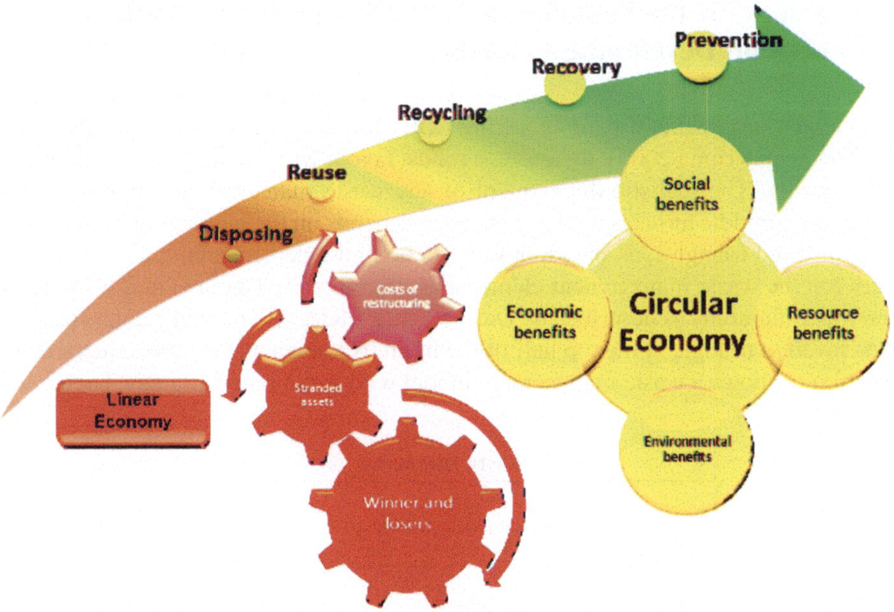

Fig. 1 The components of transition model from linear to circular economy (Salguero-Puerta et al. 2019)

The change from an economy that is linear to a CBE offers the possibility of providing high-quality and resource-efficient services, improving competitiveness and, at the same time, demanding new collaborative approaches from stakeholders throughout the value chain (Blomsma and Tennant 2020). CBE aims to prevent rapid depletion of resources and reduce waste generation (Pisitsankkhakarn and Vassanadumrongdee 2020). The CBE not only help eliminate waste but also find a new value for it (Urbinati and Chiaroni 2019). This indicates that what could otherwise be referred to as waste becomes a valuable source of materials for subsequent use, either directly or as feedstock for other productions (Gall et al. 2020). The concept of CBE draws on its competence to offer environmentally friendly solutions necessary for the sustainable development of a country. In addition, the application of circular supply chain concepts helps to use resources efficiently, which purportedly leads to added value for the society (Singhal et al. 2020). This is of significant importance to developing the economy. The traditional method of using and dumping tires has been fast discarded in developed countries through the use of some established EOLT management strategies, while most developing countries are not doing well in this regard. In this study, analysis of responses from semi-structured interviews with selected experts similar to methods used by Garnett and Cooper (2014), Ishola et al. (2019b), and Bonah et al. (2020) was used to validate the need to develop a specific strategy for EOLT management in emerging economies.

2 Factors for the Sustainable EOLT Management System: Focus on Developing Economy

Attaining a sustainable way of managing EOLT had been challenging over the years (Lebreton and Tuma 2006); therefore, a robust layout waste management strategy is of necessity. The realistic deployment of any waste management programs for a short term, middle term, as well as long term depends on the interaction between the two mirrored triangles, as shown in Fig. 2. The figure depicts a unique relationship between the waste management elements as described by Ishola et al. (2018c) and the waste hierarchy as described by Defra (2012), ERIA (2016), and Kanters (2020). The inverted triangle at the top half of the figure is known as the "waste hierarchy." The figure shows a waste management model with a combined strategy, having the

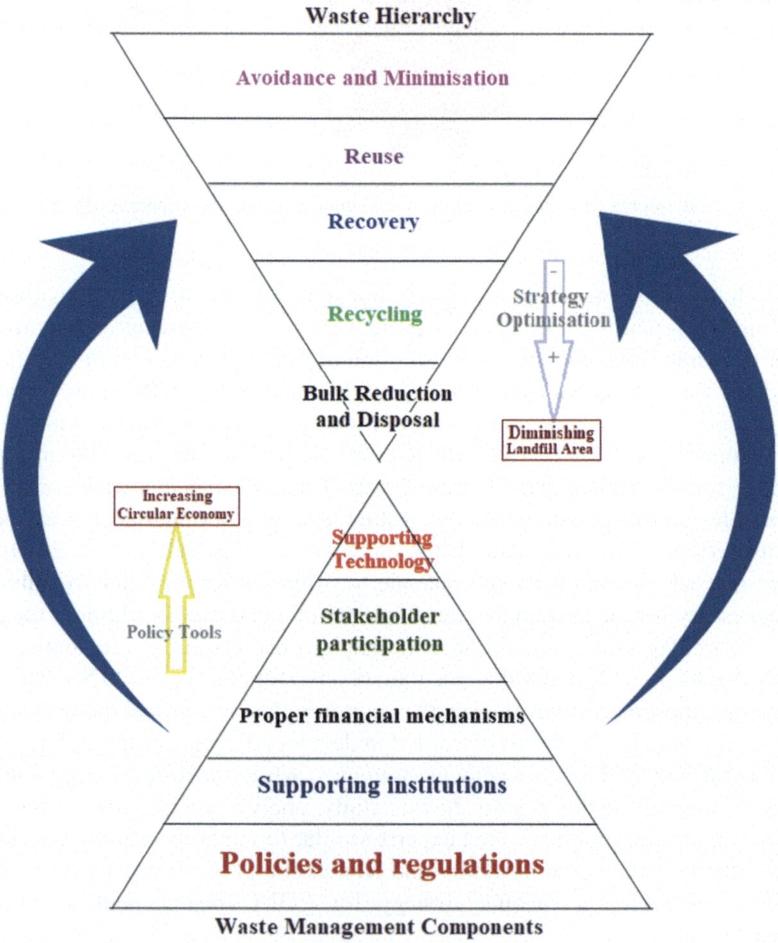

Fig. 2 An integrated waste management model (Ishola et al. 2018a)

waste management component triangle representing the "engine" while the waste hierarchy is the "track." Like any other waste management systems, for a country to have an effective EOLT management system, the United Nations recommended five key elements that must be present for such establishment (Uriarte-Miranda et al. 2018). These fundamental elements include the policy and regulations, the supporting institutional framework, the financial mechanism, the stakeholders' involvement, and the supporting technologies (Ezeudu and Ezeudu 2019).

The waste hierarchy highlights all the existing levels of wastes' life cycle from creation through recovery to disposal (Parvez et al. 2019). It is being controlled by the aggregate actions (or inactions) from all levels of the waste management apparatus. The waste hierarchy mostly describes an instruction of priority which establishes the best inclusive environmental option in waste policy and legislation (Agovino et al. 2020). The waste disposal hierarchy is in preference order of waste avoidance: reuse, recycling, recovery (including energy recovery) and only as a last resort, safe disposal (landfill) (Kanters 2020) as highlighted in Fig. 2. The approach of waste hierarchy is aimed at preventing and eliminating the origins of eco-friendly glitches (Park et al. 2018). The effectiveness and efficiency of a waste management system can largely be determined by the different management options implemented at different levels of the waste hierarchy (Sebola et al. 2018). For instance, CBE for tires begins with having a strong regulation to ensure that the quality of tires available in the market is of both optimal performance and durable (ETRMA 2019a). Appropriate waste management in accordance to waste hierarchy has advantages including reducing pollutants, avoiding greenhouse gas emissions, conserving resources, saving energy, and developing more environmentally friendly innovations (Agovino et al. 2020). The hierarchy had been used solely to address waste management issues. However, it was discovered that the waste hierarchy, as it were, is an insufficient basis for comprehensive waste management (Ewijk and Stegemann 2016).

It is essential to note that rules and regulations build the basis of any program on waste management (Marshall and Farahbakhsh 2013), (Ishola et al. 2020b); thus, it occupies the base (first level) of the base triangle. Government regulations have been proven to have the maximum relationship with other factors related to the implementation of the circular supply chain (Singhal et al. 2020). However, any sustainable waste management policy must take into account the human and environmental security and safety in addition to the economic and social sustainability of the country and other stakeholders (Olukanni et al. 2016). These goals require strong political will to set in order guidelines and regulations that will direct such desired management scheme (Keefe and Fellow 2016). Frequent regime change is a significant and recurring cause of failure of many good policies in developing countries (Makinde 2005). To avoid this kind of issue, once a waste management policy or regulation has been established, a detailed feasibility and viability study should lead to a pilot project. After the pilot project had been certified satisfactory by all stakeholders, the policy should be passed into a bill, law, or act that is autonomous without the overwhelming influence of any incoming leadership. It is crucial to have a leading institution set up for running the program autonomously while it controls other engaged institutions (Duangburong et al. 2015). These are tagged as the supporting institutions represented on the second level of the listed factors from the base of the waste management triangle in Fig. 2.

The provided institutional framework creates the platforms to facilitate the itemized objectives and tasks to achieve an effective and sustainable waste tire waste management (Wilson et al. 2013). Connecting all likely resources in terms of finances and others to generate and maintain the needed functional departments is also a significant responsibility of the bearing institution. For any proposed EOLT management program, the lead institution can either be a government tier or a constituted body. The lead institution is to harmonize the involvement of the stakeholding institutions, specifically government (in case of an independent body), private investors, tire producers, the recycled product market entities, and end-users by communicating with all of them. Failure to execute this function does not lead to the desired results (Olukanni and Nwafor 2019). The lead institution is responsible for setting up salvaging facilities in addition to overseeing its operations. The autonomy of the management of waste tires in lead institution may ensure the retainment of capable hands on the program despite the change of government regimes. It is also expected that such autonomy will engender economic stability and independence. After initial funding from the government during the commissioning period, the mechanism of cost recovery takes a full progression, limiting management contributions.

The constituted management deals unswervingly with the public besides dominated the existing markets for the goods as prescribed during the feasibility studies. In a bid for an independent waste tire management institution to tolerate the aforementioned, they should have a friendly rapport alongside community who are the primary investor being the "manufacturer" of the waste (Olukanni et al. 2018). Constant and unrelenting public awareness is not negotiable for a notable change in the culture of the populace (Strydom 2018). Tactics to motivate end-users on the use, reuse, and collection of waste tires must be deployed by the institution (Rousta et al. 2020). Choice of supporting technology is obliged to be made having in awareness the economic strong point of the participating nation. For a developing economy, a balance must be struck between using obsolete technology and a brand-new technology as the project must be able to break even quickly enough; otherwise, the process leads to the creation of a white elephant project. Depending on the business organogram, all the concerned tiers of governments and the supporting institutions contribute their quotas to these projects, with an expectation of a reasonable rate of return on their investments (Duangburong et al. 2015).

The development of institutional capacity for the CBE requires as many technological innovations of systemic value with the configuration of evolving business model which correspond to the principles of the CBE (Pieroni et al. 2019). The subsidiary expertise at the top level of the waste management triangle in Fig. 2 connotes the recycling technological innovations that would be needed for the management program. The transition from the linear economy to the CBE begins with the development of high technology and technological knowledge, thus decoupling the use of nonrenewable resources from production and creating the means to seek alternative renewable resources (Ogunmakinde 2019). Moreover, several technologies may need to be combined (Forbord and Hansen 2020). This may involve pyrolysis (Ishola et al. 2019b), (Soni and Gaikwad 2017), pulverizing and grinding techniques, and the accompanying material characterization methods among other related technologies (Šandrk-Nukić and Miličević 2019). Technology service centers are industry-scale initiatives strategically sited in locations easily accessible to tire collectors. The institution makes these decisions through expert advice after extensive feasibility studies and cost-benefit studies (Martínez et al. 2013).

This present study aims at developing an EOLT circular economy recovery system that integrates all the indispensable factors mentioned above having in mind all the intricacies of a growing economy. The management system design follows the prescribed waste hierarchy and management principles. The EOLT management system is purposed to incorporate the experts' insights based on field experience. The system design is programmed such to put in place measures to circumference all the mitigating factors like inadequate infrastructure, underfunding, and political interference.

3 End-of-Life Tire Recovery Techniques: Restructuring Toward Circular Economy

Constant evaluation and optimization of the waste management strategy also lead to a seamless improvement in economic efficiency at the different hierarchical levels. Used tires are more valuable, sustainable, and less hazardous if a well-designed management technique is used (Ferronato and Torretta 2019). Fig. 3 shows a comprehensive circular-based economy-oriented end-of-life tire management. There are lots of uses of EOLT as a source of secondary raw materials in various sectors and industries (Landi et al. 2016), starting from the energy sector to the road building and construction and also to the manufacture of chemicals. Such uses had been identified in the following forms:

- *Whole tire form* – These are uses of the tires in their whole form as they are after removing them from the rims. No cutting, crushing, or processing is required. This shape is generally suitable for windbreakers, erosion retainment walls, road embankments, artificial reefs, coastal protection, avalanche protection, waters flow breaker, slope stabilization, and construction and insulation of landfills. Their use spreads to agriculture by means of silage tongs, although to a minimal extent (Campbell 2008).
- *Shredded tire form* – The shredding of tires can be by exposing them to some power-driven strains. The shredding proportions can be between 25 and 300 mm contingent on the expectations of the application. The tire aggregate is 30–50% lighter than sand and gravel. Tire aggregates are useful for sidewalks, roads, railways, and the construction of landfills. In addition, it is suitable as subfill and backfill as well as embankments and bridges (Hu et al. 2014).

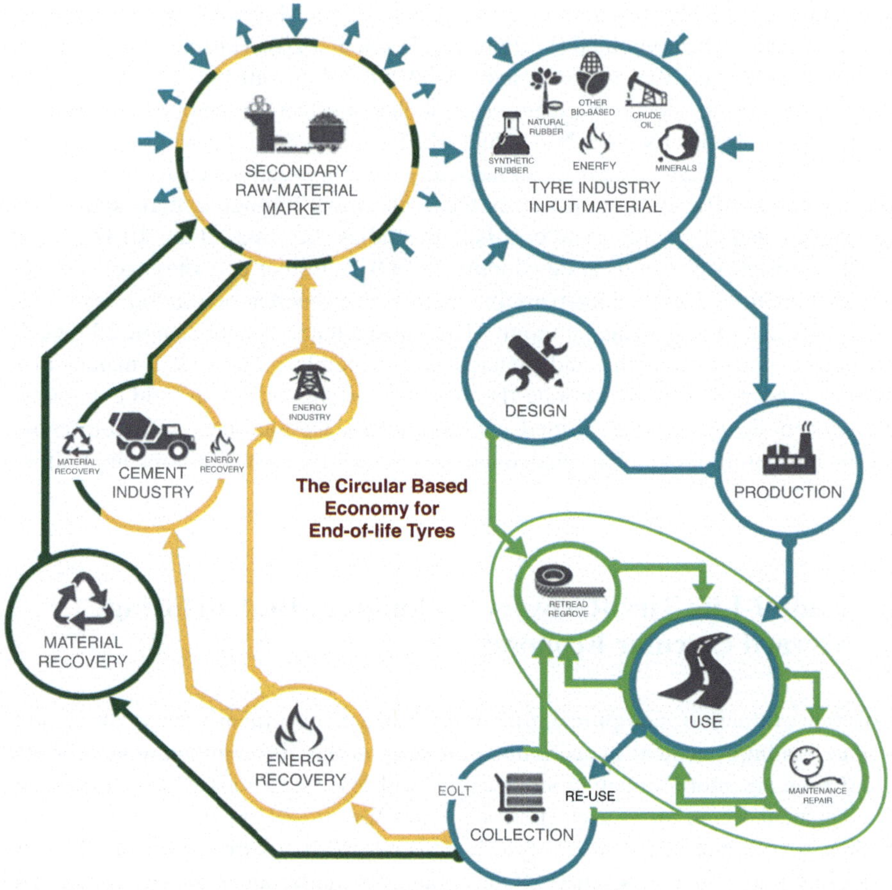

Fig. 3 A typical circular-based economy cycle for end-of -life tires (EOLT) (ERJ 2020), (Dobrot et al. 2019), (ETRMA 2017, 2019b)

- *Powdered and crumb form* – After the fabric and steel components have been removed, tires can be reduced to granular or powdered form. There are various applications of powdered tire; it covers up to flooring for stadiums and playgrounds, molded rubber products, etc. (Gomes et al. 2019).

Figure 3 is a flowchart of various end products from used tires, ranging from various physical products to the extraction of tire oil that could aid as a different energy front. During reprocessing, certain end products and by-products can be used as raw materials for other industries, thereby promoting cleaner productions. In the long run, safe and effective disposal has been achieved. Given the highlighted use of recycled used tire products, as shown in Fig. 3, two main markets could be recognized as the energy-based and material-based markets (Muzenda 2014).

3.1 Energy Retrieval

Tire-derived fuel (TDF) is a widely used energy source in cement kilns, thermal power plants, industrial boilers, steel mills, pulp, and paper mills (Duarte-Ribeiro-de-Souza and D'Agosto 2013). EOLT pyrolysis gas and/or oil has also been widely used as alternative fuels suitable for use in automotive engines (Antoniou and Zabaniotou 2015). TDF has some notable advantages. Tires burned in a controlled oven environment have a relatively lower carbon content per unit of fuel compared to some solid fuels, thereby reducing greenhouse gas emissions (Ishola et al. 2018c). Tires have a high energy content which corresponds to energy sources compared to other known biomasses which makes it a preferable for engineering applications (Svetlana et al. 2021). In Europe, cement industries are significant beneficiaries of fuel or energy from waste tires. Kilns are increasingly being equipped to use waste tires as supplement fuel and still be in compliance with some designated emission standards (Pehlken and Essadiqi 2005). The net calorific value of tires is within 26 and 44 MJ/KJ (Williams 2013), similar to that of coal and some other biomass (Islam et al. 2013). Tire disintegrates completely at 650 °C (Ishola et al. 2020a), whereas the temperature in cement kilns is usually around at 1800 °C. Tires are thus completely disposed to give carbon dioxide and water with slag and tar as residues in a controlled environment (Ishola et al. 2018a). Using whole tires as fuels in industrial kilns saves preparation (quartering or shredding) costs (Mmereki et al. 2017) (Derakhshan et al. 2017) (Jansen et al. 2016). The possibility of having TDF as the leading waste tires usage in a developing country might not be very feasible. Having such industries like cement kilns, thermal power stations, steel mills, and industrial boilers evenly located around the developing country may not be so much expected. Consequently, significant attention is on the alternative material market of waste tires as a preferable option to being engaged in for EOLT recovery plan for developing nations (Juan 2013).

3.2 Material Recovery

Solid products that can be recovered from EOLT include steel, rubber in both natural and synthetic forms, carbon black or graphite, and rubber granules (Sagar et al. 2018). Recovery may be by either mechanical, chemical, or thermochemical processes (Athanassiades 2013). EOLT material recovery can be divided into recovery for civil engineering uses and general engineering uses.

Whole or shredded wires can be used for the various application. These include road insulation, backfill for walls, embankments, field drains, rainwater runoff barriers/erosion control, marsh and wetland establishment, crash barriers, speed humps, and jetty bumpers. The usage was found appropriate due to some characteristics of the precursors such as lightweight, good permeability, shock and noise absorbing, durability, etc. (Poulikakos et al. 2017). Rubber modified asphalt can be derived from the

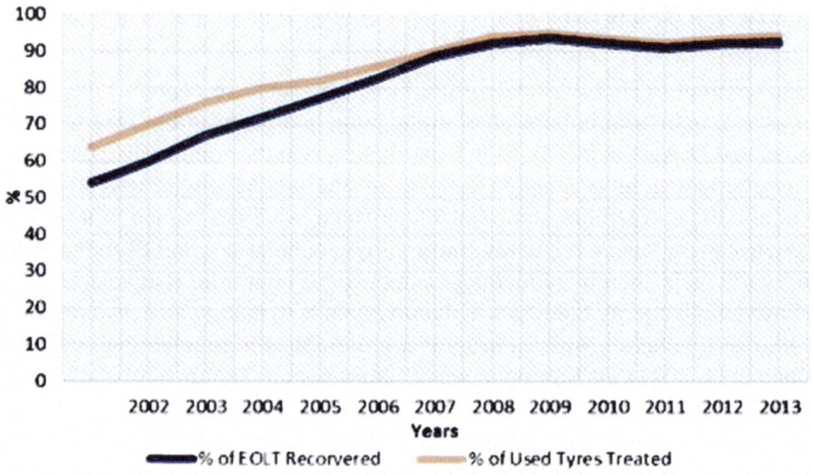

Fig. 4 World EOLT recovery evolution. (Source: ETRMA 2015)

grounded rubber. This asphalt can be used for pavements, which have advantages such as longer life span and less noise. Many kinds of research are ongoing on the recycled rubber blend with polymers and also to increase rubber use in asphalt in varied climates. Rubber granules are produced either by cryogenic (freeze) grinding or ambient grinding, the former is obtained by using liquid nitrogen to freeze the tires before file processing producing finer particles (Hu et al. 2014). Rubber materials can also be employed for road surfacing, flooring, and mats or adhesives (Muzenda 2014). A recently published report shows the recoverability of limonene from old tires by pyrolysis (Januszewicz et al. 2020). Figure 4 is a graph that depicts the EOLT recovery evolution of the industry between 2001 and 2013. More and more high-end applications for recycling materials are developing, which increases the value of recycling materials compared to recycling energy (Torretta et al. 2015).

4 End-of-Life Tire Management Systems

Maximizing resources from waste tires by preventing misuse, wastage, and contrary exploitation is a step toward harnessing waste tire potentials. In Europe, recovery or collection of waste tires generally follow three (3) significant systems, namely, producer responsibility, government responsibility, and free market system.

4.1 Extended Producer Responsibility (EPR)

Producer responsibility clause actuates total or partial operational or financial responsibility of the manufacturer for a product extended to the post-consumer stage of a product's life cycle (Oh and Hettiarachchi 2020). Under this system, the

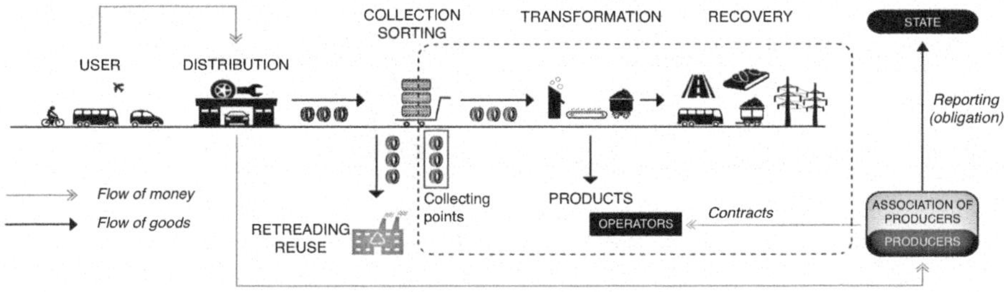

Fig. 5 Producer responsibility Flowchart (Jain 2016; Scott 2016)

manufacturers are to ensure that the wastes created by and from their products are collected and handled responsibly (Niza et al. 2014). The producers are under law to retrieve back their products as their property right after the useful life cycle has been completed (Luth and Koppejan 2020). This implies that the producer is responsible for the tire waste generated by their consumers (ETRMA 2015). Figure 5 represents a flowchart for extended producer responsibility.

Financing of the waste tire management is done by a tax rate or fee placed on tire manufacturers or tire importers (Scott 2016). These financial implications are passed down through a value chain down to the distributors and subsequently to end-users. Producer responsibility is about the most used in European regions with about 21 countries adopting a legal framework that assigns the responsibility of the post-life management of tires to the producers (Winternitz et al. 2019). Analysis of EPR situation in an emerging economy revealed that while the government imposed full financial and operational responsibility on tire manufacturers and importers, other actors in the product chain were indignantly left out of executing their tasks and responsibilities assigned to them (Park et al. 2018). This indicated that control might be a significant issue in emerging economies where appropriate check and balance may not be guaranteed. In addition, due to some identified emerging economy-associated crises such as inadequate power supply (Ishola et al. 2019a, 2020b), among the rest, developing economies do not have a favorable environment to serve as sites for producers. As a result, growing economies largely depend on goods imported by manufacturers from industrialized countries (Torrente-velásquez et al. 2020) and, therefore, may not have access to producer responsibilities at all.

4.2 *Government Responsibility System*

Under this system, the government is responsible for the management of waste tires. The funds for the management program come from the tax imposed on producers of tires as directed by government policy, also afterward conceded on to the purchaser (Šandrk-Nukić and Miličević 2019). Meanwhile, this management system engages tax system; it is likewise denoted as the "national tax system." The implication is that tax levied on the sale of tires is utilized to run the waste tire management system

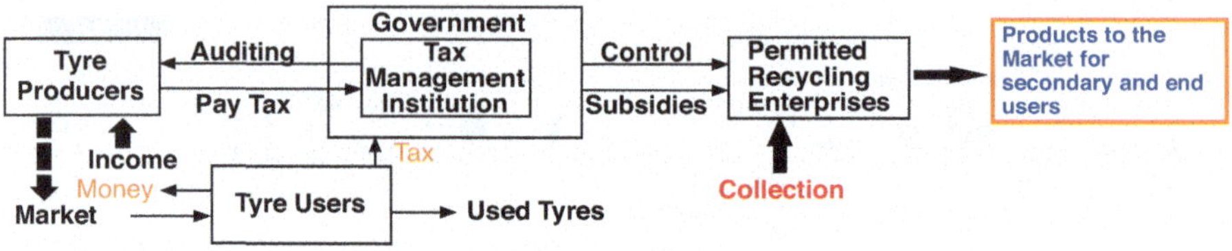

Fig. 6 Flowchart of a typical government responsibility system (Cao et al. 2016)

(ETRMA 2019a). Examples of countries which follow the national tax system are Denmark, Croatia, Slovak Republic, and Latvia (Sebola et al. 2018). As a result of the lack of indigenous tire industries, the government may have no access to company tax from the tire producers. This renders this strategy unfeasible for the most developing world. Figure 6 describes the flow pattern of this system.

4.3 Liberal System

The liberal system is also popularly known as the free market system. The system gives numerous waste tire management firms an unrestricted market so as to capitalize in the retrieval and reprocessing of tires. However, the legislation or government highlights the purposes or ethics of retrieval to be encountered but does not elect a specific institution to the organization (Mmereki et al. 2017). In this system, all the entrepreneurs in the retrieval value chain agreement under free-market situations perform in acquiescence by means of the license given (Šandrk-Nukić and Miličević 2019). Likewise, underneath the recognized objectives, contracts can be bidden for by companies, possibly for a specific province, county, or state to be in charge of the supervision of waste tires (Uriarte-Miranda et al. 2018). Partaking in the free market scheme by the companies is voluntary; nevertheless, the business has to conform with the regulations about entrepreneurship, use, transportation, storage, and disposal of wastes and other relating laws provided on a general by the government (Athanassiades 2013). Germany, Austria, Switzerland, and the United Kingdom are developed countries under the free-market system (Jain 2016). However, for developing economies, the inadequacy of supporting institutional framework may hamper the implementation of the free-market system. Figure 7 describes the flow pattern of this system.

4.4 System Evaluation Towards a Sustainable Management Model for Developing Economy

A brief systemic analysis was conducted to investigate the non-effectiveness of EOLT management strategies in developing countries. A semi-structured interview was conducted for a number (12) of waste management researchers/specialists from

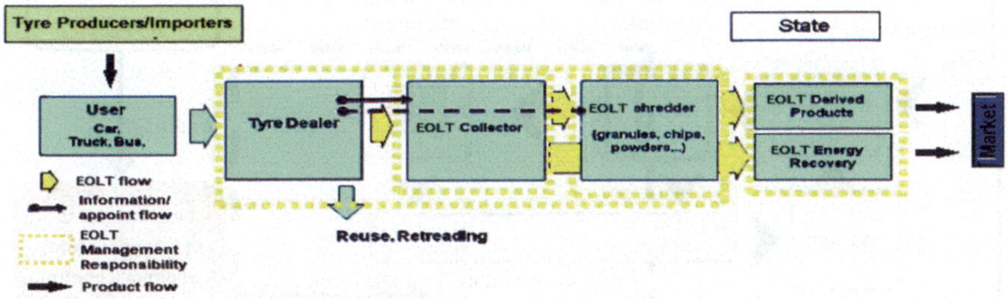

Fig. 7 Flowchart of liberal system (Jain 2016; Scott 2016)

6 developing countries to arrive at a consensus expert-based analysis. The interviewee's view of the current situation of EOLT management in their respective countries of practice was rated against the best practices. The following relevant report was extracted from the responses of the waste management researchers/specialists. While all the respondents correctly identified the EOLT established management strategies. 33.3% of the respondents feel that the total adoption of any of the established strategies in the developed countries would greatly help in solving the EOLT management issues in the developing economies. Nevertheless, 50% of the respondents believe that the peculiarity of the developing countries would be impediments to expected results. Finally, the remaining 16.7% is either indifferent or feels a specialized study is needed before arriving at such conclusions. The summary of the study also showed that 66.7% of the respondent's countries of practice are confirmed not to follow any particular EOLT management strategies other than the general (municipal) waste management. All these countries are also confirmed to engage in the ideology of four "Rs" of waste management, though, in different degrees, ranging from strong engagement to very weak. The listed Rs are reduce, reuse, recycle, and recovery of waste materials, as mentioned earlier in the report. Approximately 58.3% of the respondents believed that, if followed correctly, the general philosophy in waste management is adequate for EOLT management in developing countries. Nevertheless, all the respondents strongly agreed that a specialized EOLT management specifically designed for developing and the emerging economy will be of higher benefits.

5 The "Pseudo-Producer Responsibility" Approach

Emerging economies are most likely to face the dual threat of non-admittance to manufacturer responsibilities and simultaneously a much larger number of used tires per capita to deal with, as established earlier. This new model addressed these

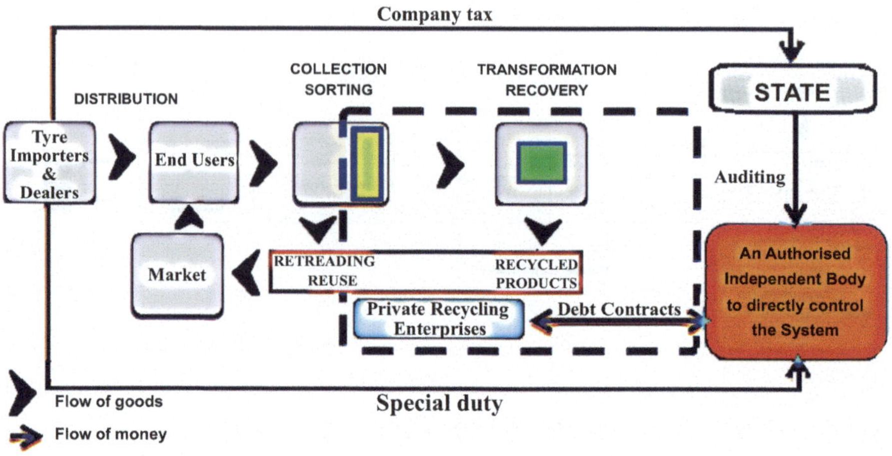

Fig. 8 Flowchart of the "pseudo-producer" responsibility system

threats by constituting an intervention agency to replace the producers in conducting the producer responsibility functions. The new model is as represented in Fig. 8. An additional duty on new and used tires on import (entry) is collected by the government to fund the agency. However, it may be impossible to eliminate the flow of used vehicles and tires into an emerging economy. However, it is particularly important to put a restriction on the importation of specific tire ages. At the ports, the rejected tires are sent directly to the waste tire collection points close to the ports. The "pseudo-producer" agency would ensure an incessant and properly managed stream of secondhand tires within the recycling system. The agency is expected to enjoy an appreciable degree of autonomy. They are to issue contracts to local private investors with an ability to of issuing debt contracts to strengthen the technological capability of the local private investors with a well-defined payback modality. However, since a lot of developing countries are battling with corruption problems, the government arm should conduct contract probes to ensure due process and monitoring as well as periodic external body auditing of the institution. This is expected to aid the sustainability of the model by blocking cash flow leakages. There is a complete circular cash flow to maintain the system and start generating profits after payback, if run as prescribed. In addition, the system has the capacity to reduce the amount of EOLT generated during the product life cycle. It is not only expected to lead to a reduction of the waste management cost, but principally, it is designed to lead to a profitable system. The government is expected to provide guidelines for the business relationship with the responsibility of providing a local content favored business conditions for the local investors.

6 Conclusions

EOLT can potentially pose serious pollution risks to the environment if they are not managed strategically. Developing countries were endangered in this regard. Developed countries are approaching zero tire waste, via the application of specific management strategies adopted. But views of some experts conclude that some primary socioeconomic and cultural status of an average developing country necessitated a need for the design of a model that will suit the emerging economies. The "pseudo-producer" responsibility system was designed for effective EOLT management for emerging economies. Measures to ensure the recycling of EOLTs have been successfully modeled to take advantage of the EOLT recycled products and at the same time avoid or reduce the hazards associated with unhealthy pilling and disposal. The system projects flow of information, materials, and cash to formulate a sustainable and formidable management model for EOLTs. The new strategy was designed to be based on well-defined policies and regulations on the way to a harmless recycling structure. The designed strategy engenders the circular economy concept by incorporating all the key stakeholders into an incentive-driven circular system while being closely monitored and controlled by the constituted authorities. The implementation of this model is expected to increase the circular economy on the local market by opening up platforms where supply meets the demand for reuse and recycling of waste tires.

The designed pseudo-producer liability systems may not be suitable for countries having challenges with managing their border activities effectively. Many developing countries suffer from corruption and organizational issues, which can be a pointer to difficulty in the sanity of border operations. This stands as a likely limitation to the proposed EOLT waste management model. Validation of the model will best be conducted using a dynamic system modeling and simulation tool. A comparative analysis of all the above-described models in simulated emerging economies using artificial intelligence has been considered as a future scope in relation to this present work.

References

Agovino M, Garofalo A, Mariani A (2020) Effects of environmental regulation on separate waste collection dynamics: empirical evidence from Italy. J Clean Prod 124:30–40. https://doi.org/10.1016/j.jclepro.2016.02.082

Ahmed V, Batool S (2015) *India-Pakistan Trade: Perspectives from the Automobile Sector in Pakistan.*

Antoniou N, Zabaniotou A (2015) Experimental proof of concept for a sustainable End of Life Tyres pyrolysis with energy and porous materials production. J Clean Prod 101:323–336. https://doi.org/10.1016/j.jclepro.2015.03.101

Athanassiades E (2013) Waste tyre pyrolysis: sustainable recovery and reuse of a valuable resource. Retrieved from http://easyaccess.lib.cuhk.edu.hk/login?url=http://search.proquest.com/docview/1780249748?accountid=10371%5Cn, http://findit.lib.cuhk.edu.hk/852cuhk/?url_

ver=Z39.88-2004&rft_val_fmt=info:ofi/fmt:kev:mtx:dissertation&genre=dissertations+%26+theses&sid=ProQ:P

Bittencourt ES, Hora C, Fontes DO (2020) Modeling the Socioeconomic Metabolism of End-of-Life Tires Using Structural Equations: A Brazilian Case Study. *Sustainability 12*.

Blomsma F, Tennant M (2020) Circular economy: Preserving materials or products? Introducing the Resource States framework. *Resources,* Conserv Recycl. https://doi.org/10.1016/j.resconrec.2020.104698

Bonah E, Nii M, Ansah S et al (2020) Nuclear energy for sustainable development: SWOT analysis on Ghana's nuclear agenda. Energy Reports *6*:107–115. https://doi.org/10.1016/j.egyr.2019.11.163

Campbell A (2008) Determining a Waste Tyre Management System for Hong Kong. *Waste Manag 04*(44).

Cao J, Lu B, Chen Y et al (2016) Extended producer responsibility system in China improves e-waste recycling: Government policies, enterprise, and public awareness. Renew Sust Ener Rev 62:882–894. https://doi.org/10.1016/j.rser.2016.04.078

Defra (2012) *The Economics of Waste and Waste Policy*. Retrieved from Department for Environment, Food and Rural Affairs website: https://assets.publishing.service.gov.uk/government/uploads/system/uploads/attachment_data/file/69500/pb13548-economic-principles-wr110613.pdf

Derakhshan Z, Ghaneian MT, Mahvi AH et al (2017) A new recycling technique for the waste tires reuse. Environ Res 158:462–469. https://doi.org/10.1016/j.envres.2017.07.003

Dobrot D, Dobrot G, Dobrescu T et al (2019) The Redesigning of Tires and the Recycling Process to Maintain an Efficient Circular Economy. Sustainability 11:5204.

Duangburong J, Tantayanon S, Bhandhubanyong P (2015) A Breakthrough Challenge with Tyre Waste Management: Thailand Perspective. Int J Soc Sci Hum 5(9):768–772. https://doi.org/10.7763/IJSSH.2015.V5.553

Duarte-Ribeiro-de-Souza C, D'Agosto M. De AD (2013) Value chain analysis applied to the scrap tire reverse logistics chain: An applied study of co-processing in the cement industry. Res, Conserv Recycl *78*:15–25. https://doi.org/10.1016/j.resconrec.2013.06.007

Dufton PW (2001) *End-of-Life Tyres – Exploiting their Value.*

ERIA (2016) *Towards a Circular Economy: Corporate Management and Policy Pathways* (V. Anbumozhi & Jootae Kim, Eds.). Jakarta: ERIA.

ERJ (2020) ETRMA Calls Policy Makers Support Sustainability Bid. *European Rubber Journal.*

ETRMA (2015) *End-of-life Tyre Report 2015.*

ETRMA (2017) *Moving Innovation That Cares.*

ETRMA (2019a) Circular Economy. Retrieved from https://www.etrma.org/key-topics/circular-economy/

ETRMA (2019b) *Sustainable and Resilient Natural Rubber Supply Chain.*

Ewijk S Van Stegemann JA (2016) Limitations of the waste hierarchy for achieving absolute reductions in material throughput. J Clean Prod 132:122–128. https://doi.org/10.1016/j.jclepro.2014.11.051

Ezeudu OB Ezeudu TS (2019) Implementation of Circular Economy Principles in Industrial Solid Waste Management: Case Studies from a Developing Economy (Nigeria). *Recycl 4*(42).

Ferronato N, Torretta V (2019) Waste Mismanagement in Developing Countries: A Review of Global Issues. *Int J Envir Res Pub Health.* https://doi.org/10.3390/ijerph16061060

Forbord M, Hansen L (2020) Enacting sustainable transitions: A case of biogas production and public transport in Trøndelag, Norway. J Clean Prod. https://doi.org/10.1016/j.jclepro.2020.120156

Gall M, Wiener M, Chagas C et al (2020) Building a circular plastics economy with informal waste pickers: Recyclate quality, business model, and societal impacts. Res, Conserv Recycl. https://doi.org/10.1016/j.resconrec.2020.104685

Garnett K, Cooper T (2014) Effective dialogue: Enhanced public engagement as a legitimising tool for municipal waste management decision-making. Waste Manag 34(12):2709–2726. https://doi.org/10.1016/j.wasman.2014.08.011

Gipser M (2000) A Tire Model for Ride & Durability Simulations. *International ADAMS/FTire User's Conference*, 1–13. Tokyo.

Godfrey L, Oelofse S (2017) Historical Review of Waste Management and Recycling in South Africa. *Resources*, *6*(4):57. https://doi.org/10.3390/resources6040057

Gomes TS, Neto GR, De-Salles ACN et al (2019) End-of-Life Tire Destination from a Life Cycle Assessment Perspective. In *New Frontiers on Life Cycle Assessment-Theory and Application*. IntechOpen.

Gravagnuolo A, Angrisano M, Girard LF (2019) Circular Economy Strategies in Eight Historic Port Cities: Criteria and Indicators Towards a Circular City Assessment Framework. Sustainab 11(3512).

Hartley K, Santen RV, Kirchherr J et al (2020) Policies for transitioning towards a circular economy: Expectations from the European Union (EU). Res, Conserv Recycl. https://doi.org/10.1016/j.resconrec.2019.104634

Hu N, Cheng CM, Wen XF et al (2014) Comparison Study of Scrap Tires Management between China and the USA. Adv Mat Res 878:90–98. https://doi.org/10.4028/www.scientific.net/AMR.878.90

Ishola, F. A., Oyawale, F. A., Inegbenebor, A. O., & Boyo, H. (2018a). Design of a high Temperature 'Anaerobic Gas-Furnace' suitable for Pyrolysis. IOP Conference Series: Materials Science and Engineering, 413, 012079. https://doi.org/10.1088/1757-899x/413/1/012079

Ishola F, Ajayi O, Oyawale F. A et al (2018b) Sustainable End-of-Life Tyre (EOLT) Management for Developing Countries – A Review. *Industrial Engineering and Operation Management (IEOM) Conference, South Africa.*

Ishola F, Oyawale F, Inegbenebor A et al (2020a) Mathematical Analysis and Thermal Modelling of a Pilot-Scale Pyrolysis Gas Furnace. J Adv Res Fluid Mech Therm Sci 65(1):81–93.

Ishola F, Towoju O, Mamudu A et al (2020b) Nigerian Oil Palm Industry as a Sustainable Renewable Energy Resource. *E3S Web of Conferences*, *152*, 02005.

Ishola FA, Azeta J, Agbi G et al (2019a) Simulation for Material Selection for A Pico Pelton Turbine ' s Wheel and Buckets. *Proc Manuf 30.*

Ishola FA, Inegbenebor AO, Oyawale, FA (2019b) Thermal Modelling for A Pilot Scale Pyrolytic Furnace for Production of Carbon Black. J Phy: Conf Ser, 1378(3):032089. https://doi.org/10.1088/1742-6596/1378/3/032089

Ishola FA, Ogunrinola I, Onwardi PN et al (2018c) 2-Dimensional CFD Simulation of A Gas-Fired Pyrolysis Furnace. Int J Mech Eng Tech 9(12):383–394.

Ishola FA, Olatunji OO, Ayo OO et al (2019c) Sustainable Nuclear Energy Exploration in Nigeria – A SWOT Analysis. Proc Manuf 35:1165–1171. https://doi.org/10.1016/j.promfg.2019.06.072

Islam MR, Islam MN, Mustafi NN et al (2013). Thermal recycling of solid tire wastes for alternative liquid fuel: The first commercial step in Bangladesh. Proc Eng. https://doi.org/10.1016/j.proeng.2013.03.162

Jain A (2016) *Compendium of Technologies for the Recovery of Materials/Energy from End of Life (EOL) Tyres.*

Jansen S, Schmeitz A, Akkermans L (2016) *Study on some safety-related aspects of tyre use.* https://doi.org/10.2832/67191

Januszewicz K, Kazimierski P, Kosakowski W et al (2020) Waste Tyres Pyrolysis for Obtaining Limonene. Materials 1–30.

Juan D (2013) Waste Tyre Pyrolysis. Renew Sust Energ Rev 23:179–213. https://doi.org/10.1016/j.rser.2013.02.038

Kanters J (2020) Circular Building Design: An Analysis of Barriers and Drivers for a Circular Building Sector. Buildings 10(77):1–16.

Karagiannidis A, Kasampalis T (2010) Resource recovery from end-of-life tyres in Greece : A field survey, state-of-art and trends. Waste Manag Res 28(6):520–532. https://doi.org/10.1177/0734242X09341073

Keefe LO, Fellow C (2016) *Investigating global best practice waste tyre management*. 1–42.

Kinobe JR, Gebresenbet G, Vinnerås, B (2012) Reverse Logistics Related to Waste Management with Emphasis on Developing Countries—A Review Paper. J Environ Sci Eng B(1):1104–1118.

Landi D, Vitali S, Germani M (2016) Environmental Analysis of Different End of Life Scenarios of Tires Textile Fibers. Proced CIRP 48:508–513. https://doi.org/10.1016/j.procir.2016.03.141

Lebreton B, Tuma A (2006) A quantitative approach to assessing the profitability of car and truck tire remanufacturing. Int J Prod Econ 104:639–652. https://doi.org/10.1016/j.ijpe.2004.11.010

Luth J, Koppejan R (2020) Extended producer responsibility for lamps in Nordic countries: best practices and challenges in closing material loops. J Clean Prod 123(2016):167–179. https://doi.org/10.1016/j.jclepro.2015.06.131

Makinde T (2005) Problems of Policy Implementation in Developing Nations: The Nigerian Experience Problems of Policy Implementation in Developing Nations. *J Soc Sci 11*(1):63–69. https://doi.org/10.1080/09718923.2005.11892495

Managing End-of-Life Tires. (2015) Retrieved from www.wbcsd.org/web/tires or from tires@wbcsd.org

Marshall RE, Farahbakhsh K (2013) Systems approaches to integrated solid waste management in developing countries. Waste Manag 33(4):988–1003. https://doi.org/10.1016/j.wasman.2012.12.023

Martínez JD, Puy N, Murillo R et al (2013) Waste tyre pyrolysis – A review. Renew Sust Ener Rev 23:179–213. https://doi.org/10.1016/j.rser.2013.02.038

Mmereki D, Machola B, Mokokwe K (2017) Status of waste tyres and management practice in Botswana. J Air Waste Manag Ass. https://doi.org/10.1080/10962247.2017.1279696

Muzenda E (2014) A Discussion of Waste Tyre Utilization Options. Eng Tech. https://doi.org/10.15242/IIE.E0314593

Niza S, Santos E, Costa I et al (2014) Extended producer responsibility policy in Portugal: a strategy towards improving waste management performance. J Clean Prod 64:277–287. https://doi.org/10.1016/j.jclepro.2013.07.037

Ogunmakinde OE (2019) A Review of Circular Economy Development Models in China, Germany and Japan. Sustainability 4(27).

Oh J, Hettiarachchi H (2020) Collective Action in Waste Management: A Comparative Study of Recycling and Recovery Initiatives from Brazil, Indonesia, and Nigeria. Recycling 5(4).

Olukanni DO, Adeleke JO, Aremu DD (2016) A Review of Local Factors Affecting Solid Waste Collection in Nigeria. Pollution 2(3):339–356. https://doi.org/10.7508/pj.2016.03.008

Olukanni DO, Aipoh AO, Kalabo IH (2018) Recycling and Reuse Technology: Waste to Wealth Initiative in a Private Tertiary Institution, Nigeria. Recycling. https://doi.org/10.3390/recycling3030044

Olukanni DO, Nwafor CO (2019) Public-Private Sector Involvement in Providing Efficient Solid Waste Management Services in Nigeria. Recycling 4(19).

Omole DO, Isiorho SA, Ndambuki JM (2016) Waste management practices in Nigeria: Impacts and mitigation. The Geolog Soc America Special Paper 2520(33):377–386. https://doi.org/10.1130/2016.2520(33)

Park J, Díaz-posada N, Mejía-dugand S (2018) Challenges in implementing the extended producer responsibility in an emerging economy: The end-of-life tire management in Colombia. J Clean Prod 189:754–762. https://doi.org/10.1016/j.jclepro.2018.04.058

Parvez N, Agrawal A, Kumar A (2019) Solid Waste Management on a Campus in a Developing Country: A Study of the Indian Institute of Technology Roorkee. Recycling 4(28):1–22.

Pehlken A, Essadiqi E (2005) *Scrap Tire Recycling in Canada* (Vol. 08). https://doi.org/10.13140/2.1.1941.8400

Pieroni MPP, Mcaloone TC, Pigosso DCA (2019) Configuring New Business Models for Circular Economy through Product – Service Systems. Sustainability 11(3727).

Pisitsankkhakarn R, Vassanadumrongdee S (2020) Enhancing purchase intention in circular economy: An empirical evidence of remanufactured automotive product in Thailand. Res Conserv Recycl 156:104702. https://doi.org/10.1016/j.resconrec.2020.104702

Poulikakos LD, Papadaskalopoulou C, Hofko B et al (2017) Harvesting the unexplored potential of European waste materials for road construction. Res Conserv Recycl 116:32–44. https://doi.org/10.1016/j.resconrec.2016.09.008

Rosendorfová M, Vybochová I, Beukering P (1998) Waste management and recycling of tyres in Europe. In *Institute for Environmental Studies*. Retrieved from http://dare2.ubvu.vu.nl/handle/1871/22424

Rousta K, Zisen L, Hellwig C (2020) Household Waste Sorting Participation in Developing Countries A Meta-Analysis. Recycling 5(6).

Sagar M, Nibedita K, Manohar N et al (2018) A potential utilization of end-of-life tyres as recycled carbon black in EPDM rubber. Waste Management 74:110–122. https://doi.org/10.1016/j.wasman.2018.01.003

Salguero-Puerta L, Leyva-Díaz JC, Cortés-García FJ et al (2019) Sustainability Indicators Concerning Waste Management for Implementation of the Circular Economy Model on the University of Lome (Togo) Campus. Int J Environ Res Pub Health 16(12):2234.

Šandrk-Nukić I, Miličević I (2019) Fostering Eco-Innovation: Waste Tyre Rubber and Circular Economy in Croatia. Inter Desc Compl Syst 17(2-B):326–344. https://doi.org/10.7906/indecs.17.2.9

Scott E (2016) End-of-life Tyre Report 2015. In *European Tyre & Rubber Manufacturers Association (ETRMA).*

Sebola MR, Mativenga PT, Pretorius J (2018) A Benchmark Study of Waste Tyre Recycling in South Africa to European Union Practice. Proced CIRP 69:950–955. https://doi.org/10.1016/j.procir.2017.11.137

Svetlana, Dabic-Miletic Vladimir, Simic Selman, Karagoz (2021) End-of-life tire management: a critical review Abstract Environmental Science and Pollution Research 28(48) 68053–68070. https://doi.org/10.1007/s11356-021-16263-6

Singhal D, Tripathy S, Kumar S (2020) Remanufacturing for the circular economy: Study and evaluation of critical factors. Res Conserv Recycl 156:104681. https://doi.org/10.1016/j.resconrec.2020.104681

Soni T, Gaikwad A (2017) Waste Pyrolysis Tire Oil As Alternative Fuel For Diesel Engines. Inter J Mech Prod Eng Res Dev 7(6):271–278.

Strydom WF (2018) Applying the Theory of Planned Behavior to Recycling Behavior in South Africa. Recycling. https://doi.org/10.3390/recycling3030043

Torrente-velásquez JM, Ripa M, Chifari R et al (2020) A waste lexicon to negotiate extended producer responsibility in free trade agreements. Res Conserv Recycl 156. https://doi.org/10.1016/j.resconrec.2020.104711

Torretta V, Cristina E, Ragazzi M et al (2015) Treatment and disposal of tyres: Two EU approaches. A review. Waste Manag 45:152–160. https://doi.org/10.1016/j.wasman.2015.04.018

Urbinati A, Chiaroni D (2019) Managing the Introduction of Circular Products: Evidence from the Beverage Industry. Sustainability 11(3650):1–12.

Uriarte-Miranda ML, Caballero-Morales SO, Martinez-Flores JL et al (2018) Reverse Logistic Strategy for the Management of Tire Waste in Mexico and Russia: Review and Conceptual Model. Sustainability. https://doi.org/10.3390/su10103398

Vanhamäki S, Virtanen M, Luste S et al (2020) Transition towards a circular economy at a regional level: A case study on closing biological loops. Res Conserv Recycl. https://doi.org/10.1016/j.resconrec.2020.104716

Wang KS, Hsu FS, Liu PP (2002) Modeling the bathtub shape hazard rate function in terms of reliability. Rel Eng Syst Saf 75:397–406.

Weissman SL, Sackman JL, Gillen D et al (2003) *Extending the Lifespan of Tires*. Berkeley, California.

Williams PT (2013) Pyrolysis of waste tyres: A review. Waste Manag. https://doi.org/10.1016/j.wasman.2013.05.003

Wilson DC, Velis CA, Rodic L (2013) Integrated sustainable waste management in developing countries. *Institution of Civil Engineers: Waste and Resource Management.*

Winternitz K, Heggie M, Baird J (2019) Extended producer responsibility for waste tyres in the EU: Lessons learnt from three case studies – Belgium, Italy and The Netherlands. Waste Manag 89:386–396. https://doi.org/10.1016/j.wasman.2019.04.023

Zedler Ł, Przybysz-romatowska M, Haponiuk J et al (2020) Modification of Ground Tire Rubber — Promising Approach for Development of Green Composites. J Compos Sci 4(2):1–11.

Life Cycle Engineering for Material Recovery: The Case of Residential Envelope Construction

Matan Mayer

1 Introduction

Single-family home construction is by far the largest and fastest-growing sector in the US building industry, with over 900,000 new buildings completed last year alone and over 73% growth in the last decade (US Census Bureau 2022). For the most part, this construction activity lacks any consideration of future end-of-use waste generation. Introducing circularity into this industry is a multifaceted challenge, which includes multiple economic, regulatory, and design-related aspects. A basic precondition to tackling this challenge is an ability to accurately measure circularity potential (Schvaneveldt 2003). Circularity potential assessment is relatively developed for consumer products in aspects ranging from evaluating unfastening effort of widely used fasteners (Sodhi et al. 2004), through estimation of product end-of-life disassembly potential during early design phases (Germani et al. 2014), quantification of end-of-life disassembly effort and cost (Das et al. 2000), and prediction of total time of disassembly operations for a given product (Gungor and Gupta 1997), to calculation of disassembly depth effectiveness (Giudice 2010). In contrast, far less attention has been given to circularity potential assessment in the building industry. Significant developments in this field include introduction of a decision-making framework for demolition technique selection for improved material recovery (Abdullah et al. 2003), evaluation scheme for recycling potential of construction materials (Thormark 2001), and a method for assessing loop potential for building materials (Rosen 2019). Those sources propose evaluating recovery potential at either the material or the system levels, but not at both.

Aiming to bridge this gap, Mayer and Bechthold (2017) propose an assessment framework that quantifies recovery potential in buildings at both the product and the

M. Mayer (✉)
IE University, Madrid, Spain
e-mail: matan.mayer@ie.edu

I. S. Dunmade et al. (eds.), *Sustainable Engineering*, Green Energy and Technology, https://doi.org/10.1007/978-3-031-47215-2_11

assembly levels. The scheme relies on an existing market and liability distinction between building products (manufacturers) and building assemblies (designers and contractors). At the product level, the scheme includes four evaluation criteria: (A) recyclability, which evaluates the availability and efficiency of existing recycling technologies from a market and an environmental perspective; (B) surface treatment, which examines the ability of coatings such as fireproofing to be easily removed during end-of-life recovery; (C) binders, which assesses the potential of adhesives and mortars to allow contaminant-free separation for reuse and recycling; and (D) material diversity, which looks at the number of different components, coatings, and binders as a potential obstacle for full material recovery. At the assembly level, the scheme includes four evaluation criteria as well: (A) average product scores for all products in the assembly; (B) connection index, which accounts for separation damage, tool type required for disassembly, and disassembly time; (C) access index, which considers the correlation between disassembly sequence and product life expectancy, as well as the existence of layers that limit the access to other layers with shorter life expectancies; and (D) component integration, which is based on the notion that prefabricated assemblies are likely to lead to higher recovery potential due to their ability to be easily removed from the building in one piece and allow full disassembly offsite. Using these criteria, the developed framework computes a material recovery potential index (MRPI) for the evaluated assembly. The index ranks products and assemblies on a scale of 0.0–1.0, where a higher result indicates stronger recovery potential. In this context, this paper describes an experimental implementation of the framework on an adaptation of commonly used assemblies for full circularity.

This chapter demonstrates using the MRPI assessment method to inform assembly design decisions. Light timber frame envelope assembly is considered as a test case, a recovery-oriented redesign process is described, and its outcomes are evaluated. Lastly, a comparative analysis is presented and discussed.

2 Method

As illustrated in the previous section, the MRPI assessment method can be used to compare the recovery potential of diverse assemblies on a single score scale. A direct extension of the comparative assessment of existing assembly designs would be the informed creation of new ones. Within that context, this chapter describes a redesign of a common envelope assembly type based on its material recovery performance. It should be noted that this operation is meant as a speculative exercise with the sole purpose of exploring the limits of recovery potential using widely available materials, products, and connection types. The resulting design, therefore, may not fully address the wide range of envelope performance considerations such as thermal resistance, acoustic insulation, and so forth.

2.1 Timber Frame Assembly as a Test Case

The growing popularity of timber frame construction in the American residential market has been a major driver of two opposing trends from a material recovery perspective. On the one hand, the constant high demand for timber frame components pushed the industry to develop highly efficient and standardized products, which in turn may contribute to greater end-of-life utilization. On the other hand, the rapid growth of this industry also rushed the demise of traditional skill-intensive timber-made connections in favor of mass-produced steel products such as nail plates and staples (Staib et al. 2013). The latter are highly problematic for material separation and recovery purposes. In addition to their dominant market presence, timber frame envelope assemblies also make a compelling test case for a recovery-oriented redesign exercise since they inherently require a separation between structure, insulation, and weather protection elements. This separation positions timber frame envelope assemblies in a preferable starting point from a material recovery standpoint.

2.2 Preassembly

While highly standardized, timber frame assemblies consist, for the most part, of discrete off-the-shelf products which are typically put together at the construction site. Although it offers contractors a certain degree of operational flexibility, this practice regularly results in ad hoc sub-assemblies and large amounts of irreversible connections such as staples and nails. In addition to encouraging inefficiency, it also results in far more waste than the comparable preassembled alternatives. Research shows that implementing a high degree of prefabrication and preassembly in standard campus concrete buildings, for example, reduced on-site waste production by over 97% (Begum et al. 2010). Although no similar studies focusing specifically on timber frame envelope assemblies have been found, it is fair to assume that increased prefabrication and preassembly in this assembly type could lead to less wasteful and ultimately more recoverable buildings.

2.3 Modularity

Modular construction systems are defined as closed systems (entirely made by a single producer) which are prefabricated and preassembled by a manufacturer independent of a particular building. For this reason, systems of modular units inherently require a well-defined aggregation and assembly logic. These qualities may in turn contribute to a smoother end-of-life deconstruction process and higher diversion rates. In the context of buildings, modular units can be classified into three categories: *frames*, including columns, beams, and trusses; *panels*, including composite products such as SIPs; and *units*, including entire buildings or large building segments (Staib et al. 2013). From a material recovery standpoint, a modular system with high complexity and large dimensions poses greater challenges than a simple and small system, mainly because the size of the unit determines the type of tools required for handling it, and its complexity affects its disassembly depth. That said, complexity is often desirable as it enables multifunctionality. Examining the three modular unit categories through this lens, frame systems are simple enough to enable high maintenance and recovery flexibility but lack complexity; panel systems offer greater complexity but are more problematic to maintain and replace due to their typically larger dimensions; and unit systems offer the highest complexity but are also the largest and not as flexible. It seems that a need exists, therefore, for a hybrid modular system, one which would be complex enough to offer a wide range of functionalities while also small enough to be easily handled by a single person for increased recovery potential. Such a system could be a smart brick of sorts, possessing the structural capabilities and discreteness of a frame with the insulation and weather protection capabilities of a panel. The following sections describe the process of designing this system.

2.4 Material Composition

Considering only their internal wall composition, residential timber frame envelopes present ample examples of good material recovery practices: they are based on a complete separation between systems; they consist of simple and widely available products; and they offer a high degree of flexibility in accommodating life cycle modifications such as building additions and reprograming. In terms of material properties, however, there seems to be much room for improvement. First, for the most part, timber frame envelope components consist of relatively soft materials. Even purely structural components such as studs are produced from softwood species. This softness often invites extensive deployment of connection types such as nails and staples, which are traditionally designed for the sole purpose of rapid assembly. Second, except for vapor barriers, most of the materials in a typical timber frame envelope assembly tend to be of a generally low recovery value. The low market value of salvaged softwood, OSB, plywood, or fiberglass insulation eventually disincentivizes deconstruction and material recovery efforts. Finally, the typical lack of order in site-constructed timber frame assemblies, combined with the extensive use of nails and the low recovery market value of most of their components, make demolition the preferred end-of-life solution for the vast majority of single-family homes in the USA. To conclude, in order to improve material recovery performance in this type of envelope assemblies, a reconsideration of materiality is required along with an adoption of general configuration principles (Fig. 1).

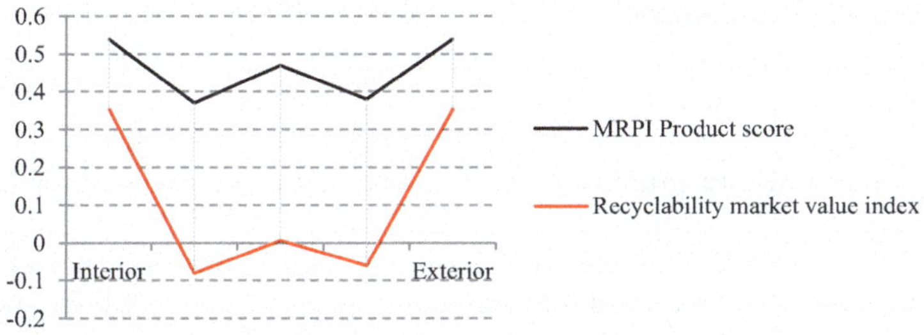

Fig. 1 MRPI product and recyclability market value index scores describing a timber frame envelope cross section

2.5 *Disassembly Depth*

Disassembly depth reflects the effort-to-gain ratio in the recovery of components from a given assembly. From a disassembly depth standpoint, timber frame envelope assemblies are problematic due to two main reasons: First, the recovery market value of most materials in the assembly is similar and low. This fact reduces the chances of a full disassembly from occurring on purely economic grounds. In other product types, where valuable components are placed deep within the assembly (e.g., like in laptops), a full disassembly is typically proven to be more economically viable. Second, in a residential context, timber frame envelope assemblies will usually feature face layers which are more difficult to remove than the inner components of the assembly. This is the case for gypsum plasterboard as an interior face layer, which is often attached to the wall's structure by a combination of adhesive and screws, as well as for vinyl or wood siding as an exterior face layer, which is usually attached to the assembly using nails. To address both of those disassembly depth challenges, a redesigned envelope assembly should (a) contain materials with high recovery value deep within the assembly and (b) rearrange connection types so that the external layers of the assembly are the easiest to remove.

2.6 *Informed Design Process*

Given the recovery-related challenges surveyed in the previous sections, existing timber frame envelope assembly types require improvement on three fronts: material composition, connection type, and level of integration (modularity). The MRPI method can be used in this case to facilitate an iterative design process, where different strategies are evaluated and compared to each other to select a highly recoverable solution. Based on the assumption that the selection of face materials largely depends on consumer taste rather than on performance, this study examines a redesign of the assembly's inner components only.

2.6.1 Material Composition

Apart from accommodating mechanical, electrical, and plumbing systems, the inner envelope assembly fulfils four main function groups – structure, thermal insulation, sheathing, and moisture protection. In the case of timber frame assemblies, all groups except moisture protection (vapor barrier and Tyvek) are typically made of materials with relatively low recovery potential. Those materials, namely, wood studs for structure, fiberglass batt for insulation, and plywood or OSB for sheathing, can easily be replaced by a number of commonly used materials with similar performance properties and higher recovery potential.

Figure 2 shows a comparison of MRPI scores and cost between standard and alternative materials for structural, insulation, and sheathing applications. In the structure category, extruded aluminum is found to have the highest recovery potential score but also the highest price per lb. In the insulation category, cotton batt achieves the highest MRPI score and also costs slightly less than the standard fiberglass batt. Lastly, in the sheathing category, soft HDPE sheet is found to have the highest recovery potential but also a substantially higher cost than the standard plywood or OSB sheet.

2.6.2 Connection Type

As established earlier in this chapter, in order to successfully facilitate greater disassembly depth, connections with a high degree of reversibility should be placed closer to the extremities of the envelope's cross section. Most existing timber frame envelopes assemblies but exhibits a reversed situation: the connections which are closer to the envelope's surface are often harder to disassemble than those which are closer to the envelope's core. Figure 3 compares the typical connection-type score distribution across a timber envelope assembly with an ideal distribution for material recovery. Note that even at its lowest point, the ideal distribution aims at a connection score of no lower than 0.85: that of a bolt and nut connection. While other

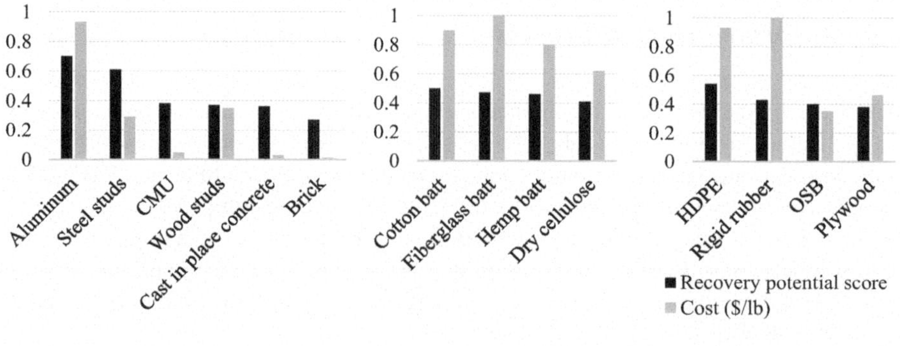

Fig. 2 Recovery potential and cost of material options for structure (left), thermal insulation (center), and sheathing (right)

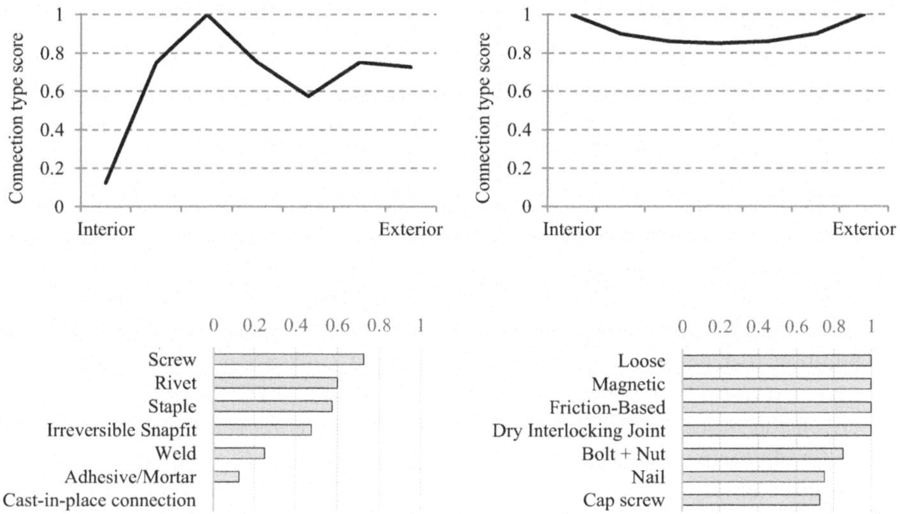

Fig. 3 Top left: Connection type scores describing a timber frame envelope cross section and a disassembly-oriented configuration. Below: Connection type scores by common types

lower scoring connection types, such as screws, may allow full disassembly, they involve a degree of damage which may prevent full reuse of the assembly's components.

2.6.3 Level of Integration

Simply put, the level of integration in an assembly largely depends on the number of component groups preassembled off-site. The rationale behind tying component integration with material recovery is that off-site preassembly ensures that much of the disassembly can be performed off-site as well, thereby reducing the duration of on-site operations and increasing the project's diversion rates. In most timber frame envelope assemblies, the components are typically assembled on site, meaning that their integration score is usually 0.00. In this context, a redesign should be based on two main assumptions: (a) the redesign keeps the common separation between envelope systems (structure, insulation, sheathing, and moisture protection) and (b) the face layers (plasterboard and siding) will be added on site for the reasons discussed earlier in this chapter. Under these assumptions, the highest possible integration score is 0.71 (5/7). Considering the previous discussion regarding modularity, along with integration and complexity, the size of each assembly module plays a major role in facilitating easy and quick full disassembly. The combination of these considerations into a design scheme is discussed in the next section.

2.7 Configuration

The analytical process described in previous sections suggests the following design operations for maximizing material recovery potential in frame-based residential envelope assemblies:

A. From a material standpoint, for the structure, softwood should be replaced by extruded aluminum or steel. Given that the target building types are mostly low rise, aluminum may make a better alternative as it is lighter and less susceptible to atmospheric corrosion even when untreated (Hung 2020). The analysis also shows that for insulation, fiberglass batt should be replaced by cotton, which in addition to having a higher recovery potential is also less expensive. Finally, plywood or OSB sheathing should be replaced by soft HDPE or rubber boards. Given the more developed infrastructure for HDPE recycling, it may make a better alternative than rubber.
B. From a connection-type standpoint, the typical nails and staples for timber products should be replaced by friction-based or bolt and nut connections. Ideally, friction-based or magnetic connections should be placed closer to the envelope face layers; however, due to the assumed use of plasterboard and common siding types, the use of bolt and nut connections may be required closer to the assembly surface as well.
C. From a component integration standpoint, the redesigned assembly should strive to incorporate all systems (except face layers) into a single module. Such module should be light and small enough to be handled by one individual but also large enough to ensure quick assembly and disassembly of the entire structure. Research suggests that 30 kg (66 lb) can be considered as an upper limit load for a recurring lifting activity by an average construction worker (Hartmann and Fleischer 2005). The width and height of the module should not exceed 149 cm (4.88 ft), which is the average arm span of an adult female (Steele and Chenier 1990).

Considering these findings, the resulting assembly design, shown in Figs. 4, 5 and 6, is an integrated preassembled unit containing the following: an aluminum square tube structural frame (with a HDPE thermal break), cotton insulation batt, friction-fit HDPE sheathing board, and stainless steel connector plates. Once aggregated, the units are covered by a whether barrier (Tyvek) and siding on the exterior, as well as plasterboard on the interior. The unit size described here is 24″ × 24″ × 7.5″ (60 cm × 60 cm × 19 cm), because at this size the weight of each unit approaches 60 lb.; however, these dimensions can vary to correspond to project-specific considerations. In order to address typical end conditions, Fig. 4 also describes a corner module where the structure and insulation accommodate a 90 degree turn. Accommodating other end conditions such as fenestration, roof assembly, and foundations is not addressed here, as solving their design goes beyond the scope of this chapter. These issues would be expected to be addressed within the context of a design project.

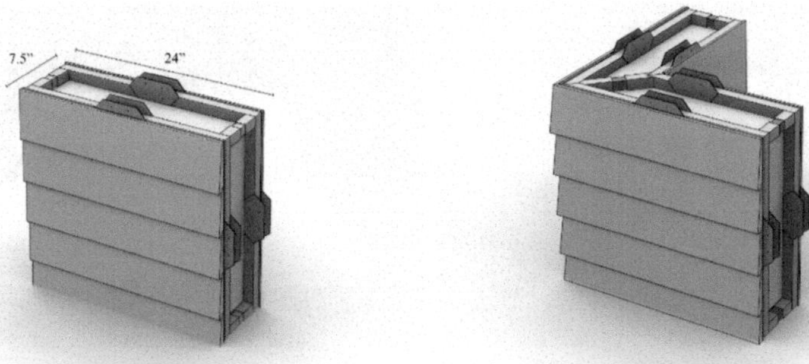

Fig. 4 Integrated wall module (left) and corner module design. Although the drawing includes exterior and interior finish layers (siding and plasterboard), they are intended to be applied to the entire envelope surface post-assembly

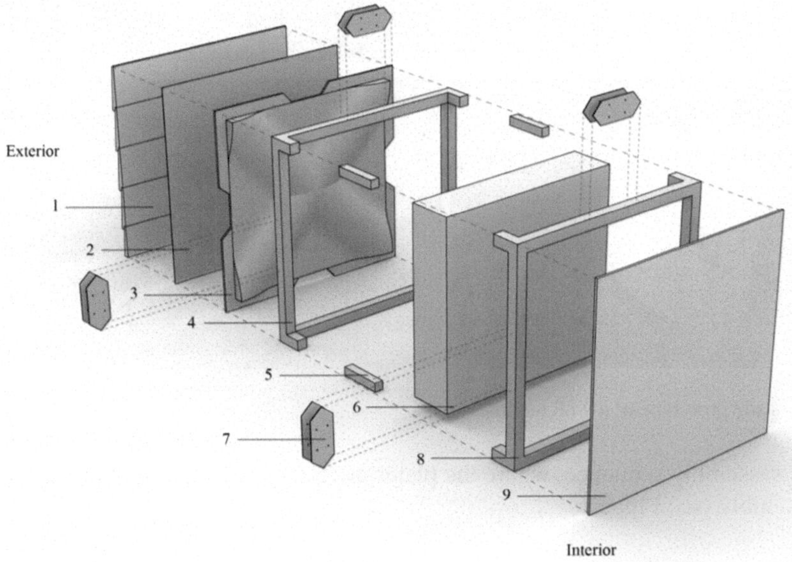

Fig. 5 Exploded view of a wall module: (1) exterior siding layer; (2) moisture barrier; (3) soft HDPE sheathing; (4, 8) aluminum structural frame; (5) HDPE thermal break; (6) cotton insulation; (7) stainless steel connection plate; (9) plasterboard

Regarding its connection strategy, all the components in the assembly are joined by friction-fit connections (insulation, sheathing, and thermal break) or by bolts and nuts. The stainless steel connection plates connect adjacent modules at frame midpoints in order to avoid weakening the frame's corners. The HDPE sheathing layer recedes at these midpoints and grants easy access to the plates in order to facilitate

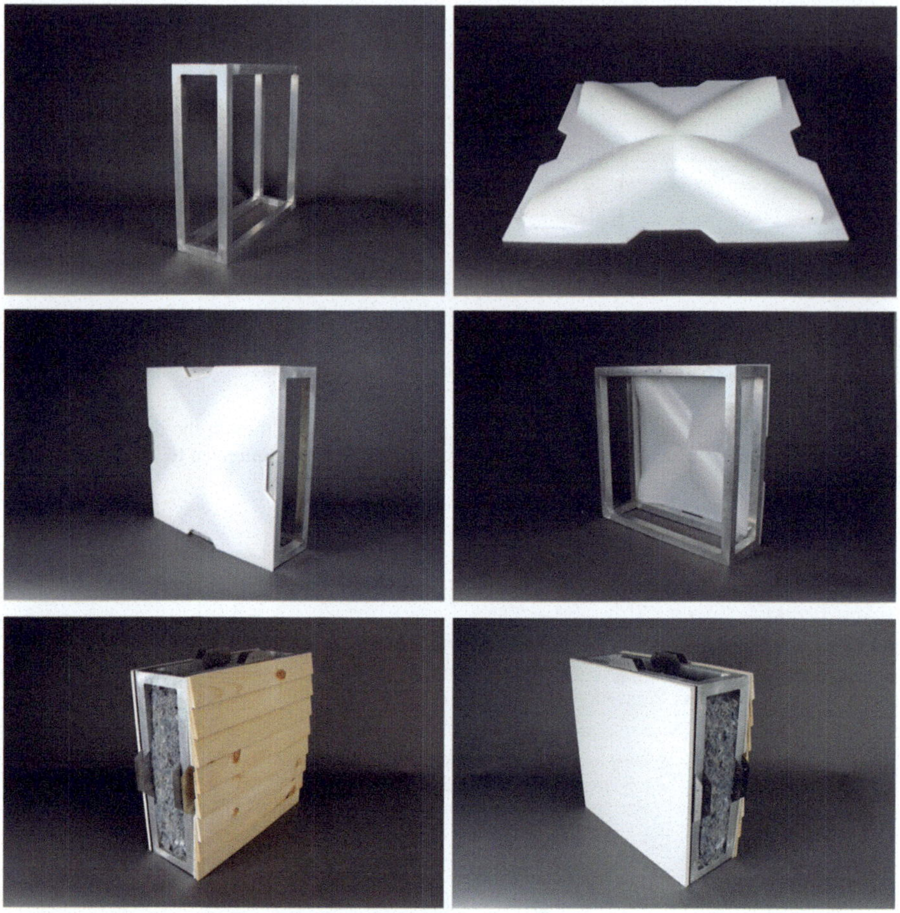

Fig. 6 Full-scale prototype of a wall module

an ideal disassembly sequence, where the plates are removed first to maximize off-site disassembly (see Fig. 7).

3 Results and Discussion

As discussed earlier in this chapter, the proposed design offers a hybrid approach which aims to combine the flexibility of frame-based assemblies with the high system integration potential in modular or unit-based assemblies. Assessing the effectiveness of the proposed design in improving material recovery potential requires a comparative study with one of the existing envelope assemblies analyzed in Mayer and Bechthold (2017). Given the hybrid nature of the proposed design, candidate

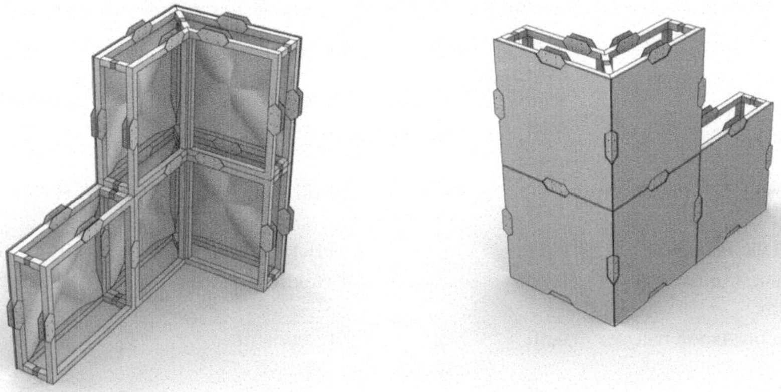

Fig. 7 Axonometric view of stacked wall modules from the interior (left) and the exterior. These drawings depict only frame, connection, and sheathing components

systems for a comparison could be either frame-based or unit-based. However, since the structural approach of the proposed design as well as its separation of systems is based on frame assemblies, it would make most sense to compare it to a frame-based envelope assembly. An ideal assembly for this purpose would be a timber frame envelope, primarily due to its dominant presence in the American residential market. Its prevalence can be translated into significant positive environmental impact with the introduction of a recovery-oriented alternative. Table 1 shows a MRPI score breakdown for the proposed design. Final scores are displayed once for the assembly including its face layers (siding and plasterboard) and once excluding those layers.

The face layers are excluded in one of the score breakdowns as the assumption is that the selection of siding and interior finishes is up to individual consumer taste rather than only performance considerations. Comparing these results to the MRPI score breakdown of a timber frame assembly with wood siding offers several observations (Table 2).

A. The proposed design achieves material recovery potential that is 35–47% higher than that of the timber frame assembly. This difference can be primarily attributed to the relatively irreversible connection types used in the timber frame assembly, as well as to the low level of integration and off-site assembly that is characteristic of its construction process.
B. Surprisingly, the product score of the timber assembly and the proposed design are similar. This is due in part to the use of the same siding and drywall. The 0.10 product score difference between the two score breakdowns for the proposed design supports this hypothesis.
C. As shown in Fig. 8, the proposed design offers more balance between the various assembly categories, mainly due to its higher integration level and higher connection reversibility.

Table 1 MRPI score breakdown for the proposed assembly design including typical face layers (left) and excluding them

Product	MRPI score	Notes	Connection type	Score
Gypsum plaster drywall sheet	0.10	Paperbacked	Bolt + nut	0.85
Stainless steel connection plate	0.57		Bolt + nut	0.85
HDPE thermal break	0.54		Friction fit	1.00
Aluminum square tube structure	0.70		Weld	0.25
Cotton insulation batt	0.50	Not paperbacked	Friction fit	1.00
Sheathing HDPE sheet	0.54		Friction fit	1.00
HDPE Tyvek sheet	0.54		Magnetic	1.00
Pine siding	0.27	Treated and stained	Magnetic	1.00
Total (average)	0.47			0.86

Including face layers			*Excluding face layers*		
MRPI assembly category	Weight	Score	MRPI assembly category	Weight	Score
Product score	0.3	0.47	Product score	0.3	0.57
Connection type	0.4	0.86	Connection type	0.4	0.82
Access	0.2	0.65	Access	0.2	0.75
Component integration	0.1	0.75	Component integration	0.1	1.00
Total score		**0.69**	Total score		**0.75**

Table 2 MRPI score breakdown of a timber frame assembly with pine siding

MRPI assembly category	Weight	Score	Notes
Product score	0.3	0.43	
Connection type	0.4	0.67	
Access	0.2	0.46	The sheathing has the lowest access score in the assembly
Component integration	0.1	0.14	Drywall is the only component group assembled off-site
Total score		**0.51**	

Beyond the higher material recovery potential it generates, this balance plays an important role in ensuring that material recovery is allowed to occur de facto. In other words, an assembly which excels in only one or two of the categories and performs poorly on the rest may hinder recovery rather than facilitate it. A SIP assembly, for example, would achieve a high integration score but would in fact be much harder to recycle than a standard timber frame assembly.

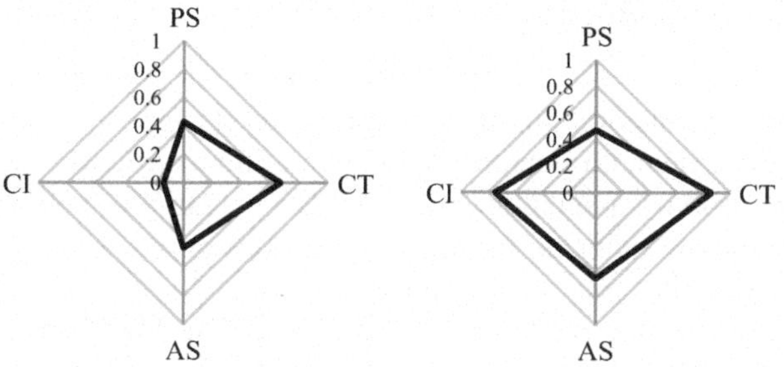

Fig. 8 MRPI assembly category scores for timber frame assembly with pine siding (left) and for the proposed design (PS product score, CT connection type, AS access score, CI component integration)

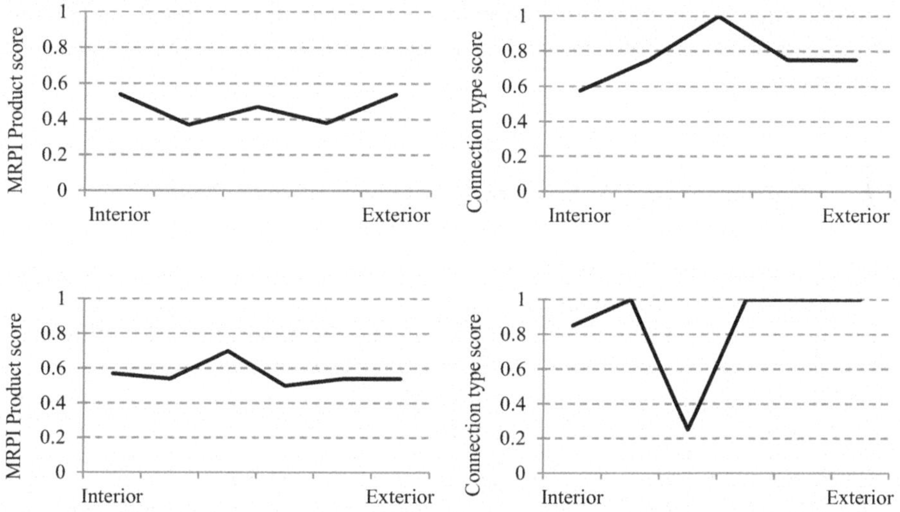

Fig. 9 MRPI product score and connection type score variation throughout the envelope's cross section for timber frame assembly (top) and the proposed design. The graphs exclude face layers

D. Considering product score and connection type score variation through the envelope's cross section, the proposed design (at the bottom of Fig. 9) exhibits more score consistency with the clear exception of the aluminum frame. The frame achieves both a higher product score due to high recyclability levels and a low connection type score due to it being welded together. Considering the fact that the frame's parts being welded together does not prevent or hinder the removal of other components from the assembly nor does it impact the recyclability of the frame, one could choose to understate its impact on the connection type graph. It should be noted that these graphs disregard the effect of siding and plasterboard since both assemblies have been modeled as having identical face layers.

4 Conclusion

In conclusion, although the proposed design may benefit from further rounds of refinement, this chapter demonstrates how the MRPI evaluation method can be used to produce assemblies with a higher potential of being reused or fully recycled. As with any other performance-driven design process, aiming for highly recoverable assemblies may compromise other objectives such as low cost, lightweight structure, seismic resiliency, and so forth. The MRPI method is therefore intended to serve professionals, researchers, and policy makers as one instrument in a larger toolbox. It goes without saying that the weighing of one performance requirement against another should be performed individually for each project and that aiming for full end-of-life recovery may not be equally ideal for all building types or life expectancies.

4.1 Attainment of UN Sustainable Development Goals

This study is positioned within the discipline of life cycle engineering, with a specific focus on end-of-life solutions in the building industry. In reducing demolition waste and the associated release of pollutants into the atmosphere, the study contributes to the attainment of two UN Sustainable Development Goals within goals 11 (*Make cities and human settlements inclusive, safe, resilient and sustainable*) and 12 (*Ensure sustainable consumption and production patterns*):

- 11.6: *By 2030, reduce the adverse* per capita *environmental impact of cities, including by paying special attention to air quality and municipal and other waste management.*
- 12.5: *By 2030, substantially reduce waste generation through prevention, reduction, recycling, and reuse.*

References

Abdullah, A., Anumba, C., & Durmisevic, E. (2003). Decision tools for demolition techniques selection. In Proceedings (pp. 55-72).

Begum, R. A., Satari, S. K., & Pereira, J. J. (2010). Waste generation and recycling: Comparison of conventional and industrialized building systems. American Journal of Environmental Sciences, 6(4), 383.

Das, S. K., Yedlarajiah, P., & Narendra, R. (2000). An approach for estimating the end-of-life product disassembly effort and cost. International Journal of Production Research, 38(3), 657-673.

Germani, M., Mandolini, M., Marconi, M., & Rossi, M. (2014). An approach to analytically evaluate the product disassemblability during the design process. Procedia CIRP, 21, 336-341.

Giudice, F. (2010). Disassembly depth distribution for ease of service: a rule-based approach. Journal of Engineering Design, 21(4), 375-411.

Gungor, A., & Gupta, S. M. (1997). An evaluation methodology for disassembly processes. Computers & Industrial Engineering, 33(1-2), 329-332.
Hartmann, B., & Fleischer, A. G. (2005). Physical load exposure at construction sites. Scandinavian journal of work, environment & health, 88-95.
Hung, F. S. (2020). Design of lightweight aluminum alloy building materials for corrosion and wear resistance. Emerging Materials Research, 9(3), 750-757.
Mayer, M., & Bechthold, M. (2017). Development of policy metrics for circularity assessment in building assemblies. Economics and Policy of Energy and the Environment, 2017(1-2), 57-84.
Schvaneveldt, S. J. (2003). Environmental performance of products: Benchmarks and tools for measuring improvement. Benchmarking: An International Journal, 10(2), 136-151.
Sodhi, R., Sonnenberg, M., & Das, S. (2004). Evaluating the unfastening effort in design for disassembly and serviceability. Journal of Engineering Design, 15(1), 69-90.
Staib, G., Dörrhöfer, A., & Rosenthal, M. (2013). Components and systems: Modular construction–Design, structure, new technologies. Walter de Gruyter.
Steele, M. F., & Chenier, T. C. (1990). Arm-span, height, and age in black and white women. Annals of human biology, 17(6), 533-541.
Thormark, C. (2001). Recycling potential and design for disassembly in buildings. Lund Institute of Technology, Lund University, Lund, Sweden.
Rosen, A. (2019). Assessment of loop potential. In: Hillebrandt, A., Riegler-Floors, P., Rosen, A. & Seggeweis, J. (2019). Manual of recycling: buildings as sources of materials. Walter de Gruyter.
US Census Bureau (MCD): Cheryl Cornish, Stephen Cooper. "US Census Bureau New Residential Construction." Census.gov, 2022. https://www.census.gov/construction/nrc/index.html.

Part III
Sustainability in Manufacturing, Infrastructure, and Engineering Design

Disassembly 4.0: A Pragmatic Approach to Achieve Sustainability in Engineering

Jenny Coenen

1 The Need for Circular Manufacturing

Prolonging a product's life cycle is a very effective way of reducing its material footprint over time and preventing the use of raw materials. This can be achieved by either maintenance, repair, refurbishing, remanufacturing, upgrading, reuse, or a combination of these processes. The benefits of circular manufacturing are threefold as it leads to a lower material footprint and less waste and it can help to overcome supply chain issues. Existing literature on circular manufacturing puts – with good reason – much emphasis on design aspects like design-for-repairability, design-for-disassembly, etc. Insights on the how to organize and innovate the disassembly and remanufacturing processes are not that easily available. But how to deal with products of which the design, history, and actual state are not or only partly known? And how to deal with the reality of lower volumes and high variety of products when "going circular"? Hence, more practical information is needed about how to increase the percentage of component and material recovery and required innovations. *Criticality* of materials is related to the perceived supply risk and its vulnerability to supply disruption (as described by Bakker et al. 2019). Many of these critical materials can be found in e-waste. The actual recycling rate of most critical materials is low.

Solar panels, for example, as well as many other (electronic) products are shredded now instead of reused. Predictions show that in 2050, the world will have a "waste mountain" of 60 million tons of solar panels (Eijsbouts and Jehee 2021). Estimates show that manufacturing industry contributes around 15% of the 58 gigaton of the worldwide footprint (Circle Economy 2021). Remanufacturing of

---

J. Coenen (✉)
The Hague University of Applied Sciences, EN Den Haag, Netherlands
e-mail: j.m.g.coenen@hhs.nl

I. S. Dunmade et al. (eds.), *Sustainable Engineering*, Green Energy and Technology, https://doi.org/10.1007/978-3-031-47215-2_12

vehicles, electronics, machines, and equipment will have the highest impact on reduction of this carbon footprint.

2 Circular Manufacturing

Circular manufacturing strategies are ranked, based on the value that can be retained in a product. As such, reuse is preferred over recycling. Different variants of reuse exist: direct *reuse*, *repair*, *refurbishment,* and *remanufacturing*, together sometimes abbreviated as *ReX*. They result in a different status of the product. For remanufacturing this status is "as new or better"; for repair and refurbish, the product functionality is restored to a certain level (but less than new). Figure 1 shows also other

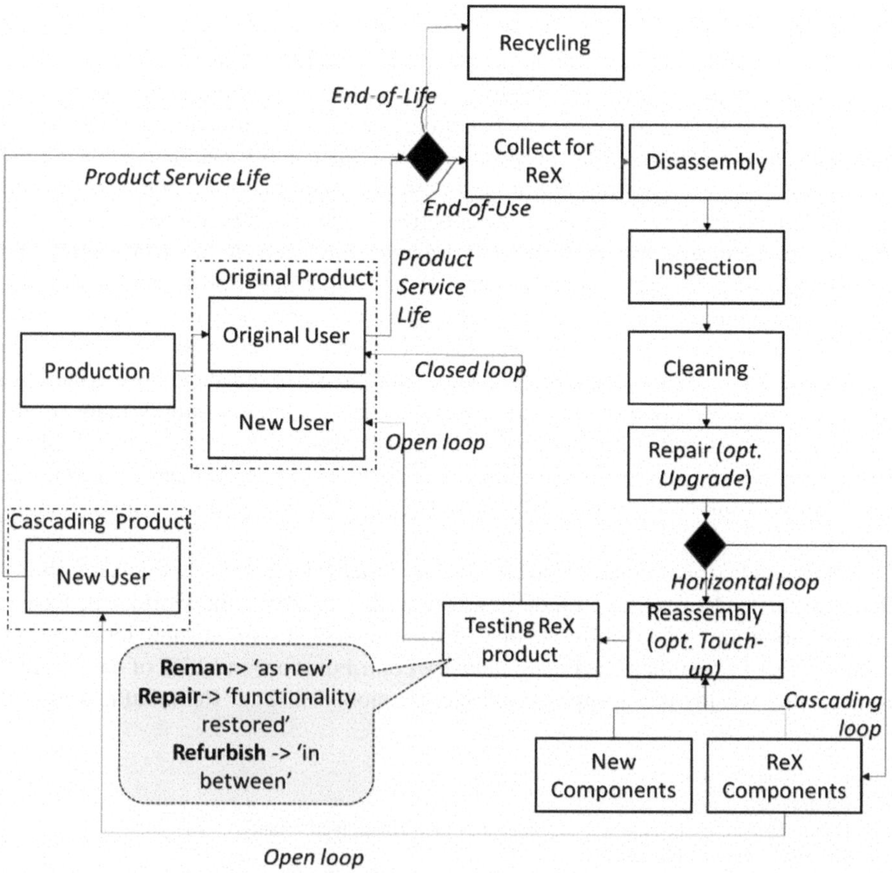

Fig. 1 ReX process including cascading. (Extension of Nasr et al. 2019)

differences like aesthetic touch-up, which is part of refurbishment and remanufacturing, but usually not of repair.

Figure 1 also illustrates other relevant definitions related to circular manufacturing (Nasr et al. 2019): the *product service life* starts at the moment a product is released for use after manufacture and ends at *end-of-life* (*EOL*). It spans product use, any repair, refurbishment or remanufacturing, as well as "hibernation" periods between when the product is not used. EOL is the point in the product's service life at which the product is beyond recovery and can no longer function as required and should be disposed of or – preferably – recycled. *End-of-use* (*EOU*) is the point in the product's service life at which the product doesn't meet the needs of the original owner anymore, due to defects (material/physical/technical obsolescence), lack of interoperability or incompatibility of software (functional obsolescence), the desire for a new version (psychological/desirability obsolescence), or because the product has become illegitimate (for instance, because the certification ends). The so-called *closed-loop* ReX occurs when the recovered product goes back to its original user, while *open loop* allows for a new user. With horizontal loop is meant the recovery of the original product. This previously sold product or module, intended for the remanufacturing process, is called a *core*. The *cascading loop* (of constituent products) contains the recovery of disassembled parts from a product. From this follows that no matter what circular strategy is chosen, the process always starts with disassembly.

Reuse by remanufacturing is the preferred circular manufacturing strategy (for instance, argumented by Stahel (2010)). Whether is this *possible* depends on several factors such as the maximum number of times a component could be effectively reused and the maximum potential service life of the product, after which no extension would be possible. Whether it is *appropriate* depends on the product's use-phase energy requirement: if this has been significantly reduced by redesign of the product, ReX is not a good alternative (Allwood et al. 2011). If salvage at EOU is *probable* depends on organization of return flow and if it is *profitable* depends on the cost of ReX versus the price of a ReX product. So to determine an optimal ReX strategy, these aspects should be weighted: environmental (or sustainable) economic and technical feasibility.

3 Feasibility Aspects of Circular Manufacturing

In Fig. 2 different dimensions and the relevant metrics for remanufacturing feasibility are illustrated. *Environmental impact* can be measured by avoided material use (kg); avoided energy use (MJ); avoided material emissions (kg CO2 eq.); avoided process energy (MJ); and avoided process emissions (kg CO2 eq.). Instruments like *life cycle analysis* (*LCA*), eco-design, and eco-efficiency can be used to calculate these metrics Instead of executing a full LCA (which is time-consuming) for ReX, the assessment can be simplified. This is done by checking if an extension in product service life justifies the additional required footprint.

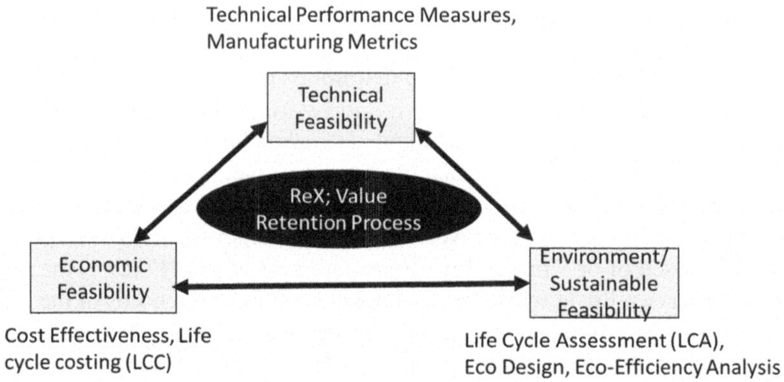

Fig. 2 Feasibility metrics. (Combining figures by Iacovidou et al. (2017), Zhang et al. (2021))

This can be done as follows: Start with the original product that has a footprint (e) that is divided by its "first" product service life (y). The ReX product has a total footprint that is the sum of the footprint of all the phases in its product service life ($p \times e + e$). p is used here to indicate the factor "footprint of the ReX process/original footprint." The ReX product has a total product service life that is the sum of its first life span (y) plus the subsequent product service life phases (y_{rm}). Dividing total footprint by total product service life should be smaller than footprint new/first life cycle (Eq. 1). When simplifying this to Eq. 2, it is clear that gained product service life (y_{rm}) should be larger than footprint factor (p) multiplied by first service life (y).

$$\frac{p \times e + e}{y_{rm} + y} < \frac{e}{y} \quad \text{(Eq. 1)}$$

$$y_{rm} > p \times y \quad \text{(Eq. 2)}$$

Xiao et al. (2018) describe this method in the context as a case study in which life cycle analysis is used to compare the environmental impacts (and cost) of a manufactured loading machine with its remanufactured counterparts. The study showed significant sustainability impact, whereas remanufacturing required only ~25% of the original footprint, while the life span of the product was doubled. This is just an example of course, but it is easy to imagine situations in which the savings can be even higher, for instance, for products with lots of waste material, like a complex machined piece of 2 kg that is manufactured from a 200 kg block of metal. If ReX leads to a prolongation of 80% of the first service life, at the cost of extra footprint (p) of 5% of the original footprint (this is a safe assumption because the huge material loss of complex machining doesn't take place), then it is obvious that ReX is sustainably feasible.

Economic feasibility depends on the cost advantage of reused products compared to new products. This can be expressed in cost-effectiveness or by means of life cycle costing. Aspects to consider are cost of materials, both the new material inputs to the process and removal (costs for disposal or recycling), and cost of return flow (Is a trade-in compensation for cores required?) Is it possible to organize return flows at an acceptable cost level? Uncertainty about the state of the components adds additional risk and could be overcome by collecting usage data, for instance, by condition monitoring. This also helps to select optimal ReX intervals: if you wait too long, this could lead to more damage and too short causes more disassembly costs. Of course, there is also the cost of the ReX process itself: the accessibility and cost of spares; accessibility to work instructions; ease-of-disassembly of broken parts; ease of troubleshooting and functionality testing; the complexity of re-assembly and recommissioning; the degree of automation; the ease and cost of cleaning, etc. These costs are mainly related to the technical feasibility and will be further discussed in the next section. Cost benefits are determined by the perceived value of the product. In general, clients expect a lower price and some experts give a rule-of-thumb that 60–90% of the initial sales price is considered a competitive price for a product after its first end-of-use (Boorsma et al. 2018). A short delivery time may give a good competitive advantage over a new product though. If the remanufacturing is not only done to reuse a product as a whole but also for "harvesting" parts, the economical evaluation becomes more complex. So a remanufacturing business should consider the most appropriate disassembly level and the best recovery option (reuse, remanufacturing, recycling, or disposal) for the components (Ramírez et al. 2020).

4 Technical Feasibility of Remanufacturing

In order to assess the technical feasibility of remanufacturing, refurbish, and repair, a closer look is needed at the relevant processes shown in Fig. 1: disassembly, inspection, cleaning, repair, reassembly, and testing. Figure 3 shows these process

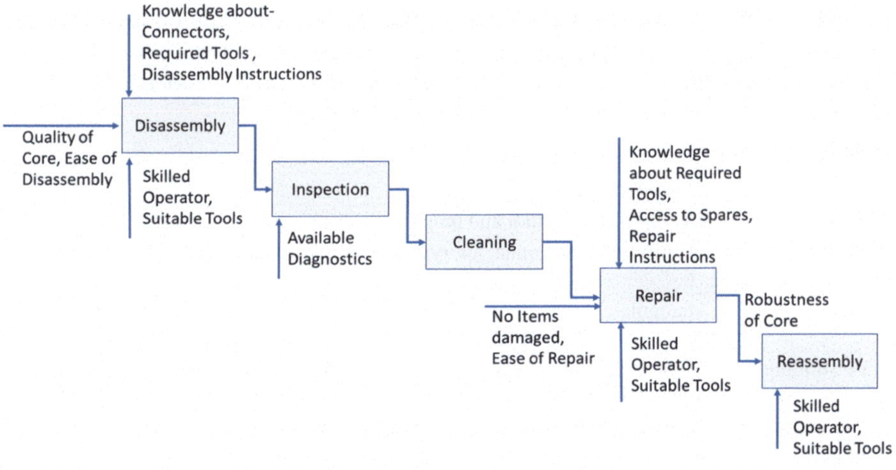

Fig. 3 ReX processes and factors

steps with relevant factors that impact the technical feasibility. These will be elaborated in the following section.

4.1 Disassembly Process

Table 1 zooms further into the basic disassembly tasks (for instance, described in Vanegas et al. 2018). If we elaborate this for an example product like a phone (see Fig. 4), the disassembly process would look like explained in Fig. 5. The main steps are given in the top row and consist of "removing the closing lid" and "removing the battery." Then the object must be turned such that top side comes in reach and subsequent steps "remove front," "remove screen," and "remove screws" can be completed. Then elaborating, for instance, for step "remove battery" (in this example executed by means of a tool with suction cups), this results in the constituent disassembly tasks "tool change," "position," "lift," and "sort." Sorting (see Fig. 6) is an important step: the removed battery should either be classified as "to recycle," or "needs recondition," or "good for reuse," or "investigate by battery diagnostics." For the case of a phone, the battery health can usually be checked by the initial diagnostics, still in some cases further investigation might be required after disassembly.

The *quality and degradation state of the incoming core* can vary highly depending on the cause of component EOU which could either be caused by failure or planned replacement. The different parts within a product can have a different stage of degradation (different load, different wear and tear), also some components can be reused for multiple service lives (for instance, metal frames), while others are limited to a single service life (like for electronic components and software). Separate instances of similar parts/products may have different degradation because of a different service life. In addition, the quality of earlier maintenance and repair plays a role here. Such factors impact disassembly tasks necessity and duration.

The *ease of disassembly* can be expressed directly in *disassembly time* (Vanegas et al. 2018). To estimate disassembly time, the eDIM metric can be used (Peeters 2018): the number of components and their disassembly sequence should be provided, and then the basic disassembly task durations can be looked up from a

Table 1 Basic disassembly tasks

Basic task	Description
1. Tool change	Pick up/put away tool and prepare for different use
2. Identify connectors	Identify location connector, type connector, required tool
3. Manipulation	Manipulate the product to access or identify a connector in order to disconnect it, e.g., flipping
4. Positioning	Position the tool relative to the fastener prior to the actual disconnection process, for instance, placing a screwdriver on top of a screw
5. Disconnection	Actual disconnection, for instance, unscrewing
6. Removal	Removing the unfastened components and putting them into bins

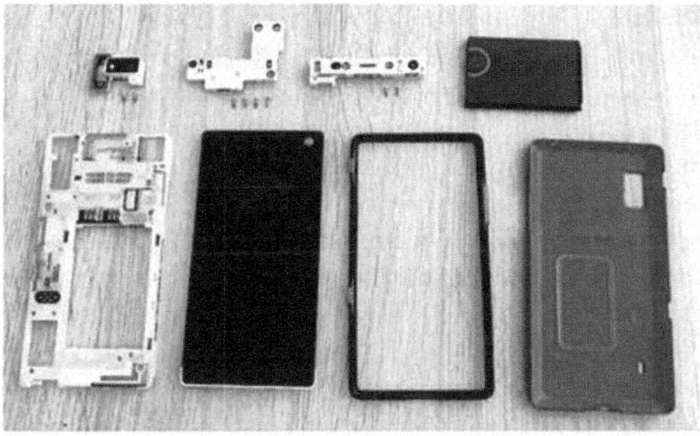

Fig. 4 Disassembled Phone

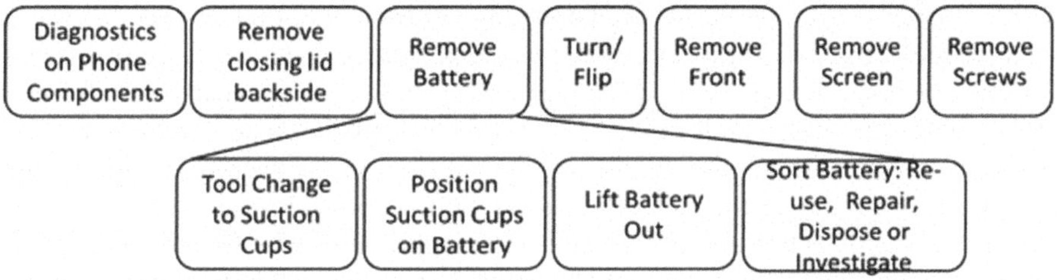

Fig. 5 Disassembly process

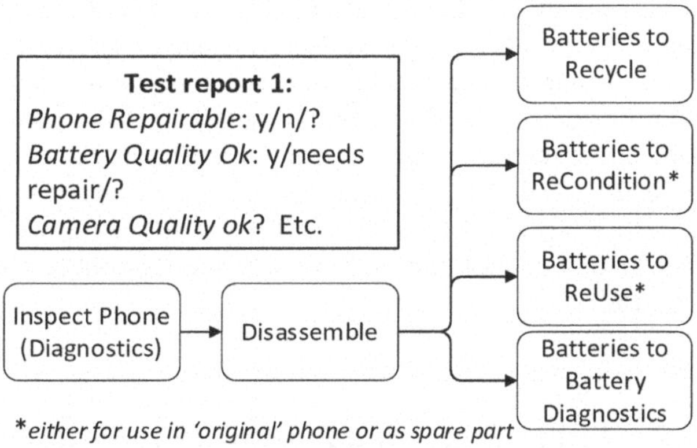

Fig. 6 Sort options disassembled parts

database and with this the duration. The six basic disassembly tasks (Table 1) are based on a work measurement system called MOST (the use of MOST in disassembly is, for instance, explained by Dangal et al. 2022). The database is filled with indications of the time required for disassembly based on MOST, under the assumed performance of an average skilled operator, under standard conditions at a normal pace. During disassembly, parts can be recovered in modules or in separate parts. Recovery in modules can increase efficiency, for instance, because certain removal actions are combined or because the subsequent recovery action is aimed at the module. When determining the *ease of disassembly*, four parameters are considered leading by de Fazio et al. (2021): disassembly sequence/depth, type of tools required, fastener reusability/reversibility, and disassembly time by analyzing the product architecture and determining the disassembly precedence relations. The joining method (welding, gluing, sewing, bolting, screwing, clicking, etc.) and use of fasteners (nuts and bolts, rivets, screws, snap-fits) play an important role. This should be a compromise between ease of assembly, sufficiently strong to withstand during service life and enable an efficient disassembly (Jeandin and Mascle 2016). However, "reversible" fasteners like bolts and screws can solidify, corrode, or be damaged such that they can only be disassembled destructively. Required tools for disassembly, access to fasteners, required force, and the probability of damaging the product (for instance, breaking a cover when trying to release a snap-fit) are relevant for the ease of disassembly.

This brings us to topic of having knowledge upfront about the connectors used, need for tools, force to apply, etc. such that *proper disassembly and repair instructions can be provided*. This is not trivial as disassembly operators can be confronted with a high variability and versions of products and the state products are in. Products may have undergone (unforeseen) modifications and service requirements may stretch for a longer time so the repair and reman portfolio can be big. In addition, a clear decision tree about when and when not to dissemble (based on diagnostics or entry criteria) and what should be recovered under what conditions is needed to have a smooth ReX process. Therefore, it is important to have access to the product and component IDs (are they known or clearly visible on the incoming products?) and the relevant technical info (like bill of materials, a wiring diagram, etc.) and relevant info about hazardous substances and waste disposal requirements.

Skills can be enhanced with the availability of good work instructions. Repair and reman are only sparsely covered in the curricula of technical and vocational education; therefore, most operators are trained in-house (confirmed in (Boorsma et al. 2018)). Other important resources are tools (sometimes specifically designed for the manufacturing process and also required for the disassembly), the aforementioned (automatic) testing, and whenever possible automation of the disassembly itself.

4.2 Inspection: Availability of Diagnostics

Diagnostics are important to provide supplementary process information and the outcomes of diagnostics determine the further ReX flow. A starting point for diagnostics is the ease of identification. Failure diagnostics can be gathered by testing before disassembly (based on the overall product) and after disassembly for supplementary diagnostics on individual parts. For electrical components these can include power/voltage tests; for mechanical components noise measurements; visual inspection; measuring temperature, pressure, etc.; and reading out "on-board" diagnostics. For this last item, the ReX party should have access to the diagnosis info and error codes of the original manufacturer, which is not always the case. Diagnostics and testing can be labor-intensive; therefore, in-line (automatically during another process step) or automatic testing can largely improve the efficiency of ReX. in addition, nondestructive test techniques (used, for instance, to test the remaining strength of a joint) like ultrasonic flaw detection or eddy current inspection are useful in ReX. Finally, dimension measurements are needed to see if old parts still meet the required tolerances (not too much wear or deflection). All these techniques also play a role in the quality assessment of the finished ReX product.

4.3 Possibilities for Cleaning

Components can also have different levels of "difficulty of cleaning," ranging from cannot be cleaned economically, difficult to clean, or special treatment required, moderate cleaning and easy cleaning (Parsa and Saadat 2021). This cleaning of dirt, grease, sediment, corrosion, etc. can be done by several methods, for instance, chemical cleaning, or mechanical (brushes, sanding, grinding), this without damaging the part and exposing operators to hazardous materials.

4.4 Repair

Access to spare parts can either be established by the OEM (who can or cannot give access to original components), by the remanufacturer who keeps salvaged parts from returned EOL on stock, or from (endorsed) third parties that deliver equivalent or reversed engineered parts. Interchangeability with third parties is an important topic. When standardized parts are used (for instance, for modular designs or product families or subsequent revisions of parts), the ReX party takes less risk by keeping replacement/spare parts in stock, as explained by Dangal et al. (2022).

Robustness of a core is about the expected degradation (by design) during the service life, to what degree it allows for repair and the expected number of service lives. Quality on the other hand is – in this context – considered to reflect the "actual state" of the core. So, a very robust core that had an unusual tough life can still have a low quality when entering the ReX process. Still, robustness is good indicator of the probability of successful ReX, mainly for the repair and reassembly tasks. Total cost of ownership plays a role here; it is preferred to invest in a more expensive robust core that has several service lives instead of a less expensive one with a single or short product service life (Goodall et al. 2014). The above covers also most important aspects for reassembly, with the exception of the additional touch-up of the product. This can be either functional or mainly aesthetic and is often a surface treatment like galvanizing, brushing electroplating, thermal spraying, laser cladding, painting, etc.

5 Innovation in Shopfloor Processes for Circular Manufacturing

All aspects of technical feasibility of disassembly and remanufacturing have potential for innovation. The required innovations will often rely on digital technologies (like tagging and tracking equipment in the field) and developments in Industry 4.0 like using cobots to enhance productivity and efficiency of the available resources.

An overview is given in Table 2. For instance, Riggs et al. (2015) and Bentaha et al. (2022) consider the design and balancing of a disassembly line with the purpose to have a very efficient process. Due to the fact that the EOL state of a product

Table 2 Innovations in disassembly

Task	Innovation required
Process design	
Reduce "waste"[a]	Short change-over times, reorganize disassembly layouts and organization. Optimal disassembly paths and line balancing
Digitization	Integration of digital tools, IoT (both for improving return logistics and the information collection for and during ReX process execution
Process execution	
Tool change	Universal tools; tool change mechanisms (redesign of joints)
Identify connectors	Both from PDM; by vision; disassembly instructions carried on digital product passport
Manipulation	Optimize position/order; generate this beforehand based on 3D CAD model, automation (robotics/mechatronics)
Positioning	Vision, sensing, other smart aids; "teach" by putting lots of sensors on hand tools and operators such that this can be fed to robotic tools
Disconnection	Automation (robotics/mechatronics)
Removal	Sorting and recognition methods, inline testing of recovered/unfastened components
Process digitization	
Inspection	Embedded or inline testing
Information collection	Condition monitoring, reversed engineering by means of 3D scanning, digital product passports (with information about materials, maintenance, manufacturing info, etc.); product life cycle management (PLM) systems that can handle open, closed, horizontal, and cascading loops in ReX; applying data science, application of MBD/PMI and extend that to disassembly and repair info, protocols for electronic data interchange)
Decision-making	Operator support
Repair/ reassembly	
Repair	Additive manufacturing both for manufacturing spare parts and for repairing damaged parts

[a]"Waste" in the meaning of lean manufacturing, i.e., unnecessary motion, processing, inventory, etc.

may vary and that multiple recovery options exist, this is a difficult optimization problem. *Internet of Things (IoT)* solutions like tagging and tracking products can play a role both for improving return logistics and the information collection for and during ReX process execution. When product instances carry unique identifiers, this can allow for storing information from *condition monitoring* on *Digital Product Passports* (DPP) with information about materials, maintenance, manufacturing info, etc. It also allows for applying data science to recognize patterns and optimize ReX decisions. Universal multi-head robotic tools can be used to facilitate automatic disassembly and decrease disassembly times. This is explained by J. Li et al. (2018), and H. Li et al. (2022) give some further description of disassembly robots developed by Apple Inc. for automatic disassembly of phones. They argue that by adding vision to the process and a database with relevant features of phone components, such a system could be expanded to develop successful and efficient disassembly plans and programs for other types as well. Other types of innovations aim purely at the disconnection task itself, like in Blümel and Raatz (2022) who propose to apply induced vibrations on fasteners. This reduces the needed loosening torque and minimizes the risk of breaking the fasteners. In addition, in repair tasks alternative technology is being developed, for example, applying additive manufacturing techniques like laser cladding and cold spraying to restore damaged parts. Moreover, replacing missing spares is within reach by means of 3D (metal) printing. In the field of smart information collection, reversed engineering by means of 3D scanning can deliver missing technical documentation of products. The increasing use of *product life cycle management (PLM)* systems helps to systematically store such information. Opportunities lie in increasing PLM functionality such that it can easily handle open, closed, horizontal, and cascading loops in ReX as well. The same holds true for CAD enhancements like *model-based design/product manufacturing information (MBD/PMI)*. MBD allows for adding manufacturing information like tolerances and welding codes directly in the engineering model. Extending that with disassembly and repair info would expand its use to ReX. Finally protocols for *electronic data interchange (EDI)* for purchase orders could be expanded to deal with remanufactured or other ReX components.

6 Reflection on Impact

One can easily argue that other aspects shall be more decisive for increasing the feasibility of circular manufacturing than those mentioned above. References like Singhal et al. (2020) conclude that *design-for-remanufacturing* is the most critical factor, followed by managing trade-in and return flow at end-of-life. Many initiatives and efforts toward systematic methods for design improvement are described in literature like in Bracquené et al. (2021) and Dangal et al. (2022). In general, these methods aim first at identifying the so-called priority parts that are functionally important and most prone to failure. For these parts a *repairability index* (Dalhammer et al. 2022) should be calculated. Then redesigning and comparing the old and new indices can be used to distil eco-design lines and evaluate the effect of using less different fastener types, taking into account fastener accessibility, clustering components that have similar expected EOS durations in subassemblies ("clumping"), etc. Another direction is that of enabling standardization and exchangeability of components. Developments in product configuration tools (for instance, described by Cannas et al. 2022) are interesting, and these could be expanded with offering clients configured products that include ReX parts (and show the difference in price and delivery time while providing the same functionality). It is also very important to manage the value perception of the user by taking away misunderstanding about the quality and safety of the ReX products. Still, it is relevant to further improve the technical feasibility of ReX processes as described in this chapter, also because design innovations won't have immediate effect on the installed base of products currently in use. Regulatory bodies can play an important role in overcoming barriers for ReX. To name a few examples: make sure that *used goods* are not legally classified as *waste*; lift bans and restrictions on the trade-in of cores; facilitate that used products can get renewed safety certification; and provided regulations for accessibility spares and instructions (also for third parties). The new Circular Economy Action Plan launched in 2020 introduces legislation that sets minimum design requirements, a legal framework in order to get access to spare parts and repair information, and a compulsory system to score repairability. France already requires a repairability index for electronics and white goods. For industrial markets similar legislation would be welcome. Moreover, public expenditure by governmental bodies can play an important role, for instance, by considering ReX products as preferred supplies.

7 Conclusion

This chapter gives an introductory overview of relevant factors and required innovations for enhancing the technical and hence economic feasibility of circular manufacturing strategies. The contribution of this overview is that it looks beyond design recommendations into shopfloor implications and how to benefit from increased digital maturity in manufacturing. Plenty of challenges still lie ahead in this field, and this chapter gives some further guidance in how to address them systematically.

References

Allwood, J. M., Ashby, M. F., Gutowski, T. G., & Worrell, E. (2011). Material efficiency: A white paper. *Resources, Conservation and Recycling*, *55*(3), 362–381. https://doi.org/10.1016/j.resconrec.2010.11.002

Bakker, C., den Hollander, M., Peck, D., & Balkenende, R. (2019). *Circular Product Design: Addressing Critical Materials through Design* (pp. 179–192). https://doi.org/10.1142/9789813271050_0009

Bentaha, M.-L., Marangé, P., Voisin, A., & Moalla, N. (2022). End-of-Life product quality management for efficient design of disassembly lines under uncertainty. *International Journal of Production Research*, 1–22. https://doi.org/10.1080/00207543.2022.2028199

Blümel, R., & Raatz, A. (2022). Research on Gentle Loosening of Solidified Bolted Joints for Complex Capital Goods. *Procedia CIRP, 105*, 541–546. https://doi.org/10.1016/j.procir.2022.02.090

Boorsma, N., Peck, D., Fischer, S., Bakker, C., & Balkenende, R. (2018). Capabilities required to tackle barriers to remanufacturing. In B. Kopacek (Ed.), *Proceedings of the Conference Going Green – CARE INNOVATION 2018,* pp. 1–12.

Bracquené, E., Peeters, J., Alfieri, F., Sanfélix, J., Duflou, J., Dewulf, W., & Cordella, M. (2021). Analysis of evaluation systems for product repairability: A case study for washing machines. *Journal of Cleaner Production, 281*, 125122. https://doi.org/10.1016/j.jclepro.2020.125122

Cannas, V. G., Masi, A., Pero, M., & Brunø, T. D. (2022). Implementing configurators to enable mass customization in the Engineer-to-Order industry: a multiple case study research. *Production Planning & Control, 33*(9–10), 974–994. https://doi.org/10.1080/09537287.2020.1837941

Circle Economy (2021) Circularity Gap Report. (2021). *Circularity-gap.world*

Dalhammer, C. , Hartman, C. , Larsson, J. , Jarelin, J., Milios, L. , & Mont, O. (2022). Moving away from the throwaway society. Five policy instruments for extending the life of consumer durables.

Dangal, S., Faludi, J., & Balkenende, R. (2022). Design Aspects in Repairability Scoring Systems: Comparing Their Objectivity and Completeness. *Sustainability, 14*(14), 8634. https://doi.org/10.3390/su14148634

de Fazio, F., Bakker, C., Flipsen, B., & Balkenende, R. (2021). The Disassembly Map: A new method to enhance design for product repairability. *Journal of Cleaner Production, 320*, 128552. https://doi.org/10.1016/j.jclepro.2021.128552

Eijsbouts, R. J. J., & Jehee, W. (2021). *Roadmap circulaire fotovoltaische industrie*. www.wecycle.nl

Goodall, P., Rosamond, E., & Harding, J. (2014). A review of the state of the art in tools and techniques used to evaluate remanufacturing feasibility. *Journal of Cleaner Production, 81*, 1–15. https://doi.org/10.1016/j.jclepro.2014.06.014

Iacovidou, E., Velis, C. A., Purnell, P., Zwirner, O., Brown, A., Hahladakis, J., Millward-Hopkins, J., & Williams, P. T. (2017). Metrics for optimising the multi-dimensional value of resources recovered from waste in a circular economy: A critical review. *Journal of Cleaner Production, 166*, 910–938. https://doi.org/10.1016/j.jclepro.2017.07.100

Jeandin, T., & Mascle, C. (2016). A New Model to Select Fasteners in Design for Disassembly. *Procedia CIRP, 40*, 425–430. https://doi.org/10.1016/j.procir.2016.01.084

Li, H., Sun, H., Fan, F., Liu, H., Li, L., & Yin, F. (2022). Research on an intelligent disassembling method for multi-type mobile phones based on rough set theory. *Procedia CIRP, 105*, 547–552. https://doi.org/10.1016/j.procir.2022.02.091

Li, J., Barwood, M., & Rahimifard, S. (2018). Robotic disassembly for increased recovery of strategically important materials from electrical vehicles. *Robotics and Computer-Integrated Manufacturing, 50*, 203–212. https://doi.org/10.1016/j.rcim.2017.09.013

Nasr, N., Russell, J., Bringezu, S., Hellweg, S., Hilton, B., Kreiss, C., & von Gries, N. (2019). *A Report of the International Resource Panel. United Nations Environment Programme.*

Parsa, S., & Saadat, M. (2021). Human-robot collaboration disassembly planning for end-of-life product disassembly process. *Robotics and Computer-Integrated Manufacturing, 71*, 102170. https://doi.org/10.1016/j.rcim.2021.102170

Peeters J.R., (2018). eDIM: further development of the method to assess the ease of disassembly and reassembly of products — Application to notebook computers.

Ramírez, F. J., Aledo, J. A., Gamez, J. A., & Pham, D. T. (2020). Economic modelling of robotic disassembly in end-of-life product recovery for remanufacturing. *Computers & Industrial Engineering, 142*, 106339. https://doi.org/10.1016/j.cie.2020.106339

Riggs, R. J., Battaïa, O., & Hu, S. J. (2015). Disassembly line balancing under high variety of end of life states using a joint precedence graph approach. *Journal of Manufacturing Systems, 37*, 638–648. https://doi.org/10.1016/j.jmsy.2014.11.002

Singhal, D., Tripathy, S., & Jena, S. K. (2020). Remanufacturing for the circular economy: Study and evaluation of critical factors. *Resources, Conservation and Recycling, 156*, 104681. https://doi.org/10.1016/j.resconrec.2020.104681

Stahel, W. R. (2010). *The Performance Economy*. Palgrave Macmillan UK. https://doi.org/10.1057/9780230274907

Vanegas, P., Peeters, J. R., Cattrysse, D., Tecchio, P., Ardente, F., Mathieux, F., Dewulf, W., & Duflou, J. R. (2018). Ease of disassembly of products to support circular economy strategies. *Resources, Conservation and Recycling, 135*, 323–334. https://doi.org/10.1016/j.resconrec.2017.06.022

Xiao, L., Liu, W., Guo, Q., Gao, L., Zhang, G., & Chen, X. (2018). Comparative life cycle assessment of manufactured and remanufactured loading machines in China. *Resources, Conservation and Recycling, 131*, 225–234. https://doi.org/10.1016/j.resconrec.2017.12.021

Zhang, X., Tang, Y., Zhang, H., Jiang, Z., & Cai, W. (2021). Remanufacturability evaluation of end-of-life products considering technology, economy and environment: A review. *Science of The Total Environment, 764*, 142922. https://doi.org/10.1016/j.scitotenv.2020.142922

Additive Manufacturing: Impact, Prospects, and Challenges in Sustainable Engineering

Cynthia Samuel Abima and Nkosinathi Madushele

1 Introduction

The ultimate objective of sustainability is to lessen the negative environmental effects of human activity while boosting economic and social advantages throughout the product's life cycle (Foster 2020; Mensah 2019; Terlouw et al. 2021). Sustainable engineering practices are becoming increasingly crucial as the global population continues to grow, requiring more resources and energy to meet the demands of modern life (Wu et al. 2022). According to its definition, sustainable engineering is the practice of designing products and operating systems to use resources and energy at a rate that does not jeopardize the natural resources for future needs while reducing environmental impact and improving living standards (Lokhande et al. 2021; Hauschild et al. 2020; Stojčí et al. 2018). This involves understanding and integrating the natural and built environments and considering a product's or system's ethical, social, economic, and ecological impacts. It aims to reduce the environmental impact of engineering practices, including using renewable energy sources, reducing waste, and minimizing pollutants.

Sustainable designs drive toward reducing the amount of materials and processes employed in a production process, thus boosting industrial efficiency and reducing the amount of resources used (Javaid et al. 2021). In doing so, companies can reduce overall costs while still producing quality products or services. This practice can lead to increased profits, more competitive prices, and a reduction of the environmental impact of the production process. Sustainable engineering also seeks to minimize energy consumption and use hazardous materials during manufacturing. The push for sustainable engineering practices will ultimately lead to the

C. S. Abima (✉) · N. Madushele
Department of Mechanical Engineering Science, University of Johannesburg, Johannesburg, South Africa
e-mail: cynthias@uj.ac.za; nmadushelee@uj.ac.za

I. S. Dunmade et al. (eds.), *Sustainable Engineering*, Green Energy and Technology, https://doi.org/10.1007/978-3-031-47215-2_13

development of new technologies, materials, and processes that can help reduce energy consumption and pollution. A major driver for sustainable engineering is the novel additive manufacturing (3D printing) technology.

2 Additive Manufacturing

Additive manufacturing (AM), often known as 3D printing, is a new technique that is swiftly becoming a key industrial manufacturing process. It is a novel, disruptive approach to creating components and finished goods (Araujo et al. 2021). It has the potential to make a long-lasting impact in the industrial world. In additive manufacturing, objects are created in a predetermined pattern by successive layer-layer deposition of materials (Raza and Zhong 2022; Godina et al. 2020), typically in the form of powdered metals, plastics, polymers, or composites. In additive manufacturing, a three-dimensional (3D) model is built using computer-aided design software. The model is broken down into individual layers using a slicer program and then exported into a format that defines the surface mesh or volume in 3D space to be printed. Its wide acceptance into the industrial production framework is due to its series of sustainability features (Stojčí et al. 2018) and high-quality technology (Kadir et al. 2020). This process allows for the creation of complex designs and shapes that would otherwise be challenging or impossible to create with traditional manufacturing techniques (Priyadarshini et al. 2022).

AM has the prospects to greatly reduce manufacturing costs and time while opening up new possibilities for product design, redesigning, remanufacturing, end-of-life recovering, reusing, and recycling (Priaone et al. 2021). These features are the intent of sustainable engineering. Another remarkable benefit of AM is its ability to fabricate high-strength materials, superalloys, and smart materials such as inconel, titanium, and stainless steel in fewer manufacturing steps (Nirish and Rajendra 2020; Attar et al. 2019). To promote the technologies and applications, the American Society of Testing Materials (ASTM 2015) categorized AM processes into seven methods (material extrusion, directed energy deposition, powder bed fusion, vat photopolymerization, binder jetting material jetting, and sheet lamination).

2.1 Material Extrusion

Material extrusion technique deposits a continuous filament of materials such as acrylonitrile butadiene styrene (ABS), polylactic acid (PLA), or composites to build a part. To build a part, the material is fed from a spool and heated and extruded via a nozzle in a predetermined layer-by-layer pattern (Srivastava et al. 2022). This technique is well-suited for prototyping and end-use part as it is capable of creating complex geometries with exact dimensional accuracy and stability, making it appropriate for most industrial applications (Mohamed et al. 2015). It is easy to operate

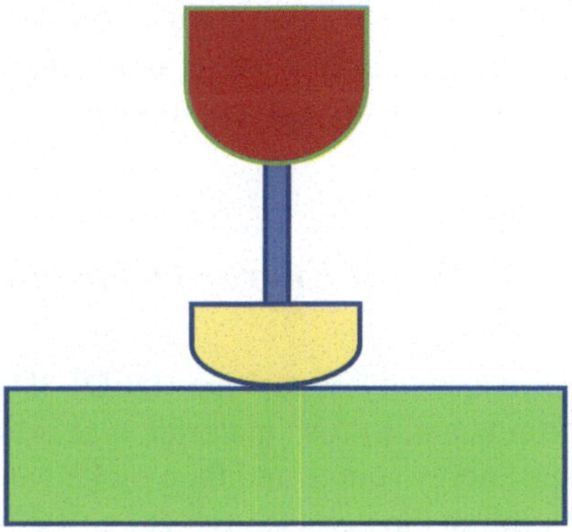

Fig. 1 Material extrusion

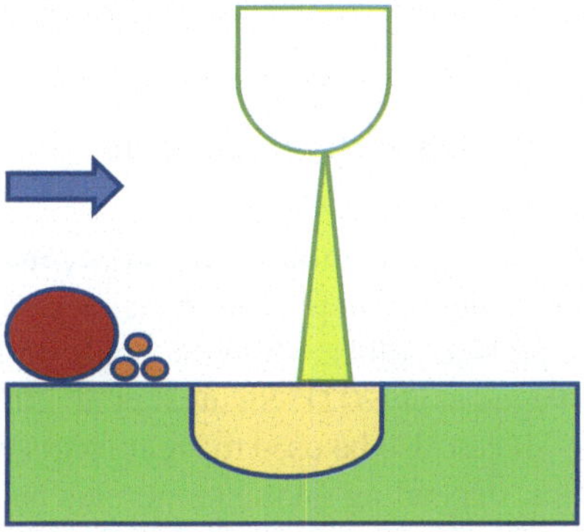

Fig. 2 Powder bed fusion

and has great operating speed. The materials used in this process are readily available with relatively low cost. Figure 1 shows the schematic representation of material extrusion process. This technique is also called fused deposition modeling (FDM).

However, the coding time for complex structures is extensive, and the print quality is determined by the maximum temperature of the nozzles and the capacity of the machine.

2.2 *Powder Bed Fusion*

Part building in the powder bed fusion method begins with applying a thin layer of powder to the bed, after which it is selectively melted and fused using a laser or electron beam (Aboulkhair et al. 2019; Kaushik et al. 2022). This process is also called selective laser melting (SLM), layer-wise laser melting (LLM), or selective laser sintering (when the powder is partially melted). Layer upon layer, the procedure is repeated until the print is completed. When an electronic beam is used to achieve fusion, the process is often referred to as electron beam melting (EBM) (Fig. 2).

Powder bed fusion is well-suited for producing high-precision, complex parts with high accuracy and fine details. It is capable of printing a variety of materials.

The process is often used to produce parts requiring high strength and toughness properties because of its uniform melting temperature, making it ideal for aerospace, medical, and automotive applications.

2.3 *Direct Energy Deposition*

Direct energy deposition (DED) 3D printer uses either a plasma arc, laser, or electron beam, to fuse material as it is added onto a substrate. The material is fed into the energy source and then melted and deposited onto the substrate layer by layer to build up the parts (Singh et al. 2022). Figure 3 shows the schematic representation of the direct energy deposition process.

2.4 *Vat Polymerization*

Vat polymerization is commonly used to produce photopolymer resins. It involves exposing liquid photopolymer resins to ultraviolet (UV) light to convert them from liquids to solids, while the platform lowers the object after each new layer is cured (Pagac et al. 2021; Shaukat et al. 2022) (Fig. 4).

It can also be used to create parts with a variety of mechanical and physical qualities, depending on the monomers and initiators utilized in the process.

2.5 *Material Jetting*

With the help of a print head, viscous or liquid substance droplets are selectively deposited onto a build platform in material jetting procedure (Gülcan et al. 2021). The material solidifies as it is deposited, and the print head moves layer by layer to build up the final part (Fig. 5).

Fig. 3 Direct energy deposition

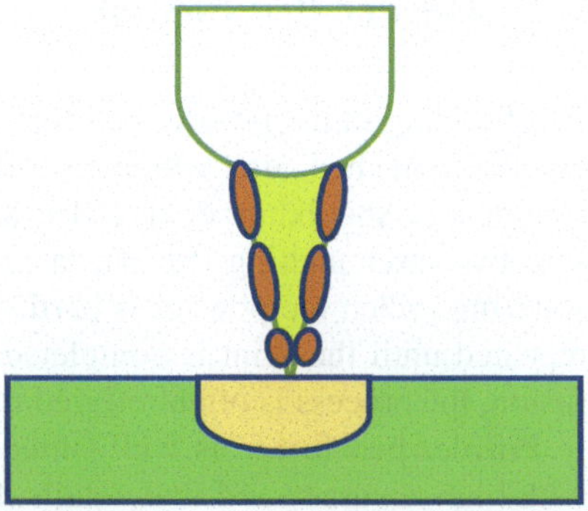

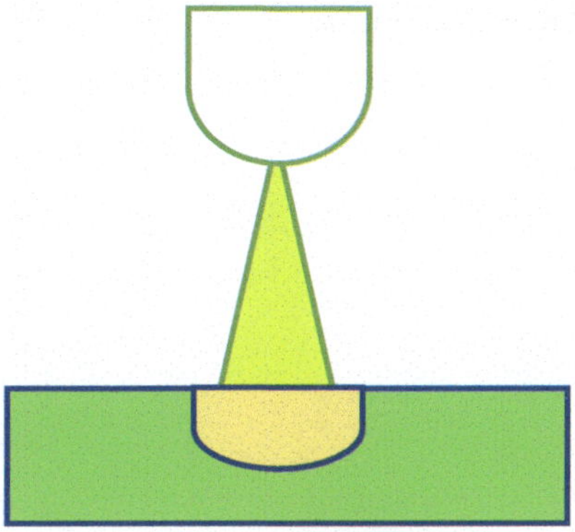

Fig. 4 Vat polymerization

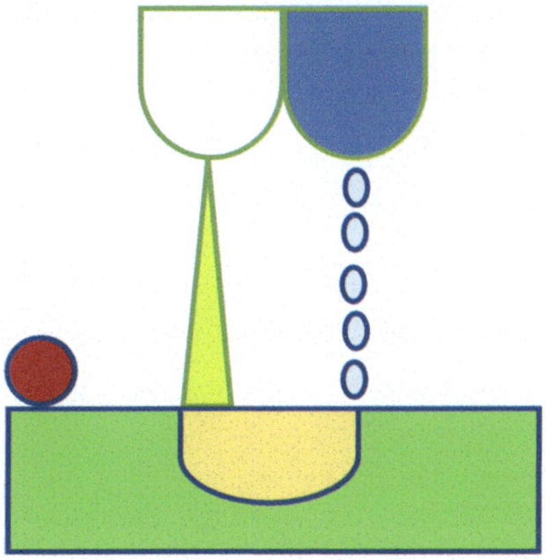

Fig. 5 Material jetting

One of the benefits of material jetting is its versatility, as it can be used to print a wide range of materials, including photopolymers, wax-like materials, and bio-based materials. Material jetting can also produce parts with high resolution and sharp details.

2.6 *Binder Jetting*

In binder jetting technique, a binding agent is selectively deposited onto a layer of powder material to fuse the powder together (Al-Gawhari and Mohammed Ali 2022). The procedure is repeated until the desired part is completed (Fig. 6).

This process can handle multiple materials with diverse colors because of its multiple print head. It also has the advantage of printing without the need for support structures. However, binder jetting may have limitations in terms of the surface finish and quality of the final part. It may also require additional postprocessing steps, such as sintering, to improve the properties of the final print (Mostafaei et al. 2021). Additionally, the process may have limitations in terms of the level of detail that can be achieved and may require a controlled environment.

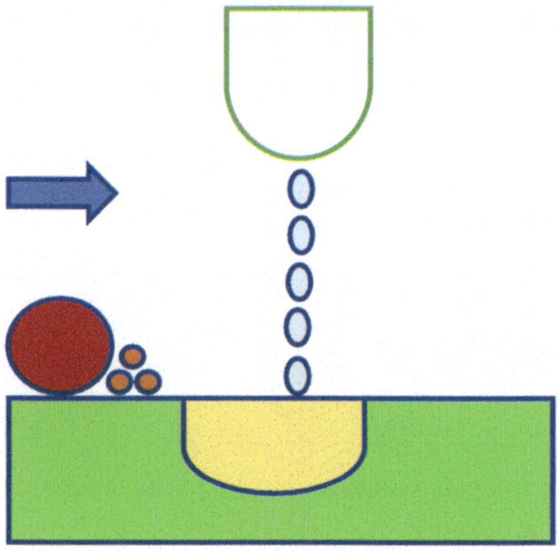

Fig. 6 Binder jetting

Fig. 7 Sheet lamination

2.7 Sheet Lamination

The sheet lamination method involves stacking and bonding layers of flat sheets of material fed to the system by rollers to build up the final part. The process typically uses a flatbed platform and a blade or laser to cut the sheets into the desired shape and then bonded together with an adhesive or bonding agent (Harun et al. 2018) (Fig. 7).

3 Impact of Additive Manufacturing on Sustainable Engineering

The design process of additive manufacturing is seen to promote sustainability in a number of ways. The impact of additive manufacturing as an industrial process will be discussed with respect to the three key areas: economic, environmental, and social facets.

3.1 Economic Impact of AM on Sustainable Engineering

A life cycle analysis, according to Gebler et al. (2014), revealed that AM has the potential to save expenses between $113 and $370 billion by the year 2025 due to a decrease in material usage. The following design features of the AM processes promote economic sustainability.

3.1.1 Product Design Freedom and Optimization

The use of advanced software during the design process of AM further promotes sustainability. Using robust computer-aided design (CAD) software, designers can quickly and accurately create complex designs and shapes tailored to their applications (Buonamici et al. 2020). Not only does this promote efficient production, but it also reduces the amount of time and energy which would otherwise be wasted in the design process. In addition, the use of advanced simulation software allows designers to test and validate their designs before the actual production process begins. More so, rapid prototyping also allows engineers to try various designs and materials to find the most sustainable solution. This reduces the amount of time (Dircksen and Feldmann 2020) and energy wasted in the design process and allows for the optimization of the design for the best possible performance. These attributes help eliminate repair and/or remanufacturing costs (Dircksen and Feldmann 2020). It is important to note that additive manufacturing opens up new possibilities for research and development. By promoting technological innovation and development and supporting sustainable development goals (SDG) 9 and 4 (Industry, Innovation, and Infrastructure) and quality education, respectively, it provides new learning possibilities in engineering, design, robotics, and manufacturing.

3.1.2 Zero Tooling and Highly Automated Process

In contrast to conventional processing techniques, AM procedures enable the fabrication of components without the need of equipment like molds, drills, dies, etc., hence lowering total cost (Khorasani et al. 2022). Productivity rises due to lower total cost of manufacturing, which eventually drives down the price of the finished good and makes it more affordable.

3.1.3 Waste Reduction

AM's layer-by-layer construction method enables the exact and precise fabrication of intricate forms and shapes with reduced material use and waste output (Nirish and Rajendra 2020). Figure 8 shows the process flow for subtractive and

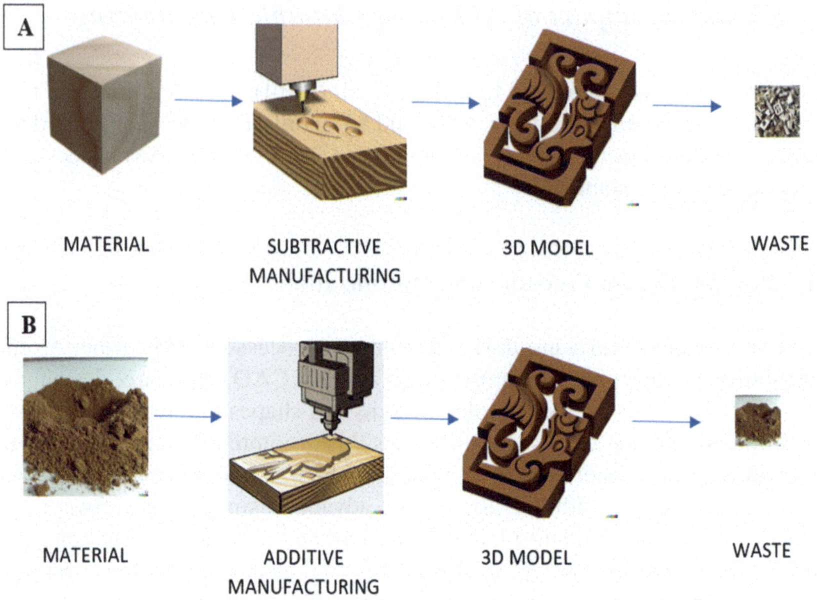

Fig. 8 Process flow: (**a**) subtractive manufacturing process, (**b**) additive manufacturing process

additive manufacturing processes. Compared to traditional manufacturing, minimal material waste is generated with AM process, contributing to SDG 12 (responsible consumption and production by reducing resource use and environmental impact).

3.1.4 Energy Consumption

AM lowers total energy usage by eliminating the energy needed to remove material from the workpiece (Watson and Taminger 2018; Peng et al. 2018).

3.1.5 Shorter and Smarter Value Chain

Additive manufacturing also enables shorter and more intelligent supply chains due to the ability to build parts on demand and with reduced lead times (Omairi and Ismail 2021). Businesses no longer need to keep huge stockpiles of parts. Additionally, additive manufacturing can lower transportation costs since parts can be produced locally rather than shipped to the end user (Godina et al. 2020).

3.1.6 Material Recovery and Recycling

AM makes it possible to reuse scrap materials, such as metals and biodegradables, which further cuts costs and reduces the environmental impact on resource usage and waste (Colorado et al. 2020).

3.1.7 End-of-Product Life Recovery

The amount of energy and resources required to replace components and parts can be decreased by using AM to repair them. Companies may increase the life of their products by using AM repair techniques. Figure 9 illustrates the end-of-product life recovery of AM processes.

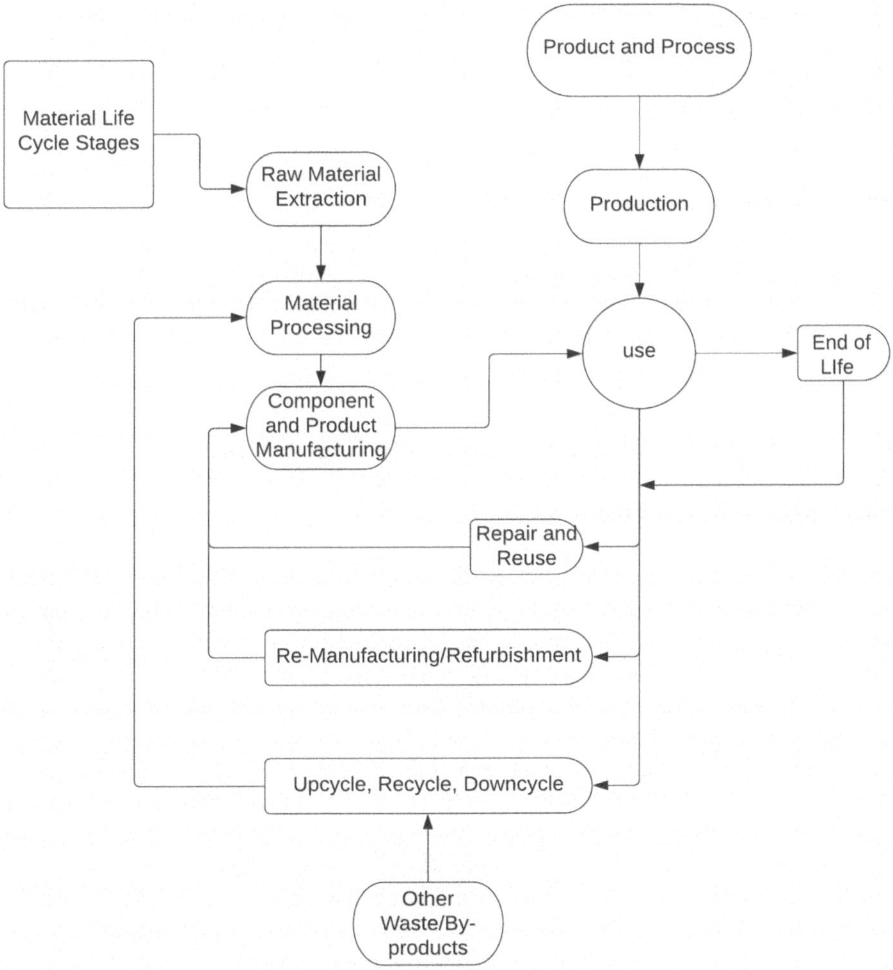

Fig. 9 Material and product life cycle

3.2 Environmental Impact of AM on Sustainable Engineering

3.2.1 Carbon Footprint

AM processes have a smaller environmental impact compared to conventional manufacturing techniques due to the limited need for labor and raw materials (Raza and Zhong 2022). AM methods enable end users to print the part themselves after receiving the 3D model from the manufacturer. This lowers shipping expenses and related greenhouse gas emissions.

3.2.2 Sustainable Energy Production

AM can be used to manufacture intricate parts for renewable energy systems like solar panels and wind turbines. Optimizing the design prior to printing using simulation software is feasible, thus improving design efficiency (Rasaki et al. 2021). This supports SDG 7, which strives to guarantee everyone access to cheap, dependable, sustainable, and modern energy.

3.2.3 Control Atmosphere

The majority of AM operations are enclosed and furnished with dust and fume extractors, preventing the machine operator from breathing in dangerous chemicals, gases, or particles, thus providing safe working conditions.

3.3 Social Impact AM on Sustainable Engineering

3.3.1 Accessible Healthcare

The ability to produce custom-made parts and the simplicity of the supply chain are considered the main social benefits (Ford and Despeisse 2016). The creation of customized parts has a great impact on the customers' well-being. It meets individual needs and promotes loyalty and market acceptance. AM has been used to produce custom dentures, human organs, tissues, scaffolds, blood vessels, prostheses, and orthosis implants such as hand ankle-foot, elbow, wrist, and hip orthopedics (Afshar et al. 2016; Chua et al. 2021; Jha et al. 2022; Li et al. 2020; Alqahtani et al. 2020; Jiao et al. 2022) and prosthetic limbs tailored to individual patients. All these improve the well-being and life span of humans. Still, on the biomedical aspect, the biodegradable resin has made it possible to print tissue-engineered scaffolds with precisely crafted pore architectures that support cell attachment and tissue regeneration (Gibson et al. 2021). This can improve the effectiveness of medical treatments, reduce recovery times, and improve patient outcomes. AM may support SDG 3's goal of ensuring healthy lives and promoting well-being for all people of all ages by improving access to and the quality of healthcare.

4 Prospect of Additive Manufacturing Toward Sustainable Engineering

Gebler et al. (2014) envisage a 5% reduction in CO_2 emission and energy intensities in industrial manufacturing companies through AM by the year 2025. AM industries and application sectors forecast shows that AM will have a market of 230–550 billion USD by 2025 (Ford and Despeisse 2016), meaning AM will have a big share in the manufacturing industry. It is estimated that the adoption of AM can cut down manufacturers' costs and increase savings up to 113–370 billion USD by 2025 (Gebler et al. 2014). According to Gillian Foster (2020), the latest data from the International Energy Agency show that building-related emissions have increased by 45% since 1990. By implementing AM procedures, these environmental effects of buildings, particularly greenhouse gas emissions, may be reduced to mitigate climate change and global warming.

Lakhdar et al. (2021) alluded that AM processes can replace injection molding for low-volume production of ceramic components with complex-shaped. By 2025, AM's commercial aerospace applications are anticipated to generate $430.87 billion in revenue (Blakey-milner et al. 2021). The deposition of conductive materials through the printing of passive circuit components, such as condensers and coils, resistors, diodes, and organic light emitting diodes (OLEDs), can be achieved with AM (Jiménez et al. 2019). Once new materials that exhibit flexibility coupled with strength for very thin thickness are fully developed, there is a chance for commercial-scale production of footwear and clothing (Jiménez et al. 2019). Presently, only a few motorist parts have been successfully produced via AM route. New materials development will pave the way for additive manufacturing of most motor parts and components. Lastly, the combination of building information and additive manufacturing has the potential to produce prototype construction machines (Jiménez et al. 2019). AM will also lead to sustainable agriculture by producing specialized tools such as sensors and drones for irrigation systems. These can help farmers improve their crop yields and contribute to SDG 2 which aims to end hunger, achieve food security and improved nutrition, and promote sustainable agriculture. Furthermore, AM can improve disaster resilience of a country by producing emergency response equipment, such as drones, robots, and shelters, quickly and on site (Belhadi et al. 2022). This supports SDG 11, which calls for inclusive, secure, resilient, and sustainable cities and human settlements. Last but not least, AM may be used to create parts for goods that are intended for disassembly and reuse (Hettiarachchi et al. 2022). By fostering sustainable production and consumption practices and minimizing waste, this can aid in the development of a more circular economy (Colorado et al. 2020; Sauerwein et al. 2019). AM may support the SDGs 12 (responsible consumption and production) and 13 (climate action) by creating a circular economy.

5 Challenges of Additive Manufacturing on Sustainable Engineering

Despite the numerous benefits of AM, there are still a number of obstacles to overcome. The high energy consumption is a significant drawback of this novel technique. The required high temperatures to melt and extrude the materials utilized in the process contributes to the high energy demand, especially for complicated and high-precision components. More so, finishing and postprocessing procedures add to AM processes' energy consumption. The small variety of materials that are readily available is another difficulty for AM. The range of components that may be manufactured is currently limited because most AM methods are only compatible with a small number of plastics, metal alloys, and polymers. Additionally, the materials used in AM are often expensive, making the process cost-prohibitive for some applications. A third challenge of AM is the lack of standardization across different systems. Each AM system currently has its own set of parameters and settings, making it difficult to transfer parts between different machines. This lack of standardization also makes it difficult to compare the performance of different AM systems. Fourth, for superior quality, postprocessing and finishing operations is often required. These processes are time-consuming and expensive, resulting in increased costs. The fifth limitation is the object size that can be printed. The printing size is limited to the printer size. Printers with the larger bed are capital intensive. Lastly, the lack of expertise in AM technology has limits its adoption in most manufacturing industries, especially in Africa and the overall product quality. In order to overcome these challenges and make AM a more sustainable engineering technology, there is a need for further research and development in the area.

6 Recommendation

Specifically, research is needed into more cost-effective and energy-efficient ways to create parts. More so, there needs to be a push for more standardization across different AM systems to make transferring parts between different machines easier. The design process of additive manufacturing can be further optimized by integrating sustainable renewable energy sources, such as solar and wind power, to power the process.

Finally, there needs to be an effort to reduce the cost and lead times associated with postprocessing and finishing operations. AM can become a more sustainable engineering process to allow engineers to create more complex parts with greater accuracy and repeatability while reducing the environmental impact and creating more value.

7 Conclusion

Sustainable engineering helps to create a more sustainable and equitable society for which AM is a major driver. AM incorporates the key sustainable manufacturing strategies in the industrial community known as the 6Rs (reuse, reduce, recycle, redesign, reverse, and repair). Additive manufacturing can revolutionize how we design and manufacture products, leading to a greener environment, eco-friendly products, extended life span, and more economic value. As technology continues to develop and new materials, such as carbon fiber and reinforced polymers, continue to emerge, the possibilities of additive manufacturing are endless.

References

G. Foster, "Circular economy strategies for adaptive reuse of cultural heritage buildings to reduce environmental impacts," *Resour. Conserv. Recycl.*, vol. 152, 2020.

J. Mensah, "Sustainable development: Meaning, history, principles, pillars, and implications for human action: Literature review," *Cognet Soc. Sci.*, vol. 5, no. 1, pp. 1–21, 2019, doi: ensah, Cogent Social Sciences (2019), 5: 1653531. https://doi.org/10.1080/23311886.2019.1653531.

T. Terlouw, C. Bauer, L. Rosa, and M. Mazzotti, "Life cycle assessment of carbon removal technologies: a critical review," *Energy Environ. Sci.*, vol. 14, pp. 1701–1721, 2021, [Online]. Available: https://phys.org/news/2021-11-life-carbon-capture.html

H. Wu, H. Mehrabi, P. Karagiannidis, and N. Naveed, "Additive manufacturing of recycled plastics: Strategies towards a more sustainable future," *J. Clean. Prod.*, vol. 335, pp. 1–17, 2022, doi: https://doi.org/10.1016/j.jclepro.2021.130236.

A. Lokhande, C. Venkateswaran, M. Ramachandran, C. Vidhya, and R. Kurinjimalar, "EST Journal on Emerging trends in Modelling and Manufacturing," *EST J. Emerg. trends Model. Manuf.*, vol. 7, no. 2, pp. 63–69, 2021.

M. Z. Hauschild, S. Kara, and I. Røpke, "CIRP Annals - Manufacturing Technology Absolute sustainability : Challenges to life cycle engineering," *CIRP Ann. - Manuf. Technol.*, vol. 69, pp. 533–553, 2020, doi: https://doi.org/10.1016/j.cirp.2020.05.004.

M. Stojčí, E. K. Zavadskas, D. Pamučar, Ž. Stevi, and A. Mardani, "Application of MCDM Methods in Sustainability Engineering : A Literature Review 2008 – 2018," *Symmetry (Basel).*, vol. 11, no. 350, pp. 1–24, 2018, doi: https://doi.org/10.3390/sym11030350.

M. Javaid, A. Haleem, R. P. Singh, R. Suman, and S. Rab, "Role of additive manufacturing applications towards environmental sustainability," *Adv. Ind. Eng. Polym. Res.*, vol. 4, no. 4, pp. 312–322, 2021, doi: https://doi.org/10.1016/j.aiepr.2021.07.005.

N. Araujo, V. Pacheco, and L. Costa, "Smart Additive Manufacturing The Path to the Digital Value Chain," *Technologies*, vol. 9, no. 88, pp. 1–13, 2021.

M. H. Raza and R. Y. Zhong, "A sustainable roadmap for additive manufacturing using geopolymers in construction industry," *Resour. Conserv. Recycl.*, vol. 186, pp. 1–18, 2022, doi: https://doi.org/10.1016/j.resconrec.2022.106592.

R. Godina, I. Ribeiro, F. Matos, B. T. Ferreira, H. Carvalho, and P. Peças, "Impact assessment of additive manufacturing on sustainable business models in industry 4.0 context," *Sustainability*, vol. 12, no. 17, pp. 1–21, 2020, doi: https://doi.org/10.3390/su12177066.

A.Z.A.Kadir,Y.Yusof,andM.S.Wahab,"AdditiveManufacturingCostestimationmodels-aclassificationreview,"*Int.J.Adv.Manuf.Technol.*,vol.107,pp.4033–4053,2020,[Online].Available:https://www.nano-di.com/blog/2019-additive-manufacturing-cost-drivers-4-key-considerations

J. Priyadarshini, R. Kr Singh, R. Mishra, and M. Mustafa Kamal, "Adoption of additive manufacturing for sustainable operations in the era of circular economy: Self-assessment framework with case illustration," *Comput. Ind. Eng.*, vol. 171, pp. 1–14, 2022, doi: https://doi.org/10.1016/j.cie.2022.108514.

P. C. Priaone, G. Campateli, A. R. Catalano, and F. Baffa, "AM life cycle.pdf," *CIP J. Manuf. Sci. Technol.*, vol. 35, pp. 943–958, 2021.

M. Nirish and R. Rajendra, "Suitability of metal additive manufacturing processes for part topology optimisation – A comparative study," *Mater. Today Proc.*, vol. 27, pp. 1601–1607, 2020, doi: https://doi.org/10.1016/j.matpr.2020.03.275.

H. Attar, M. J. Bermingham, S. Ehtemam-Haghighi, A. Sehghan-manshadi, D. Kent, and M.S.Dargusch, "Evaluation of the mechanical and wear properties of titanium produced by three different additive manufacturing methods for biomedical application," *Mater. Sci. Eng. A*, vol. 760, pp. 339–345, 2019.

M. Srivastava, S. Rathee, V. Patel, A. Kumar, Praveennath, and G. Koppa, "A review of various materials for additive manufacturing: Recent trends and processing issues," *J. Mater. Res. Technol.*, vol. 21, pp. 2612–2641, 2022.

O. A. Mohamed, S. H. Masood, and J. L. Bhowmik, "Optimisation of fused deposition modeling process parameters: a review of current research and future prospects," *Addit. Manuf.*, vol. 3, pp. 42–53, 2015, doi: https://doi.org/10.1007/s40436-014-0097-7.

N. T. Aboulkhair, M. Simonelli, L. Parry, I. Ashcroft, C. Tuck, and R. Hague, "3D printing of Aluminium alloys: Additive Manufacturing of Aluminium alloys using selective laser melting," *Prog. Mater. Sci.*, vol. 106, pp. 1–50, 2019, doi: https://doi.org/10.1016/j.pmatsci.2019.100578.

Kaushik V, N. Kumar B, S. Kumar S, and V. M, "Magnesium role in additive manufacturing of biomedical implants – Challenges and opportunities," *Addit. Manuf.*, vol. 55, no. November 2021, p. 102802, 2022, doi: 10.1016/j.addma.2022.102802.

R. Singh, K. K. Saxena, and P. Singhal, "Role of additive manufacturing in dental applications using ceramics: A review," *Mater. Today Proc.*, vol. 56, pp. 2359–2364, 2022, doi: https://doi.org/10.1016/j.matpr.2021.12.157.

M. Pagac *et al.*, "A review of vat photopolymerisation technology: Materials, applications, challenges, and future trends of 3d printing," *Polymers (Basel).*, vol. 13, pp. 1–20, 2021, doi: https://doi.org/10.3390/polym13040598.

U. Shaukat, E. Rossegger, and S. Schlögl, "A Review of Multi-Material 3D Printing of Functional Materials via Vat Photopolymerization," *Polymers (Basel).*, vol. 14, no. 12, 2022, doi: https://doi.org/10.3390/polym14122449.

O. Gülcan, K. Günaydın, and A. Tamer, "The state of the art of material jetting—a critical review," *Polymers (Basel).*, vol. 13, pp. 1–19, 2021, doi: https://doi.org/10.3390/polym13162829.

F. J. Al-Gawhari and A. A. Mohammed Ali, "Types of 3D Printers Applied in Industrial Pharmacy and Drug Delivery," *Tech. Biochem.*, vol. 3, no. 2, pp. 1–14, 2022, doi: https://doi.org/10.47577/biochemmed.v3i2.6064.

A. Mostafaei *et al.*, "Binder jet 3D printing—Process parameters, materials, properties, modeling, and challenges," *Prog. Mater. Sci.*, vol. 119, pp. 1–138, 2021, doi: https://doi.org/10.1016/j.pmatsci.2020.100707.

W. S. W. Harun *et al.*, "A review of powdered additive manufacturing techniques for Ti-6al-4v biomedical applications," *Powder Technol. 331*, vol. 331, pp. 74–97, 2018.

M. Gebler, A. J. M. Schoot Uiterkamp, and C. Visser, "A global sustainability perspective on 3D printing technologies," *Energy Policy*, vol. 74, pp. 158–167, 2014, doi: https://doi.org/10.1016/j.enpol.2014.08.033.

F. Buonamici *et al.*, "A practical methodology for computer-aided design of custom 3D printable casts for wrist fractures," *Vis. Comput.*, vol. 36, no. 2, pp. 375–390, 2020, doi: https://doi.org/10.1007/s00371-018-01624-z.

M. Dircksen and C. Feldmann, "Holistic evaluation of the impacts of additive manufacturing on sustainability, distribution costs, and time in global supply chains," *Transp. Res. Procedia*, vol. 48, no. 2019, pp. 2140–2165, 2020, doi: https://doi.org/10.1016/j.trpro.2020.08.272.

M. Khorasani, A. H. Ghasemi, B. Rolfe, and I. Gibson, "Additive manufacturing a powerful tool for the aerospace industry," *Rapid Prototyp. J.*, vol. 28, no. 1, pp. 87–100, 2022, doi: https://doi.org/10.1108/RPJ-01-2021-0009.

J. K. Watson and K. M. B. Taminger, "A decision-support model for selecting additive manufacturing versus subtractive manufacturing based on energy consumption," *J. Clean. Prod.*, vol. 176, pp. 1316–1322, 2018, doi: https://doi.org/10.1016/j.jclepro.2015.12.009.

T. Peng, K. Kellens, R. Tang, C. Chen, and G. Chen, "Sustainability of additive manufacturing: An overview on its energy demand and environmental impact," *Addit. Manuf.*, vol. 21, pp. 694–704, 2018, doi: https://doi.org/10.1016/j.addma.2018.04.022.

A. Omairi and Z. H. Ismail, "Towards Machine Learning for Error Compensation in Additive Manufacturing," *Appl. Sci.*, vol. 11, pp. 1–27, 2021.

H. A. Colorado, G. E. I. Velasquez, and S. N. Monteiro, "Sustainability of additive manufacturing: the circular economy of materials and environmental perspectives," *J. Mater. Res. Technol.*, vol. 9, no. 4, pp. 8221–8234, 2020.

S. A. Rasaki, C. Liu, C. Lao, H. Zhang, and Z. Chen, "The innovative contribution of additive manufacturing towards revolutionising fuel cell fabrication for clean energy generation: A comprehensive review," *Renew. Sustain. Energy Rev.*, vol. 148, pp. 1–25, 2021, doi: https://doi.org/10.1016/j.rser.2021.111369.

S. Ford and M. Despeisse, "Additive manufacturing and sustainability: an exploratory study of the advantages and challenges," *J. Clean. Prod.*, vol. 137, pp. 1573–1587, 2016, doi: https://doi.org/10.1016/j.jclepro.2016.04.150.

M. Afshar, A. P. Anaraki, H. Montazerian, and J. Kadkhodapour, "Additive manufacturing and mechanical characterisation of graded porosity scaffolds designed based on triply periodic minimal surface architectures," *J. Mech. Behav. Biomed. Mater.*, vol. 62, pp. 481–494, 2016, doi: https://doi.org/10.1016/j.jmbbm.2016.05.027.

K. Chua, I. Khan, R. Malhotra, and D. Zhu, "Additive manufacturing and 3D printing of metallic biomaterials," *Eng. Regen.*, vol. 2, no. November 2021, pp. 288–299, 2021, doi: 10.1016/j.engreg.2021.11.002.

M. K. Jha, S. Gupta, V. Chaudhary, and P. Gupta, "Material selection for biomedical application in additive manufacturing using TOPSIS approach," *Mater. Today Proc.*, vol. 62, pp. 1452–1457, 2022, doi: https://doi.org/10.1016/j.matpr.2022.01.423.

C. Li, D. Pisignano, Y. Zhao, and J. Xue, "Advances in Medical Applications of Additive Manufacturing," *Engineering*, vol. 6, no. 11, pp. 1222–1231, 2020, doi: https://doi.org/10.1016/j.eng.2020.02.018.

M. S. Alqahtani, A. Al-Tamimi, H. Almeida, G. Cooper, and P. Bartolo, "A review on the use of additive manufacturing to produce lower limb orthoses," *Prog. Addit. Manuf.*, vol. 5, pp. 85–94, 2020, doi: https://doi.org/10.1007/s40964-019-00104-7.

C. Jiao *et al.*, "Additive manufacturing of Bio-inspired ceramic bone Scaffolds: Structural Design, mechanical properties and biocompatibility," *Mater. Des.*, vol. 217, p. 110610, 2022, doi: https://doi.org/10.1016/j.matdes.2022.110610.

I. Gibson, D. Rosen, B. Stucker, and M. Khorasani, *Additive Manufacturing Technologies*, Third. Springer, 2021.

Y. Lakhdar, C. Tuck, J. Binner, A. Terry, and R. Goodridge, "Additive manufacturing of advanced ceramic materials," *Prog. Mater. Sci.*, vol. 116, pp. 1–50, 2021, doi: https://doi.org/10.1016/j.pmatsci.2020.100736.

B. Blakey-milner *et al.*, "Metal additive manufacturing in aerospace : A review," *Mater. Des.*, vol. 209, pp. 1–33, 2021, doi: https://doi.org/10.1016/j.matdes.2021.110008.

M. Jiménez, L. Romero, I. A. Dom, and M. Dom, "Additive Manufacturing Technologies : An Overview about 3D Printing Methods and Future Prospects," *Complexity*, vol. 2019, pp. 1–30, 2019.

A. Belhadi, S. S. Kamble, M. Venkatesh, C. J. Chiappetta Jabbour, and I. Benkhati, "Building supply chain resilience and efficiency through additive manufacturing: An ambidextrous perspective on the dynamic capability view," *Int. J. Prod. Econ.*, vol. 249, pp. 1–20, 2022, doi: https://doi.org/10.1016/j.ijpe.2022.108516.

B. D. Hettiarachchi, M. Brandenburg, and S. Seuring, "Connecting additive manufacturing to circular economy implementation strategies: Links, contingencies and causal loops," *Int. J. Prod. Econ.*, vol. 246, p. 108414, 2022, doi: https://doi.org/10.1016/j.ijpe.2022.108414.

M. Sauerwein, E. Doubrovski, R. Balkenende, and C. Bakker, "Exploring the potential of additive manufacturing for product design in a circular economy," *J. Clean. Prod.*, vol. 226, pp. 1138–1149, 2019, doi: https://doi.org/10.1016/j.jclepro.2019.04.108.

Application of Reconfigurable System Thinking in Mining and Mineral Processing Environment: Toward Sustainable Mineral Beneficiation

Makinde Olasumbo, Khumbulani Mpofu, and Boitumelo Ramatsetse

Notations

C_c	Convertibility associated with configuration
C_m	Convertibility associated with machine
C_H	Convertibility associated with material handling
I	Convertibility associated with configuration
R	Convertibility associated with machine
X	Convertibility associated with material handling
N	Individual machines in the system
$Q1$	Equipped with a screen panel replacement device
$Q2$	Easily reprogrammed with flexible software
$Q3$	The system is designed to allow addition and subtraction of modular components
$Q4$	Flexible fixturing capability
$Q5$	Large capacity magazine
$t1$	The time necessary to place an order for the precise number of RVS machine subsystems
$t2$	The amount of time needed to build exactly how many RVS machine subsystems the clients have ordered

M. Olasumbo (✉)
Department of Quality and Operations Management, University of Johannesburg, Johannesburg, South Africa
e-mail: olasumbom@uj.ac.za

K. Mpofu
Department of Industrial Engineering, Tshwane University of Technology, Pretoria, South Africa
e-mail: mpofuk@tut.ac.za

B. Ramatsetse
Department of Mechanical & Mechatronics Engineering, Stellenbosch University, Stellenbosch, South Africa
e-mail: ramatsetse@sun.ac.za

I. S. Dunmade et al. (eds.), *Sustainable Engineering*, Green Energy and Technology, https://doi.org/10.1007/978-3-031-47215-2_14

$t3$	The length of time needed to examine (to find defects) the precise number of RVS machine subsystems manufactured
$t4$	The time needed to rework or repair the damaged RVS subsystems or spare part(s)
$t5$	The time needed to transport the manufactured RVS subsystem(s) or spare part(s) (Qi) to the users of the RVS machine
$t6$	The time required to load and transport the RVS subsystem(s) or spare part(s) (Qi) to the users
$t7$	The amount of time needed to unload the RVS components or spare parts (Qi) off the truck and place them in the authorized storage space for the RVS machine operator
$Lt - ord$	Lead time
Qi	Total number of RVS spare parts or subsystems
$Rt1$	Reconfiguration time
TPL	Total production loss
PG	Productivity gain
PC	Productivity capacity
CD	The current demand for mineral concentrates
OD	The demand for old mineral concentrations

1 Introduction: Background and Overview

Mining and mineral processing industries have the ability to directly contribute to economic growth, employment, and profit especially in low-income nations. Such industries do this by ascertaining the sustainability of various operational activities throughout the life of the mine or plant. This is frequently accomplished by working together with the government and society to make sure that operations have a more positive impact on the environment, climate change, and social capital. Likewise, mining and mineral processing industries in order to align with the Sustainable Development Goals (SDGs) are urged to practice responsible extraction, generate less waste, use safer methods, apply new sustainable technologies, promote the improvement of the well-being of neighboring inhabitants, decrease emissions, and improve environmental stewardship (Monteiro et al. 2019; Cole and Broadhurst 2021; Mesquita et al. 2021). One of the most popular machines used in mineral beneficiation is called screens. With functions in sizing, grading, scalping, dewatering, desliming, and media recovery, they are among the most established and frequently used physical size separation techniques (Wills and Napier-Munn 2006). Vibrating screen machines are the most important part of the comminution process in the mining and mineral processing. Figure 1 depicts the position of the screen in a typical comminution process.

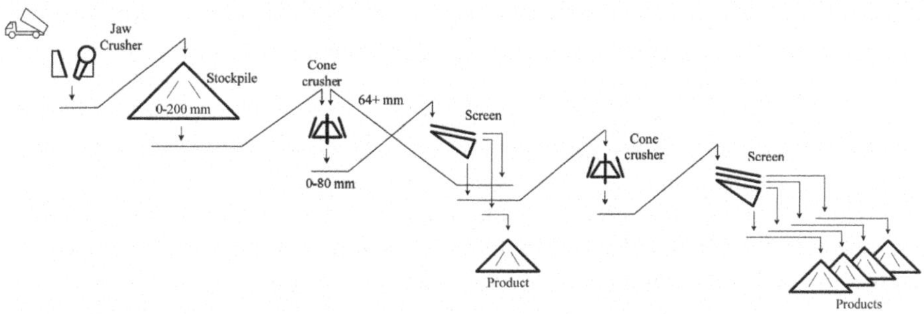

Fig. 1 Screening plant operation (Asbjörnsson 2013)

These machines are mainly used to classify and grade various mineral particle samples into various sizes as per customer requirements. Other authors **define** vibrating screens as:

> Using particle size as the primary variable effecting separation, screens divide the feed particles into two or more distinct size classes (Drzymala 2007). Makinde (Makinde et al. 2015a, b) defined vibrating screens as devices which divide a material flow into grades, for further processing to an intermediary product or a finished product.

Nevertheless, the modern mining and mineral processing markets are driven by changing customer requirements which influence the nature of operations in the global markets. A report by Kesler (2007) on mineral demand and supply into the twenty-first century revealed that the demand for mineral concentrate products has increased exponentially, and depending on the commodity of interest, projected world mineral reserves are 20 to about 1000 times greater than current annual production. In order to accomplish the goals established by the mining sector, it was also necessary to build new systems that could adapt to changing needs in mining and in the mineral processing industry (Ramatsetse et al. 2013). According to Moss (2011), "it is necessary to continually improve the technologies employed in mining operations, to render mining processes more efficient, effective and safety-enhanced, potentially increasing production." To address this, many authors have made attempts to develop innovative screening solutions capable of dealing with these aforementioned challenges as well as contribute toward efficient screening operations. These operations can further be classified into dry and wet screening. In wet screening, water is added to a screen to increase its efficiency as well as its effectiveness. Adding water to the feed material or spraying on the screen deck while the material is being screened are two ways to introduce it. Moving and static vibrating screens can also be distinguished from each other as well as by structural characteristics like horizontalness or inclination. The traditional screening equipment used in processing industries has an unfavorable degree of vibration and noise. Additionally, they have had unacceptably low screening effectiveness, high energy usage, high maintenance costs, low productivity, and poor worker safety. Almost all processing plants have employed these traditional vibrating machines. The majority of today's material separation technology produces the shaking using large, ineffective electric motors with an imbalanced spinning mass. These vibrating devices are frequently the bottleneck in the entire process in addition to being tremendously noisy, ineffective, and expensive maintenance. Additionally, the frequent usage of these machines in meeting the fluctuating mineral concentrate demands is a drawback in many conventional machines, thus requiring improvements in their design and operations in responding to these challenges. In addressing this challenge, various researchers have proposed new techniques to transform conventional vibrating screen into other forms of vibrating screens such as inclined, horizontal, grizzly, resonance, dewatering, high frequency, modular, banana, super-heavy (Guo et al. 2011), combined (Guo et al. 2011), multidimensional (Wang et al. 2011), dual frequency (Hou et al. 2012), low noise (Hua et al. 2004), etc. Most of these screens are developed based on the long years of experienced mining machine managers. These machines come in a wide range of designs and driving options, such as electromagnetic forces, eccentric shaft rotation, direct motor vibration, and others. Customers are given the best system possible by combining mesh, capacity, and shaking in accordance with their particle size, shape, and characteristics, as well as flow rate, purpose, location, and operating conditions.

There is an interest from mining machine managers and operators to acquire additional machines that will assist in their operations partly due to the unstable demand and to the rapid advances in technology, which force companies and plants to source for additional machineries that can be used to supplement the operations of conventional machines which are limited in their scalability attributes. These preventive measures are often put in place as to create a podium that will aid quick response to customer demands to ensure that processed minerals are available when required. To take advantage of this chance, researchers from Tshwane University of Technology (TUT) are increasingly coming up with new innovative ways of improving the current screening process to more sophisticated equipment that can handle changes in demand for mineral concentrates. Yoram Koren, an Emeritus Professor at the University of Michigan, presented the idea of RMS, which was used to accomplish this project. By definition, RMS "is one designed at the outset for rapid change in its structure, as well as its hardware and software components, in order to quickly adjust its production capacity and functionality within a part family in response to sudden market changes or intrinsic system change" (Koren et al. 1999). By adopting RMS principles, traditional screening machines become reconfigurable machines, forming a new type of equipment(s) designed to function with a certain product and allow quick changes to their structure, whether in terms of functionality or scalability (Katz 2007). It is clear from several government studies and media briefings that many mining companies continue to struggle with intricate chains of operations and labor-intensive processes that are heavily impacted by external variables, restricting their profitability and capacity to compete on a global scale. Hence, it is essential to implement novel innovative solutions and strategies that will transform mining and beneficiation operations. It is also apparent that only companies that have capability to adopt these aforementioned strategies will endure and survive in the competitive world. Thus, decision-makers in the mining organization should be aware of these technologies and maximize their usage and implementation.

The role of mining machine managers includes management of company assets and insuring the availability of these machines when needed for execution of operations. Based on these aforementioned duties of mining machine managers, the mining industry has experienced the developments of different cohorts of vibrating screens. In view of this, this book chapter discusses in detail the distinction between conventional screens and the proposed reconfigurable vibrating screens, application of RMS principle in different parts of mining machineries, features of these machines, and case studies that have emerged around the world.

2 Overview of the Screening Process and Its Related Subsystems

Screening is a process for dividing a mixture of solid particles of different sizes into different fractions based on differences in size and form. It entails pushing the mixture through a screen with a predetermined aperture size (Wills and Napier-Munn 2006). Depending on whether the particles are less than or greater than the size of the screen aperture when they are put onto the screen surface, they will either flow through the apertures or be kept on the screen surface. The terms "oversize particles" and "undersize particles" are used to describe particles that are larger than the aperture and smaller than the aperture, respectively. The oversize particles move down the length of the screen and are discharged into the oversize collecting chamber, while the undersize particles pass through the screen surface and are collected in the undersize stream. Figure 2 illustrates the process of screening in detail.

Figure 3 presents the various subsystems of a typical vibrating screen machine. These components can be listed and described as follows:

- *Screen deck frame*: This describes a steel structure that mounts and supports the screen panels, side plates, and back plates needed in screening mineral particles.
- *Screen mesh/panels*: The sieve surface, also known as the screen mesh, can be made of a variety of materials with varying aperture sizes, such as metal, rubber, or plastic. This allows undersized and oversized mineral particles to flow over and through this screening media.
- *Suspensions*: Its main function is to dampen the shock and vibration that the vibrating screen system encounters when screening the mineral particles.
- *Vibrating motor*: Technical devices that generate and transmit mechanical vibrations by supplying energy. These vibration motors are mostly used in bulk material handling processes, for conveying and screening of various mineral particles into required sizes.

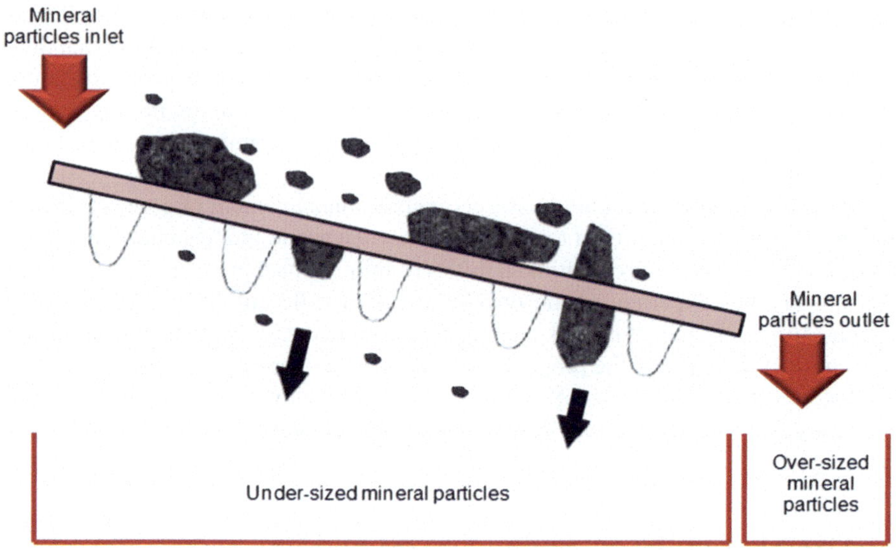

Fig. 2 Process flow for classification of mineral concentrates into diverse sizes. (Modified from Gupta and Yan 2006)

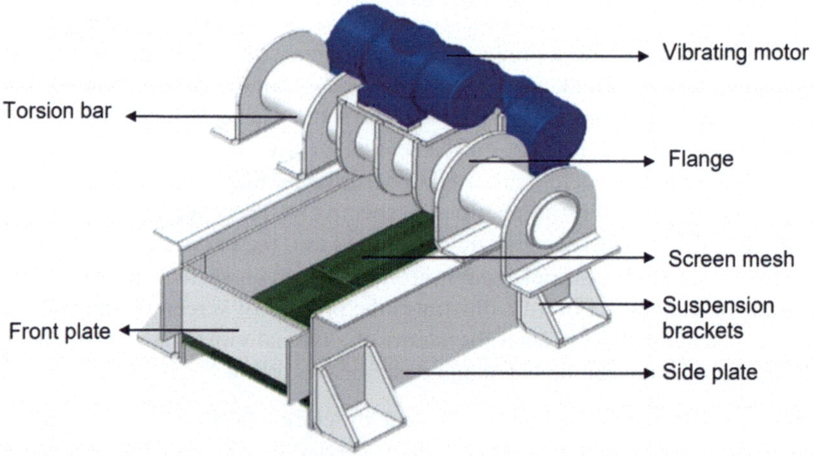

Fig. 3 Vibrating screen and its various subsystems

- *Torsion bar:* The center of the screen has a torsion bar, also known as a structural beam, which minimizes twisting and guards against side plate failure. Two connection flanges that are mounted onto either side of the screen make up the torsion bar.

- *Side plates:* The side plates are in charge of keeping the mineral particles contained inside the screen's surface and apertures. They are welded to the suspension support system for efficient vibration reduction and to maintain structural balance.

3 Evolution of Vibrating Screen Paradigms

3.1 Drivers of Change of Vibrating Screen Developments

The requirement to process higher-grade mineral particles, separate fine mineral particles, employ large scale equipment, and remove surplus wet overburden particles led to modifications in production setup and differences in vibratory screening product obtained. Due to this, the primary causes of the development of the vibratory screen trends depicted in Fig. 4 are explored.

Customer demand for various grades of mineral particles led to design revisions throughout the 1800s and 1900s. Various ore grades can be achieved with a single screen in a laboratory setting or many screens stacked on top of one another. Screen meshes of lower, medium, and bigger opening diameters are required for the shift from mineral granules of high-quality to smaller mineral granules. Due to consumer demands for diversity in sizing of mineral particles in nonmechanized as well as small-scale mining industries, these development trends were precipitated (Makinde et al. 2015a, b). The necessity to transition from laboratory size to large industrial scale applications drove design advancements between 1900 and 2000. The layer is distinguished by an increase in the machines' handling capability. Following the processing of mineral particle size, consumers' demand for mineral concentrates

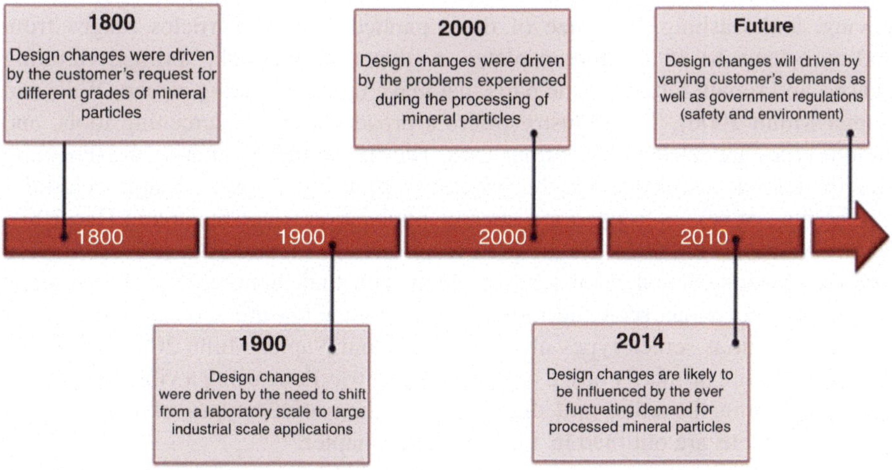

Fig. 4 Vibrating screen evolution trends

has increased as a result of these development trends. In light of this, a single deck vibrating screen was built to produce the appropriate mineral concentrates. It has a large screen surface structure, high efficiency, and capacitive motors. This phase's design adjustments between 2000 and 2014 were prompted by problems with the screening of mineral particles, such as the deterioration and wearing out of vibrating screen components like screen plates. Additionally, issues like noise levels have made it necessary to upgrade these equipment. In order to obtain finer mineral particle separation, the mining machinery industry has also created a multi-deck frame system with a variety of screen surface patterns and aperture sizes known as multi-deck vibrating screens. As a result, these enhanced beneficiation machine designs have produced smaller mineral particles made of platinum, diamond, and gold. The design changes from 2014 are expected to be influenced by the fluctuating market demand for processed mineral particles. At this point, the movement analysis and structural changes to the standard machines are prescribed as a future development path in screening technology, while on the other hand new constructive goals are pursued. Future machines will likely need to be reconfigurable because of traits like scalability, modularity, and integrability, which are expected to inject uncertainty into the requirements for mineral processing. Future screening technologies will also be largely influenced by the variances of the manufactured machines. Customers will be able to customize their own screening equipment and include features specific to their modifications, thanks to this feature.

3.2 *Types of Vibrating Screens*

The rectangular screening surface of vibrating screens has infeed and oversize discharge located at opposing ends. In addition to separating mineral particles based on size, this processing device also conducts classifying, scooping, dewatering, wet sieving, and washing. The size of the separated mineral particles ranges from 300 mm down to 45 m. Screen plates, a support framework, springs, vibrating motors, and screen fabric are the parts that make up the vibrating screen (Wills and Napier-Munn 2006). The industry offers a broad variety of screening tools, and various types are selected for various uses. The size of the ore charge, the screening aim, the intended capacity, and the screening efficiency all play a role in determining the type of screen to use for a specific application (Fuerstenau and Han 2003; Napier-Munn et al. 2005). As adopted from Fuerstenau and Han (2003), Table 2 lists the various categories of screens along with their applications. The mineral processing sector has seen the most success with vibrating screens, which have rendered previous screen types obsolete (Wills and Napier-Munn 2006). This section's themes are primarily based on research that was done using a vibrating screen. The functioning, benefits, and drawbacks of the various types of conventional vibrating screens are outlined in Table 1 of this chapter.

Table 2 Evolution of machines based on system characteristics (Abdi et al. 2018)

	Type of machine system		
System characteristics	Dedicated machines	Flexible machines	Reconfigurable machines
Process technology over time	Fixed	Adaptable	Responsive
Market conditions	Stable	Predictable	Uncertain
Product volume	High	Low/medium	Medium
Product variant	None (single product)	Medium	Medium/high
Process design	Continuous flow	Semi-connected or disconnected flow	Semi-connected/ connected flow
Manufacturing policy	Pushing	Pulling	Customizing
Present/future level of the gap between conventional and demand variations	High/very high	Medium/high	Low/very low (expected)
Integration into existing plant	Low integration	Low and rigid integration	High and flexible integration
Performance measurement indicators	Production output/input	WIP and cost	Changer over cost and time

4 Distinct Features between Dedicated, Flexible, and Reconfigurable Screening Machines

Conventional machines also known as dedicated machines were mainly designed to operate in a stable market conditions; thus, their production capacity is limited. Contrarily, flexible machines were created to cater to certain product kinds under predictable market circumstances in order to meet consumer needs. Flexible machines offer various advantages over dedicated machinery which includes greater machine efficiency, increased labor productivity, improved quality, and shorter lead times. As the deficiencies between conventional machines due to fluctuating in mineral concentrate demands are evident, the next generation of mining machines called reconfigurable machines are flooding the markets since they offer customized flexibility by adjusting their system components to suit new working conditions. Table 2 illustrates the distinctions between the various system types and their associated characteristics (adapted from Abdi et al. 2018).

5 Application of RMS Principles in Mining Machineries

According to Yoram Koren, RMS offers a podium to reach the precise capacity and functionality required, at the precise time it is required. RMS as applied in the mining industries involves the integration of modules within its hardware structure of the machine and integration of novel features and accessories needed for effectively enhancing the functionality of these machines in order to cope unstable market

Table 1 Types of conventional vibrating screens and their respective advantages and disadvantages Makinde (2014)

Type of screen	Description and application	Operating mechanism
Inclined	Crushing circuits, scalping, and high capacity	Utilization of mechanically generated vertical circular or elliptical vibration caused by the spinning of unbalanced wheels or flywheels connected to a single drive shaft
Grizzly	Scalping of coarse rock preceding crushers, bins, belts	Use of grizzly gaps inclined at approximately 20° and set at a fixed distance
Horizontal	Productivity is increased and great screening efficiency is ensured by horizontal screening equipment	Use of a linear or an elliptical vibration produced by a triple-shaft vibrator
Resonance	Applied to screening coal, limestone, sintered ore, iron ore, copper ore, and sintered ore	Utilize a dynamically balanced frame whose inherent frequency is the same as the vibrating screen body, has a balanced mass design with equal and opposing movement for the top and lower decks
Dewatering	Wet scalping and dewatering from 10 mesh and finer	Use of an inclined slight up-hill for water from wet mineral particles
Oscillating	0.5 in. + 60 mesh, light, free flowing	Use of linear stroke vibrator as well as variable slope of around 30–40° at the feed end of the machine, reducing to around 0–15° in increments of 3.5–5° for effective separation (Beerkircher 1997)
Banana	In most size applications, screens provide better throughputs than similar horizontal screens	These screens minimize the bed depth created at the feed end of the screens by producing high material travel velocities on the top deck slopes
Modular	Screens are created specifically for each operation, with special mounting locations and feeding configurations	Uses several separate screen cloth or panel modules connected in series to create a larger screen unit that is then driven by an electric vibrator
Revolving	Low capacity and efficiency; placer mining; scrub, wash, and rough size	A screen for sizing coarse material that consists of a rotating frame holding a cylindrical (or occasionally conical) screening surface
High frequency	Used for achieving fine and ultrafine screening.	The separation of mineral particles is accomplished by using frequencies up to 3600 rpm and electrical solenoids or motors

conditions caused often by changes in customer demands and machine downtime. Reconfigurable machine tools, manufacturing cells, process planning, as well control systems have all been the subject of extensive research, according to Makinde (Makinde 2014), and it is comparable to how RMS has been applied in the mining equipment sector in South Africa. Reconfigurable machines exhibit six (6) fundamental RMS principles, i.e., *integrability, scalability, customized flexibility, convertibility, modularity*, and *diagnosability*. Mehrabi et al. (2002) provided a thorough explanation of the fundamental characteristics of RMS: *convertibility*, which refers to the capacity to quickly modify the operation of current systems and equipment to meet new production needs; *scalability,* which refers to the capability to quickly alter capacity of the production by introducing or eliminating manufacturing resources (such as machines) and/or modifying system components; *modularity*, which refers to the partition of operational tasks into manageable parts that may be switched between several production systems to achieve best possible arrangement; *integrability*, which refers to the capacity to swiftly and accurately integrate modules using a variety of informational, mechanical, and control interfaces that support integration and communication; *customization*, which refers to the capacity to produce a certain product depending on the needs and specifications to satisfy the customer design requirements; and *diagnosability*, which refers to the capacity to automatically detect a system's present status to identify and treat the underlying causes of output product faults and promptly fix operational issues. How each of these ideas is applied to develop the RVS machines is discussed in the section that follows.

Convertibility is the capacity to change a machine's present operating function into another operating function in order to accommodate changing consumer needs. This means that the functionality of the RVS machine is currently screening outflowing 10 mm granite particles, as requested by customers, using the 10 mm opening screen decks fitted to the RVS machine. Processing of granite particles can be changed from the first configuration to the last configuration (26 from 15 mm, this is the new customer requirement). This is achieved by removing the screen plates (10 mm opening) currently fitted to the machine and replacing them with screen plates with a 15 mm opening as shown in Fig. 8.

Scalability is the capacity to modify the machine's capacity for beneficiation by adding or removing parts. This RMS principle is implemented on the machine by adding a side plate module, by adding a back plate module if machine scalability is to be achieved in the y-direction, by adding a front plate module if machine scalability is to be accomplished either in the x-direction, or by combining both methods as indicated in Fig. 9.

Modularity is a characteristic that defines how replaceable a system's parts or modules are. A modular system is made up of a number of components that each do a variety of minor tasks that together fulfill the machine's main function. Modules

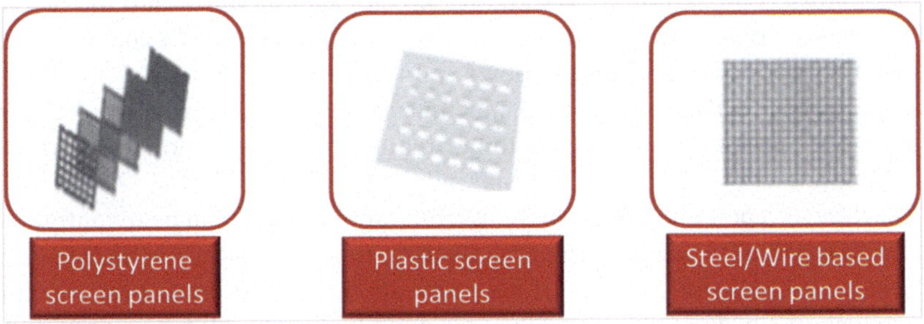

Fig. 8 Varying screen panel sizes (Makinde et al. 2018)

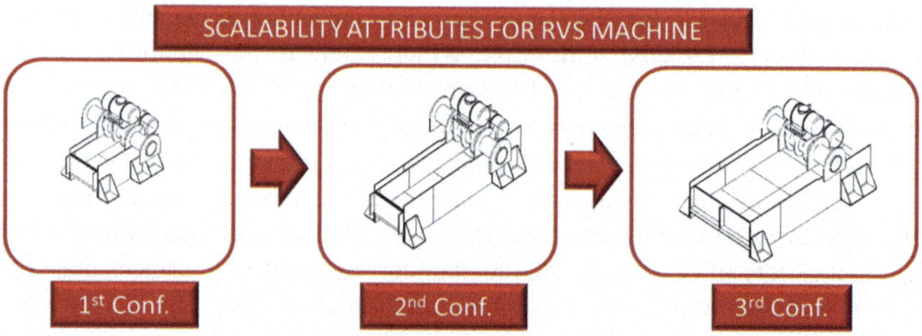

Fig. 9 Scalability attributes for the RVS machine

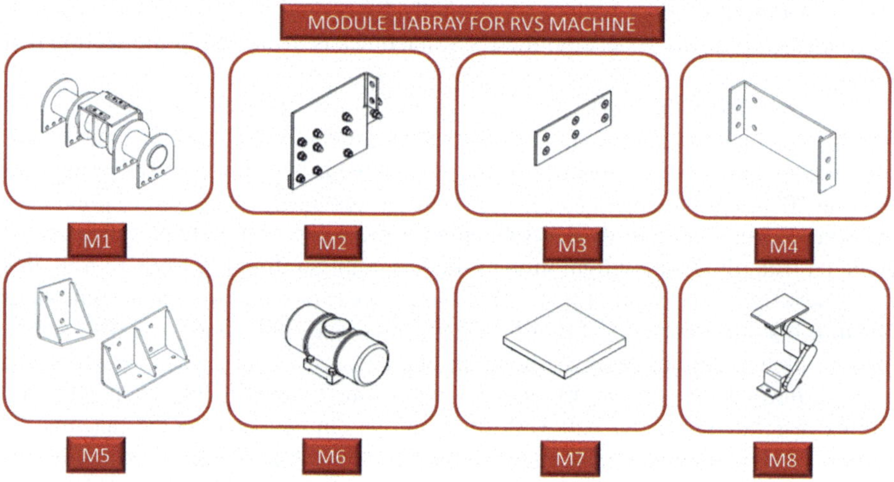

Fig. 10 Subtraction and addition of modules on the RVS machine

can be removed, replaced, or upgraded without affecting other components to meet customer needs. This means that in the module library of this machine, screen plate modules are provided with different opening sizes based on customer requirements as shown in Fig. 10.

Integrability is the ability to fit in more screen panel modules into the reconfigured screen side plate structure using mechanical and automated control tools that facilitate the integration process as shown in Fig. 11.

Customization is the ability of the machine system to be reconfigured to a new configuration capable of meeting customer requirements. As reported by Makinde et al. (Makinde et al. 2015a, b) and Makinde et al. (Makinde et al. 2016a, b), the RVS machine is a potential enrichment machine solution that can be reconfigured to meet the sporadically changing mineral needs of new customers, on the one hand.

Diagnosability is the capacity of a system application to predict the machine's performance in steady state using other machine life cycle-related metrics and the failure and repair history of the machine's numerous subsystems.

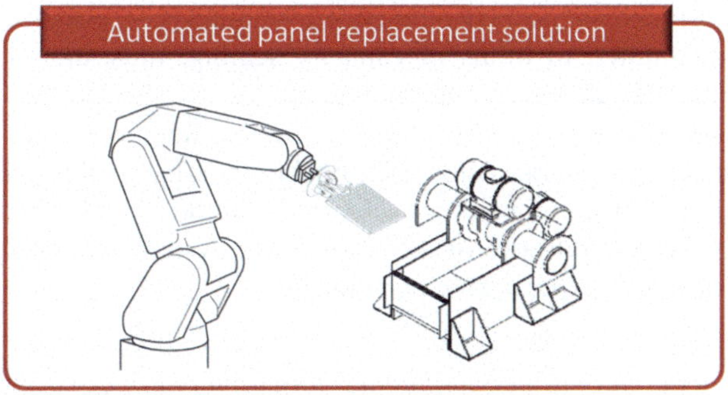

Fig. 11 Automated robotic solution for replacement of screen panel modules on the RVS machine

6 Case Studies of RMS Concept Application in Mineral Processing Industries

The RVS machine is a novel beneficiation machine that was created using the RMS concepts introduced by Koren et al. (1999) for screening various mineral particles of various sizes and production quantities required by customers in the surface and underground mining industry (Ramatsetse et al. 2017). The innovative design improves the sieving of mineral particles through the built-in geometric sieve structure, which ensures a change in the surface structure of the sieve, so that even in the event of unexpected events such as machine downtime due to malfunctions as well as other failures, sufficient quantities of mineral concentrates can be sieved (Makinde et al., Makinde et al. 2016a, b). The RVS operates in its first position much like a standard vibrating screen. The screen may be adjusted to the necessary capacity as the output of mineral particles rises. The panel's maximum configuration outperforms that of a typical screen with the same characteristics. Designed for smaller and larger screening operations, the RVS machine may be set from a capacity of 2500 mm × 1500 mm to other capacity sizes (3000 mm × 1800 mm, 3500 mm × 2000 mm, 4000 mm × 2000 mm and 4700 mm × 2500 mm), while the RVS machine designed for small-scale screening operations can be configured from a capacity size of 300 mm × 600 mm to other capacity sizes (300 mm × 1200 mm and 600 mm × 1200 mm). The RVS machine's major subcomponents are necessary to achieve that machine's best performance, according to Ramatsetse et al. (2022).

6.1 *Scalability Planning for RVS Machine Upkeep*

In order to address or plan for the changes in customer requirements as well as varieties of products in the market, the machine be capable to quickly be reconfigured in a cost-effective manner with the correct capacity requirements. According to Wang and Koren, (Wang and Koren 2012) "capacity scalability of a manufacturing system is a necessary characteristic for rapidly adjusting the production capacity in discrete steps, thereby allowing the system throughput to be adjusted from one yield to another to meet changing market demands." Conventional bending press machines are limited in terms of capacity scaling, thus short fall of addressing the ever-changing demands and product ranges. As alluded earlier, in the chapter, this issue can only be solved by adoption of either flexible or reconfigurable machines. These machines can be scaled with smaller increments through integration and disintegration of machine modules selected form the module library when the need arises.

For the purpose of this case study, system scalability is given by the following formulae, expressed in percentage:

Machine system scalability = Maximum production capacity (100) – smallest incremental capacity in percentage.

The system is highly scalable if the smallest capacity increase can be made to the system output in order to adapt it to meet changing market demand. For instance, Table 3 presents the various system scalability values for different part geometries of different length configurations.

The energy consumption for various configurations will differ because the machine has been designed to accommodate a range of mineral particle concentrates. Adenuga et al. (2020) investigated the prediction of energy requirements based on measurement data and a statistical regression model utilizing a variable frequency drive (VFD) to operate a fixed-speed electric agitating motor on the RVS machine. The electrical control system is made up of the following subsystems:

(a) **Electrical circuit board.** The electrical circuit delivers optimal electrical power to the electric motor for the vibration of the RVS screen panel structure.
(b) **Electric vibrating motor.** The vibration device agitates the loaded reconfigurable screen structure and ensures a uniformly distribution of mineral particles across the screen surface.
(c) **PLC and PID control system.** The programmable logic controller (PLC) is the generic structured controller, and proportional integral derivatives (PID) is the control software platform. These are used in configuring and programming the control system for the machine. However, the control system must be correctly tuned by the control designer to ensure optimal vibration of the RVS screen by the electric motor.

Table 3 Scalability planning for the RVS machine for different part geometries

Suitable machine configuration (mm)	System scalability
305 mm × 610 mm	100%
305 mm × 1220 mm	79.8%
610 mm × 1220 mm	66.67%

7 Convertibility Assessment for Reconfigurable Manufacturing Systems

In this case study, the convertibility assessment was conducted using the methodologies proposed by Maler-Speredelonl et al. (Maler-Speredelozzi et al. 2003). System convertibility includes contributions due to machines, how they are configured and maintained to operate. To calculate the convertibility measure, three key parameters are required:

C_c = convertibity associated with configuration
C_m = convertibity associated with machine
C_H = convertibity associated with material handling

The general formula for calculating convertibility measures can be obtained as follows:

$$C_s = w_1 C_c + w_2 C_m + w_3 C_H$$

I = convertibity associated with configuration
R = convertibity associated with machine
X = convertibity associated with material handling

The general formula for calculating convertibility measures can be obtained as follows:

$$C_c = \left[\frac{R ? X}{I} \right]$$

Incremental Conversion

For the purpose of this case study, the incremental conversion factor of the RVS machine was calculated considering the various production volumes which can be processed at a particular time interval. For instance, if the demand increases from the initial screening production, it implies that the machine has to be fully halted before any reconfiguration can happen. Thus, the minimum increment of conversion of that particular screening machine is 1.00 or 100%. Similarly, the incremental conversion values for machine configuration "b" to "c" of the RVS machine can be determined.

7.1 Routing Connections

According to Maler-Speredelonl et al. (2003), in manufacturing systems, a higher degree of convertibility of a machine is measured by a greater number of routing connections that occur on that particular machine. In this instance, the connections between the various modules of the RVS machine are counted as part of the total number of routing connections for each configuration. For instance, machine configuration "c" has one routing connection, while machine configuration "d" has five routing connections. This is due to the fact that configuration "c" can only accommodate an interface of one module to perform a bending of specific part type; however, configuration "d" which allows interface of three additional modules has also an advantage.

7.2 Routing Modules

The different mineral concentrate sizes that may be screened without necessitating changeovers are determined by the minimum number of duplicated modules at a certain stage in the process plan.

The results of Table 4 revealed that RVS machine configuration "*e*" with a C_c scoring of 136.36 is the first ranked configuration option (Fig. 12).

Table 4 Configuration convertibility for RVS machine

	I	R	X	C_c	Rank
a	0.33	7	4	84.84	2
c	1.00	1	1	1	5
b	0.50	3	2	12	4

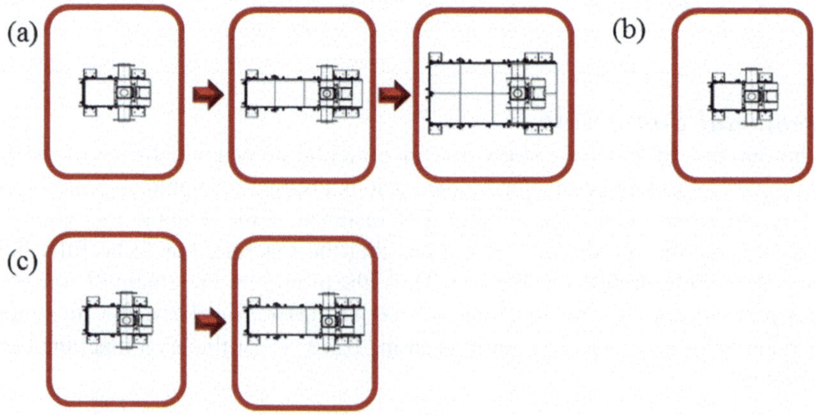

Fig. 12 Assessment tree analysis for RVS machine convertibility

7.3 Convertibility of the Machine

System convertibility is influenced by the configuration chosen as well as machine convertibility C_M which is calculated according to the following equation:

$$C_m = \frac{?_{i=1}^{N} C_m}{N}$$

The assumption that certain machines have traits and attributes that make them intrinsically more convertible underlies the machine system convertibility for each of the N distinct machines in the system. Among these characteristics is whether the device is:

$Q1$ = equipped with a screen panel replacement device
$Q2$ = easily reprogrammed with flexible software
$Q3$ = the system is design to allow addition & subtraction of modular components
$Q4$ = equipped with flexible fixturing capability
$Q5$ = equipped with a large capacity magazine

Figure 13 presents an assessment tree analysis for the RVS machine convertibility rating score. This assessment is also essential to determine the critical path of this machine. The results on the assessment tree analysis revealed that RVS has a convertibility scoring of 4 as the best critical path for the machine.

8 Modularity Assessment for Reconfigurable Systems

Table 5 presents the modularity assessment scores for the various mineral concentrate products that can be produced through the different configurations of the RVS machine. These mineral concentrate products can further be classified into various grades depending on their processing requirements. Mineral concentrates, which can be lumpy or powdery in texture, are refined ores from which the majority of waste components have been removed. Research conducted so far has resulted in providing different perspectives for the identification of efficient screening technologies that can be used to beneficiate the various minerals demanded by the customers.

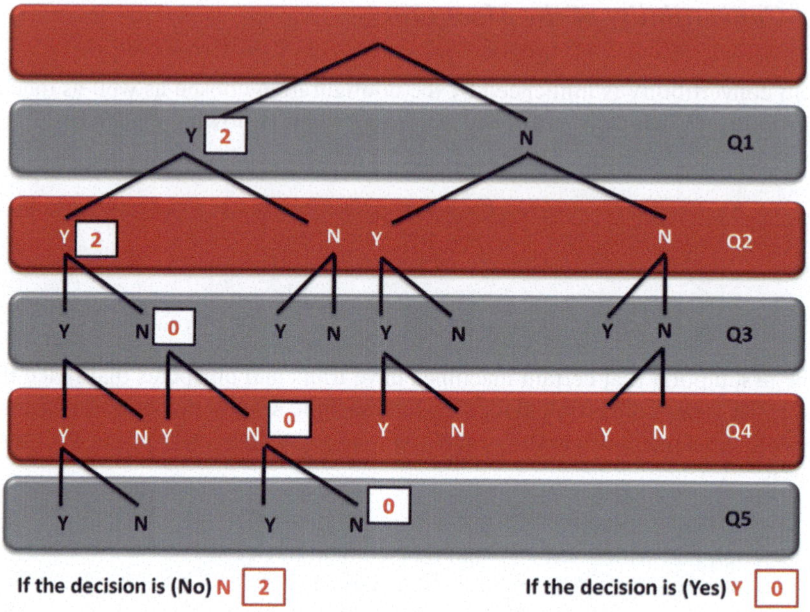

Fig. 13 Assessment tree analysis for RVS machine convertibility

Table 5 Modularity assessment for the RVS machine at different configurations

Machines	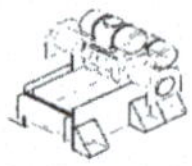 Screening small production volumes	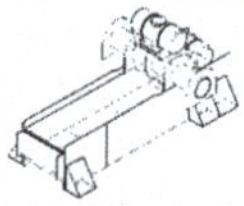 Screening medium production volumes	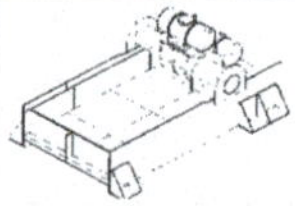 Screening high production volumes
Production size	Small demands	Medium demands	High demands
Grade A			
Grade B			
Grade C			
Screen panel modules required	**2**	**4**	**8**

Fig. 14 Common machine module relationship

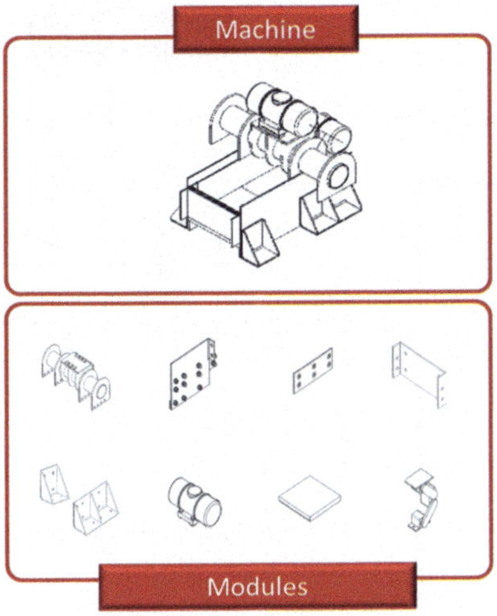

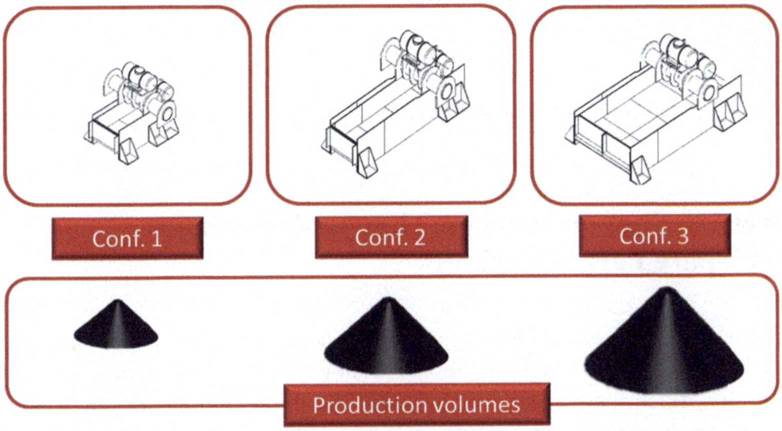

Fig. 15 Machine-product volume relationship various RVS machine configurations

(i) *Machine-Module Relationship*

This feature, as seen in Fig. 14, demonstrates the similarity in modules among the system's chosen machines. More specifically, it is the amount of shared modules among all of the system's machines, i.e., which modules are most essential or often utilized in terms of their compatibility with the machines. In terms of cost and complexity of reconfiguration, a module performs better the more widely used it is among machines.

(ii) *Machine-Product Volume Relationship*

The product/product family also has an impact on the system modularity, so to integrate the product information, we have the following formula, which quantifies the part-module relationship between the chosen modules and the product under consideration, as shown in Fig. 15.

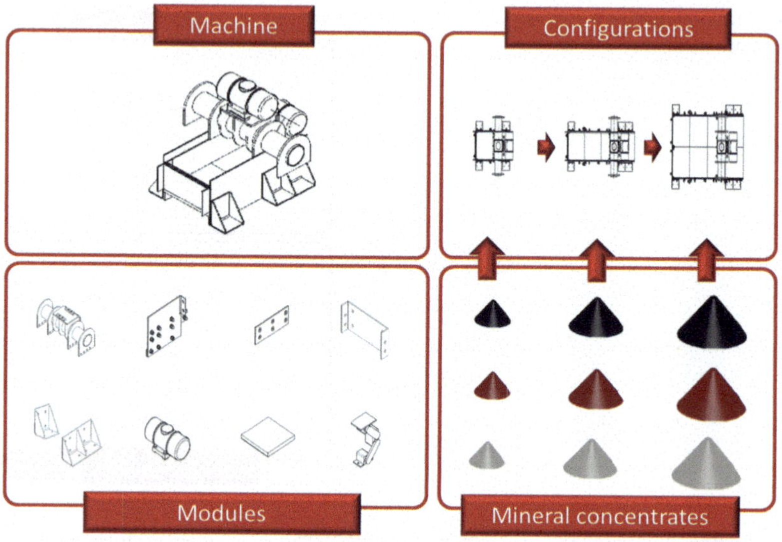

Fig. 16 Machine-component relationships for RVS machine

(iii) Machine-component relationship (Fig. 16)

9 Decision Support System (Time to Reconfigure and Spare Part Ordering)

Building a new plant demands a significantly different set of talents and abilities than modernizing an old one to the most recent production, efficiency, and standard requirements. Each renovation project is distinct in its complexity due to the difficulties experienced by plant operators. It can only be effective if the customer's partner and contractor are well-versed in the complexity of mining and mineral processing procedures and have the relevant expertise to put these novel ideas into practice within a rigidly scheduled downtime schedule. For RVS machine management, a highly effective maintenance planning system is necessary. When the RVS machine is working at an additional configuration capacity, this system estimates the time (i.e., reconfiguration time) needed to satisfy new customer needs or make up for production losses sustained during maintenance of the RVS machine. In light of this, the maintenance planning model presented below might be utilized to establish the best maintenance system for the RVS machine when it is deployed in the mining sector. The model assumes that the vendors of the replacement parts for RVS machines employ a make-to-order (MTO) supply chain method to satisfy customer demand.

RVS machine maintenance planning model formulation the parameters used for this model formulation are highlighted as follows

The total number of RVS subsystem(s) or spare part(s) (Qi) ordered by the users of RVS machine is given by

$$Qi = \sum \text{RVS} - \text{Subi}\, ni = 1 \tag{1}$$

where i is the quantity of spare components or RVS machine subsystems that the machine's maintenance managers will require to reconfigure the RVS machine or replace any worn-out subsystems with spare parts. Equation 2 provides the lead time needed to receive the RVS subsystem(s) or spare part(s) requested from the supplier, assuming there isn't any rework or replication of RVS machine spare parts.

$$Lt - ord = t1 + t2 + t3 + t4 + t5 + t6 + t7 \tag{2}$$

Equation 3 provides the lead time needed to receive the RVS subsystem(s) or spare part(s) requested from the supplier if reproduction or rework is necessary due to the existence of faults during the manufacture of the RVS subsystem(s) or spare part(s).

$$Lt - ord = t1 + t2 + 2t3 + 2t4 + t5 + t6 + t7 \tag{3}$$

In light of the facts provided above, Eq. 4 provides the total lead time needed to repair the RVS machine as well as to reconfigure the RVS machine in order to satisfy new client requests or make up for production losses caused by machine maintenance.

$$TLt = Lt - ord + Lt - replace \tag{4}$$

where *Lt–replace* is the time needed to install new subsystems in the case of machine reconfiguration, or the time needed to maintain the RVS machine and replace or repair worn-out subsystems. Additionally, on the one hand, the time it takes to make up for output losses caused by maintenance on machines or other unanticipated events {(i.e., reconfiguration time ($Rt1$)} when the RVS machine is operating at a new configuration capacity can be determined using Eq. 5.

$$\text{Rt1=Total Production Loss (TPL) the period of RVS machine maintenance Productivity Gain (PG) of the RVS machine} \tag{5}$$

where TPL = TLt × Productivity Capacity (PC) of the RVS machine at its first configuration ($PC\ RVS\ at\ 1st\ Config$), and PG =$PC\ RVS\ at\ n$th $Config$. - $PC\ RVS\ at\ 1st\ Config$. Note that "n" is the new configuration status of the RVS machine in recovering the production loss. However, the time required to meet new customer demands {(i.e., reconfiguration time ($Rt2$)}, when the RVS machine is operating at a new configuration capacity, can be determined using Eq. 6.

$$Rt2 = \frac{\text{Productivity gain-expectant}\left(\text{Pgain-expectant}\right)}{\text{Productivity Gain}\left(\text{PG}\right)\text{ of the RVS machine}} \quad (6)$$

where *Pgain−expectant* = Current Mineral Concentrates Demand (CD) − Old Mineral Concentrates Demand (OD) + Total Production Loss (TPL) that emanates during the reconfiguration of the RVS machine.

The reconfiguration time equations (*Rt*1) and (*Rt*2), to be used by the RVS machine to recover production loss that emanates during machine maintenance and meet new customer demands, were formulated, in order to ensure effective, just in time (JIT), and lean maintenance of the RVS machine as well as lean beneficiation of mineral particles on the RVS machine. However, an agent-based maintenance planning system that is capable of autonomously planning the maintenance of the RVS machine must be designed and developed to assure real-time maintenance planning of the RVS machine while it is being used in the mining sector.

10 Conclusions

To meet client expectations, the mining machinery industry must continually improve technology. Vibrating screen manufacturing has evolved owing to a variety of development requirements, including price, quality, competition, size, productivity, etc. These variables influence the economic value and advantages of increased vibrating screen output. As industries become more competitive on the global market and produce more cost-effective vibratory screens for consumers, the entire economy benefits. The separation of mineral particles efficiently and effectively utilizing more efficient screen materials has been developed because of treatment technology advancements in the construction of vibrating screens, which have increased productivity and improved treatment operations. Reconfigurable vibrating screens with adjustable screen decks and plate designs have also been proposed for use in the mining sector in the future to address these issues. This is due to the voracious nature of mineral concentrate users as well as the need to meet production targets, most commonly hampered or unmet as result of vibrating screen failures. To satisfy the requirements of mineral concentrate users, the proposed design and development of the reconfigurable vibrating screen will inherit RMS principles such as scalability, convertibility, modularity, customization, and integrability suggested by Koren et al. (1999) and Mehrabi et al. (2002). Lastly, the RVS machine has been designed to fulfil Sustainable Development Goal (SDG) 12. This goal emphasizes on "responsible consumption and production." This machine achieves this goal by producing what the customer needs at a particular time in the industry, due to its scalability abilities.

References

Abdi, M.R., Labib, A.W., Edalat, F.D. and Abdi, A., 2018. Integrated Reconfigurable Manufacturing Systems and Smart Value Chain: Sustainable Infrastructure for the Factory of the Future. Springer.

Adenuga, O.T., Mpofu, K. and Ramatsetse, B.I., 2020. Exploring energy efficiency prediction method for Industry 4.0: A reconfigurable vibrating screen case study. Procedia Manufacturing, 51, pp.243-250.

Asbjörnsson, G., 2013. Modelling and simulation of dynamic behaviour in crushing plants. Chalmers Tekniska Hogskola (Sweden).

Beerkircher, G. (1997). Banana screen technology. Comminution Practices. Kawatra, S.K. (ed.). SME, Litleton, CO. pp. 37–40.

Cole, M.J. and Broadhurst, J.L., 2021. Measuring the sustainable development goals (SDGs) in mining host communities: A South African case study. The Extractive Industries and Society, 8(1), pp.233-243.

Drzymala, J., 2007. Mineral processing. Foundations of theory and practice of minerallurgy, pp.381-382.

Fuerstenau, M.C. and Han, K.N. eds., 2003. Principles of mineral processing. SME.

Guo, N., Lin, J. and Huang, W., 2011, July. Development of 2YAC2460 super-heavy vibrating screen. In 2011 Second International Conference on Mechanic Automation and Control Engineering (pp. 1225-1227). IEEE.

Gupta, A. & Yan, D. 2006. Mineral Processing Design and Operation, Elsevier Science and Technology.

Hou, Y.J., Fang, P. and Zeng, L., 2012. Finite element analysis of dual-frequency vibrating screen. In Advanced Materials Research (Vol. 479, pp. 2124-2128). Trans Tech Publications Ltd.

Hua, Z., Shirong, G. and Xiaoyan, Y., 2004. A novel low-noise vibrating, screen. Coal Preparation, 24(1/2), pp.85-96.

Katz, R., 2007. Design principles of reconfigurable machines. The International Journal of Advanced Manufacturing Technology, 34(5–6), pp.430–439.

Kesler, S.E., 2007. Mineral supply and demand into the 21st century. In Proceedings for a Workshop on Deposit Modeling, Mineral Resource Assessment, and Their Role in Sustainable Development. Circular (Vol. 1294, pp. 55–62).

Koren, Y., Jovane, F., Heisel, U., Moriwaki, T., Pritschow, G., Ulsoy G., and VanBrussel, H.: Reconfigurable Manufacturing Systems. A keynote paper. CIRP Annals, Vol. 48, No. 2, pp. 6-12, November 1999.

Makinde, O.A., Ramatsetse, B.I. and Mpofu, K., 2015a. Review of vibrating screen development trends: Linking the past and the future in mining machinery industries. International Journal of Mineral Processing, 145, pp.17-22.

Makinde, O.A., 2014. Functionality assessment of a reconfigurable vibrating screen (Doctoral dissertation, Tshwane University of Technology).

Makinde, O.A., Mpofu, K. and Ramatsetse, B., 2016a. Establishment of the best maintenance practices for optimal reconfigurable vibrating screen management using decision techniques. International Journal of Quality & Reliability Management.

Makinde, O.A., Mpofu, K. and Ramatsetse, B., 2016b. Mining business optimisation through a reconfigurable vibrating screen design. African Journal of Science, Technology, Innovation and Development, 8(1), pp.88-96.

Makinde, O.A., Mpofu, K., Ramatsetse, B.I., Adeyeri, M.K. and Ayodeji, S.P., 2018. A maintenance system model for optimal reconfigurable vibrating screen management. Journal of Industrial Engineering International, 14, pp.521-535.

Makinde, O.A., Ramatsetse, B.I., Mpofu, K. and Oyesola, M.O., 2015b, March. Holistic methodologies for optimal reconfigurable vibrating screen design, development, production, utilization and management Part I. In SAIMM Conference (pp. 11-12).

Maler-Speredelozzi, V., Koren, Y.H.S.J. and Hu, S.J., 2003. Convertibility measures for manufacturing systems. CIRP annals, 52(1), pp.367-370.

Mehrabi, M. G., A. G. Ulsoy, Y. Koren, and P. Heytler. 2002. "Trends and perspectives in flexible and reconfigurable manufacturing systems." Journal of Intelligent Manufacturing13 (2): 135–146. doi:10.1023/A:1014536330551

Mesquita, M.J., Corazza, R.I., Souza, M.C.O., Gomes, G.N., Noronha, I. and Macedo, D., 2021. Mining and sustainability. In Environmental Sustainability (pp. 155–179). CRC Press.

Monteiro, N.B.R., da Silva, E.A. and Neto, J.M.M., 2019. Sustainable development goals in mining. Journal of Cleaner Production, 228, pp.509-520.

Moss, S., 2011. Development and Testing of a Scale Model Clamshell Mucker, and Validation by Discrete Element Modelling.

Napier-Munn, T. J., Morrell, S. & Morrison, R. D. 2005. Mineral Comminution Circuits Their Operation and Optimisation, Queensland: Julius Kruttschnitt Mineral Research Centre.

Ramatsetse, B.I., Matsebe, O., Mpofu, K. and Desai, D.A., 2013. Conceptual design framework for developing a reconfigurable vibrating screen for small and medium mining enterprises. SAIIE25 Proceedings, Stellenbosch, South Africa, p.1.

Ramatsetse, B., Mpofu, K. and Makinde, O., 2017. Failure and sensitivity analysis of a reconfigurable vibrating screen using finite element analysis. Case Studies in Engineering Failure Analysis, 9, pp.40-51.

Ramatsetse, B., Mpofu, K., Daniyan, I. and Makinde, O., 2022. Assessment of Reconfigurable Vibrating Screen Technology for the Mining Industries. In Flexible Automation and Intelligent Manufacturing: The Human-Data-Technology Nexus: Proceedings of FAIM 2022, June 19–23, 2022, Detroit, Michigan, USA (pp. 79-87). Cham: Springer International Publishing.

Wang, C.J., Ma, L.Z. and Li, Y.M., 2011. Multi-Dimensional Vibrating Screen Design and Motion Analysis. In Advanced Materials Research (Vol. 201, pp. 1672–1677). Trans Tech Publications Ltd.

Wang, W. and Koren, Y., 2012. Scalability planning for reconfigurable manufacturing systems. Journal of manufacturing systems, 31(2), pp.83-91.

Wills, B.A. & Napier-Munn, T. 2006. Mineral Processing Technology 7th ed., UK: Elsevier Science and Technology Books.

Evoking Design-As-Agency for Sustainable Engineering: The Art of Jeffrey Smart

Graeme Byrne

1 Evoking the Transforming Powers of Technology

The Australian artist Jeffrey Smart (1921–2013) depicted our engineered world in allegorical or metaphorical terms—where various meanings are possible. And so his art can open up engineering design and technical systems to critical and sustainable re-evaluations. Yet, in the first instance, it may appear that his paintings of autostradas, construction sites, shipping containers, and the like simply celebrate systems-centric design, not least as engineered objects have regular shapes and mathematical patterns. Indeed, as bright colors combine with perfect or powerful geometries, his art is partly about the productive intensities of infrastructure and its ability to remake our lived environs. As physically robust and beautiful forms combine through pure geometries, we can sense that we are in a new realm, one transformed by the technology. For instance, in *Guiding Spheres II (Homage to Cezanne)* (1979) the bright, spherical road markers recede at regular intervals, as do the container trucks (Fig. 1). As the yellow orbs shrink rapidly away from the eye, his art therefore expresses how engineering design breathes new life into global logistics. As such, Smart's geometric symbols express how our experience of place and space is transformed when mediated through new technology (also see Ihde 1990; Fry 2019).

But another form of hyper-realism (or surrealism) soon intrudes: the colors are too bold, the slick surfaces leave us adrift in a smooth, semi-automated desert, and the gray atmosphere suggests a brooding—perhaps even bruised—reticence. Along

Note: The (few) paintings not reproduced below may be found online or by consulting the references mentioned in brackets after the title of the painting.

G. Byrne (✉)
Previously at Engineering College, Australian National University,
Acton, Australian Capital Territory, Australia
e-mail: byrnegraeme@hotmail.com

I. S. Dunmade et al. (eds.), *Sustainable Engineering*, Green Energy and Technology, https://doi.org/10.1007/978-3-031-47215-2_15

Fig. 1 Jeffery Smart, *Guiding Spheres II (Homage to Cezanne)* (1979)

with the feelings of transformation, therefore, we are also discomfited. Indeed, the figures in Smart's paintings often appear to be confused by their "new violent environment" (Smart quoted in McGrath 1969, p. 32). A freeway or new housing estate, for example, can displace people and communities. Ultimately, his art evokes the ways in which our technologies simultaneously create new ways of life (or ontologies) *and* initiate painful restructuring.

In his lecture "As I See It," Smart sought to put such divergent visual ideas into words. Initially, geometrical forms resemble a "Chinese jar" as they express technology's transformative or even "metaphysical" qualities. Or, as already mentioned above, perfect geometries can be used to symbolize how a technology can modulate how we experience space, mass, and time and hence bring desired potential into reality. Communications technologies, for example, are highly valued as they annihilate how we experience aspects of space or distance. Yet evoking how technology alters our experience of the world provides only a first step toward realizing a more critical or environmentally aware approach. With our "blind faith in onwards, ever onwards," further meditation is required before we might discern the "crisis in our concept of progress" (Smart in Quartermaine 1983, pp. 121–3).

The inclusion of human figures then helps with the next stage. When an artist invokes geometry as a symbol of technological transformation, there is a danger that an implied authority—the technical expert, for example—will become an unassailable presence. Nevertheless, as Smart includes human figures, his paintings elicit feelings that are not only rooted in locality. Even if his figures are glimpsed only through a rear-view mirror, present merely in road signs, or pinned to a plain of logistics, they bear ironic witness. The emergent qualities of the built environs are then subject to variation assessments, possibly even a sense of crisis. The transformative power of technology is then no longer determined exclusively by geometries, technologies-in-themselves, or by experts alone. With experts, publics, ecologies, and many other interests all involved, a distributed and non-determinant capacity is needed instead—one that requires imagination and judgment about technological transformation or design-as-agency.

2 The Importance of Inclusive Process

Indeed, Smart well understood that efforts to remake our environs would gain if inclusive or empowering processes existed too. Consider, for example, the unresolved tension between the looming forces and minor individuals in *Control Tower* (1969) (Fig. 2). In the first instance, two organized forces vie for dominance. The large letter "C" points to the control tower—or maybe a large organization—that looms above. Second, the "C" sign is physically attached to another massive

Fig. 2 Jeffery Smart, *Control Tower* (1969)

structure—the red brick silo—that bulges across the lower part of the image. As such, the letter "C" and its accompanying arrow might refer to another force: an organized mass of red coming from another direction. Yet, consistent with Smart's characteristic approach, such geometric (and autonomous) symbols of power are offset by the human figures. As the day trippers to the airport freely roam about on an intermediate level, they seem to be mediating the larger forces or have their own interpretive frame. Therefore, with various vantage points from which to appreciate the engineered structures, design is entwined with extra-technical considerations. Rather than being technically determined by the powers-that-be, design becomes a question of efficient for *whom*, or for *what matters*. A participatory process, one prepared to discuss the design permutations, then becomes part of the design process (especially in liberal societies). Experts need to be able to discuss "facts and values together in ways that illustrate what configurations of the two would mean for…policy decisions pertaining to specific situations" (Fischer 2017, p. 273). And some economists argue along similar lines. Between autocratic stratagems—left, right, or expert—pathways need to be found that motivate the many (Acemoglu and Robinson 2013, pp. 73–83). As such, nation building—or building sustainable technologies and systems—will need to find enough support and good governance for extensive and ongoing implementation (Owens 2023).

Smart sets in motion a distributed and therefore contingent design process at the microlevel of the economy too. In *Factory staff, Erehwyna* (1972), ominous clouds gather above a Tuscan landscape (Fig. 3). But Smart has taken care with the formal structure of the painting to imply relations between engineers, employees, community, and landscapes may help or hinder outcomes. Initially, the rectangle of the engineering office expands into the rectangle of the factory proper according to the Golden Ratio of 1:1.618, a ratio often adopted by Western artists to epitomize proportional beauty (Gordon 2020). The staff are then assembled for their group photo

Fig. 3 Jeffery Smart, *Factory staff, Erehwyna* (1972)

in front of the rectangle that initiates the geometric sequence. This implies that their skills and relations with management are as important as the factory layout itself.

The surreal form on the hill then signposts how another techno-future might be in the offing. On the one hand, the factory could be a global "Anywhere" (Erehwyna spelled backward) of production. But just as Smart has given the town a name that involves a reversal, the assumptions of global industrial engineering might be undergoing a reversal at the hands of a local Tuscan or Florentine ideal. In fact, just as Smart has sought to celebrate how a sense of the beautiful may enhance production, this is what such an ideal—called *disegno*—also involves. A concentrated process of drawing and art making puts the designer more in tune with the dynamics of the world and so can feed innovation or creative knowledge synthesis (Ciammaichella and Menchetelli 2022; Edgerton 1984). The painting might therefore translate today as a visual version of ESG (environmental, social, and governance) *disegno*. While motivated by the dangers, close study of the multidisciplinary potentials in combination might happily resynthesize an industrial region.

The paintings of Smart also help engineers to appreciate a fundamental commonality between Settlers and Indigenous peoples and so better partner for the sustainable cultivation of "Country" as well. However, in the first instance, the painting *Mount Tom Price* (1966) might be thought to depict an insurmountable cultural dichotomy. Under one mountain, a mining encampment is in the early stages of development, along with loading gantries, both of which are necessary before the ore can be shipped to Japan, South Korea, and China. In contrast, the lack of Settler technology on the second mountain, with natural stratifications visible, signifies the presence of deep time and Indigenous peoples. Yet this juxtaposition of cultures under the ominous sky could be interpreted as something other than a clash between cultures that are innately more technological than the other. If it were depicting such a clash, it would constrain Indigenous peoples to being part of the landscape. It would then deny them the full suite of human capacities—expressive, social contractual, and technological—by which they can form their own design-as-agency and actively negotiate the mines of Settler nationalism (plus their own economic prospects).

Smart's painting, however, depicts a firestick-farmed landscape (with partially burned trees in the foreground). Because the trees in the foreground appear too burned, it is possible to question how well Smart has done this. Yet the capacity of Indigenous peoples to transform yet maintain the country by utilizing that most Promethean of technologies—controlled fire—is being acknowledged. As such, even in the tightly knit (mechanical) solidarity of traditional Indigenous peoples, who simultaneously sing and negotiate their technological practices through their Dreaming or cosmological stories, they still assert their presence (Forrest and Senuik Cicek 2021, p. 3). Indigenous technological innovation tends to be constrained by cultural conventions such that bark from a tree is used to fashion a canoe only in-so-far as the tree remains alive. On the other hand, Settler cultures can be thought of as flailing about because of the opposite sociology of knowledge production: over specialization or separation between disciplines. This painting, however, can enable us to move beyond what might appear unbridgeable differences. Rather

Fig. 4 Jeffery Smart, *The Construction Fence* (1978)

than debating who has design-as-agency, or a capacity to bring various disciplines and then judgment to bear, both cultures have such a capacity. This may then give rise to potential partnership in sustainable engineering endeavors on Country (Kutay and Leigh 2017).

One final example of an empowering process, especially as it can facilitate sustainable engineering, has to do with gender inclusion. Sustainable systems benefit by finding ways to literally empower the domestic economy. They could benefit by employing more female engineers too. *The Construction Fence* (1978) engages our imagination in such terms (Fig. 4). As the near-airborne figure running in front of the concertina wall is a youthful female, she might fly above the barrier to become part of what's happening on the other side of the green curtain. As a gay man, Smart was keenly aware of the engineering world's masculine culture and chose a girl as his herald of change. As Smart draws together geometric, surreal, and contesting colors, cross-cutting but also enabling levers and hooks can be imagined as well. While the red stripes against the green curtain dramatize the barriers, another interpretation suggests that sustainable engineering might gain if it were more open to half the population.

3 How Environmental, Technical, and Social Sustainabilities Need to Be Reworked Together

So far, we have seen how engineering needs to cultivate not only physical potentials but the social and visionary aspects too. When it comes to energy transitions, regional resources and ecosystems need to be re-networked in conjunction with ideas for economies and regional businesses. Nowadays, power line easements are particularly contentious. Again, therefore, an ability to consider diverse criteria can facilitate sustainable engineering (especially if technical systems are to be substantially remade while maintaining liberal or social democracies). But with a re-networking of technical, social, and ecological relations, painful choices often arise.

Fig. 5 Jeffrey Smart, *Cahill Expressway*, 1962, National Gallery of Victoria, Melbourne

And this difficult process is captured in the painting *Cahill Expressway* (1962) (Fig. 5). In fact, while making art about engineering where human and ecological risks were mutually implicated, his distinctive moods crystallized (Allen 2008, p. 23). As already considered, he would then elaborate on the need to think about technical, political, and environmental matters conjointly for decades to come (Quartermaine 1983, p. 57).

At first glance, *Cahill Expressway* may appear to be about how a hapless pedestrian has become lost or cut off by a roadway. In fact, during the late 1950s, many Sydney-siders were alarmed by a plan to cut a deep trench between the Botanical Gardens and the neighboring Domain. But while it might allow access to a freeway, it might destroy the Moreton Bay fig trees of Fig Tree Avenue. Nonetheless, if engineers favored a tunnel, the stately trees might survive. Many continued to worry, however, that engineers preferred to cut down the trees. George Molnar, a cartoonist for the *Sydney Morning Herald*, soon turned his attention to the matter (Molnar 1959). At first glance, the technical experts appear to be presenting options. These include the poisoning, burning, or blasting the trees. On further reflection, however, this meant tree removal remained the only option.

Smart's *Cahill Expressway* may have then drawn inspiration from Molnar's cartoon. It certainly provides another take on the theme of lost limbs. Just as a favorite stroll has now been cut off by the feeder road, Smart presents us with a hapless pedestrian whose arm is missing from a jacket sleeve. As the isolated figure sports a missing arm, it laments the absence of a beloved avenue of trees too. A photograph of the threatened avenue may have contributed to the visual metaphor. This depicts a visiting, wartime sailor perambulating down Fig Tree Avenue with a coated arm tucked into a pocket (State Library, New South Wales 1941). The phantom limb(s) and the loss of a tree-lined walk are then synthesized in a bereft figure, unable to embrace the new road. A haunting sense of loss is emphasized by his position, silhouetted against a dark cavern. Just as Jane Jacobs was critiquing how freeways and car culture would dissect human-scaled neighborhoods in Manhattan, Smart has arrived at his own "Noplace" in Sydney, and a sense of melancholy arises (Jacobs 1961, p. 338).

4 Going Deeper into the Sources of a Lack of Interdisciplinarity

To better express what a technological response to urban and environmental issues entails, Smart has hit upon a combination of symbols that would become his artistic style. Evocative geometries, strange skies, traffic signs, and mute—yet somehow inquiring—individuals help us to imagine the transformations in the offing yet the quandaries and injuries of infrastructure, which then encourages us to design with the people affected, their institutions of governance, plus environments all in mind. But just as films such as *The Matrix* or *Blade Runner 2049* doubt whether diverse people and multiple disciplines can synthesize enough agency to find a way out of technical cages, his paintings come to fear that diverse perspectives and criteria might be lost from view. As he turns to consider mass-produced housing blocks, his paintings still hint at a designer's ability to manipulate architecture to build a community and express personality. Yet his concerns seem to deepen, not least as mass housing estates came to be designed by the doyen of such architecture (plus planning and civil engineering): Le Corbusier.

Le Corbusier certainly started well. He incorporated internal corridors into his high-rise apartment blocks, wide enough to be considered pedestrian streets, conducive to encounters between families. Yet Le Corbusier's innovations soon faded. Projects prioritized narrow definitions of efficient construction over individual and community expression. Smart's *Approach to a City III* (1968–1969) then deftly captures the ambivalence that comes with living in hyper-modern environments (Fig. 6). He depicts a lone couple struggling to walk across a windswept bridge toward a high-rise apartment as refugees from the countryside, suitcase in hand. Capturing the feelings of many—including those facing massive urban renewal in China—the couple struggles to surmount a wave of change that has displaced them into a nature-less limbo. Nevertheless, just as a bright container truck is on the horizon, the canyons of modernity have the potential to lead toward a brighter future.

In the main, however, a fear of a singular, dominant criteria increasingly presents itself. As such, Smart may have been following the historian of infrastructure and the city, Lewis Mumford. Mumford had observed a mono or "megatech" tendency throughout human history (Mumford 1934, p. 237). Cities planned in accordance

Fig. 6 Jeffery Smart, *Approach to a City III* (1968–1969)

with Renaissance traditions are structured by roads or lines-of-sight emanating from a central plaza. A rational planning and perspectival order can then expand such as to overwhelm local conviviality. At the extreme, a reductive or homogenous trend can take hold, a destructive tendency Mumford put down to humanity's death instinct (Swer 2003). Indeed, when Smart returns to a car-related theme, the painting *Motor Dump, Pisa II* (1971), this pessimistic view of an unbounded or uncontextualized rationality comes to the fore (Pearce 2011, p. 112).

Pisa was Galileo's birthplace and, by extension, the birthplace of modern mechanics and physics. During the 1600s, Galileo had used the town's bell tower—now known as the Leaning Tower of Pisa—to conduct experiments into whether falling bodies accelerate at the same rate. Over 400 years later, however, Smart's painting depicts a tragic form of mechanics. Disused cars are heaped up at the bottom of the tower, which is itself reduced to a measuring stick. A pile of waste has therefore come to obscure the potential of shared public space, such as with pedestrians and cyclists. In a manner similar to that described by Mumford in his *City in History* (1961), space and place are mangled by the car:

> The private motor car, whose extension has devoured the one commodity the suburb could rightly boast: space. Instead of buildings set in a park, we now have buildings set in a parking lot. (p. 576)

Fig. 7 Jeffery Smart, *Monument and Car Park* (1972)

However, in Smart's *Monument and Car Park* (1972), a sense of contested space makes a comeback, if only eventually (Fig. 7). This time, urban form radiates out from the central pinnacle (based on Rome's Piazza del Popolo) whereby techniques of visual perspective emphasize how shared space has been drained away by overly rationalized planning for what has become, again, a car park. In particular, Smart foregrounds a drainage grill on the Piazza's edge. But not least because of a typically shy presence, the single-use symbolism is disrupted. A small pedestrian presence makes a return. The Piazza then remains open to various forms of use. Smart is therefore re-asserting that over-arching rationalities are never complete. Recent scholarship also stresses how the overlays of the political, cultural, and local "situations" established by Roman forms were then elaborated by Renaissance and later periods (Vesely 2017, pp. 151–4). Or, today, sustainable systems can benefit from co-existing possibilities. Modest and autonomous economies and communities may enhance the larger networks of sustainable engineering.

5 How Technological Scale or "Reaction Time" Further Compounds Risk

Smart's *Autobahn in the Black Forest II* (1979–1980) continues to wonder how outmoded rationalities, this time as technical momentums, are a problem (Fig. 8), in particular, why we might fail to appreciate the long reaction time needed to reform incumbent systems. Professor Schellnhuber, from the Potsdam Institute for Climate Impact Research, explains the issue succinctly as it compounds the risk of climate change:

> [In any situation] if the reaction time and the intervention time becomes equal then you have lost control…[And] this can be applied…to climate change. And given the tipping points we have in store, we have still an intervention time of probably thirty years. But the reaction time is similar, actually. So we are about to lose control." (Schellnhuber in Pottsdam Institute for Climate Impact Research 2021, 19:15 to 19:33)

But why have we left it so late? For Smart, a propensity for blindness arises as a result of incumbent forces. As the chevrons quickly follow one after another, we are

Fig. 8 Jeffery Smart, *Autobahn in the Black Forest II* (1979–1980)

Fig. 9 Jeffery Smart, *Container Train in the Landscape* (1983–1984)

forced to focus on our immediate situation. But this means it is easy to push other criteria out of the frame such that the fate of the forest is pushed to the periphery of our vision. Or, today, just as incumbent carbon systems are adept at appealing against the potential disruptions to regular business, the costs of delay are pushed away. Smart includes the rust stain on the leading chevron column to perhaps highlight this point of view: when our attention is focused on what is immediate, we might suffer from a degradation of awareness. The forest is (already) black; simultaneously, the danger is (already) upon us. But we have difficulty grasping the risk. Our negligence of the natural world has reached a code red, this time as incumbent systems demand we ignore what is economic in the long term.

Smart's *Container Train in the Landscape* (1983–1984) considers another systematic oversight (Fig. 9). As the bright containers blare their way through the landscape, the imports or exports mean more goods and/or taxes for the Australian economy. As the train gets the go-ahead from the electromechanical signal, the gap between the containers also allows us to glimpse the local landscape as both destination and source of goods. This gap exists, however, only as a result of an inconsistency. A close inspection of the painting reveals the wagon below the gap is 50 percent longer than the others. The perspectival view exists because of an induced absence and which enables us to view the world in depth (so to speak). A coupling mechanism, necessary to join two other wagons together, is also absent. Adding to the artistic interventions, the iron rails are obscured so the train floats over a sea (of grass).

All in all, once we have inspected the scene for eccentricities, the momentum of the train appears increasingly fantastical. But just as the coverage and coupling devices of global logistics are no longer totally encompassing, so a different perspective of Australia's conveyor-belt economy can be gained. Today, indeed, at a time of no-less intense trading networks, sustainable solution may suffer from an intensification of the problem: an increase in consumption thence a drop in availability of natural resources. Especially as new wind and solar technologies seek to underpin sustainable economies, some wonder whether enough copper and rare earth metals will be available to complete the task (Michaux 2022). Instead of cohabiting with the pastels of the Australian ecology, a flawed ghost train still intrudes, with linear economics even more decoupled from material limits. Even as Japan and other countries are concerned to maintain the train, there is an implied need for yet another way of assembling engineering. In stark contrast to the linearities of the painting, there is a need to act in concert with the evaluations of a circular economics.

Fig. 10 Jeffery Smart, *Ring a Rosy* (1988–1989)

Smart further highlights the need for sustainable designers to remain alive to other systemic/conceptual risks in his *Ring a Rosy* (1988–1989) (Fig. 10). In the painting, a small group of children play a game in a park, oblivious to the unstable ground beneath their feet. Smart was, of course, unaware of the COVID-19 pandemic that would occur 30 years after the creation of this painting. But while engineering systems enable much in the way of wealth generation, risk is now manufactured on an industrial scale. Witness the Bhopal tragedy in India, the radioactive fire at Chernobyl, and the plastics as well as carbon dioxide pollution of today. And so while a group of youngsters—future generations—happily play in their park, the by-products of industrial wealth are eating away at their foundations. While ubiquitous yet ghostly high-rise apartments hover in the background, risks are accumulating, even to the extent that "we all fall down." The unstable ground, which forms the primary metaphor of the painting, can then become a reference to the systemic risks considered by Ulrich Beck in his *Risk Society* (1992). A blanket risk can be imposed indiscriminately, with debilitating effects not least for innocent nations. Again, therefore, if engineers are attuned to the chronic and global extent of who and what is at risk, a need to collaborate with global institutions can gain in urgency as well.

6 Conclusion: Evoking Design-as-Agency to Secure Sustainable Engineering

In conclusion, Jeffrey Smart evokes the dilemmas, dislocations, and even crises of infrastructure across a series of paintings. But he also evokes how urban and natural environments realize their emergent qualities because of syntheses between diverse disciplines and visions. Technological, economic, environmental, social-contractual, and visionary elements are all in play. This might imply that energy transitions or sustainable engineering are hard to achieve. Nonetheless, rather than retreating into economically determined or systems-centric design, a responsiveness to multiple criteria remains key for Smart. Simultaneously invoking counterfactual symbols of bruised skies and surreal landscapes, the dilemmas but then choices are brought to life. By including "David-like" individuals (as in Michelangelo's David), otherwise alienating landscapes are reclaimed by a diversity of actors and perspectives. By enabling us to imagine how the presence of multiple faculties creates contingent circumstances, he is concerned to cultivate design-as-agency. A potentially contingent or free and open design process is enlivened by the mutual or co-evolving partnerships between our imaginative, analytic, and technological capacities. Rather than lamenting the "wicked" nature of design situations, contingency plus choice becomes its paradoxical essence.

Again, however, some might argue that Smart sometimes emphasizes a (supposed) historical tendency toward mono-technics. As his art dramatizes technological change, his Prometheanism may be as tragic as it is transformative. Yet a multi-perspectival—thence participative—design sensibility remains his escape route. Smart once commented that the role of people in his work was to instill a sense of "responsibility and failure at the same time" into our built environs (Pearce 2011, p. 136), which can then form the basis of a facilitative design method. As metaphors multiply the influences in play, art can provide the means by which to discuss several design options. The design process then becomes difficult to corrupt or less prone to enclosures of technocratic or autocratic politics. Further, a diversity of participants produces a greater range of insights which can be enabling of creative outcomes. Considered processes can also enable enlistment into a program of sustainable transition for the long haul.

Smart's evocations of diverse forces and faculties also assuage shortfalls in technical drawing (or computer-aided design). Such rationalistic practices of rendering and modeling are, of course, a boon for engineering and materials science, etc. But they can obscure the ontologies at stake and truncate the creative process too. As used by architects, product designers, and engineers, they do not express how a structure may challenge or change our ways of living (Dovey 1993, pp. 256–8). Again, however, rather than wholly collapsing engineering poetics into an alliance with sleek surfaces and industrial processes, his art includes ironic figures. His metaphors therefore allow ontological choices to gain expression or raise questions about "lived spatiality and meaning" within the built environment (Motycka Weston

2017, p. 197). Overall, his poetics of technical systems establish a metaphorical resource amenable to open processes and multiple criteria. With the help of his art, engineers are more able to contemplate the possibilities of local sites and so pursue innovative and sustainable goals together.

References

Acemoglu, Daren and Robinson, James A. (2013). *Why Nations Fail. The Origins of Power, Prosperity and Poverty*. London: Profile Books.

Allen, Christopher. (2008). *Jeffrey Smart: Unpublished Paintings 1940–2007*. Melbourne: Thames and Hudson.

Beck, Ulrich. (1992). *Risk Society: Towards a New Modernity*. Trans. Mark Ritter. London: Sage Publications.

Ciammaichella, Massimiliano and Menchetelli, Valeria. (2022). Drawing and Design. Declensions of Terms and Practices Actualizations. *Disegno*, 11(2), 7–13. https://disegno.unioneitalianadisegno.it/index.php/disegno/article/view/474/730. Accessed 18 April 2022.

Dovey, Kim. (1993). Putting Geometry in its Place: Towards a Phenomenology of the Design Process. In David Seamon (Ed.), *Dwelling, Seeing, and Designing: Towards a Phenomenological Ecology* (pp. 247–269). Albany, NY: State University of New York Press.

Edgerton, Samuel Y. Jr. (1984). Galileo, Florentine "Disegno," and the "Strange Spottednesse" of the Moon. *Art Journal*, 44(3), 225–232.

Fischer, Frank. (2017). *Climate Crisis and the Democratic Prospect. Participatory Governance in Sustainable Communities*. Oxford: Oxford University Press.

Forrest, Reed and Senuik Cicek, Jillian. (2021). Rethinking the Engineering Design Process: Advantages of Incorporating Indigenous Knowledges, Perspectives, and Methodologies. In: Proceedings 2021 Canadian Engineering Association Conference, 20–23 June, 2021. DOI: https://doi.org/10.24908/pceea.vi0.14911. Accessed 2 April 2023.

Fry, Tony. (2019). Design as an Ontological Question. In Anne-Marie Willis (Ed.), *The Design Philosophy Reader* (pp. 27–29). London: Bloomsbury Visual Arts.

Gordon, Nick. (2020). Jeffrey Smart's Geometry. Factory Staff , Erehwyna. https://www.theshortcourse.org/blog/blog-post-jeffreysmartgeometry. Accessed October 4, 2022.

Ihde, Don. (1990). *Technology and the Lifeworld: From Garden to Earth*. Bloomington: Indiana University Press.

Jacobs, Jane. (1961). *The Life and Death of American Cities*. New York: Random House.

Kutay, Cat and Leigh, Elyssebeth. (2017). Aboriginal Engineering for an enduring civilization. In: Proceedings. Australian Association for Engineering Education 2017 Conference. http://hdl.handle.net/10453/121485. Accessed 5 April 2023.

McGrath, Sandra. (1969). Jeffrey Smart. *Art and Australia* 7(1), 32–39.

Michaux, Simon. (2022). Discussion: The quantity of metals required to manufacture just one generation of renewable technology to phase out fossil fuels. https://www.youtube.com/watch?v=MBVmnKuBocc&list=LL&index=23&t=2028s. Accessed October 27, 2022.

Molnar, George. (1959). Our engineers did consider many alternatives to cutting down trees, but found them impractical [cartoon]. *Sydney Morning Herald*, February 6. https://nla.gov.au/nla.obj-147529400/view. Accessed October 4, 2022.

Motycka Weston, Dagmar. (2017). Corporeal Spatiality and the Restorative Fragment in Early Twentieth Century Art and Architecture. In Henriette Steiner and Maximilian Sternberg (Eds), *Phenomenologies of the City: Studies in the History and Philosophy of Architecture* (pp. 195–210). London: Routledge.

Mumford, Lewis. (1934). *Technics and Civilization*. New York: Harcourt, Brace and co.

Owens, Kate. (2023). Harnessing the transformative potential of climate governance. The 2023 Iain McCalman Lecture, Sydney Environment Institute, University of Sydney [Video]. Retrieved from https://www.sydney.edu.au/sydney-environment-institute/events/2023/the-2023-iain-mccalman-lecture.html.

Pearce, Barry. (2011). *Jeffrey Smart*. Sydney: The Beagle Press.

Pottsdam Institute for Climate Impact Research. (2021). Tipping Points for Sustainability Transitions. https://www.youtube.com/watch?v=AAGQHPbsvTY. Accessed May 7, 2021.

Quartermaine, Peter. (1983). *Jeffrey Smart*. South Yarra: Gryphon Books.

State Library New South Wales. (1941). Fig trees in Domain (taken for AUP) [photograph]. https://collection.sl.nsw.gov.au/record/9O4DXEpn. Accessed October 4, 2022.

Swer, Gregory Morgan. (2003). The Road to Necropolis: Technics and Death in the Philosophy of Lewis Mumford. *The History of the Human Sciences,* 16(4), 39–59.

Vesely, Dalibor. (2017). Between Architecture and the City. In Henriette Steiner and Maximilian Sternberg (Eds), *Phenomenologies of the City: Studies in the History and Philosophy of Architecture* (pp. 151–165). London: Routledge.

Driving Effective Sustainable Housing Infrastructure Delivery in South Africa Through Incorporation of Socioeconomic Development factor

Oluwatobi Mary Owojori and Emem O. Anwana

1 Introduction

The need for sustainable infrastructure and its associated profound implications on socioeconomic development is one of the most pervasive and difficult issues facing African cities in the wake of the twenty-first century. According to Thacker et al. (2017), infrastructure is defined as needed support, facilities, instruments, and systems required for an individual's physical, mental, or social well-being. Among these are sewage systems, sanitary facilities, roads, power, drainage, waste management, and other public transit systems (Nubi 2002). In other terms, the infrastructure that supports housing is what gives housing its functionality.

Housing infrastructure further refers to the fundamental facilities, services, systems, and installations required or suitable for the operation of a housing development, including those for energy, communications, water, sewerage, and transportation (law insider). It is characterized as "the physical elements of interconnected systems supplying the services and merchandise necessary to enable, maintain, or improve society livelihood" which is a right in many constitutions.

The original version of the chapter has been revised. A correction to this chapter can be found at https://doi.org/10.1007/978-3-031-47215-2_29

O. M. Owojori (✉)
Department of Finance and Investment Management, University of Johannesburg, Johannesburg, South Africa
e-mail: oluwatobi.owojori@nithecs.ac.za

E. O. Anwana
Department of Applied Law, Faculty of Management Sciences, Durban University of Technology, Durban, South Africa
e-mail: emema@dut.ac.za

I. S. Dunmade et al. (eds.), *Sustainable Engineering*, Green Energy and Technology, https://doi.org/10.1007/978-3-031-47215-2_16

Housing infrastructure development is thus described as the establishment and maintenance of habitable, social, and sustainable public and private residential environments to ensure viable households and communities in areas allowing convenient access to economic opportunities and social amenities to which residents will have access by the South African Housing Act of 1997. Section 26 of the 1996 South African Constitution recognizes the right to housing and states that everyone has the right to access adequate housing. Section 26(2) further states that "the state is required to take reasonable legislative and other measures, within the limits of its resources, to ensure that the right to housing is gradually realized."

The constitution further stated that "the rights in our Bill of Rights are interrelated and mutually supportive." There is no question that the fundamental social ideals of human dignity, freedom, and equality, as well as the right to adequate housing, cannot be seen apart. It and the other socioeconomic rights are closely related. Social and economic rights must be interpreted together within the context of the entire Constitution (The Constitution of South Africa 1996). Therefore, to provide sustainable communities with greater access to economic opportunities and social facilities, housing infrastructure must do more than just construct housing.

The right to adequate housing is inextricably linked to many other intersecting rights (Farha et al. 2008) such as the rights to public participation, equality, human dignity, just administrative action, access to information, and access to justice, as well as a variety of socioeconomic benefits and conveniences These include having access to housing, water, power, sanitation, livelihoods, transportation, medical facilities, educational institutions, and cultural and recreational amenities (Behr et al. 2021) including libraries, parks, swimming pools, sports facilities, and religious institutions. Together, these rights and socioeconomic benefits reduce poverty, promote equality, and enhance people's quality of life.

The national development plan (NDP) for South Africa, which is organized into 13 chapters, provides sector-specific targets and a vision for the country that aims to guarantee the attainment of a "decent standard of life" for all South Africans by 2030. Unfortunately, given its objective of promoting "sustainable human settlements," the government has not yet adopted the kind of all-encompassing approach to such development challenges that would fundamentally address the persistent socioeconomic variance among South African cities and towns.

For the realization of Vision 2030 to become a reality during the next 10 years, the NDP and its measures must be implemented holistically. Every development must strike a balance between social, environmental, and economic concerns to be considered sustainable (Owojori and Okoro 2022a). The National Housing Code's updated national housing programs and the DHS's social/rental housing-related initiatives are optimistic steps in this direction. However, there is evidence that the environmental aspect of sustainability is receiving more attention at the implementation level. As a result, sustainability plans typically fall within the purview of environmental issues than social and economic ones (Weingaertner and Moberg 2014).

There is a strong link between housing delivery and socioeconomic development, and many studies have explored this relationship from various perspectives. One common approach is to examine the perspectives of individuals and communities toward housing and its impact on their socioeconomic status. This has spurred researchers to study various factors that can influence housing delivery and socioeconomic development. One example of this is a study conducted in South Africa by Nnadozie (2013); the study tested the hypothesis "the relative odds of access to basic services in post-apartheid South Africa, for different demographic segments of the society." The study recommends the establishment of a service delivery model that would redress the substantial unevenness in access to and delivery of basic services, according to the demographic segmentations (population group and income level).

Swope and Hernández (2019) also presented a vision for healthy housing that is based on a social justice perspective and argues that social policy should not only focus on redistribution but also on raising the productive capacity of the people and promoting social cohesion. It underscores the importance of a comprehensive approach to addressing social inequality. These inequalities are often reflected in divergent perspectives toward housing and development, with some researchers and stakeholders prioritizing access to basic services such as water and electricity, while others prioritize the creation of jobs and economic opportunities. For instance, researchers such as Moyo and Mambo (2022) have highlighted the importance of incorporating socioeconomic development factors into housing delivery programs in order to ensure that they are responsive to the needs and perspectives of local communities.

Similarly, Ramafamba and Mears (2012) explored how service delivery can be a strategy for local economic development in South Africa. It investigates how service delivery in Mamelodi has contributed to employment and business development emphasizing the need to enhancing the linkages and synergies between housing delivery and other socioeconomic factors to facilitate access to basic services to ensure sustainability over the long term. Therefore, the relationship between housing delivery and socioeconomic development is complex and multifaceted, and researchers continue to explore this topic from a range of perspectives in order to better understand the challenges and opportunities associated with providing affordable and adequate housing for all.

To achieve this, robust frameworks need to be put in place to concretely measure all the dimensions of social and economic development metrics. It is necessary to take a multifaceted approach that considers a range of factors to further address the broader issue of housing by identifying socioeconomic development factors that can be incorporated for sustainable housing infrastructure delivery.

There is a need for the significant study, advocacy, and action at the socioeconomic levels with regard to strategy, initiatives, and housing priorities. The objective of the study is to explore the socioeconomic development (SED) components that can be integrated into housing infrastructure for sustainable development. The socioeconomic metrics considered in this study for sustainable infrastructure delivery are (1) infrastructure accessibility, (2) social equity, (3) social cohesion and inclusivity, (4) economic investments, (5) health and well-being, and (6) community services. By integrating socioeconomic development factors into housing infrastructure delivery, there is maximization of the positive social and economic impacts of engineering projects for sustainability. This can further contribute to improving the quality and quantity of housing for low-income people in well-located areas, enhancing their living conditions and opportunities, and advancing socioeconomic development and spatial transformation in South Africa.

The six factors are related to different SDGs that aim to improve various aspects of sustainable development: *infrastructure accessibility* (SDG 9, resilient infrastructure and innovation), *social equity* (SDG 10, reduced inequalities and inclusion), *social cohesion* and inclusivity (SDG 16, peaceful and inclusive societies and institutions), *economic investments* (SDG 8, sustainable economic growth and decent work), *health and well-being* (SDG 3, healthy lives and well-being), and *community services* (SDG 11, sustainable cities and communities with basic services relevant for sustainable housing infrastructure delivery, as they reflect the economic and social dimensions of sustainability). The provided information will assist those involved in the built environment in integrating socioeconomic strategies to deliver sustainable housing infrastructure.

1.1 Rationale for the Study

The study is motivated by the recognition that housing infrastructure has a significant role in ensuring an adequate quality of human life and advancing socioeconomic development and reducing poverty, unemployment, and inequalities. The study is also informed by the challenges and gaps in the current housing infrastructure delivery system in South Africa such as the dearth of social and economic services (Sobantu 2020), as a function of factors not limited to political interference, bureaucracies, corruption, and lack of funding.

There haven't been precise considerations on how housing could contribute to social development despite a rise in the level of literature emphasizing the state of South Africa's housing delivery and associated problems. In other words, the connection between social and economic development and housing has not been well studied in both literature and practice. This requires innovative solutions that leverage social and economic service provision within the context of social and economic sustainability in the infrastructure delivery process.

There is a need to examine the context and scope in which SED factors can be integrated into housing infrastructure delivery in SA as part of sustainable engineering process. The study aims to provide insight into incorporating socioeconomic development (SED) factors into the housing infrastructure delivery process in South Africa (SA). The study responds to the call for sustainable development goals by addressing several aspects of sustainable development, such as adequate housing, basic services, social and economic development, and alignment with the national and global agendas for the communities in South Africa. The study intends to assist stakeholders in mapping out strategies for delivering sustainable housing infrastructure in South Africa that are aligned with the national development plan and the sustainable development goals.

2 Methodology

The study adopted a qualitative review while using content analysis to address the overall goal and the specific objective of this chapter. An initial search of the Scopus academic database produced scarce relevant documents using the key phrases "housing infrastructure *and* socioeconomic integration" *or* "socioeconomic integrated housing infrastructure," which easily established the peripheral position of housing infrastructure in South Africa concerning social and economic integration. A larger set of publications were found through similar searches conducted using Google Scholar. With some notable exceptions, though, most of these publications merely made brief mention of housing when they were examined more closely.

An investigation of the research cited in these instances revealed that the concepts underpinning the SED components may appear in the literature singly or collectively under a variety of designations. Due to this and the large number of terms included in the "housing infrastructure socioeconomic indicators," rather than utilizing simple keyword searches, identifying literature for this evaluation required a continuous, iterative approach of reference extraction. Gray literature (reports and working papers, government documents), academic journal articles, book chapters, websites, and other publications were gathered into an EndNote database. Although the idea of housing infrastructure and SED considerations in the South African context was of particular interest, this study also draws on a wealth of foreign literature on housing infrastructure and sustainable development.

The review is divided into four main parts. Section 1 sets the scene for infrastructure in South Africa from precolonial times to the present. In Sect. 2, the relationship between housing infrastructure, socioeconomic development, and sustainable development goals (SDGs) is discussed. We examine how housing infrastructure relates to socioeconomic development in light of sustainable development objectives in the international literature. The obstacles to the delivery of sustainable housing infrastructure are covered in Sect. 3. In Sect. 4, which includes recommendations and a conclusion, the study asserts that incorporating SED elements has significant but mostly unexplored promise for all sustainable housing delivery in South Africa, especially in light of the fragmented state created by earlier apartheid regimes. The review's scope does not include all topics that can be relevant to socioeconomic indicators. The study presented here does not purport to be exhaustive; rather, it serves as an introduction to the various SED indicators that might guide the construction of sustainable housing infrastructure.

3 Infrastructure in the Precolonial South Africa

Before colonial times, most black Africans endured severe oppression and injustice that were committed by European colonizers, who eventually became apartheid bureaucrats after settling in the nation (Mubangizi 2008; Maylam 2017). The black African population was denied opportunity and subjected to discrimination due to unjust, racist laws and practices during the time of colonial and apartheid authority, which was marked by egregious human rights violations (Modise and Mtshiselwa 2013). The Native Land Acts of 1913, which distributed 87% of the land to white settlers, was at the foundation of the suffering of the black African community (Beinart and Delius 2014). Afterward, under apartheid, they were not permitted to live in some places and were forcefully transported to other areas which were classified as being for black people. The Unjust Group Areas Act of 1950, which allowed the construction of townships for the housing developments of blacks, colored people, and Indians, was the primary piece of legislation utilized to accomplish this. These townships are intrinsically linked to the problem of cheap migrant labor and oppression of the primarily black African population since they were required by South Africa's increasing urbanization.

4 State of South Africa Housing Infrastructure Today

The township, a legacy of colonialism and apartheid, was unintentionally strengthened and spread throughout South Africa after it gained democracy in 1994 as a result of the Reconstruction and Development Programme, the new administration's mass housing scheme (RDP) (Mar and Edmonds 2010). South Africa had a significant housing crisis as a result of colonialism and apartheid, which harmed social advancement and residents' access to basic human rights. Human rights, service delivery, housing, and social development challenges are all intertwined in South Africa. This is so that the inhabitants' access to other rights, such as water, power, safety, and security, can be ensured. Since 1994, according to the 2011 National Development Plan (NDP), the government has constructed more than 3 million housing units with subsidies for low-income families (National Planning Commission [NPC] 2011).

Access to essential services has also increased, with 97% of homes now having access to water and nearly 75% having access to power and sanitary facilities (NPC 2011). Despite these advancements, the availability of adequate housing, consistent water and power, a supply of safe water, accessible public transportation, and hygienic and respectable sanitation facilities continues to be a daily challenge for many South Africans, especially those in underdeveloped rural and suburban areas neighborhoods (NPC 2011). South Africa has long been one of the continent's most developed nations, and it has also significantly boosted global economic growth.

More so than many African countries, the region has prospered economically. Nonetheless, it is unlikely to be able to achieve the Sustainable Development Goals (SDGs) by 2030. According to a World Bank assessment, South Africa is also the most unequal nation in the world, with race being a deciding factor in a society where 10% of the population owns more than 80% of the wealth (Victor et al. 2022). The fact is that South Africa hasn't had inclusive development, and due to the country's growing urbanization, there is an enormous need for social and economic infrastructural development. There have been relatively few concrete suggestions on how housing could specifically result in social development rewards, despite an increase in the literature highlighting South Africa's housing delivery progress and related issues. In other words, the connection between social development and practical housing has not been studied in both literature and practice. It is significant that the African National Congress (ANC)-led government recognized housing delivery as a plan to advance social rights and economic progress in 1994. Government-initiated reforms were launched as soon as they came into office intending to address the housing crisis and address broader social and economic equality shortfalls for the underprivileged (Charlton 2013); discourse on housing, social justice, and social development is required in the nation. There have been growing concerns for a people-centered approach that may improve the housing, social, and economic rights of the underprivileged. A more sustainable housing delivery system that is able to douse the negative social and economic effects of apartheid and colonization is required to improve South Africans' quality of life.

5 Housing Infrastructure in the Context of Socioeconomic and Sustainable Development

The importance of infrastructures in fostering and sustaining economic growth has been recognized throughout the last 20 years (Calderon and Seven 2004). Infrastructure projects have been promoted as engines for socioeconomic growth and sustainable development. Infrastructure is the framework on which a community is created, according to Anderson and Agrawal (2011). The provision of essential services is crucial in ensuring that unemployment is minimized, inequality is mitigated, and social value is strengthened. According to Fourie (2006), infrastructure improvements lead to greater socioeconomic development. Infrastructure is vital to the growth of any society. The realization of the socioeconomic SDGs depends on sustainable infrastructure. Given that 3.5 billion people currently live in poverty, infrastructure's importance in achieving SDG 1's goal of eradicating poverty is perhaps the most significant (Atamanov et al. 2018).

Among the primary goals of infrastructure development is to enhance access to basic services, and integrated sustainable infrastructure that provides access to electricity, transportation, potable water, and adequate sanitation is closely linked to eradicating inequality by expanding economic opportunities and streamlining communication channels in addition to increasing access to basic services (Bhattacharya et al. 2016). The United Nations (UN) created targets for socioeconomic and environmental development in the form of the Millennium Development Goals (MDGs) and Sustainable Development Goals (SDGs) (Ji et al. 2021).

Sustainable housing infrastructure enables governments and the private sector to deliver services that support individual livelihoods and broad economic growth while boosting the quality of life and human dignity. Examples include networks of good housing, roadways, and power and water networks as outlined in the SDGs. The 2030 Agenda for Sustainable Development has a significant section on housing infrastructure as shown in Table 1, which is also a key factor in accomplishing many of the SDGs. The sustainability of essential services like water and power during a crisis is ensured by sustainable housing infrastructure, which also provides communities with greater stability and lessens the interruption to their living conditions.

Table 1 Housing infrastructure as an integral part of the SDGs

Goal 1 No poverty	Sustainability is facilitated through equitable land access, housing, and the economic ability they supply
Goal 3 Good health and Well-being	Housing infrastructure that is accessible, sustainable, secure, and resilient is crucial for both physical and mental health
Goal 6 Clean water and sanitation	A fundamental need and essential element of healthy and sustainable housing is having access to adequate, efficient water and sanitation services
Goal 7 Affordable and clean energy	Energy poverty is a subtype of housing infrastructure poverty that affects surrounding populations' social, environmental, and health conditions
Goal 8 Decent work and economic growth	Sustainable home infrastructure affords communities more time for useful activities, which improves economic potential
Goal 9 Industry, innovation, and infrastructure	Emerging solutions provide an economically resilient, cost-effective, and ecologically responsible housing strategy
Goal 10 Reduced inequalities	Housing infrastructure equity assists communities in reducing inequality
Goal 11 Sustainable cities and communities	Considering housing has a transformative effect on socioeconomic and environmental results, placing housing infrastructure at the center fosters the development of better towns and cities

Adapted from Rust (2020) Centre for housing in Africa

The SDGs recognize the right to sufficient housing as a key component of these tenets, which has long been a top priority also for UN-Habitat. A covering over one's head is only one aspect of adequate housing. Additionally, it means providing accessibility, adequate security, adequate basic infrastructure including water supply and sanitation facilities, suitable environmental quality, health-related factors, and adequate and accessible location concerning work and basic facilities (UN-HABITAT 2015). Table 1 shows the contribution of housing infrastructure to the sustainable development goals.

5.1 Socioeconomic Integrated Housing Infrastructure Delivery

The dearth of social and economic services in the surrounding locality is one of the challenges that many new and existing housing developments impose (Sobantu 2020). There is a greater need to make sure that social and economic facilities are constructed and coordinated with housing projects moving forward to realize the concept of sustainable human settlements. Despite its natural and human resources, South Africa nevertheless faces many socioeconomic problems, including high unemployment rates, poverty, social inequality, and limited access to public services.

These issues still plague South Africans and degrade their living conditions. Even though it is a developing nation, improving the housing infrastructure must be a major priority. A good housing infrastructure can ensure that people's situations improve and aid the economy. Social and economic infrastructure is critical to the development of sustainable communities (Noyoo and Sobantu 2019). While the provision of housing, potable water, and electricity is vital for meeting basic human needs, other services such as schools, transport, and health care are important for ensuring the long-term satisfaction of residents (Atamanov et al. 2018). These infrastructure types work together to build a framework that enables societies with opportunities for social and economic well-being to be established by local citizens.

Housing Infrastructure projects offer a lot of benefits as shown in Fig. 1. It can open up employment opportunities for formerly unemployed individuals, support specialized supply chains, enhance the urban environment, lower social inclusion obstacles, and ultimately improve the well-being of people and communities (Sobantu 2020). This is referred to as social capital. However, it can be developed at all stages of a project's lifecycle from the earliest planning through designing, procurement, delivery, maintenance, and eventually decommissioning. Currently, it is predominantly considered during procurement and construction.

As stewards of the environment, engineers must not only produce functional designs but also contribute to sustainable development, given the underlying economic equity and contextual justice needs of the communities they support and manage (Cruickshank 2007). To adequately grasp the contextual factors, developers will need to consider some metrics, including those from social and economic sciences. This serves as the impetus for the current study, which presents key socioeconomic parameters that are generally crucial for the development of sustainable housing infrastructure.

Measurement of broader social and economic effects is becoming more popular, especially for major projects that affect communities in both positive and negative ways. For instance, the Public Services (Social Value) Act, which went into effect in the United Kingdom in 2013, mandates that individuals who design and deliver basic infrastructure consider how they might achieve additional social, economic, and environmental advantages (Public Service Act 2012). However, this is not commonly done on nongovernment projects, and there is currently no consistently

- Improved quality of life and dignity of residence,
- Affordable access to housing,
- Improved health and lower incidents of illness, fatalities and material losses, better labour productivity,
- Better conditions for human development, employment, creativity and economic growth,
- Durability and low maintenance cost,
- Protection against natural hazards,
- Improved efficiency and savings on the use of energy, water and other physical resources,
- Better environmental protection and sanitary conditions,
- Contribution towards climate adaptation and mitigation,
- More sustainable and socially inclusive urban growth,
- Social cohesion and political stability.

Fig. 1 Benefits of housing infrastructure (Adapted from Golubchikov and Badyina 2012)

applied set of indicators. As such, the stakeholders in this process who have experiences in economics and the social sciences must comprehend and accurately reflect the diverse needs, values, and cultures of the community including deciding which socioeconomic indicators to utilize.

According to Ritchie and Gill (2011), they identified seven capitals—"built, social, human, cultural, economic, political, and natural"—as critical to any sustainable society. Two forms of capital, the social and economic, are considered in this chapter. The socioeconomic metrics considered in this study are shown in Table 2 as (1) infrastructure accessibility, (2) social equity, (3) social cohesion and inclusivity, (4) economic investments, (5) health and well-being, and (6) community services. For engineering and construction projects to be integrated into communities in a way that enhances socioeconomic development and sustainable housing infrastructure, rather than worsening socioeconomic concerns, all six socioeconomic metric descriptors are essential. We chose these dimensions because they reflect the most fundamental infrastructural components that societies should have to promote local living standards and are frequently employed when assessing social and economic factors for societal development.

Table 2 Socioeconomic descriptors for sustainable housing infrastructure

Socioeconomic indicator categories/ dimensions	*Socioeconomic descriptor sets*	*Sustainable outcomes*
Infrastructural accessibility	Ease of access to clean water Ease of access to electricity Ensure design that compliments the character of the community Design mixed-used places where local amenities are easily accessible Design network structure of connected areas to link major transport nodes Ensure houses are designed well connected to jobs, clinics, shops, schools, and other services	Improvement in all transport modes as per time, security, and accessibility Improvement in economic productivity, Well-being levels, and environmental safety Functions as essential components to enhance inclusivity and Well-being in cities
Social equity	Designing areas that are suitable for people of various socioeconomic backgrounds Designing neighbourhoods with a suitable range of housing kinds lowering the cost of some of this housing so that it is accessible to those with incomes below the median	Enhances equity in cities and enable social services Lowers crime Reduced levels of social exclusion
Social cohesion	Design of housing units, including spaces that allow occupants to socialize (garden, meeting space, and relaxation facilities) Expanded horizontal mobility areas that can be used for community interaction, diverse resident-led events, and shared spaces for recreation and entertainment	Facilitates transformative social connections Provides a strong feeling of community while fostering a healthy environment
Economic investments	Developing adequate housing to increase labour productivity Ensuring housing and employment are integrated Forming partnerships for the construction of houses geared at young people promoting indigenous business ventures housing for the most vulnerable groups should be prioritized	Enable improvement of people's lives Improved job prospects for the less skilled Influence on the community's economic performance
Health and Well-being	Design to encourage social interaction, civic participation, physical activity, and recreational time Infrastructure integrating design, inclusivity, and sustainability with convenient public transportation, walking, and cycling options Expanded areas for cycling and walking and landscaping Well-delineated public and green spaces	Facilitate strengthened networks Improve the autonomy and sustainability of the community

(continued)

Table 2 (continued)

Socioeconomic indicator categories/ dimensions	*Socioeconomic descriptor sets*	*Sustainable outcomes*
Community services	Realigning streets to integrate city centers, services, and urban green space closer together Increasing the number of designated pedestrian streets Create extra parking spots with better designs close to commercial areas. transportation system improvements More public places for organizations Pedestrianization of additional local streets	Facilitates improved mobility Improves the quality of life Improve the autonomy and sustainability of the community

5.2 *Barriers to Sustainable Housing Delivery in the SA Context*

Sustainable housing infrastructure remains one of the most important infrastructure needs in South Africa and most of the developing world as it encourages growth, improves welfare, and generates jobs (du Plessis 2010). However, the creation of infrastructure is complex, and the road from conception to construction and operation is a long one full of obstacles like corruption, political and bureaucratic interference, and lack of funding. Because of corruption, those tasked with enforcing the constitutionally given right to obtain housing have routinely misappropriated or corruptly diverted the funds intended for the progressive realization of sustainable housing (Rotberg 2019).

The delivery of housing infrastructure in South Africa is plagued by an endemic kind of corruption, which harms economic growth, unemployment, inequality, and poverty levels (Manomano et al. 2016). Through inflationary pressures, extortion from contractors, and other corrupt activities, corruption impedes the supply of housing infrastructure. Although the state's objective is to be hierarchical and bureaucratic to maintain accountability, this has proven difficult to achieve, particularly in SA and other parts of Africa, necessitating other measures to combat such vices (Nathan 2013). A long tradition of corruption in South Africa's housing delivery system is just one of several possible causes for this. Examples of historical narratives of rent-seeking, corruption, and clientelism in the pre-and post-apartheid South African state can be found in Hyslop (2005) and Lodge (1998). They posited that the vast corruption that existed under the apartheid system continues in South Africa under its current form.

Additionally, there is the problem of political interference in the management of service delivery, which is a major concern in many states. Municipalities have significant difficulties in managing the interaction between politicians and officials. Relationships between important political and administrative authorities in the government units appear to be strained, and there is inappropriate poor governance in administrative problems (Mngomezulu 2020). It cannot be overstated how important it is to assess how service delivery generally works in South Africa.

The issue of bureaucratic procedures is another major barrier. The current housing delivery challenges are caused by a highly complex bureaucratic, administrative, financial, and institutional structure that was left over from the previous administration as well as the massive housing backlog and the desperate and cynical nature of the poor. According to Gumbo (2014), the delayed and complex procedures involved in providing land and land services, such as planning, basic service provision, the building of housing, and occupation, have slowed down the delivery of housing infrastructure.

Another barrier is the lack of funding, expertise, and human capital. These issues have serious implications for the successful delivery and sustainability of housing infrastructure projects. Restricted opportunities for loans and other debt financing sources, a lack of local tax and duty revenues, and shaky planning and procurement procedures used to conduct public projects are further infrastructure development barriers (Owojori and Okoro 2022b). Prioritizing social and economic concerns will ensure that the delivery of infrastructure remains a crucial component of the economy's resilience and economic output.

5.3 *Mainstream Examples of Socioeconomic Development Integrated Projects*

There are typically a handful of successful cases that result in fascinating outcomes when it comes to socioeconomic integrated housing infrastructure. In Brazil, two cities in the state of Bahia, Feira de Santana and Salvador, underwent construction as part of the Bahia Urban Areas Integrated Development Project, which was funded by the International Bank for Reconstruction and Development (IBRD). The goal of the project was to provide access to basic services, better housing options, and social support services in Salvador's least developed and most vulnerable neighborhoods to eliminate urban inequalities in a sustainable manner. The project comes after several prior urban development initiatives in Bahia that were supported by the World Bank and aimed to integrate the provision of basic infrastructure and social services in low-income neighborhoods (World Bank 2018).

According to the project's theory of change, integrated physical interventions (such as infrastructure, housing, and equipment) and social interventions that promote civil society, enhance social services for local communities, and offer job training and income-generating opportunities will strengthen physical, human, and social capital, which is a prerequisite for sustainable housing infrastructure delivery and mitigating urban poverty. By providing access to basic services, improved housing, and social support services, the project was intended to improve sustainable housing provision with the socioeconomic development of the area of influence in a sustainable manner in the most vulnerable areas of Salvador and some other cities in the borrower's territory.

Another initiative with a comparable impact is Carlos Moreno's "15-minute city" model, which was developed in 2016 and has been tested in Paris. It is based on the ideas of convenience and mobility. The model provided people with social, economic, and cultural activities that were less than 15 minutes away (Moreno et al. 2021). The idea is also influenced by New Urbanism, a development strategy that focuses on creating diverse neighborhoods that are walkable and environmentally sustainable. According to Cliff Ellis, New Urbanism integrates a variety of spatial ideas to create high-quality, human-centered plans that support sustainable growth and urban rehabilitation.

Sweden expanded on the approach by emphasizing the construction of streets and crossings. Their bottom-up approach to urban design, spearheaded by innovation agency Vinnova, attempts to link residents to their built environment. The key feature of this idea is that it is feasible to restructure the urban setup so that all of the city's essential services and amenities may be reached by foot in less than 15 minutes. According to Moreno (2021), urban dwellers would have enough time for themselves as well as forge and nurture stronger social ties, resulting in more urban sustainability.

Since individuals are frequently "stuck" in traffic, and important infrastructures like road networks have historically been positioned to favor automotive flow over human movement, this has not been practicable in present metropolitan settings. Only a few cities have bike lanes or walking paths specifically designated for this. In addition to the human and social aspects, the model is thought to open up new opportunities for cities to pursue environmental sustainability agendas, particularly the reduction of emissions from cars, and this has a positive bearing on adhering to the Paris Agreement as well as achieving SDG 13 on Climate Action (Allam 2020). On its economic aspect, several advantages were realized, such as the creation of job possibilities, social equity, and the reduction of excess economic loss resulting from lost productive time in traffic jams.

Colombia's capital, Bogotá, located in South America, likewise invested in this strategy to advance sustainable infrastructure promptly throughout the pandemic. The resurgence of a human-centered approach to urban planning, with thriving, sociable neighborhoods linked by first-rate public transportation and cycling infrastructure, has improved accessibility to essential facilities and services, as well as the standard of living for disadvantaged areas (Gutierrez 2020). A crucial component of the 15-Minute City is proximity to resources and services, as well as a reinforced, equitable urban strategy that provides opportunities for people from different social strata that are simple to access and located within proximity.

5.4 *Summary and Recommendations*

This study highlights important housing infrastructure elements linked to socioeconomic growth, shedding insight into the relationship between housing infrastructure and sustainable development. Socioeconomic and racial issues have historically and now affected the housing options available to varied communities in South Africa. Additionally, gaps in access to sustainable housing infrastructure did not develop naturally; rather, they were socially constructed and enforced as previously mentioned. Access to sufficient housing infrastructure is crucial, given the effects of inadequate housing infrastructure on numerous socioeconomic and well-being factors. Action on housing as a measure to minimize socioeconomic inequities is urgently needed from stakeholders, government and public housing providers, and professionals.

It is imperative to change current rhetoric so that the underlying housing circumstances that promote health, well-being, and a feeling of community are adequately considered. To achieve health and well-being, housing must be acknowledged as a significant source of health and well-being among all stakeholders, not just those involved in the built environment but also those in the many other sectors that intersect with housing. It will take interdisciplinary cooperation between built environment professionals, city and regional planners, housing developers, and other relevant personnel to progress this discourse. They must consider the socioeconomic effects of housing supply, risks, stability, and the local community.

Sustainability should be the main principle in designing housing infrastructure as it contributes to the quality of life. This chapter makes the case that contemporary innovations in sustainable housing delivery and practices should give more weight to social and well-being concerns. To achieve this, suggestions are provided.

Improved infrastructure accessibility: The accessibility concept is crucial to nearly all types of service planning. The main rationale for its inclusion in housing infrastructure planning models is that it is essential to fostering territorial, economic, and social cohesion, which in turn fosters sustainable development. Many components go into improving service access, for instance, more consideration must be given to the creation of targeting methods that are efficient and practical in fostering inclusive service delivery and access. Because South Africa is such a diverse country, social equity is crucial. It is important to assess how well individuals of different backgrounds can get along Housing measures based on these concepts of justice and inclusion are thus important in achieving UN's envisioned SDGs, which strive to develop inclusive and sustainable settlements by 2030. This fosters a unity and sense of *Ubuntu*. Initiatives that make it easier to construct and design homes as well as neighborhoods that are adaptable and take into account the various demands of communities can help achieve this.

Economic investments: If a concerned socioeconomic group's access to employment is impeded, the delivery of housing infrastructure will not be sufficient. Government and the other agencies should collaborate to adequately address the interests and needs of the marginalized in terms of employment and training.

Financial mobilization: For the execution of sustainable housing policies and initiatives, financial resources must be mobilized. This can be done by working with the private sector, public-private partnerships, and organizations. Financial resources must be procured for the implementation of sustainable housing strategies and projects. This can be accomplished by collaborating with businesses, organizations, and public-private partnerships.

Housing infrastructure as part of National Development Strategies: Strategies for socioeconomic development and housing infrastructure must be purposefully weaved and incorporated into the National Development Plan, especially the crucial economic infrastructure that underpins the 2030 National Development Plan, to be compatible with new infrastructure and take long-term socioeconomic development into account.

Improved governance: Enhancing governance and combating corruption to the point when these goals are fully institutionalized in the relevant governmental and nongovernmental structures and practices. Reforming governance structures and launching strategic investment, research, and training programs are necessary for this to occur gradually.

Research and knowledge: Advancement in research and knowledge related to sustainable housing infrastructure is required. The field still has a lot to learn about the best ways to build and execute sustainable housing infrastructure as a track to socioeconomic development. Discussions about sustainable housing should take place in academics, government, business, and nonprofit activities, as well as political and policy debates.

References

Allam Z (2020). The rise of autonomous smart cities: technology, economic performance, and climate resilience. Springer International Publishing, Cham

Andersson, K. and Agrawal, A. (2011). "Inequalities, institutions, and forest commons". *Global Environmental Change*, 21(3), 866–75

Atamanov, A., Azevedo, J. P., Aguilar, R. A. C., Chen, S., Rodas, P. A. C., Dewina, R., et al. (2018). April 2018 PovcalNet Update: What's New. Available at: http://documents.worldbank.org/curated/en/173171524715215230/April-2018-Povcalnet-Update-What-s-New

Behr, Daniela M., Chen, L., Goel, A., Haider, K., et al. (2021). Introducing the Adequate Housing Index: A New Approach to Estimate the Adequate Housing Deficit within and across Emerging Economies. Policy Research Working Paper; No. 9830. World Bank, Washington, DC.

Beinart W., and Delius, P. (2014). "The Historical Context and Legacy of the Natives Land Act of 1913". *Journal of Southern African Studies*, 40(4), 667–688.

Bhattacharya, A., Meltzer, J., Oppenheim, J., Qureshi, Z. and Stern, N. (2016). Delivering on Sustainable Infrastructure for Better Development and Better Climate. The Brookings Institute.

Calderón, C. and Serven, L. (2004). The effects of infrastructure development on growth and income distribution. elibrary.worldbank.org/doi/book/10.1596/1813-9450-3400

Charlton, S. (2013). *State ambitions and peoples practices: An exploration of RDP housing in Johannesburg* (Unpublished Doctoral thesis). University of Sheffield, Sheffield, United Kingdom.

Cruickshank, H. (2007). "The evolving role of engineers: towards sustainable development of the built environment". *Journal of International Development,* 19, 111–121

Du Plessis, A. (2010). Local environmental governance and the role of local government in realising section 24 of the South African Constitution.

Farha, L., Meers, J., and Roca, Z. (2008). Right to Adequate Housing.

Fourie, D. (2006). "The application of good governance in public financial management". *Journal of public administration*, 41(2.2), 434–443.

Golubchikov, O. and Badyina, A. (2012). Sustainable housing for sustainable cities a policy framework for developing countries. UN-Habitat. United Nations Human Settlements Programme 2012. ISBN: 978-92-1-132488-4

Gumbo, T. (2014). Housing the Urban Poor: Experiences of Un-Serviced Housing Plots. Pretoria: Human Sciences Research Council (HRSC).

Gutierrez, G. (2020). Bogotá: sustainability at the core. How the Colombian capital is transforming lives through sustainable development.

Hyslop J (2005). "Political corruption: Before and after apartheid". *Journal of Southern African Studies* 31(4), 773–789

Ji, X., Zhang, Y., Mirza, N., Umar, M. and Rizvi, S. K. A. (2021). "The impact of carbon neutrality on the investment performance: Evidence from the equity mutual funds in BRICS". *Journal of Environmental Management*, 297, 113228

Lodge T (1998). "Political corruption in South Africa". *African Affairs,* 97(387), 157–187.

Manomano, T., Tanga, PT and Tanyi, P. (2016). "Housing Problems and Programs in South Africa: A Literature Review". *Journal of Sociology and Social Anthropology*, 7(2),113–122

Mar, and Edmonds, P.E. (2010). *Making Settler Colonial Space: Perspectives on Race, Place, and Identity* (pp. 1–24). New York, NY: Palgrave Macmillan of Southern African Studies, 40(4), 667–688.

Maylam, P. (2017). *South Africa's racial past: The history and historiography of racism, segregation, and apartheid.* Routledge

Mngomezulu, S. (2020). "Political Interference in the Administration of Service Delivery in Umlalazi Local Municipality of KwaZulu-Natal, South Africa". *Journal of Economics and Behavioral Studies* 12(1)38–45.

Modise, L., and Mtshiselwa, N. (2013). "The Natives Land Act of 1913 engineered the poverty of black South Africans: A historical ecclesiastical perspective". *Studia Historiae Ecclesiasticae*, 39(2), 359–37

Moreno, C., Allam, Z., Chabaud, D., Gall, C., and Pratlong, F. (2021). "Introducing the 15-Minute City: Sustainability, Resilience and Place Identity in Future Post-Pandemic Cities". *Smart Cities, 4*(1), 93–111

Moyo, T., and Mambo, J. (2022). "The use of Systems Thinking for Housing Delivery in South Africa". *African Journal of Public Sector Development and Governance* 5(1), 1–18.

Mubangizi, B. C. (2008). "Responses to poverty in post-apartheid South Africa: some reflections. *International Journal of Social Welfare*, 17(2), 174–181

Nathan, O. (2013). Building revolutionary working class counter power: Municipalities, Service Delivery and Protest. Zabalaza, 19 March.

National Planning Commission (NPC). (2011). National Development Plan (NDP): Vision 2030. Pretoria: NPC

Nnadozie, R. C. (2013). "Access to basic services in post-apartheid South Africa: What has changed? Measuring on a relative basis". *African Statistical Journal*, 16, 71–99

Noyoo, N., and Sobantu, M. (2019). *Deconstructing and decolonising spatiality: Voluntary and affordable housing for a transforming Johannesburg*. In M. T. Myambo (Ed.), Reversing urban spatiality (pp. 35–42). London: Routledge.

Nubi, T.O (2002). Procuring, Managing and Financing Urban Infrastructure: Towards an Integrated Approach Land Management and Property Tax Reform in Nigeria, in `Omirin et al., (ed.) Department of Estate Management, University of Lagos, Akoka

Owojori, O. and Okoro C. (2022a). "Making the 'Available Desirable' Using Adaptive Reuse (AR) In Sustainable Construction: A Systematic Review and Directions from case studies ". *The Twelfth International Conference on Construction in the 21st Century (CITC-12Amman, Jordan | May 16–19, 2022.* ISBN: 979-8-9865406-0-3

Owojori, O.M.; Okoro, C.S. (2022b). Overcoming Challenges Associated with Circular Economy in Real Estate Development. In *Sustainable Education and Development—Making Cities and Human Settlements Inclusive, Safe, Resilient, and Sustainable*; Mojekwu, J.N., Thwala, W.,

Public Service (Social Value) Act. (2012). Chapter 3 The Stationery Office Ltd London, UK

Ramafamba, E., and Mears, R. (2012). "The role of service delivery in local economic development: A case study of Mamelodi Township. African Journal of Business Management" ,6(4), 1564–1572

Ritchie, L.A. and D.A. Gill (2011) "Considering Community Capitals in Disaster Recovery and Resilience," PERI Scope, Public Entity Risk Institute, 14(2).

Rotberg, R.I. (2019). The corruption cure: How citizens and leaders can combat graft. Princeton of University Press, New Jersey, USA

Rust, K. (2020). Housing is central to achieving the SDGs. Available from: https://housingfinanceafrica.org/documents/housing-is-central-to-achieving-the-sdgs/

Sobantu, M. (2020) "Housing and Social Development: Linking social and economic aspects of development through housing delivery," *Social Development Issues*, 42(2).

Swope, C. B., and Hernández, D. (2019). "Housing as a determinant of health equity: A conceptual model". *Social Science & Medicine*, 243, 112571.

Thacker, S., Adshead, A., Fay, M. and Hallegatte, S. (2017). "Infrastructure for sustainable development". *Springer Nature*. https://doi.org/https://doi.org/10.1038/s41893-019-0256-8

The Constitution of South Africa. (1996). *The Constitution of the Republic of South Africa, 1996*. Retrieved from https://www.gov.za/documents/constitution/constitution-republic-south-africa-1996-1

UN-HABITAT (2015). *Housing at the Center of New Urban Agenda*, pp. 3–4.

Weingaertner, C., and Moberg, A. (2014). "Exploring social sustainability: learning from perspectives on urban development and companies and products". *Sustainable Development*, 22(2), 122–133

Victor, S., Precious, Z., Facundo, C. (2022). *Inequality in Southern Africa: An Assessment of the Southern African Customs Union (English)*. Washington, D.C.: World Bank Group.

World Bank. 2018. Country— Bahia Poor Urban Areas Integrated Development Project: Viver Melhor II. Independent Evaluation Group, Project Performance Assessment Report 127250. Washington, DC: World Bank.

Green Infrastructure: Planning for Sustainable and Resilient Small Towns – Evidence from the Seine Valley in France

Marie Asma Ben Othmen, Mai Laila, Lukas Madl, Felix Schachenmayr, and Gabriella Trotta-Brambilla

1 Introduction

Sustainable urbanization is part of the United Nations' Sustainable Development Goals. Mainly, SDG 11: Sustainable Cities, which relies on achieving targets such as preserving cultural and natural heritage and tackling climate change, is being addressed by 196 countries worldwide through the New Urban Agenda (UN-Habitat 2020). Although cities are crucial in achieving a sustainable future for a world becoming urban, not all people living in urban areas live in large cities. Small towns – with less than 100,000 inhabitants – are crucial for connecting 46% of the world's population in rural areas and 35% in cities. They are also cornerstones in improving urban-rural linkages (Roberts 2016) as well as the economic development of their wider area (Sietchiping et al. 2014). According to the recent UN-Habitat report "Unpacking the Value of Sustainable Urbanization," there are solid interdependent relationships between small towns, cities, and rural areas that are cornerstones to achieving sustainable urbanization.

Achieving SDG 11 in small towns is challenging, as projections expect their population to increase by 170% by 2030 (Lambe 2012). This would result in the

M. A. Ben Othmen (✉)
INTERACT Research Unit–Innovation, Land Management, Agriculture, Agro-Industries, Knowledge, and Technology, UniLaSalle, Rouen, France
e-mail: marie-asma.benothmen@unilasalle.fr

M. Laila · L. Madl · F. Schachenmayr
Urban Agriculture and Green Cities, UniLaSalle, Rouen, France
e-mail: mai.laila@etu.unilasalle.fr; lukas.madl@etu.unilasalle.fr; felix.schachenmayr@etu.unilasalle.fr

G. Trotta-Brambilla
National School of Architecture, Normandie, Rouen, France
e-mail: gabriella.trotta@rouen.archi.fr

I. S. Dunmade et al. (eds.), *Sustainable Engineering*, Green Energy and Technology, https://doi.org/10.1007/978-3-031-47215-2_17

expansion and the reclassification of rural areas into urban annexes, exacerbating unsustainable urbanization trends (Report World Cities 2020).

Small towns face the same sustainability challenges as large cities and must adapt to increasing and unpredictable vulnerability to climate change and address its induced risks. However, these towns' current urban development model fails to promote sustainability and ensure their resilience (Mallick et al. 2021). Scholars have attempted to rethink sustainable urbanization in small towns by considering restoring the co-evolutionary relationships between towns and their surrounding ecological and agricultural systems. With this regard, Fanfani (2020) suggested the *urban-bioregion* paradigm to design functional, cultural, and environmental rules. These are fundamental tenets of achieving resilience in small towns and, more broadly, enhancing practices for urban, peri-urban, and rural integrated planning policies (Fanfani et al. 2022). Another facet of the literature utilized the concept of green infrastructure (GI) to promote sustainable planning interventions in small towns and enhance their resilience. King and Shackleton (2020), for instance, underscore the role of green infrastructure in providing people with opportunities to advocate for ecosystem services and empowering individuals and communities to become responsible stewards of the natural environment, which could, ultimately, help achieve sustainable urbanization.

In this chapter, we suggest reconsidering the approaches to enhancing the resilience of small towns through a more functional and sustainable urban planning perspective. To do so, we explore the potential of combining green infrastructure and bio-region paradigm approaches by considering a case study from the Seine Valley in northwestern France: *Rives-en-Seine*.

This chapter does not consider this small town merely as an administrative entity. It views it instead as an ecological unit where the built environment is embedded into a more extensive natural system and an "opportunity to seek a new balance between established communities, the surrounding environment, and local heritage" (Fanfani 2018). The primary focus of this chapter is to investigate the potential of green infrastructures in providing a diverse array of ecosystem services that are crucial to human well-being, urban sustainability (O'Brien et al. 2017), and resilience (Bush and Doyon 2019) in small towns.

2 Setting the Stage: Defining Green Infrastructure (GI)

Green infrastructure is one of the many nature-based solutions and ecosystem service strategies deployed across the globe to contribute to human well-being and environmental protection (Fedele et al. 2018). Based on landscape principles, the green infrastructure concept evokes a multi-scalar network of ecological elements providing a variety of functions and benefits (Benedict MA and McMahon E. 2001) and reconnecting urban environments with the biosphere (Seddon et al. 2021).

The green infrastructure concept has prevailed in the European tradition of spatial planning, land conservation, and landscape design since the nineteenth century (Benedict MA and McMahon E. 2001). Even if still relatively new, it remains a guiding principle for planning efforts to provide equitable access to high-quality green spaces. Moreover, what is interesting about green infrastructure is that they have gained increased attention from planners and architects but also from ecologists, environmental groups, and even politicians (Mell 2017).

Parkways, green belts, or garden cities entail what is known to be the earlier examples of green infrastructure (Ignatieva et al. 2011) that expanded to all forms to include several forms of integrating environmental management and ecological conservation approaches leading to a more holistic approach to green infrastructure as we know it today. However, beyond its original purpose of improving the aesthetic value of urban areas and building cultural identity, green infrastructure is now recognized for its multifunctional role and equal importance as an urban planning tool alongside other planning instruments contributing to people's quality of life. For instance, Monteiro et al. (2020) emphasize the green infrastructure's role in carbon sequestration, cooling, mitigating air pollution, and offering recreational opportunities and health benefits. On the other hand, Dong et al. (2017) point to the relevance of green infrastructure in flood risk management. According to the authors, the flooding risk is no more to be tackled through the sole mean of gray infrastructure (e.g., civil engineering protective infrastructure, such as dykes and seawalls). The authors even call for "greening the gray infrastructure" to ensure the ecological enhancement and urban drainage resilience reinforcement.

Similarly, Alves et al. (2019) argue that a mix of green, blue, and gray infrastructures will likely result in the best adaptation strategy to face climate change. These alternatives complement each other, as while gray infrastructure efficiently reduces flooding risks, green spaces offer additional benefits that gray infrastructure cannot offer (Brink et al. 2016). Moreover, not only Jaffe (2010) argues that green infrastructure is more cost-effective than traditional stormwater management methods, but Dell'Anna et al. (2022) emphasize their benefits in terms of raising property values and generating economic benefits in the areas where it is implemented.

In recent years, while large cities have made notable progress in implementing green infrastructure to address or anticipate future trends associated with climate change, small towns have been slow to follow suit due to financial constraints and the prioritization of other economic development initiatives over climate change adaptation (Salas Tobón and Barton 2019). This chapter addresses this gap and demonstrates that carefully designed green infrastructure considering multiple functions and ecological continuities – taking into account the wider bioregion – can bring multiple environmental, social, and economic benefits, thereby enhancing small towns' resilience.

3 Case Study Presentation: Rives-En-Seine (Northwestern France)

Rives-en-Seine is a small town in northwestern France that is part of the Seine Valley landscape and the Natural Regional Park: *Les Boucles de la Seine Normande* (Fig. 1). This small town is relevant as a case study because even if ecological elements are still robust markers in the area, over decades, successive spatial transformations caused significant degradation to the local environment and reinforced its vulnerability to climate change-induced hazards (e.g., increased heatwaves and floods caused by the river levels rising, erosive runoffs, and water quality deterioration). Rives-en-Seine also has the peculiarity of being a relatively newly born town from the merger in 2016 of three villages: Caudebec-en-Caux (mostly urbanized), Villequier, and Saint Wandrille-Rançon (both primarily rural). The Seine River is a natural central axis that structures the territory, links the three villages, and even gives its name to the newly formed municipality. Nonetheless, each of the three municipalities has a recognized history and cultural and natural heritage, making it challenging to build a common identity for the newly formed small town (Ben Othmen et al., 2022).

Despite its urbanization, Rives-en-Seine is not disconnected from conservation efforts. Not only is it a link but also a habitat itself, surrounded by several other valuable habitats, and can play a significant role in mitigating the loss of species. Indeed, Rives-en-Seine is located within and between several important biotopes and biodiverse habitats, including wet meadows on the Seine side that are home to a diverse range of aquatic, amphibian, insect, and bird species. The small town is surrounded by forests in the west and east, while urban areas border the classified wetland of the Marais de Caudebec-en-Caux in the north making Rives-en-Seine akin to an island between crucial biodiversity reservoirs. On the other hand, the urban infrastructures – residential areas, streets, and artificial zones – stand in significant contrast to their more natural surroundings and form a barrier of

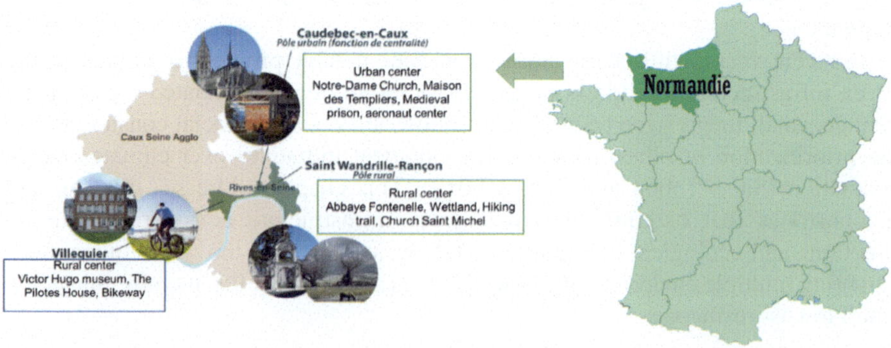

Fig. 1 Localization of Rives-en-Seine's three municipal entities (Caudebec-en-Caux, Villequier, and Saint Wandrille)

connectivity between important habitats. Similarly, the artificial structures present along the Seine riverbanks within the small town create a second barrier for the blue corridors that connect the Seine River and its marshes with the Marais, as shown in Fig. 2. Additionally, Rives-en-Seine is highly vulnerable to flooding, given its location and the anticipated rise in the Seine's water level by 2050, which may result in most of the town being submerged (IPCC 2022). Moreover, the agricultural areas of the plateaus are subject to land management pressure and intensive farming practices, further compounding the risks associated with runoffs.

4 Methodology

This chapter applies an ecosystem-based approach to spatial planning practices, which spans from investigating ecosystem properties to the social frameworks and personal values that drive and shape human interaction with nature and could effectively inform the green infrastructure implementation (Andersson et al. 2014). Accordingly, the methodology is grounded in an interactive group workshop involving the students and the professors of two institutions: the National School of Architecture of Normandy and the Earth and the Life and Earth Sciences Polytechnic Institute UniLaSalle (Rouen Campus), as well as other local stakeholders, including the regional natural park *Les Boucles de la Seine Normande* and the public establishment for land management as well as the municipal team members and local officials and representatives from Rives-en-Seine municipality. The workshop involved first a scoping phase to identify the goals of ecosystem enhancement through green infrastructure and threats to their achievement. Second, in addition to observations and species inventory of local fauna and flora, a risk analysis of the whole area spanning the three municipalities was performed. Interviews were conducted with major stakeholders involved in the town's ecological restoration

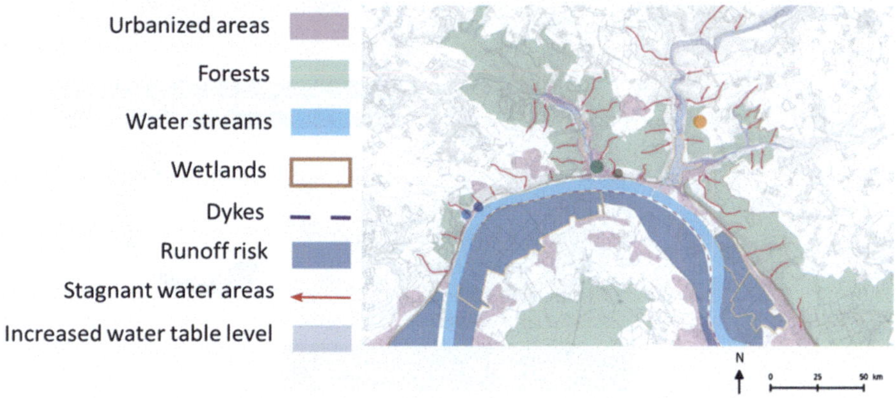

Fig. 2 Mapping flood, runoff risks, and major greenways in Rives-en-Seine

projects (watershed management agency, green spaces management services, etc.) and urban planning (land use departments). Finally, focus groups were organized with residents to understand their perception of ecosystem services benefits (e.g., environmental, social, and economic) that ecological restoration and green infrastructure could achieve.

5 Results

Despite the significant loss of natural habitats in Rives-en-Seine, there is still potential for restructuring public spaces based on the green and blue corridors, wetlands, marshes, and wooded and agricultural areas. It is crucial to reconnect these natural and urban spaces to form a dense and functional network. Our findings emphasize the importance of ecological restoration and the significant role that green infrastructure can play in providing the small town with sustainability and resilience capacity. The diagram in Fig. 3 depicts the green infrastructure options discussed with local stakeholders and residents during the workshop. It displays how these solutions can be executed and their intended outcomes (e.g., environmental, social, and economic sustainability). Each of these options was developed according to three guidelines: (1) increasing the native urban biodiversity across urbanized areas,

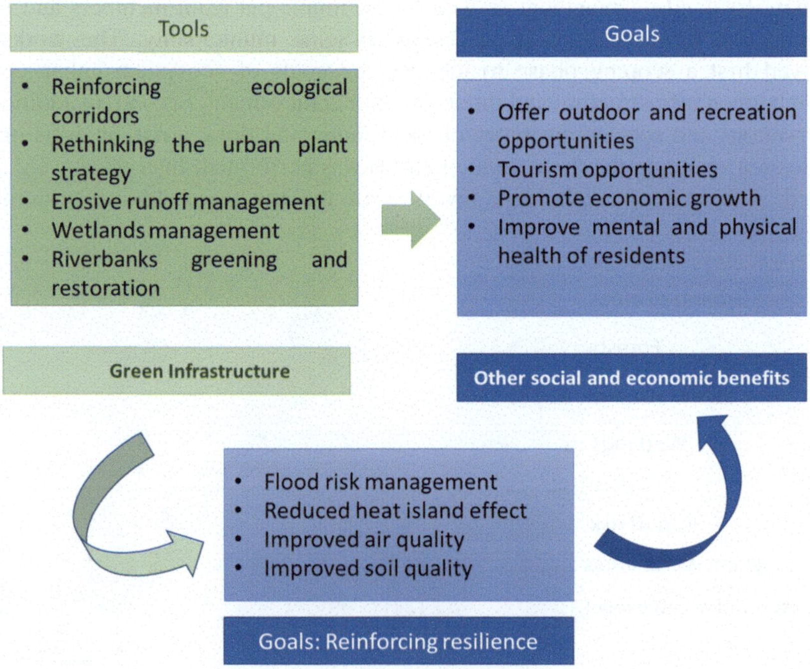

Fig. 3 Green infrastructure as tools for achieving sustainability goals (Ben Othmen & Trotta-Branbilla., 2022)

(2) improving the ecological connectivity through green and blue corridors, and (3) maximizing the potential ecosystem services productions to achieve environmental, social, and economic benefits.

Viewing the urban environment through an ecological lens enables the creation of strategies that address multiple issues and target several economic, social, and environmental goals, including improving the quality of living spaces, reducing hydrogeological risks, and promoting economic growth. Figures 4 and 5 visually

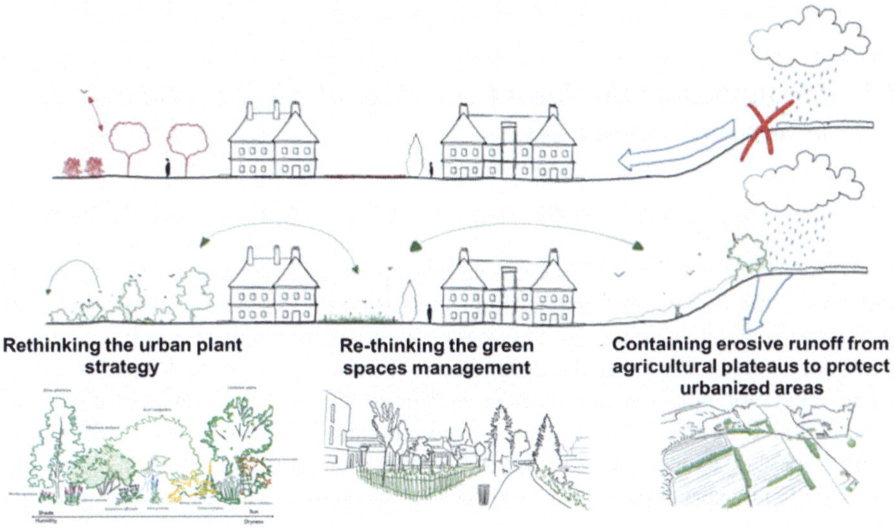

Fig. 4 Potential use of green infrastructure at the Caudebec-en-Caux (urban) – Saint Wandrille (rural) interface

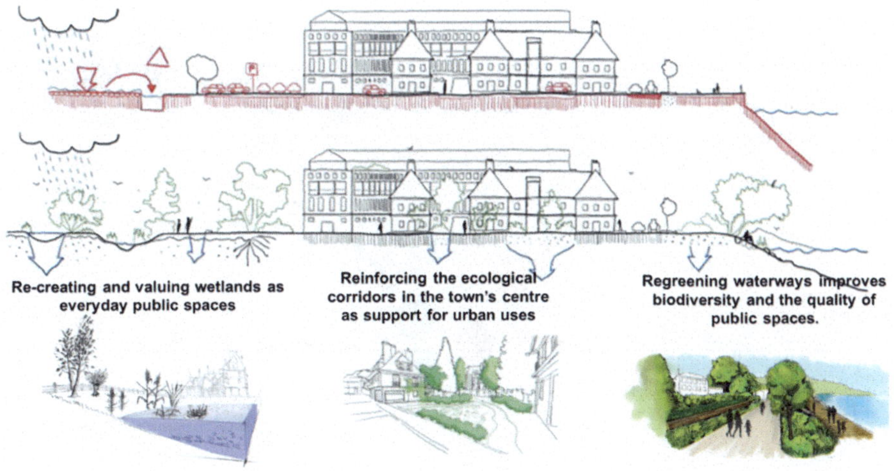

Fig. 5 Potential use of green infrastructure at Villequier (rural) – Caudebec-en-Caux (urban) interface

represent the green infrastructure options selected and articulated during the workshop. These options range from rethinking the green spaces management and urban plant strategy, containing erosive runoff from agricultural plateaus, to re-recreating and valuing wetlands as everyday public space and regreening waterways and riverbanks to improve the public space quality along with the small-town's resilience. As discussed during the workshop, these solutions are not just aesthetic or recreational but serve as a starting point for reinforcing the small town's resilience and sustainability.

5.1 *Strengthening the Town Center's Ecological Corridors to Support Urban Uses*

The natural habitats in Rives-en-Seine are highly fragmented due to the overdevelopment of urban spaces. The town center's main square, Caudebec-en-Caux, can be ecologically improved by creating new green areas, such as ecological parks, and transforming courtyards and parking spots (Fig. 6). Replacing hard paving with permeable vegetated surfaces will increase soil infiltration, reduce runoff, and direct excess rainwater to recharge the groundwater. In addition to its ecological benefits, urban vegetation contributes to human well-being by creating a sense of place and reducing the urban heat island. These developments are also well received by residents as they are expected to improve the organization of public spaces in a way that creates safe pedestrian paths and local economic activities such as farmers' markets and restaurants.

It is worth noting that for these ecological corridors to support sustainable urban uses successfully, it's essential to involve various stakeholders from the start of the

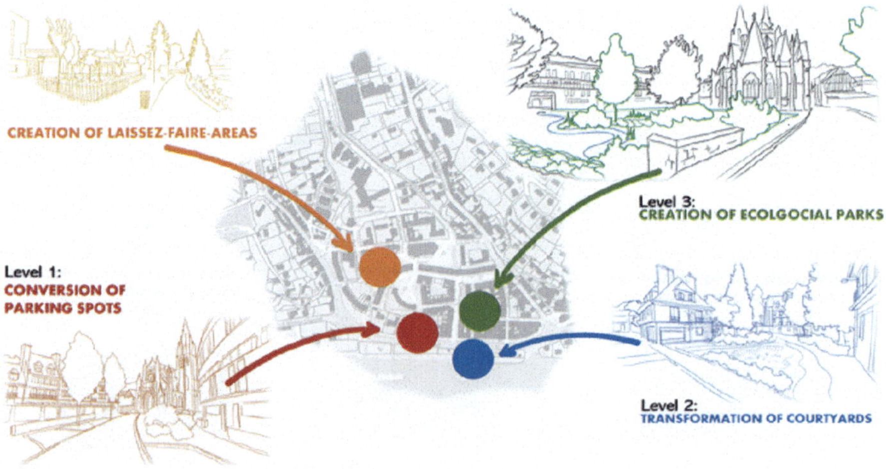

Fig. 6 Intervention strategy in the small-town center: Caudebec-en-Caux

project. Engaging urban planners, natural resource managers, ecologists, etc., in a mutual effort to integrate green spaces' ecological function into the urban functions of gray and blue infrastructure in the small town of Rives-en-Seine, is essential for the project's success over the short mid and long term.

5.2 Revising the Plant Strategy in the Town Center

The current plant strategy in Rives-en-Seine primarily includes species unsuited for the local environment. It features exotic species with less or no functions in the local ecosystem and no relation to native species. Therefore, we propose a total rework of the plant strategy toward a list mainly consisting of local trees, bushes, shrubs, and flowers (Fig. 7). We also suggest actively discouraging the use of any classified invasive species, such as *Buddleia davidii* or *Cotoneaster horizontalis*, both of which were found in the flauristic inventory. Although improving the relationship between plants and their surroundings is essential, green spaces in the town center – Caudebec-en-Caux – undergo intense management, regularly cutting bushes, mowing lawns, and removing spontaneous growth. While maintenance is necessary, these practices was found to hinder the diversity of plants life throughout the small town.

Reducing the intensity of green space management is essential to allow natural processes to occur and boost urban biodiversity. One way to achieve this is to designate "*hands-off*" or "laissez-faire" areas in some of the town's vacant areas. These areas can serve as microcosms of urban biodiversity, where vegetation can thrive, providing a natural habitat for local wildlife.

Creating such areas can also serve as an educational tool to raise environmental awareness among residents. By showcasing the importance of biodiversity and the

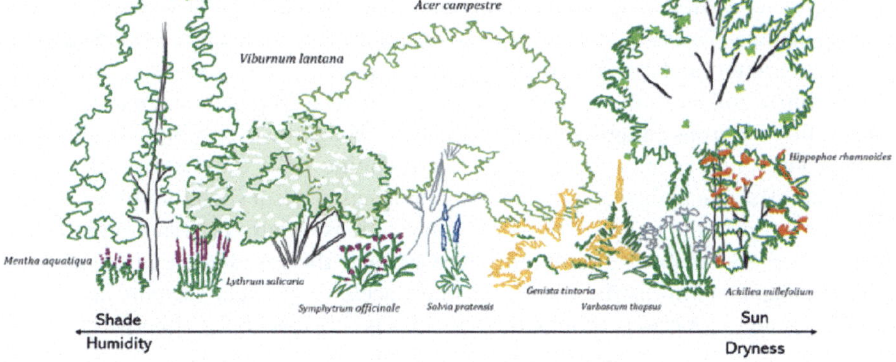

Fig. 7 Suggestions for the future plating palette, sorted by context

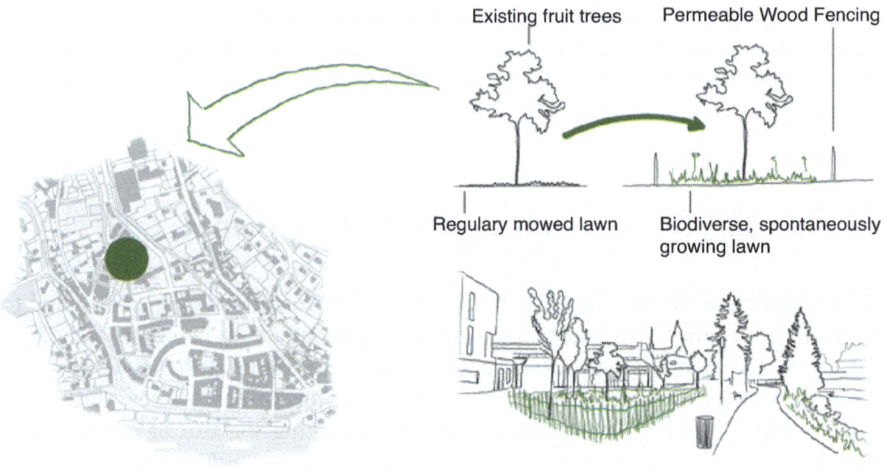

Fig. 8 Example of a laissez-faire site

benefits of allowing nature to take its course, residents can better understand the value of preserving natural habitats and promoting biodiversity in urban areas. Figure 8 can be used as an example to illustrate the potential benefits of such laissez-faire green spots.

5.3 Preventing Erosive Runoff from Agricultural Plateaus to Protect Urban Settlements

Rives-en-Seine is located amid multiple agricultural plateaus in a region affected by the persistent, even aggravated, problem of erosive runoff caused by various hydrological disturbances. To mitigate this risk, adopting appropriate cultivation techniques and vegetation distribution in these areas is crucial. Vegetative barriers, such as shrub hedges, fascines, and hedgerows, are significant in catchment management to decrease runoff and erosion effects (Richet et al. 2017). Hedgerows, in particular, can minimize soil and soil organic carbon loss, enhance crop yield, increase nutrient availability for crops in agricultural plots above hedgerows, and offer habitats for herbaceous forest species, particularly thermophilic ones (Graham et al. 2018). However, effective hedgerow implementation necessitates a spectrum of strategies. These encompass the reorganization of crop rotations, reintroduction of grasslands, adoption of sustainable agricultural practices, and the installation of small-scale water management infrastructure. These multifaceted approaches require collaborative efforts among farmers to maximize yields and mitigate risks effectively.

5.4 *Valuing Wetlands as Everyday Public Spaces*

Recognizing wetlands as public spaces vitalizes residents' daily experiences by ensuring accessibility and providing diverse recreational and leisure activities. This fosters heightened public awareness and appreciation for these areas, nurturing their preservation for future generations. Our wetland analysis found that wetland size and connectivity are essential for maintaining biodiversity, species distribution, and water quality in Rives-en-Seine. Hence, to restore wetlands effectively, a landscape approach that considers the local context and how different habitats are interconnected is necessary (Allen et al. 2020). Accordingly, a comprehensive wetland management strategy for Rives-en-Seine should consider the dynamics of these areas, particularly in Caudebec-en-Caux and Saint-Wandrille, where urbanization is expanding into water-saturated and flood-prone areas in the valley during extreme rainfall events. Thus, developing a network of wetlands crossing Caudebec-en-Caux to the Seine can play a crucial role in promoting the migration of species from marshes (located in the north) to wetlands (located in the south).

Particularly the Marais located in the north should be protected, and options to enlarge it should be considered, given that biodiversity and water quality increase with the size of wetland patches. Because Rives-en-Seine is characterized by the presence of single, small patches of water retention areas scattered around the small town, the wetland development plan has to include corridors between existing and potential new wetland patches and retention/detention areas. In particular, creating a wetland network through Caudebec-en-Caux to the Seine could promote species migration from the Marais in the north to the humid zones in the south. One way to increase the number of wetland patches is to promote and subsidize gray water treatment wetlands in new building projects or renovation processes of existing buildings. These green infrastructures serve multiple functions, such as positively influencing the microclimate and even improving residents' mental health (Mahmoudi et al. 2021). Moreover, we identified brownfields bordering the Seine shoreline as potential locations for renaturation processes (Fig. 9). The direct proximity of wetland patches to the Seine River could further encourage hydrophilic species distribution from the north to the south and vice versa.

Fig. 9 Visual representations of restored wetlands

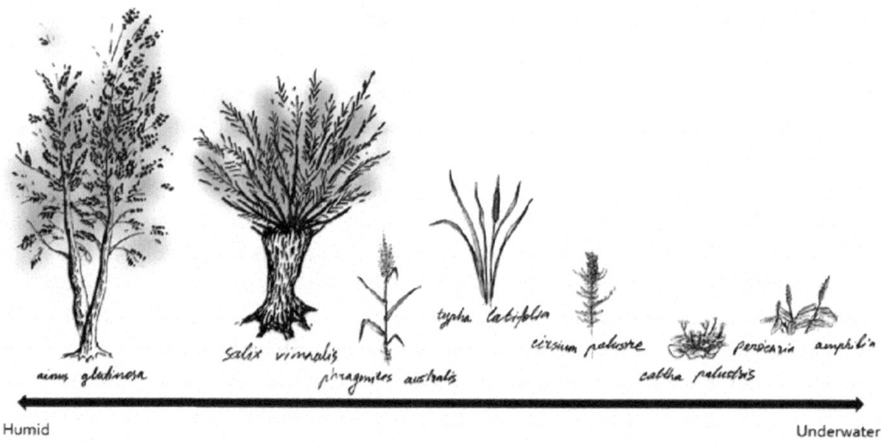

Fig. 10 Examples from a future planting palette, sorted by soil humidity

To conclude, expanding the potential of wetland networks as ecological corridors could be advanced by integrating them with greening initiatives along the Seine riverbanks. Replacing the current concrete borders with gravel ones not only enhances capacity but also elevates the aesthetic appeal of the riverbanks. In addition, using native species (Fig. 10) during the wetland restoration works, as suggested earlier in this chapter, could be an interesting option.

5.5 *The Seine Riverbank Restoration Boosts Species Diversity and Economic and Social Uses of the Public Space*

As stated previously, the Seine River is a structuring element of the landscape and economic activities in Rives-en-Seine. It also provides essential water sources, influences the microclimate, provides and affects habitats of multiple species, and interlinks the whole area of the Seine Valley in Normandy (Martinez 2009). However, the Seine banks fail to provide a suitable ecological habitat as they are dominated by concrete slabs, preventing natural vegetation growth and shelter for aquatic species and deteriorating the river's ecosystem. This is a result of past civil engineering practices that have dominated Seine's morphological restoration, along with other water streams crossing urban areas (diking, concreting, etc.).

Green infrastructure could significantly help restore the Seine riverbank and its ecological significance while boosting ecosystem services, such as flood protection. This could be achieved by allowing the stream to erode the banks in less vulnerable areas or developing flood expansion areas using riparian wetlands. Developing public spaces along water streams is also recommended, where access can be temporarily restricted during flooding events. By combining gray infrastructure with green infrastructure, such as using wooden structures and planting native species, the

Fig. 11 A vision for the riverbank restoration

Fig. 12 Image of the Seine riverbank sections that have been kept natural

riverbanks' development can be more environmentally sustainable and cost-effective (Zeng et al. 2021) (Figs. 11 and 12).

In more technical terms, we suggest a three-layer design: (a) a foundation of wooden piles with a base of stone and gravel, (b) a first layer, which involves removing concrete and reshaping the slope using geogrids filled with gravel at the bottom, and (c) a cover of woody plants planted in the upper part and an optional wire mesh for tidal protection.

In contrast, this option focuses on enhancing the natural edge of the river and does not allow easy access to the water. As a remedy, we suggest that certain areas will have a floodable wooden pathway to allow people to experience the Seine up close. Therefore, the second layer (c) is intended to feature a row of trees and permeable pavement, transforming the riverbank into a gathering space for the community and a potential redevelopment opportunity (Fig. 11, right). Thus, beyond its ecological functions and protection against floods, regreening the riverbanks could embody social functions by providing a space for people to connect with the Seine ecosystem and its functions. The design also incorporates aesthetic considerations, as exemplified by the *Le Jardin de Victor Hugo* simulation (Fig. 11, right). With the diverse range of plants it includes (as emphasized earlier in this chapter), it has the potential to attract not only residents but visitors from all over France, boosting local economic growth in the small town.

6 Conclusion

Green infrastructure has emerged as a crucial strategy to achieve sustainable development by effectively coordinating environmental, social, and economic development (Ying et al. 2022).

Envisioning effective green infrastructure plans that withstand climate change impacts depends on a deep understanding of the dynamics of the inhabitant and the natural environment. This chapter considered the specific case of implementing green infrastructure in small towns that are deeply connected to their unique local characteristics and natural resources. It highlighted the importance of considering local context when developing principles for including ecological, social, and economic components in green infrastructure design that could achieve sustainable urbanization outcomes.

So far, the use of green infrastructure has been broadly promoted in cities and has been less or not adopted to the scale of small towns. From a practice and urban planning and design perspective, such an approach risks ignoring the uniqueness of small towns and the natural environment in which they are embedded. In this chapter, we demonstrate that small towns like Rives-en-Seine, have untapped potential for incorporating ecosystem services into the built environment, restoring degraded ecosystems, and improving ecosystem function through innovative spatial design and planning. Such an approach help achieve environmental, economic, and social outcomes and promote transformative adaptation processes to climate change.

Acknowledgments The authors would like to acknowledge Rives-en-Seine stakeholders, mainly the Parc Naturel Regional des Boucles de la Seine Normande, the Mayor, the municipal team, and the residents of Rives-en-Seine for their cooperation and involvement in this research. The authors also acknowledge the Master Architecture, City and Territory students from the National School of Architecture, Normandie for their collaboration.

References

Allen, C., Gonzales, R., & Parrott, L. (2020). Modelling the contribution of ephemeral wetlands to landscape connectivity. *Ecological Modelling*, *419*, 108944. https://doi.org/10.1016/j.ecolmodel.2020.108944

Alves, A., Gersonius, B., Kapelan, Z., Vojinovic, Z., & Sanchez, A. (2019). Assessing the Co-Benefits of green-blue-grey infrastructure for sustainable urban flood risk management. *Journal of Environmental Management*, *239*(February), 244–254. https://doi.org/10.1016/j.jenvman.2019.03.036

Andersson, E., Barthel, S., Borgström, S., Colding, J., Elmqvist, T., Folke, C., & Gren, Å. (2014). Reconnecting cities to the biosphere: Stewardship of green infrastructure and urban ecosystem services. *Ambio*, *43*(4), 445–453. https://doi.org/10.1007/s13280-014-0506-y

Benedict, M. A., & McMahon, E. (2001). *Green infrastructure: conservation for the 21st century*. Sprawl Watch Clearinghouse.

Ben Othmen, M. A., Bigault-Dam, A., Léglise, N., & Trotta-Brambilla G. (2022). Quel futur pour les petites villes normandes ? Rives-en-Seine. Hors série, pp. 64, Etudes Normandes. ⟨hal-04199483⟩

Ben Othmen, A., & Trotta-Brambilla, G. (2022). Circularity and Resilience: from multidisciplinary experimentation to urban transformation. Challenges for the Next Gen Built Environment Conference, Bologna University, May 2022, Bologna, Italy. ⟨hal-04213893⟩

Brink, E., Aalders, T., Ádám, D., Feller, R., Henselek, Y., Hoffmann, A., Ibe, K., Matthey-Doret, A., Meyer, M., Negrut, N. L., Rau, A. L., Riewerts, B., von Schuckmann, L., Törnros, S., von Wehrden, H., Abson, D. J., & Wamsler, C. (2016). Cascades of green: A review of ecosystem-based adaptation in urban areas. *Global Environmental Change*, *36*, 111–123. https://doi.org/10.1016/j.gloenvcha.2015.11.003

Bush, J., & Doyon, A. (2019). Building urban resilience with nature-based solutions: How can urban planning contribute? *Cities*, *95*(October). https://doi.org/10.1016/j.cities.2019.102483

Dell'Anna, F., Bravi, M., & Bottero, M. (2022). Urban Green infrastructures: How much did they affect property prices in Singapore? *Urban Forestry & Urban Greening*, *68*, 127475. https://doi.org/10.1016/j.ufug.2022.127475

Dong, X., Guo, H., & Zeng, S. (2017). Enhancing future resilience in urban drainage system: Green versus grey infrastructure. *Water Research*, *124*, 280–289. https://doi.org/10.1016/j.watres.2017.07.038

Fanfani, D. (2018). The urban bioregion as form and project of the co-evolution between urban and rural domain. the case of the Florence metropolitan area. *International Journal of Engineering & Technology*, *7*(1.4), 61. https://doi.org/10.14419/ijet.v7i1.4.9264

Fanfani, D. (2020). Co-evolutionary Recovery of the Urban/Rural Interface: Policies, Planning, and Design Issues for the Urban Bioregion BT - Bioregional Planning and Design: Volume I: Perspectives on a Transitional Century. In D. Fanfani & A. Matarán Ruiz (Eds.), *Bioregional Planning and Design: Volume I* (pp. 129–150). Springer International Publishing. https://doi.org/10.1007/978-3-030-45870-6_8

Fanfani, D., Duží, B., Mancino, M., & Rovai, M. (2022). Multiple evaluation of urban and peri-urban agriculture and its relation to spatial planning: The case of Prato territory (Italy). *Sustainable Cities and Society*, *79*(March 2021). https://doi.org/10.1016/j.scs.2021.103636

Fedele, G., Locatelli, B., Djoudi, H., & Colloff, M. J. (2018). Reducing risks by transforming landscapes: Cross-scale effects of land-use changes on ecosystem services. *PLoS ONE*, *13*(4), 1–21. https://doi.org/10.1371/journal.pone.0195895

Graham, L., Gaulton, R., Gerard, F., & Staley, J. T. (2018). The influence of hedgerow structural condition on wildlife habitat provision in farmed landscapes. *Biological Conservation*, *220*, 122–131. https://doi.org/10.1016/j.biocon.2018.02.017

Ignatieva, M., Stewart, G. H., & Meurk, C. (2011). Planning and design of ecological networks in urban areas. *Landscape and Ecological Engineering*, *7*(1), 17–25. https://doi.org/10.1007/s11355-010-0143-y

IPCC. (2022). *Climate Change 2022 Impact, Adaptation and Vulnerability, a summary for policy-makers*. https://www.ipcc.ch/report/ar6/wg2/

Jaffe, M. (2010). Environmental reviews & case studies: reflections on Green Infrastructure economics. *Environmental Practice*, *12*(4), 357–365.

King, A., & Shackleton, C. M. (2020). Maintenance of public and private urban green infrastructure provides significant employment in Eastern Cape towns, South Africa. *Urban Forestry and Urban Greening*, *54*(February), 126740. https://doi.org/10.1016/j.ufug.2020.126740

Lambe, W. H. (2012). *Small Town Development Approaches*. UN-HABITAT. http://unhabitat.org/publications-listing/small-town-development-approaches-the-global-urban-economic-dialogue-series-series-title/

Mahmoudi, A., Mousavi, S. A., & Darvishi, P. (2021). Greywater as a sustainable source for development of green roofs: Characteristics, treatment technologies, reuse, case studies and future developments. *Journal of Environmental Management, 295*, 112991. https://doi.org/10.1016/j.jenvman.2021.112991

Mallick, S. K., Das, P., Maity, B., Rudra, S., Pramanik, M., Pradhan, B., & Sahana, M. (2021). Understanding future urban growth, urban resilience and sustainable development of small cities using prediction-adaptation-resilience (PAR) approach. *Sustainable Cities and Society, 74*, 103196. https://doi.org/10.1016/j.scs.2021.103196

Martinez, E. (2009). La vallée de la Seine en aval de Rouen. *Études Normandes, 58*(3), 11–14. https://doi.org/10.3406/etnor.2009.1770

Mell, I. C. (2017). Green infrastructure: reflections on past, present and future praxis. *Landscape Research, 42*(2), 135–145. https://doi.org/10.1080/01426397.2016.1250875

Monteiro, R., Ferreira, J. C., & Antunes, P. (2020). Green infrastructure planning principles: An integrated literature review. *Land, 9*(12), 1–19. https://doi.org/10.3390/land9120525

O'Brien, L., De Vreese, R., Kern, M., Sievänen, T., Stojanova, B., & Atmiş, E. (2017). Cultural ecosystem benefits of urban and peri-urban green infrastructure across different European countries. *Urban Forestry and Urban Greening, 24*, 236–248. https://doi.org/10.1016/j.ufug.2017.03.002

Report World Cities. (2020). *Unpacking the Value of Sustainable Urbanization*. https://doi.org/10.18356/c41ab67e-en

Richet, J.-B., Ouvry, J.-F., & Saunier, M. (2017). The role of vegetative barriers such as fascines and dense shrub hedges in catchment management to reduce runoff and erosion effects: Experimental evidence of efficiency, and conditions of use. *Ecological Engineering, 103*, 455–469. https://doi.org/10.1016/j.ecoleng.2016.08.008

Roberts, B. H. (2016). Rural urbanization and the development of small and intermediate towns. *Regional Development Dialogue, 35*(October 2016), 1–23.

Salas Tobón, A., & Barton, J. R. (2019). Assessing climate risk in small and intermediate towns and cities: A preliminary rapid appraisal tool and its application in Florencia, Colombia. *Urban Climates in Latin America*, 379–406. https://doi.org/10.1007/978-3-319-97013-4_15

Seddon, N., Smith, A., Smith, P., Key, I., Chausson, A., Girardin, C., House, J., Srivastava, S., & Turner, B. (2021). Getting the message right on nature-based solutions to climate change. *Global Change Biology, 27*(8), 1518–1546. https://doi.org/10.1111/gcb.15513

Sietchiping, R., Kago, J., Zhang, X. Q., Tuts, R., & Reid, J. (2014). The role of small and intermediate towns in enhancing urban-rural linkages for sustainable urbanization. *Regional Development Dialogue, 35*(March 2019), 48–62.

UN-Habitat. (2020). The World Cities Report 2020 The Value of Sustainable Urbanization. In *United Nations Human Settlments Programme (UN-Habitat)*.

Ying, J., Zhang, X., Zhang, Y., & Bilan, S. (2022). Green infrastructure: systematic literature review. *Economic Research-Ekonomska Istrazivanja* , *35*(1), 343–366. https://doi.org/10.1080/1331677X.2021.1893202

Zeng, J., Lin, G., & Huang, G. (2021). Evaluation of the cost-effectiveness of Green Infrastructure in climate change scenarios using TOPSIS. *Urban Forestry & Urban Greening, 64*, 127287. https://doi.org/10.1016/j.ufug.2021.127287

Sustainable Engineering of Future Urban Systems: An Inclusive Approach Toward Livable, Climate-Neutral, and Productive Smart Cities

Rizal Sebastian

1 Introduction

Making cities inclusive, safe, resilient, and sustainable is one of the 17 Sustainable Development Goals (SDGs) which have been adopted by all United Nations member states. At present, more than 50% the world's population live in cities. By 2050, the share would reach or exceed 70%. Globally, cities contribute more than 80% of gross domestic product (GDP). They are the drivers of economic growth, but they are also responsible for more than 70% of energy consumption and greenhouse gas emissions (UN Habitat 2022; United Nations 2022). These facts show the significance of cities for the goal to achieve a climate-neutral society.

In the European Union (EU), cities are the home of 75% of the population and the predominant locations of the building stocks to be decarbonized. Therefore, cities are important in the European Green Deal that gives a special attention to decarbonization of the existing building stock that is responsible for 40% of energy consumption and 36% of greenhouse gas emissions (European Commission, 2019). The Green Deal has the potential to empower a city to tackle challenges locally and in partnership with other cities. The Eurocities network, which gathers more than 200 cities from 38 countries with over 130 million population, is committed to implement the Green Deal. It is particularly concerned with climate, energy, buildings, circular economy, mobility, food systems, biodiversity, zero pollution, digital transformation, and social sustainability, including gender equality (EUROCITIES 2020).

In order to bring concrete solutions to the great challenges, realizing 100 climate-neutral and smart cities is highly prioritized and has become one of the five EU

R. Sebastian (✉)
Research Group Future Urban Systems, The Hague University of Applied Sciences, The Hague, The Netherlands
e-mail: R.Sebastian@hhs.nl

I. S. Dunmade et al. (eds.), *Sustainable Engineering*, Green Energy and Technology, https://doi.org/10.1007/978-3-031-47215-2_18

Missions for 2030 (European Commission 2022). An important objective of the smart city mission is to drive the sustainable transition using five key enablers: (1) a model for transforming cities into innovation hubs; (2) participative governance forms; (3) an economic and funding model for climate action; (4) an integrated urban planning model; and (5) smart systems and data platforms (European Commission 2020). In the Netherlands, the cities of Amsterdam, The Hague, Rotterdam, Utrecht, Groningen, Eindhoven, and Helmond have given a commitment to the EU Mission. The Dutch government has launched a policy to accelerate the sustainable transformation of the built environment under the Dutch Climate Deal. As a first step for implementing this policy, 1.5 million existing residential buildings will be renovated by 2030 through a neighborhood approach (Ministry of Economic Affairs and Climate Policy 2022).

This book chapter focuses on sustainable engineering of smart cities and has three objectives. The first objective is to articulate the concept of sustainable smart city from a practical perspective. A smart city comprises a broad range of aspects, such as the governance, social and healthcare aspects; the mobility, transport, and logistics aspects; and the environmental aspects. This book chapter highlights the built environment aspects of smart cities and particularly addresses sustainable engineering in the construction knowledge domain. The second objective is to clarify the challenges of sustainable engineering for smart cities, especially regarding energy transition, climate neutrality and climate change resilience, and circular economy in construction. These are the main challenges for sustainable transformation of urban environments as indicated by the innovation agendas described in the previous paragraphs. The third objective is to define a systemic approach to tackle the challenges of smart cities, with digitalization as a key enabler. This proposed approach is designated for applied research and higher professional education.

In the next sections of this book chapter, the practical concept of smart cities and the challenges for sustainable engineering are described. Subsequently, a practical analysis is performed on an empirical case to explicate the sound practice, ongoing research, and wider relevance. Afterward, the areas and directions for applied research are identified, and a strategy to create synergy between research, living labs, and educational curriculum for construction students is proposed. A particular emphasis is given to the quadruple-helix collaboration between the academia, industry, governments, and citizens. Such a collaboration allows for an inclusive approach toward livable, climate-neutral, and productive smart cities.

2 Practical Aim and Concept of Sustainable Smart Cities

A city is not a city without the people. In general, a city can be recognized as a large human settlement. There is no global agreement on what a city is, yet the European Commission and the Organisation for Economic Co-operation and Development are committed to developing a global, people-based definition of cities (Joint Research Centre 2019). This book chapter underscores the main aim of sustainable smart cities from a people-centric view toward the built environment, which is to enhance (a) the livability of cities as human settlements; (b) the climate neutrality and climate change resilience of cities at different geographical locations; and (c) the productivity of cities as business environments. "Livability, climate neutrality, and productivity" of smart cities respectively reflect the United Nations' sustainability action plan for "people, planet, and prosperity" (United Nations 2015).

This book chapter introduces a practical concept of smart city in view of the Big Data revolution that affects the technological and social developments in the digital era. A smart city is considered as engineering opportunities to manage people and resources in a more effective and efficient way (Ortman et al. 2020). The practical concept of smart cities is inspired by the characteristics of smart buildings in the European Commission's research framework as summarized by Moseley (2017). Three pillars of smartness are used with reference to the technological readiness of a smart building to (1) manage the building systems efficiently; (2) interact with the building occupants and users; and (3) interact with the surrounding environment and energy infrastructure. Similarly, smart cities can be recognized by their abilities to (1) efficiently enhance their sustainability performance through data-driven self-optimization of the urban systems; (2) interactively adapt to the changing societal needs and placemaking desires; and (3) effectively connect the inhabitants, buildings, surroundings, urban infrastructures, and energy networks with each other.

3 Challenges for Sustainable Engineering of Smart Cities

In dealing with the challenges for sustainable transformations of cities as addressed in the Paris Agreement and the United Nations' SDGs, the European Commission and the Organization for Economic Cooperation and Development pointed out three transitions: (1) the transition to sustainable energy; (2) the transition to a climate neutrality; and (3) the transition to circular economy (OECD 2020). This book chapter takes on a systemic approach and considers these transitions as the interrelated challenges for sustainable engineering of smart cities. In the following paragraphs, each challenge is articulated.

Energy transition can be understood as a shift from finite and fossil-based energy sources toward renewable and low-carbon energy sources while maximizing the energy-efficiency and improving the management of energy demand. The greatest progress on transition toward renewable energy can be made in cities as they are the primary energy users and they can become a catalyst for the globally needed shift to a low-carbon future (UIA 2018; IRENA 2021). In smart cities, the energy transition in neighborhoods and districts can be accelerated and scaled up. For this purpose, sustainable engineering should deal with the emerging challenges to develop and implement (1) Positive Energy Districts (PEDs); (2) integrated solutions for decentralized energy systems; and (3) inclusive energy communities. These challenges are further explained in the next paragraph.

A Positive Energy District (PED) has net-zero energy consumption, net-zero CO_2 emissions, and a surplus production of renewable energy annually. The district energy system is integrated in an urban and regional energy system (JPI Urban Europe 2020; Derkenbaeva et al. 2022). In PEDs, the focus shifts from energy-positive individual buildings to energy-positive building blocks and toward energy-positive neighborhoods. A wider impact of the energy transition process needs to be created with cities as the driving forces. In the emerging concept of PED, the priority is placed on the flexibility of the energy systems considering that the availability of renewable energy sources (RES) is not entirely schedulable and dispatchable. Such a flexibility can be enabled by digitalization and decentralization of energy systems. For this purpose, sustainable engineering needs to consider smart integrated decentralized energy (SIDE) systems. SIDE are subsets of microgrids that can increase the flexibility and resilience of neighborhood energy systems through a bottom-up approach (Graaf 2018). Finally, the acceptance and active participation of the urban communities are key in sustainable energy transition. A holistic transformation that simultaneously incorporates socioeconomic, technological, environmental, political, and institutional changes is needed (Derkenbaeva et al. 2022).

Along with energy transition, the transition toward climate neutrality and climate-change resilience is a key challenge for sustainable engineering of smart cities. Climate neutrality needs to be achieved while cities are confronted with resilience challenges due to the climate change. The United Nations Intergovernmental Panel on Climate Change (IPCC) and the Royal Netherlands Meteorological Institute highlighted the challenges of heat and precipitation in cities. Cities are usually warmer than rural areas due to the urban heat islands (UHIs). In addition, extreme rainfalls and droughts are increasingly present due to global warming (KMNI 2021). To create climate resilience, the challenge for sustainable engineering of smart cities is to improve the robustness and livability of the built environment by adopting nature's principles and implementing nature-inclusive construction. By mimicking natural organisms or ecosystems, sustainable solutions can be developed for climate change adaptations and ecological improvements in the built environment (Zari and Hecht 2020; Stevens 2021). Nature-inclusive urban regeneration can create a better living environment for humans and accommodate a wider range of animal and plant species. This should be done by deliberately creating room for biodiversity in, on, and around the buildings and public spaces in the cities (Dijkshoorn-Dekker 2022).

Finally, the transition to circular economy is part of the integral challenges for smart cities along with the transitions to renewable energy and climate neutrality. This book chapter focuses on circular construction to develop circular resource flows in the urban areas and to minimize the environmental footprints of the built environment with regard to carbon and nitrogen emissions. In the Circular Economy Guidebook for Cities, Dhawan and Beckmann (2019) envision modular and flexible design approaches for the built environment, minimizing the use of virgin materials and increasing the use of construction materials with positive impacts for the health and life quality of the inhabitants. A coherent strategy of urban mining is needed since the existing buildings and urban infrastructures contain huge volumes of valuable materials that can be extracted, reused, recycled, or upcycled with a minimum use of energy compared to the original amount of energy required to mine and process raw materials. In the Netherlands, the government has called for innovative engineering as part of the roadmap towards clean and zero-emission construction in 2030 that addresses sustainable materials, logistics, and digitalization (SEB 2021).

4 Empirical Case and Practical Analysis

This section reviews the sound practice, ongoing research, and wider relevance of an urban neighborhood project that addresses the challenges for sustainable engineering of smart cities. As a sound example, a Positive Energy District (PED) named the Groene Mient in the city of The Hague, the Netherlands, is shown in Fig. 1 and further described in this section. This case has been selected as it demonstrates practical innovations related with energy transition, climate neutrality, climate change resilience, and circular construction. It is a real urban project based on an inclusive approach driven by the local community. The project serves as a model for living lab methodology in applied research.

The Groene Mient is a social-ecological residential neighborhood characterized by energy self-sufficiency; sustainable water management; use of renewable energy as well as natural, sustainable, and recycled materials; and customizable design for

Fig. 1 The Groene Mient neighborhood in The Hague, the Netherlands. (Reproduced with consent from www.groenemient.nl)

each dwelling. The project was delivered in 2017 (https://www.groenemient.nl/ and Groot 2018). It covers an area of 7653 m^2 and consists of 33 dwellings. There is a common ecological garden where the local community can grow herbs and vegetables. As part of climate-resilience measures, at several places in the garden, the system for "water disposal, drainage and infiltration" is installed. This system can temporarily contain rainwater and let the water be absorbed by the ground gradually so that flooding at heavy rainfalls can be prevented and some water can be retained during the drought season. The materials for the building façades are natural stones and sustainable wood. As a real example of circular construction, the community hall is constructed based on reused building components and recycled materials from the demolition of an old school building at the project's location.

This project is the first PED in The Hague that was realized based on collective self-organized (CSO) housing by the local community. In a CSO housing project, a group of households work together to organize the planning, design, implementation, and maintenance of the building complex or neighborhood where they live (Piaia et al. 2017). CSO housing is emerging in Europe. In the Netherlands, successful pilot projects in various municipalities have been conducted over the past decade. In the UK, the government claimed that self-built homes are greener and believed that 25,000 self-built homes each year would save around 100,000 tons of CO_2. In Italy, the share of CSO in the housing market has exceeded 50%.

Decentralized energy management is implemented in the Groene Mient and organized by a local energy cooperative, i.e., Sterk op Stroom (https://sterkopstroom.nl/). The energy cooperative sets up a smart electricity grid for the locally generated, used, distributed, and stored electricity. It collaborates with another energy cooperative that manages a smart heat grid in the district. Together, they ensure the energy provision for the district without using natural gas or other fossil fuels. Initiatives for establishing local energy cooperatives, which are collective initiatives of the local inhabitants to set up sustainable energy systems based on a democratic governance mechanism, are supported by the Dutch government. In 2021, 676 energy cooperatives were counted in the Netherlands, and the number is growing (Schwencke 2023).

Benefitting from the Groene Mient as a living lab, the local energy cooperative currently conducts research to optimize the smart grid. Energy data from smart meters is continuously collected and monitored to gain an insight into the energy production, demand, and peak consumption within the district and in the surrounding areas. The aim is to use the locally generated solar energy directly and efficiently, for instance, through privacy-proof smart sharing between the dwellings and the nearby shops which vary in energy demands throughout the day. The research also results in a valuable insight for considering the cost-effectiveness of local energy storages in district batteries and bidirectional charging stations for electric vehicles (https://sterkopstroom.nl/).

The International Energy Agency (IEA) believes at decentralized and local energy solutions. At microgrid level, the distribution system can balance the energy flows. At city level, different parts of the energy system can be linked together. The advantages are balanced energy demand response, shaved peaks in energy demand, lowered impact on the uncertainties of RES, and optimized self-consumption of the electricity and thermal energy that is generated within the district. Overall decentralized and local energy solutions can contribute to drive renewable deployment, decarbonize the energy systems, and offer opportunities for new players in the energy markets (IEA 2017).

The Groene Mient serves as an example that is relevant for replication in other neighborhoods and cities. The feasibility of such a CSO-based PED project depends heavily on the trust-based collaboration among the citizens joined in the local energy cooperative, as well as the collaborations between the citizens and the local governments, enterprises, and academic partners. In the Groene Mient project, such collaborations manifested in the support from the provincial and municipal governments through subsidies and a consensus on the regulatory framework. Furthermore, the experimental decentralized energy networks were set up in partnership with an energy company and a distribution network operator. In addition, research on smart energy data and artificial intelligence was initiated in collaboration between the local energy cooperative, two universities, and an ICT company.

Understanding how trust is built in collective initiatives for smart urban transitions is essential for the ability and feasibility to replicate and scale up the sound practices. Zoller and Kok (2022) performed an analysis on the Groene Mient in The Hague and a similar case in Terheijden, a town in the Netherlands, using actor-network theory (ANT) as a framework. ANT envisages several phases in the interactive process of trust building along with three factors that are intertwined for establishing trust in a local energy cooperative, namely, the individuals, the organization, and the technologies. For instance, the individuals need to have confidence in the operation of the smart grid and digitalization technologies by the local energy cooperative when they commit themselves to become members of this organization.

5 Direction for Applied Research and Curriculum Development

This section presents an overview of the main subjects for applied research on sustainable engineering of smart cities. It continues by outlining the living lab method for applied research and the interrelations between applied research and the curriculum of higher professional education for construction students.

To accelerate and scale up the transitions toward sustainable energy, climate neutrality, and circular economy in the urban environments, applied research on digitalization for smart cities is urgently needed. Digitalization is the driver behind the ongoing "Industry 4.0" or the Fourth Industrial Revolution, especially through the cyber-physical systems that facilitate the development and deployment of digital twins in the built environment (Sebastian et al. 2021). Digitalization continues to evolve alongside and in between industrial revolutions. Ballon (2016) gave an overview of the evolution of the Internet as acknowledged by a broad public. It started from the "e-trend" in the 1990s when terms such as "e-mail," "e-commerce," and "e-government" became popular during the era of the Web 1.0 – also known as the "dot.com." About a decade later, the "i-trend" arrived when Apple introduced iPod, iPhone, and iPad, and the Internet could be further personalized along with the rise of social media platforms that characterized the Web 2.0. In recent years, "smart" has become the new trend along with the widespread of Internet of Things (IoT) that allows for connecting human and "smart objects," as well as the Blockchain, semantic web, and artificial intelligence (AI) technologies that characterize the Web 3.0.

This book chapter envisions practical research on digitalization for smart cities related with three themes: (a) digital data, (b) digital tools, and (c) digital processes. Regarding digital data, the focus is on digitization of the information and future-proof solutions for Big Data of the built environment. Big Data can be recognized by their three high-Vs characteristics, i.e., high volume, variety, and velocity (Ballon 2016, referred to Gartner's definition of Big Data). Regarding digital tools, the emphasis is on the hardware and software that are needed for designing, renovating, operating, and maintaining smart cities. Regarding digital processes, digital transformation of organizations in a smart city ecosystem is prioritized. The main areas for further in each theme to address the challenges for sustainable engineering of smart cities are mapped in the following matrix.

In anticipation of their technology and market readiness in the coming years, digital twins can be considered as the overarching digital innovation for sustainable engineering of smart cities. Digital twins operate on bidirectional dataflows. This means that a change in the physical object affects the digital object and vice versa (CDBB 2018). Digital twins have emerged for three primary usages with a high relevance for smart cities, in particular (1) for monitoring the actual conditions of cities; (2) for predicting future changes through simulations based on digital models and artificial intelligence (AI); and (3) for supporting decision-making to manage the cities throughout their lifecycle. Applied research is needed to progress from building information modeling (BIM) toward digital twins by developing solutions for capturing and processing dynamic data and connecting it with BIM and analytical models (Sebastian et al. 2021).

Going further from outlining a direction for applied research, this book chapter seeks ways to integrate applied research and professional education. Living labs offer an excellent opportunity for such an integration. Living labs are real environments at local or regional scale where multiple stakeholders representing the industry, government, academia, and citizens – known as quadruple helix – are involved in a design-driven methodology to foster and apply innovation (Popa, 2021). Living labs employ co-creation, real-life experimentation, and prototyping as their primary methods. Such labs strive to make an impact by embedding their results in the existing contexts, translating or scaling-up their learning methods to the other contexts, or influencing policy and regulations (Overdiek and Geerts 2023).

Ballon (2016) claimed that there is no single blueprint for making a smart city. He envisioned that creating a smart city as a permanent living lab is a recipe for success. Many of the technologies for smart cities are available for implementation, yet the biggest concern is the uncertainty about when to do it and which configuration is the most preferred amid continuous societal and political changes. A smart city living lab where new systems can be proven in a real environment attracts industry, government, academia, and citizens altogether into an innovation testbed. It will thus allow for simultaneously enhancing the technology, market, and policy readiness level (TRL, MRL, and PRL).

Aligned with the concept of living labs, the built environment knowledge domain in Dutch higher professional education system emphasizes cocreation between interdisciplinary research led by professors of applied sciences and education courses by lecturers with professional practices. Defining the research questions and facilitating the applications of research results in education and practice are to be done based on a close collaboration between the stakeholders in the quadruple helix (Glabbeek et al. 2022).

Finally, a possible element of professional education to stimulate outside-of-the-box thinking on a contemporary topic is organizing group hackathons. The term "hackathon" was first used in the software industry, and it became popular in the last decades along with the widespread of digitalization. At present, a hackathon is not solely meant for solving software programming challenges through intensive work in a short time but also for innovation challenges in various disciplines. An example of hackathon in relation to the concept of smart city was the national challenge organized by the Dutch association for engineering firms in 2022 (TechniekNederland 2022). The city of Amersfoort provided the challenge to propose creative solutions for the transformation of the Hoefkwartier, an existing district consisting of vacant office buildings, into a smart and sustainable residential district with 3500–4000 dwellings. The city promoted sustainability in energy transition, circular economy, smart mobility, climate adaptation, and nature-inclusive urban development. In this hackathon, seven teams from across the Netherlands were involved. Each team consisted of four bachelor students from a university of applied sciences, two students from a secondary vocational education institute, and two young professionals from a design/engineering/construction company. The practical impact was ensured by the team composition promoting interdisciplinary knowledge, complementary roles, and combined skills, as well as the quadruple-helix collaboration in organizing the hackathon and the use of a potential living lab.

6 Conclusions and Recommendations

Capitalizing on the increasing acceptance, availability, and affordability of digitalization, smart city developments have been growing rapidly across the globe. While digitalization is a key enabler, it is not the main aim of sustainable engineering of smart cities. This book chapter underlines the livability, climate neutrality, and productivity aspects of smart cities aligned with the people, planet, and prosperity dimensions of the UN's sustainability plan of action.

A city is a complex system where different aspects work interactively as a whole. Dealing with the complexity of engineering smart cities needs systems and network thinking. In systems thinking, a smart city is seen as an intertwining of (1) the building and civil infrastructure systems, (2) the urban energy and utility systems, and (3) the business and social systems. A systemic approach should be implemented for the sustainable transformation of the buildings and civil infrastructure throughout their lifecycle. The proposed systemic approach for smart cities consists of people-centric design, construction, and renovation; performance-based operation and predictive maintenance; and circular (re)use of building materials and sustainable transformation of neighborhoods.

The practical body of knowledge for sustainable engineering of smart cities is expanding with more and more best practices and lessons learned, yet there are still urgent demands for impact-oriented research and education, especially in the construction knowledge domain. In the previous section of this book chapter and in Table 1, a non-exhaustive overview of areas for further research on digital data, tools, and processes to tackle the smart city challenges was presented. Finally, standardization is essential for extending the replicability and scalability of the innovative solutions, as well as for measuring the progress against a coherent set of indicators that are consolidated in the ISO series for Sustainable Cities and Communities (ISO 37120: 2018, 37,122: 2019, 37,123: 2019).

Table 1 Overview of applied research areas for sustainable engineering of smart cities

Applied research themes	Areas for further research to address the challenges of smart cities		
	Transition to sustainable energy	Transition to climate neutrality and resilience	Transition to circular construction
Digital data	Digital twins of smart grids at district scale for data-driven monitoring and AI-based self-optimization	District digital twins based on 3D BIM, 3D GIS, weather and climate data for design and renovation of urban neighborhoods	Digital twins with construction material data in the lifecycle of buildings and civil infrastructures
Digital tools	Efficient monitoring and control systems at end-user, building and district level with edge computing	Immersive technologies, thermal sensors, and 3D scanning tools for climate-adaptive design simulations and performance monitoring	Interoperable LCA tools for BIM at design, renovation, maintenance, reuse, and demolition
Digital processes	Coherent procedures on data privacy and data sharing to support trust and business models in energy cooperatives	Standardization in data interoperability and data analytics for climate impact modeling in neighborhood planning and regeneration	Regulatory framework for zero emission and resource-efficient construction, renovation, and demolition processes

BIM building information modeling, *GIS* geospatial information system, *LCA* lifecycle analysis

References

Ballon P (2016) Smart cities: Hoe technologie onze steden leefbaar houdt en slimmer maakt. Lanno Campus, Leuven

CDBB (2018) The Gemini principles. Centre for Digital Built Britain, Cambridge.

Derkenbaeva E, Vega SH, Hofstede GJ, Leeuwen E van (2022) Positive energy districts: Mainstreaming energy transition in urban areas. Renewable and Sustainable Energy Reviews 153:111782

Dhawan P, Beckmann J (2019) Circular economy guidebook for cities. CSCP, Wuppertal

Dijkshoorn-Dekker M (2022) Climate resilient cities: Nature-inclusive construction will become the norm. Kennis Online Magazine, 27 June 2022

EUROCITIES (2020) The European Green Deal: Delivering results for citizens with Europe's cities. Policy paper, Brussels, February 2020

European Commission (2019) The European Green Deal. Communication from the Commission, Brussels, 11 December 2019

European Commission (2020) 100 Climate-Neutral Cities by 2030 – by and for the citizens: Interim report of the Mission Board for Climate-Neutral and Smart Cities. European Commission Directorate-General for Research and Innovation, Brussels

European Commission (2022) EU Mission: 100 climate-neutral and smart cities. Publications Office of the European Union, Luxembourg

Glabbeek N van, Meer Y van der, Willemse E (eds.) (2022) Domeinprofiel Built Environment: traditie en transitie. Domein Built Environment, Leidschendam

Graaf F de (2018) New strategies for smart integrated decentralised energy systems. Report of Project TESI117006 led by Metabolic, Amsterdam

Groot H de (2018) Studie in bruin en grijs: Groene Mient Den Haag. Het Houtblad 3, May 2018

IEA – International Energy Agency (2017) Scaling-up renewables through decentralised energy solutions. Report of workshop on 28 March 2017, Paris

IRENA – International Renewable Energy Agency (2021) Renewable Energy Policies for Cities. IRENA, Abu Dhabi

ISO 37120 (2018) Sustainable cities and communities: Indicators for city services and quality of life. ISO, Geneva

ISO 37122 (2019) Sustainable cities and communities: Indicators for smart cities. ISO, Geneva

ISO 37123 (2019) Sustainable cities and communities: Indicators for resilient cities. ISO, Geneva

Joint Research Centre (2019) The future of cities. Publication Office of the EU, Luxembourg

JPI Urban Europe (2020) SET Plan Action 3.2 – White paper on PED: Reference framework for positive energy districts and neighbourhoods. Austrian Research Promotion Agency, Vienna

KNMI (2021) Climate signal '21. KNMI, De Bilt

Ministry of Economic Affairs and Climate Policy (2022) Ontwerp Beleidsprogramma Klimaat. Rijksoverheid, The Hague

Moseley P (2017) EU Support for Innovation and Market Uptake in Smart Buildings under the Horizon 2020 Framework Programme. Buildings 7(4):105

OECD – Organisation for Economic Co-operation and Development (2020) Managing Environmental and Energy Transitions for Regions and Cities. OECD Publishing, Paris

Ortman SG, Lobo J, Smith ME (2020) Cities: Complexity, theory and history. PLoS ONE 15(12): e0243621

Overdiek A, Geerts H (2023) Innoveren met Labs 2.0: Ruimte maken voor duurzaamheidstransities. De Haagse Hogeschool, The Hague

Piaia, E, Di Giulio R, Sebastian R, Damen T (2017) Collective Self-Organised Housing: Methods, procedures and tools for new buildings and retrofits. Journal of Technology for Architecture and Environment 14

Popa E (ed) (2021) Quadruple helix collaborations in practice: Stakeholder interaction, responsibility and governance. RICONFIGURE project deliverable report D6.4

Schwencke AM (2023) Local energy monitor 2022. HIER and Energie Samen, Utrecht

SEB (2021) Routekaart schoon en emissieloos bouwen. SEB, The Hague
Sebastian R, Böhms HM, Luiten GT (2021) Developing a standardised approach of Asset Lifecycle Information Modelling for Semantic Digital Twins in the built environment. Proceedings of the 38th International Conference of CIB W78, 11–15 October 2021, Luxembourg, pp 946–955
Stevens LL (2021) Analogical Reasoning in Biomimicry Design Education. PhD Dissertation, Delft University of Technology
TechniekNederland (2022) Techathon 2022 – Herontwikkeling wijk Hoefkwartier in Amersfoort. Project booklet, 10 November 2022, Amersfoort
UIA – Urban Innovative Actions (2018) Energy transition in action: Report from European Sustainable Energy Week. UIA, Lille
UN Habitat (2022) World Cities Report 2022. UN Human Settlements Programme, Nairobi
United Nations (2015) Transforming our world: the 2030 Agenda for Sustainable Development. Resolution adopted by the General Assembly, 25 September 2015, New York
United Nations Department of Economic and Social Affairs (2022) The Sustainable Development Goals Report 2022. United Nations, New York
Zari MP, Hecht K (2020) Biomimicry for regenerative built environments: Mapping design strategies for producing ecosystem services. Biomimetics 5(18)
Zoller F, Kok O (2022) Network development and trust in local energy initiatives. Report of research on Future Urban Systems, The Hague University of Applied Sciences

Dr. Rizal Sebastian is a full professor of applied science. He is the chair of the research group Future Urban Systems at The Hague University of Applied Sciences in the Netherlands. His expertise focuses on smart cities based on the knowledge of architecture, civil engineering, and ICT. Complementary with his academic role, he works as a senior advisor on digitalization for sustainable built environments at RVO – an executive agency of the Dutch government. He also serves as a scientific expert in the EU innovation programs.

Prior to his current position, Rizal has accumulated more than 23 years of professional experience at a national institute for applied scientific research, a global engineering company, and a private software firm. He obtained a BSc degree in Architecture, an MSc degree in Construction Management, and a PhD degree in Design Management. He has become a guest professor at various universities and an author of a large number of scientific publications. His publication list can be found on: http://www.researchgate.net/profile/Rizal_Sebastian/publications/ and https://orcid.org/0000-0003-1714-8418.

Automated Shotcrete: A More Sustainable Construction Technology

Geoff Isaac, Paul Nicholas, Gavin Paul, Nico Pietroni, Teresa Vidal Calleja, Mike Xie, and Tim Schork

1 Introduction

The most common building material is concrete making it a major contributor to the ecological footprint of the construction sector, accounting for between 5 and 8% of global CO2 emissions (Anton et al. 2021, pp. 1–2). Global production of concrete totalled 4.4 billion tons in 2021, with output projected to reach 5.5 billion tons by 2050. Traditional concreting techniques rely on formwork, a mould or cast in which to pour concrete. Conventional formwork is often made using timber, but it is also possible to use other materials. Developing formwork is labour and material intense and accounts for between 35 and 60% of the total cost of concrete work (Bedarf et al. 2021, p. 3). It is also extremely wasteful, as most formwork is temporary and discarded after use, often having been used once only.

G. Isaac (✉)
School of Architecture, University of Technology Sydney, Sydney, NSW, Australia
e-mail: Geoffrey.isaac@uts.edu.au

P. Nicholas
Centre for Information Technology and Architecture, Royal Danish Academy of Fine Arts, Copenhagen, Denmark

G. Paul · T. Vidal Calleja
School of Mechanical and Mechatronic Engineering, University of Technology Sydney, Sydney, NSW, Australia

N. Pietroni
School of Computer Science, University of Technology Sydney, Sydney, NSW, Australia

M. Xie
Centre for Innovative Structures and Materials, RMIT University, Melbourne, VIC, Australia

T. Schork
School of Architecture and Built Environment, Queensland University of Technology, Brisbane, Australia

I. S. Dunmade et al. (eds.), *Sustainable Engineering*, Green Energy and Technology, https://doi.org/10.1007/978-3-031-47215-2_19

Conservative practices together with technological challenges make construction one of the least digitised and least efficient industry sectors, with concrete construction still heavily reliant on manual processes (Anton et al. 2021, p. 2). Construction is accountable for 37% of CO2 emissions related to energy (UN Global Status Report for Buildings and Construction 2021, p. 15). Industry pressures including a desire to improve worker safety, productivity and efficiency, combined with market pressures caused by urbanisation, increasing labour shortages and the challenge of climate change, are driving the growing interest in advanced technologies such as shotcreting to challenge established practices (ABB Robotics 2022, p. 5). Extending and accelerating the application of shotcrete has the potential to reduce carbon emissions and increase the efficiency of concrete construction by reducing the quantity of used material, therefore supporting the construction industry in addressing UN Sustainable Development Goal 12.

2 3D Concrete Printing

Additive manufacturing (AM) processes offer potential efficiency through lower labour demands and reduced construction waste (Bedarf et al. 2021). AM also offers the potential for leaner and more sustainable structures to be developed, reducing waste by placing material only where it is needed to meet structural demands (Anton et al. 2021, p. 1). However, cement materials suitable for 3D printing can exhibit distinct mechanical properties compared with traditional concrete (Lu et al. 2019b, p. 478). Significant challenges to improve various characteristics of the material used for 3D printing including shape retentivity and extrudability remain unresolved (Nair et al. 2020). Other limitations of 3D printing with concrete (3DPC) are its inability to build overhangs and the vertical building rate. Mixtures must remain sufficiently soft to be extrudable and to mix with the layer deposited previously, the substrate, which needs to have cured sufficiently to support itself together with the weight of the most recently deposited material without significantly sagging (Wangler et al. 2016, p. 70). Achieving a high-quality surface finish and bonding weaknesses (cold joints) caused by the layering process are issues that limit current applications and require further investigation (Neudecker et al. 2016, p. 335).

Those limitations of 3D printing specifically when applied to concrete have motivated the investigation of alternative technologies and solutions with the aim of reducing the GHG impact of the construction industry.

3 Shotcrete

Shotcrete was developed (from early in the twentieth century) to address the needs of the tunnelling and mining industries. In contrast to conventional 3D printing, shotcrete is not extruded in a bead but combined with pressurised air and sprayed to create a 3D structure. Concrete is pumped through a tube to a nozzle through which it is sprayed under pressure (and thereby compressed) on to newly blasted surfaces to stabilise the excavation. 'Conveying, compacting and application of concrete material are performed in one operation, which is distinctive for this technique' (Lindemann et al. 2017, p. 4). Manual application of shotcrete is an arduous task with physical demands on the 'nozzle man' restricting productivity. The drive for improved productivity combined with the hazardous nature of tunnelling and mining automation of the shotcreting process has been an industry priority for many decades.

Shotcrete has the potential to be significantly more time and materially efficient than 3DCP, and therefore more sustainable, for both new construction and repair, thereby also delaying end-of-life impacts (Rispin et al. 2005). A brief history of the automation of shotcrete is detailed in (Rispin et al. 2005). Early innovations were restricted to nozzle-holding devices using cranes and lifts to move and direct the application of concrete. Specialised shotcrete manipulators began to appear in the 1980s offering greater accuracy and extension possibilities. Within a decade spraying manipulators became commonplace on all large construction projects where shotcrete was used (Rispin et al. 2005, p. 4).

As demands for increased productivity grew, the need to quickly and accurately apply large volumes of shotcrete led to the development of remote controls, with radio remote quickly becoming the industry standard. Autonomous spray mobiles containing all the equipment needed to deliver quality shotcrete were tailored to meet the needs of specialist construction tasks (Rispin et al. 2005, p. 4). Computer-controlled multi-axis robots capable of scanning and spraying an area have become increasingly common since the start of the twenty-first century.

Automation has delivered significant productivity enhancements and offers the potential to create complex geometries, a significant advantage compared with 3DCP which is restricted to vertical builds. Hand-held nozzles can cover 7–9 cubic metres per hour, while mechanised spraying can cover more than double of the output up to 20 cubic metres per hour (Rispin et al. 2005, p. 5). Automation has delivered improvements in safety and reduced set-up times, requiring less people in the process.

Academic attention has recently focused on developing shotcrete as an alternative to conventional 3DCP, combining the benefits of shotcrete with AM processes, shotcrete-based 3D printing or SC3DP. Researchers at Technische Universität Braunschweig led the development of 'a robotically controlled additive manufacturing process that builds concrete components layer by layer through the controlled addition of compressed air' (Kloft et al. 2020a, p. 609). The concrete layer's height is inversely proportional to the speed of the robot-controlled nozzle, while width can be determined through adjusting the distance of the nozzle from the surface. Complex geometries including overhangs and the integration of built-in parts or reinforcement become achievable (Kloft et al. 2020, p. 2).

One of the main benefits of SC3DP is that interlayer bonding is improved using high air pressure which 'tears up' the concrete in the nozzle and creates a high contact surface area with the air stream facilitating the intermixing of additives (Kloft et al. 2020, p. 2). Layer interlocking and hence interlayer bond strength are enhanced by SC3DP compared with traditional additive manufacturing due to the high kinetic energy of the sprayed material, which reduces the prospect of cold joints forming. 'In general, SC3DP specimens show better mechanical performance than extrusion-based 3D printed materials and cast specimens of the same mixture design' because of improved interlaying bonding (Heidarnezhad and Zhang 2022, p. 7). Performance can be enhanced by precisely controlling layer thickness (Northcroft and Ziegler 2008).

3.1 Formwork

Developing traditional formwork is labour and material intense accounting for between 35 and 60% of the total concrete work cost (Bedarf et al. 2021, p. 3). It is also extremely wasteful, as most formwork is temporary and discarded after use, often having been used once only. Reducing or eliminating formwork saves not only the natural resources and labour required to build it but also the transport and disposal impacts, highlighting a significant sustainability advantage of shotcrete (Schokker 2010).

Work at Braunschweig has led to the development of techniques to SC3DP without conventional single-use formwork. A 2016 paper outlined their proposed experiments progressing from a flat wall with opposing formwork to creating complex curved walls (Neudecker et al. 2016, p. 336). Among the many challenges, the research team needed to develop simulation tools, an automated injection tool and a control system. Long-range scanners supplemented with a 3D laser triangulation scanner and a 3D vision system were used to monitor performance (Neudecker et al. 2016, p. 335). Although experiments were not complete at the time of publishing, the researchers identified 'a water-to-cementitious ratio of 0.4, with a pressure of 6.5 bar in the pneumatic cylinder of the pumping system' produced the best spray (Neudecker et al. 2016, p. 338). Pressure at the nozzle tip of 5.2 bars resulted in the highest-quality results with a rebound rate of just 8%.

The Block Research Group combined a reusable cable net with a fabric overlay to create formwork for an anticlastic mesh-reinforced sandwich shell roof (Block et al. 2017). This lightweight solution eliminated the need for both falsework and foundations for the formwork. An evolutionary design process was employed to identify the most optimal geometry for the roof, allowing the thickness of the shells to be reduced to between just 5 cm and 3 cm (Block Research Group 2018).

3.2 Reinforcing Agents

Combining the traditionally separate requirement of formwork and reinforcement into a single robotic fabrication process has potential to 'produce an additive and waste-free, material-efficient, and geometrically unconstrained method of fabricating complex non-standard concrete constructions' (Hack and Lauer 2014, p. 52). Hack and Laurer report on using acrylonitrile butadiene styrene (ABS) to print mesh mould formwork that can also act as a reinforcement for concrete structures. Concrete is sprayed and protrudes through the mesh mould with surfaces manually trowelled to smooth and level (Hack and Lauer 2014, p. 49). Using polymers developed for conventional 3D printers permitted 'precise control over the material's hardening behaviour. Pinpoint cooling during the extrusion process, for example, gives such a high level of control that free spatial extrusions become possible and, consequently, the "knitting" of structures freely in space' (Hack and Lauer 2014, p. 49). Moving to spatial extrusion, in contrast to layer deposition, significantly reduces fabrication time and can be deployed at a large scale. However, the mesh structure is not sufficiently strong to resist structural loads, limiting its application to non-structural components (Wu et al. 2022, p. 13). Hack and Lauer also highlight the potential for carbon, glass, bamboo or basalt to be co-extruded to develop constructions that can withstand high tensile forces. Hack then continued to experiment with the technique for his PhD dissertation, transitioning from polymer to structurally superior steel meshes suitable as a loadbearing construction system, which was used at the DFAB (NEST) house (see below) (Hack 2018).

Inspired by the ferrocement technique (developed in the 1940s) of manually throwing concrete against a dense, self-supporting reinforcement mesh, researchers at ETH Zurich investigated robotic spraying of glass fibre-reinforced concrete on a permeable reinforcement mesh made from carbon fibre (Taha et al. 2019). This approach allowed the researcher to move away from the limitation of only spraying horizontal layers (as with the work at Braunschweig). Square, 38 mm glass-fibre mesh was bent and stabilised into the desired shape. Glass fibre was mixed with the concrete in the nozzle of the spraying gun. Through experimentation the optimal fibre length (42.5 mm) in relation to mesh opening size was identified as crucial to ensure the material clogged the openings of the mesh and adhered to it, 'while also assuring that excess of material penetrating through the mesh during the fabrication was minimized' (Taha et al. 2019, p. 248). A double-curved structure with an average thickness of just 3 cm was successfully developed. The researchers conclude that their approach could be compatible with mesh mould in the application of surface finishing.

The DFAB (NEST) house was conceived as a multi-technology demonstrator of digital fabrication techniques (Fig. 1). A densely reinforced load-bearing concrete wall was built in situ at the house. The steel reinforcing mesh was constructed more densely than traditional steel rebar cages to prevent the concrete mix from flowing through the mesh. To build the mesh accurately, the fabricator sensed its position within the construction site and continuously monitored the shape of the mesh. A 12 m wall, 2.8 m high, requiring over 20,000 weld points was constructed using 6 mm steel rods over a period of 125 hours.[1] Undulations were incorporated into the design to stiffen the wall to compensate for its relative thinness.

[1] A video of this project can be viewed at: https://www.youtube.com/watch?v=Fi3SyfQ3hnc.

Fig. 1 The in situ fabricator building the mesh mould at the NEST house. The mesh mould process unifies the reinforcement and formwork production into a single and robotically controlled on-site fabrication system. (Image courtesy: NCCR Digital Fabrication)

In 2019, Hack et al., inspired by lattice structures traditionally constructed using steel or aluminium, began to experiment with using reinforced concrete for geometrically complex spatial structures suitable for long-spanning, column-free construction (Hack et al. 2019). In this approach, the spatial structures were modularised into planar components (Fig. 2). Using identical, planar truss girders reduced material use and eliminated the need for formwork. Planar components were 3D printed using three layers of shotcrete reinforced with a (manually placed) carbon fibre grid. After milling the edges, the planar elements were cured and then assembled. It was noted that more sophisticated module typologies need to be developed 'to allow for improved connectivity and multidirectional load transfer between neighbouring modules' (Hack et al. 2019, p. 371).

In 2020, Kloft and Hack successfully produced a fully reinforced, double-curved concrete wall over 5 square meters and 18cms thick using a 6-axes Stäubli robot (Hack and Kloft 2020). In this process horizontal and vertical (10 mm B500B steel) rebars were positioned manually as the structure was printed (Fig. 3 and described in more detail in (Kloft et al. 2020b)). A second three-centimetre layer of concrete was applied embedding the reinforcements, while creating a foundation for the surface finishing, using a trowelling process (a 3-axes Omag milling application) (Hack and Kloft 2020, p. 1130).

Mike Xie, from RMIT in Melbourne, led a project to construct a 4.2 m wall using fibre reinforced ultra-high-performance concrete sprayed over 80 moulds, 3D

Fig. 2 Assembled prototype element demonstrating proof of concept for a project at Technische Universität Braunschweig. (Image courtesy of Norman Hack)

printed from PETG (Fig. 4). After printing and assembly moulds were sprayed with concrete in layers. The demoulded polished concrete components were transported for on-site assembly (Centre for Innovative Structures and Materials 2021; Dingwen 2022).

The German government-financed Carbon Concrete Composite project is approaching completion of the construction of The Cube, a building made, in part, from precast panels with sections shotcreted onto a carbon fibre mesh. The carbon fibre reinforcement allowed double-curved geometry walls just 4 cm thick to be built without the need for conventional formwork. It is claimed that the construction will have four times the strength of a regular reinforced concrete building and contain 70% less embodied carbon (the use of clinker has been avoided). The use of flexible (rust proof) carbon reinforcement mesh has allowed for new geometric forms to be explored in the design of the building. The 220 square meter building on the grounds of the Technische Universität Dresden has a predicted lifespan nearly three time the standard 70–80 years for concrete buildings reinforced conventionally (Cousins 2021; Fearson 2021). The Cube showcases the potential for shotcrete to create longer-lasting buildings, a significant sustainability advantage compared with traditional construction techniques as end-of-life impacts are delayed.

Fig. 3 Threading unbent vertical reinforcement into the shotcrete core structure. (Image courtesy Norman Hack)

Fig. 4 Intelligent form. (Image courtesy: Dingwen 'Nic' Bao, Xin Yan, Yi Min 'Mike' Xie, Wei Qiu and Jianan Peng)

Additionally, shotcrete can be used to economically repair or rehabilitate structures, extending their lives and delaying potential impacts caused by demolition and reconstruction.

3.3 Control Systems

The accuracy and consistency of the sprayed results is determined by the sophistication of the control system which determines the location and quantity of material and calculated the desired compaction. Adaptive control algorithms can be used to incorporate feedback from sensors monitoring material distribution in real time. The ability to precisely control the thickness of a section to match its structural requirements eliminates wasted material, one of the main sustainability advantages of shotcrete. Traditional processes such as trowelling and milling can be used to improve surface finish (Hack and Kloft 2020, p. 1129).

Synchronising the measuring system with the robot-held nozzle to obtain consistent results within acceptable tolerances remains a significant challenge. Irregular environmental characteristics and fluctuations in the composition of the concrete mix can result in 'vast deviation between designated and printed geometry' (Lachmayer et al. 2023). A 2018 state-of-the-art review of in situ measurement technologies identified laser triangulation as more accurate and reliable compared with 2D camera-based monitoring (Lindemann et al. 2019). Informed by this review, the authors concluded that model-based offline planning alone was insufficient to deliver the accuracy required and implemented a series of monitoring and control initiatives using a Beckhoff control system.

Layer width and height were controlled by two algorithms developed to compensate for material displacements causing geometric inaccuracies. By controlling the velocity of deposition and the distance from the nozzle to the surface, the required layer height and width could be achieved, although the authors note that control of the layer width measurement is indirect, which they planned to address by direct measurement to deliver closed-loop control of layer width (Lindemann et al. 2019, p. 294). This project resulted in SC3DP constructing a wall with complex concrete geometries, featuring a significant overhang and integrated reinforcement for the first time (Lindemann et al. 2019, p. 296).

Researchers at Cambridge University developed trajectory planning algorithms to produce doubly curved ribbed concrete shells using the C# plug-in for Grasshopper's 3D visual programming environment (Nuh et al. 2022). Two prototypes were produced using this process: a 4.5 m nine-segment shell and a deep ribbed thin shell. The authors note that the benefit of the Grasshopper plug-in is that 'it can be applied to any robotic assembly system with a concrete sprayer attached'.

3.4 Material Mixtures

Developing suitable cementitious materials specifically for SC3DP is another area of enquiry (Lu et al. 2019a, p. 1074). Incorporating supplementary cementitious materials such as fly ash or recycled aggregate into the concrete mix avoids landfill, highlighting another sustainability advantage of SC3DP. Material must be specifically developed to ensure it is suitable to be pumped through the hose and propelled through the nozzle. Coarse aggregates have not been used to date. Pumping performance can be improved if the mixture has low plastic viscosity and low dynamic yield, both of which can be manipulated by incorporating additives (Heidarnezhad and Zhang 2022, p. 6). Of particular note Yun et al. report that adding up to 4.5% of silica fumes by weight improved pumpability by lowering viscosity (Yun et al. 2015). Yun also notes shootability is improved with higher yield stress and increased viscosity resulting in improved build-up thickness while reducing rebound rates. Neudecker et al. recommend a water to cement ratio of 0.4 applied with a pressure of 6.5 bar to achieve the best shootability performance (Neudecker et al. 2016, p. 338). Researchers at Zurich found that the addition of steel fibres led to significant improvements in ductility and strength (Pfändler et al. 2019).

To reduce the density of concrete and to improve the accuracy of spraying, researchers at the Singapore Centre for 3D Printing experimented with adding an air-entraining agent (AEA) and incorporating lightweight aggregate (fly ash cenosphere – FAC) in the mixture (Lu et al. 2019a). Adding AEA reduces yield stress, improving pumpability (Lu et al. 2019a). Experiments showed that the introduction of FAC and AEA lowered the spread diameter and slump values of freshly sprayed cement, suggesting the material could better retain its shape, but negatively impacted its pumpability (Lu et al. 2019a, p. 1075). A mixture containing 100% FAC aggregate and 0.1 grams/litre of AEA presented the lowest plastic viscosity and dynamic yield stress requiring the lowest calculated pumping pressure, achieving the best performance in delivery and deposition.

Computer-controlled dosage of accelerator creates a more uniform shotcrete quality. Adding accelerator and silica fume to shotcrete means it hardens more quickly, achieving high static yield stress more quickly (Heidarnezhad and Zhang 2022, p. 7). Material with 6% accelerator had a deformation modulus about 14 times higher than the material without accelerator (Dressler et al. 2020, p. 16). However, increasing the accelerator dosage can reduce interlayer bonding, potentially compensated by increasing the air volume flow (Lachmayer et al. 2023, p. 13).

3.4.1 Functionally Graded Concrete

Functionally graded concrete is produced by adding a gas-releasing foaming agent, such as aluminium powder, to react with alkaline hydration products or by mixing wet foams with cement paste. The later approach enables different density levels to be achieved, from 200 to 1900 kg/m^3, compared with ~2500 kg/m^3 for standard concrete. The ability to vary density creates the opportunity to develop monomaterial manufacturing processes that address demand for varying the mechanical properties of elements within a structure. Functionally graded materials offer the potential for improvements in material usage, strength and functionality while reducing weight when compared with their homogenous equivalents (Keating 2011, p. 1). Researchers at MIT produced a graded cylinder beam weighing 9% less than a solid cylindrical beam of the same dimensions capable of supporting the same load, illustrating the sustainability benefits of this approach (Keating 2011, p. 5). Functionally graded concrete can contribute to improved material efficiency in construction. In addition, 'foam concrete has lower thermal conductivity (~0.065 $Wm^{-1}K^{-1}$ at ~250 kg/m^3) compared to regular concrete (~0.5 $Wm^{-1}K^{-1}$)', decreasing the need for insulation, yet another sustainability benefit (Bedarf et al. 2021, pp. 7, 10). The internal composition of structural components can be aligned with their 'specific structural and thermal performance requirements' (Herrmann et al. 2018, p. 54).

The researchers at Universität Stuttgart combined two concrete mixes using two pumps and spray nozzles, continuously varying the quantities of mixtures: mixture one is a high-density fine-aggregate concrete, and mixture two has a lower density with a higher porosity. Significantly the two concretes were not mixed before spraying but applied at the same time through two separate nozzles. Topology optimisation algorithms were refined to achieve optimal material distribution. The pumps' volumetric flow control allows for seamless gradation across a wide spectrum of characteristics, ranging 'from low to high strength, heavy to ultra-lightweight, and low to high heat insulation properties' (Herrmann and Sobek 2017, p. 57). Experimental tests revealed that the bulk density of the specimen decreased progressively over its height, and this change was reflected in the mean compressive strength. Steel-reinforced beams with a mass reduction of 34% were successfully produced using functionally graded components (Herrmann and Sobek 2017, p. 62).

4 Future Research

Many topics require further investigation before SC3DP reaches the level of maturity required to achieve widespread deployment by the construction industry. Effectively guiding the development and deployment of in situ robotic fabrication processes presents a complex challenge that spans multiple disciplines and domains, requiring interdisciplinary collaboration and expertise (Buchli et al. 2018). Solutions require the intense collaboration of architects, materials scientists, roboticists, civil engineers and mechanical engineers, among others. Areas of research required are summarised by Heidarnezhand and Shang under three headings: (1) Establishing the correlation between the operational process parameters, material properties and the resulting printed layer geometry is crucial to enhance printing precision. We note that while it has been proven that laser triangulation is superior to 2D cameras to monitor performance, investigation of other measuring techniques is needed. (2)

Examine combining shotcrete with 3D extrusion printing, to help overcome the shortcomings of both technologies (e.g. 3D printing formwork to be over sprayed by shotcrete). (3) Innovate to develop superior printing mixtures, particularly with the addition of fibres, potentially as a replacement for reinforcement (Heidarnezhad and Zhang 2022, p. 9). Alternative reinforcement agents including plastics, carbon, glass and hemp require further experimentation. The effects of combining these reinforcement agents with other additives and accelerators remain an area ripe for further inquiry (Ivanova et al. 2022). Continued investigation into reinforcement structures that are more efficient than traditional steel rebar is needed. Offering faster build times and more efficient use of resources shotcreting is a more sustainable construction technology, but more research is required to quantify these environmental benefits. In particular, Saade et al. note a lack of comparable life cycle assessment studies (Saade et al. 2020).

Research priorities include varying the nozzle geometry to obtain more precision, as the layer geometry is determined by a variety of additional factors including pumping speed, air pressure and stand-off distance. Increasing the nozzle diameter decreases the spray velocity and results in lower compaction rates (Burak et al. 2018). Rebound rates are affected by nozzle positioning and speed, mix proportion, additives and rheological properties. Clear relationships between viscosity and yield stress and their impact on rebound rates have not yet been identified (Yun et al. 2015). Rebound not only wastes resources but also negatively impacts placement of material and potentially mechanical properties. Investigations to reduce rebound or compensate for its effects are needed. In their review of reinforcement technologies, Wu et al. emphasise the need for further development of design standards specific to printed concrete, as well as the establishment of code recognitions and/or guidelines that can verify equivalency to reinforced concrete through comprehensive testing (Wu et al. 2022, p. 21). They also call for standard testing procedures for safety-related performance to be established. SC3DP produces a rough textured surface, and more efficient, fully automated post-processes need to be developed to refine finished products.

Finally, the opportunity to continuously vary the density of the mixture (functionally graded concrete) during construction, investigated by researchers at MIT and Stuttgart, offers the possibility of achieving significant material and energy efficiencies and suggests a rich area for further investigation. The Stuttgart experiment used two nozzles to mix concrete materials at the point of application. Alternatively, more accurate methods to combine and deliver functionally graded concrete on demand require further experimentation. SC3DP functionally graded concrete creates the potential to deliver significant GHG reductions for both the concrete and construction industries through optimising structural designs, minimising the use of materials and significantly reducing waste while improving the sustainability profile of the construction industry and helping it meet the UN Sustainable Development Goal 12.

References

ABB Robotics. (2022). Building the future – how robotic automation can transform the construction industry. https://new.abb.com/products/robotics/industries/transforming-the-future-of-construction. Accessed 25 March 2023

Anton, A., Reiter, L., Wangler, T., Frangez, V., Flatt, R. J., & Dillenburger, B. (2021). A 3D concrete printing prefabrication platform for bespoke columns. Automation in Construction, 122, 103467. https://doi.org/10.1016/j.autcon.2020.103467

Bedarf, P., Dutto, A., Zanini, M., & Dillenburger, B. (2021). Foam 3D printing for construction: A review of applications, materials, and processes. Automation in Construction, 130, 103861. https://doi.org/10.1016/j.autcon.2021.103861

Block, P., Schlueter, A., Veenendaal, D., Bakker, J., Begle, M., Hischier, I., Hofer, J., Jayathissa, P., Maxwell, I., Echenagucia, T. M., Nagy, Z., Pigram, D., Svetozarevic, B., Torsing, R., Verbeek, J., Willmann, A., & Lydon, G. P. (2017). NEST HiLo: Investigating lightweight construction and adaptive energy systems. Journal of Building Engineering, 12, 332–341. https://doi.org/10.1016/j.jobe.2017.06.013

Block Research Group. (2018). HiLo Research & innovation unit for NEST. https://brg.ethz.ch/hilo. Accessed 25 March 2023

Buchli, J., Giftthaler, M., Kumar, N., Lussi, M., Sandy, T., Dörfler, K., & Hack, N. (2018). Digital in situ fabrication— Challenges and opportunities for robotic in situ fabrication in architecture, construction, and beyond. *Cement and Concrete Research, 112,* 66–75. https://doi.org/10.1016/j.cemconres.2018.05.013

Burak, E. E., Vorob'eva, Y. A., & Sheps, R. F. (2018). Investigation of the Strength Characteristics of Shotcrete as a Function of the Technological Parameters of the Application. IOP Conference Series: Materials Science and Engineering, 463(4), 042015. https://doi.org/10.1088/1757-899X/463/4/042015

Centre for Innovative Structures and Materials. (2021). Intelligent Forms. Centre for Innovativ. https://www.cism.org.au/intelligentforms. Accessed 25 March 2023

Cousins, S. (2021). Carbon concrete Cube cuts emissions and expands design options. RIBA Journal. https://www.ribaj.com/products/c3-carbon-concrete-composite-cube-exhibition-centre-dresden-germany-embodied-co2-curved-design. Accessed 25 March 2023

Dingwen, B. (2022). Performance-driven Digital Design and Robotic Fabrication Based on Topology Optimisation and Multi-agent System. Dissertation, RMIT.

Dressler, I., Freund, N., & Lowke, D. (2020). The Effect of Accelerator Dosage on Fresh Concrete Properties and on Interlayer Strength in Shotcrete 3D Printing. Materials, 13(2), Article 2. https://doi.org/10.3390/ma13020374

Fearson, A. (2021, July 30). The Cube will be "world's first building made of carbon concrete." Dezeen. https://www.dezeen.com/2021/07/30/henn-tu-dresden-carbon-fibre-concrete-building/. Accessed 25 March 2023

Hack, N., & Kloft, H. (2020). Shotcrete 3D Printing Technology for the Fabrication of Slender Fully Reinforced Freeform Concrete Elements with High Surface Quality: A Real-Scale Demonstrator. In F. P. Bos, S. S. Lucas, R. J. M. Wolfs, & T. A. M. Salet (Eds.), Second RILEM International Conference on Concrete and Digital Fabrication (pp. 1128–1137). Springer International Publishing. https://doi.org/10.1007/978-3-030-49916-7_107

Hack, N., & Lauer, W. V. (2014). Mesh-Mould: Robotically Fabricated Spatial Meshes as Reinforced Concrete Formwork. Architectural Design, 84(3), 44–53 https://doi.org/10.1002/ad.1753

Hack, N., Lindemann, H., & Kloft, H. (2019). Adaptive Modular Spatial Strutures for Shotcrete 3D Printing. 363–372. Paper presented at International Conference of the Association for Computer-Aided Architectural Design Research in Asia, Victoria University of Wellington, 15–18 April 2019

Hack, N. P. (2018). Mesh Mould: A Robotically Fabricated Structural Stay-in-Place Formwork System [Doctoral Thesis, ETH Zurich]. https://doi.org/10.3929/ethz-b-000263345

Heidarnezhad, F., & Zhang, Q. (2022). Shotcrete based 3D concrete printing: State of art, challenges, and opportunities. Construction & Building Materials, 323. https://doi.org/10.1016/j.conbuildmat.2022.126545

Herrmann, E., Mainka, J. L. C., Lindemann, H., Wirth, F., & Kloft, H. (2018). Digitally Fabricated Innovative Concrete Structures. In proceedings of the International Symposium on Automation and Robotics in Construction, IAARC Publications, Waterloo.

Herrmann, M., & Sobek, W. (2017). Functionally graded concrete: Numerical design methods and experimental tests of mass-optimized structural components. Structural Concrete, 18(1), 54–66. https://doi.org/10.1002/suco.201600011

Ivanova, I., Ivaniuk, E., Bisetti, S., Nerella, V. N., & Mechtcherine, V. (2022). Comparison between methods for indirect assessment of buildability in fresh 3D printed mortar and concrete. Cement and Concrete Research, 156, 106764. https://doi.org/10.1016/j.cemconres.2022.106764

Keating, S. (2011). Functionally Graded Rapid Prototyping. MIT Media Lab. https://www.media.mit.edu/publications/functionally-graded-rapid-prototyping/ Accessed 25 March 2023

Kloft, H., Empelmann, M., Hack, N., Herrmann, E., & Lowke, D. (2020a). Bewehrungsstrategien für den Beton-3D-Druck. Beton- Und Stahlbetonbau, 115(8), 607–616. https://doi.org/10.1002/best.202000032

Kloft, H., Empelmann, M., Hack, N., Herrmann, E., & Lowke, D. (2020b). Reinforcement strategies for 3D-concrete-printing. Civil Engineering Design, 2(4), 131–139. https://doi.org/10.1002/cend.202000022

Kloft, H., Krauss, H.-W., Hack, N., Herrmann, E., Neudecker, S., Varady, P. A., & Lowke, D. (2020). Influence of process parameters on the interlayer bond strength of concrete elements additive manufactured by Shotcrete 3D Printing (SC3DP). Cement and Concrete Research, 134, 106078. https://doi.org/10.1016/j.cemconres.2020.106078

Lachmayer, L., Böhler, D., Freund, N., Mai, I., Lowke, D., & Raatz, A. (2023). Modelling the influence of material and process parameters on Shotcrete 3D Printed strands—Cross-section adjustment for automatic robotic manufacturing. Automation in Construction, 145, 104626. https://doi.org/10.1016/j.autcon.2022.104626

Lindemann, H., Fromm, A., Ott, J., & Kloft, H. (2017). Digital Prefabrication of freeform concrete elements using shotcrete technology. Paper presented at International Association for Shell and Spatial Structures Annual Symposium, HafenCity University Hamburg, 25–28 September 2019

Lindemann, H., Gerbers, R., Ibrahim, S., Dietrich, F., Herrmann, E., Dröder, K., Raatz, A., & Kloft, H. (2019). Development of a Shotcrete 3D-Printing (SC3DP) Technology for Additive Manufacturing of Reinforced Freeform Concrete Structures. In T. Wangler & R. J. Flatt (Eds.), First RILEM International Conference on Concrete and Digital Fabrication – Digital Concrete 2018 (pp. 287–298). Springer International Publishing. https://doi.org/10.1007/978-3-319-99519-9_27

Lu, B., Qian, Y., Li, M., Weng, Y., Leong, K. F., Tan, M. J., & Qian, S. (2019a). Designing spray-based 3D printable cementitious materials with fly ash cenosphere and air entraining agent. Construction & Building Materials, 211, 1073–1084. https://doi.org/10.1016/j.conbuildmat.2019.03.186

Lu, B., Weng, Y., Li, M., Qian, Y., Leong, K. F., Tan, M. J., & Qian, S. (2019b). A systematical review of 3D printable cementitious materials. Construction & Building Materials, 207, 477–490. https://doi.org/10.1016/j.conbuildmat.2019.02.144

Nair, A., Aditya, S. D., Adarsh, R. N., Nandan, M., Dharek, M. S., Sreedhara, B. M., Prashant, S. C., & Sreekeshava, K. S. (2020). Additive Manufacturing of Concrete: Challenges and opportunities. 814(1), 12022. https://doi.org/10.1088/1757-899X/814/1/012022

Neudecker, S., Bruns, C., Gerbers, R., Heyn, J., Dietrich, F., Dröder, K., Raatz, A., & Kloft, H. (2016). A New Robotic Spray Technology for Generative Manufacturing of Complex Concrete Structures Without Formwork. Procedia CIRP, 43, 333–338. https://doi.org/10.1016/j.procir.2016.02.107

Northcroft, I., & Ziegler, C. (2008). Computerised control for robotically applied sprayed liners providing technical, economic, environmental benefits as well as additional safety features. Paper presented at 6th International Symposium on Ground Support in Mining and Civil Engineering Construction, Cape Town, 30 March to 3 April 2008.

Nuh, M., Oval, R., Orr, J., & Shepherd, P. (2022). Digital fabrication of ribbed concrete shells using automated robotic concrete spraying. Additive Manufacturing, 59, 103159. https://doi.org/10.1016/j.addma.2022.103159

Pfändler, P., Wangler, T., Mata-Falcón, J., Flatt, R. J., & Kaufmann, W. (2019). Potentials of Steel Fibres for Mesh Mould Elements. In T. Wangler & R. J. Flatt (Eds.), First RILEM International Conference on Concrete and Digital Fabrication – Digital Concrete 2018 (pp. 207–216). Springer International Publishing. https://doi.org/10.1007/978-3-319-99519-9_19

Rispin, M., Gause, C., & Kurth, T. (2005). Robotic Shotcrete Applications for Mining and Tunneling. Shotcrete, Summer, 4–9.

Saade, M. R. M., Passer, A., & Mittermayr, F. (2020). (Sprayed) concrete production in life cycle assessments: A systematic literature review. The International Journal of Life Cycle Assessment, 25(2), 188–207. https://doi.org/10.1007/s11367-019-01676-w

Schokker, A. (ed) (2010). The Sustainable Concrete Guide Applications. US Green Concrete Council, Washington

Taha, N., Walzer, A. N., Ruangjun, J., Bürgin, T., Dörfler, K., Lloret-Fritschi, E., Gramazio, F., & Kohler, M. (2019). Robotic AeroCrete – A novel robotic spraying and surface treatment technology for the production of slender reinforced concrete elements. Proceedings of the 37th ECAADe and 23rd SIGraDi Conference, 3, 245–254. https://doi.org/10.3929/ethz-b-000387276

UN Global Status Report for Buildings and Construction. (2021). 2021 Global Status Report for Buildings and Construction | Globalabc. https://globalabc.org/resources/publications/2021-global-status-report-buildings-and-construction. Accessed 25 March 2023

Wangler, T., Lloret, E., Reiter, L., Hack, N., Gramazio, F., Kohler, M., Bernhard, M., Dillenburger, B., Buchli, J., Roussel, N., & Flatt, R. (2016). Digital Concrete: Opportunities and Challenges. RILEM Technical Letters, 1, 67–75. https://doi.org/10.21809/rilemtechlett.2016.16

Wu, Z., Memari, A. M., & Duarte, J. P. (2022). State of the Art Review of Reinforcement Strategies and Technologies for 3D Printing of Concrete. Energies (Basel), 15(1), 360. https://doi.org/10.3390/en15010360

Yun, K.-K., Choi, P., & Yeon, J. H. (2015). Correlating rheological properties to the pumpability and shootability of wet-mix shotcrete mixtures. Construction and Building Materials, 98, 884–891. Scopus. https://doi.org/10.1016/j.conbuildmat.2015.09.004

Part IV
Sustainable Engineering in Biotechnology and Nanotechnology

Role of Nanoparticles in the Suppression of Diseases in Fruits and Vegetables to Improve Agricultural and Environmental Sustainability: A Bibliometric Analysis

Rahul Islam Barbhuiya, Prabhjot Kaur, Charles Wroblewski, Abdallah Elsayed, Jayasankar Subramanian, Gopu Raveendran Nair, and Ashutosh Singh

1 Introduction

Fruits and vegetables make up an integral part of a diet that is nutritionally balanced and has a positive influence on the vitality and health of the body. They contribute to cellular cleansing and regeneration, as well as the treatment of a variety of diseases (Mritunjay and Kumar 2015). On the other hand, fresh vegetables and fruits can be contaminated with pathogens at any stage of the production process, including post-harvest storage, pre-harvest handling, transit, retail, and either commercial or domestic consumption. The sources of pre-harvest contamination include compost, insects, dust, insecticides, irrigation water, soil, fungicides, faeces, wild animals, and domestic animals. Similarly, post-harvest contamination can be caused by a variety of factors, including human handling, insects, harvesting equipment, transport vehicles, processing equipment, ice, and rinse water (Olaimat and Holley 2012). Therefore, pesticides are utilised to lessen or get rid of the infections, and fertilisers are distributed so that the crops receive more nourishment and are better able to withstand disease. However, the active chemicals in conventional formulations of pesticides and fertilisers often have a limited solubility in water, which restricts crop access. Therefore, to manage infections and obtain an acceptable

R. I. Barbhuiya · P. Kaur · C. Wroblewski · A. Elsayed · A. Singh (✉)
School of Engineering, University of Guelph, Guelph, ON, Canada
e-mail: asingh47@uoguelph.ca

J. Subramanian
Department of Plant Agriculture, University of Guelph, Guelph, ON, Canada

G. R. Nair
Department of Agricultural and Biological Engineering, University of Illinois at Urbana-Champaign, Champaign, IL, USA

I. S. Dunmade et al. (eds.), *Sustainable Engineering*, Green Energy and Technology, https://doi.org/10.1007/978-3-031-47215-2_20

yield, the growers need to apply significantly more of these formulations in bigger volumes and amounts. In addition, the formulations of current fertilisers and pesticides are influenced by processes such as leaching, precipitation by soil components, and volatilisation leading to significant environmental concerns. Because of this, traditional techniques of preventing disease and supplying plants with nutrients become not only inefficient but also prohibitively expensive (Elmer and White 2018). Consequently, to achieve agricultural and environmental sustainability, we need to discover new means of delivering nutrients, treating diseases, and adjusting existing sources of nutrients.

The term "nanotechnology" refers to the application of a technology that involves the use of nanoparticles with diameters ranging from 1 to 100 nanometres. It has the possibility to improve both the quality and quantity of agricultural products by making possible the intelligent management of agricultural inputs such as irrigation systems, herbicides, and fertiliser applications. The usage of nanoparticles has the potential to help mitigate the majority of the environmental concerns caused by conventional modern agriculture's continued reliance on chemical pesticides and fertilisers. It is now being utilised in a variety of industries and has the potential to transform the methods that are utilised in modern agriculture. These materials have a high reactivity because of their vast surface area-to-volume proportion as well as distinctive physical and chemical characteristics (Sirelkhatim et al. 2015). The agriculture and food industries can be revolutionised by nanotechnology because it can provide new means for enhancing plants' capability to absorb nutrients, resisting molecular disease, and rapid disease detection. As a direct consequence of this, numerous investigations into the role that nanotechnology plays in the avoidance of disease have been carried out. However, there is a lack of comprehensive information on the role of nanoparticles in disease prevention specifically related to fruits and vegetables.

Bibliometric study is a rigorous methodology for examining publications, utilising quantitative techniques such as co-word analysis, social network analysis, and cluster analysis. The primary aim of bibliometric analysis is to furnish a succinct summary of progressions in a specific research domain, detect prevalent domains or nascent patterns, and evaluate the contributions of nations, journals, or writers utilising quantitative metrics (Agarwal et al. 2016; Peng et al. 2022). VOSviewer is a piece of software that creates a visual map of the co-occurrence of research and keywords (van Eck and Waltman 2010). In recent years, a number of bibliometric studies have been conducted that look at the published works in the fields of nanoparticles and nanotechnology (Zhu et al. 2021; Bodnariuk and Melentiev 2019; Aleixandre-Tudó et al. 2020). Therefore, this study was conducted to assess the current status of research fields in the role of nanoparticles in the disease suppression of fruits and vegetables and to investigate the trend. Our goal was to pinpoint the most active research areas in this field so that we could provide a glimpse into potential future study areas. Publications are derived from the Scopus database between 2000 and 2021 in terms of the distribution of yearly publications, co-occurrence of keywords, source journals, and countries. This study would be helpful to both new specialists and researchers in the field of fruit and vegetable disease suppression with nanoparticles in planning research direction, determining the range of research topics, and identifying novel themes.

2 Data Collection and Methodology

The study of bibliometrics will focus on the year the articles were published, the journals published, the subjects of those articles, the number of articles, the countries and institutes that participated in the publishing of papers, and how these factors were thoroughly analysed using a variety of quantitative research techniques. This study looks at the patterns that have emerged in studies on the role of nanoparticles in the suppression of vegetable and fruit diseases since 2000. The scope of the investigation is limited to papers published between 2000 and 2021. Due to the fact that the data for 2022 is still lacking, it was disregarded.

2.1 Information Gathering and Retrieval Methods

In the present study, the data pertaining to publications was collected on July 7, 2022, and was procured in the format of a "CSV File" from the Scopus database.

The publications were required to meet the following requirements:

1. The terms searched (ST) were determined by ST ("topic", which includes the Abstract title, Abstract, Keywords) as ST = ("Nanoparticles" OR "Disease Suppression" OR "Nanotechnology") AND ST = ("plant diseases" OR "plant pathogens" OR "fruits" OR "vegetables" OR "fruits diseases" OR "vegetables diseases" OR "agriculture diseases" OR "disease suppression"). Table 1 lists the key terms that were initially used to describe the research area, together with the corresponding number of articles that have been published.

Table 1 Initial search terms for the research area and the number of publications published

S. No.	Keywords	No. of published papers (from 2000 to 2021)
1	Nanoparticles and plant diseases	2579
2	Nanoparticles and plant pathogens	1332
3	Disease suppression and fruits	713
4	Disease suppression and vegetables	479
5	Nanoparticles and fruit diseases	370
6	Nanoparticles and agriculture diseases	365
7	Nanoparticles and vegetable diseases	148
8	Nanotechnology and fruit diseases	110
9	Nanotechnology and vegetable diseases	64
10	Nanoparticles and disease suppression	1
11	Nanotechnology and disease suppression	1

2. The type of the document was “Article”, “Review”, and “Book Chapter”.
3. Dates ranged from 2000 to 2021, inclusive of both the publishing and the corresponding years.
4. The following details were gathered: keywords, journals, establishments, nations, authors, publication, and citations.

The literature review showed that about 6162 articles have been written, but only a small number of articles have been written about how nanoparticles could be used in disease suppression of fruits and vegetables. Furthermore, the number of articles was lowered by deleting duplicates and by removing ostensibly unrelated publications, i.e. papers unconnected to higher education and engineering.

2.2 *Network Mapping and Data Analysis*

The study of bibliometrics can be helpful in contribution to the monitoring of the evolution and trends of successful publications. In recent years, bibliometric visualisation software has gained widespread usage as a tool for the extraction of publication data, as well as for the analysis of such data and the creation of knowledge maps (van Eck and Waltman 2010). To build the visual network maps utilised in this investigation, we analysed the CSV files that were obtained from the Scopus and extracted bibliographic data on keywords, citations, countries/regions, research institutions, and researchers using VOSviewer version 1.6.18 (van Eck and Waltman 2010). Co-citation analysis, co-authorship analysis, and co-occurrence analysis are the three most widely used bibliometric methods for extracting themes and locating hotspots in the literature, respectively. Scholars can benefit from citation analysis in a number of ways, including discovering the knowledge base of a field and determining its scope (Romero and Portillo-Salido 2019). Collaboration patterns among writers, institutions, and nations are revealed through co-authorship analysis (Newman 2004). The method of co-occurrence analysis involves examining the frequency of multiple words within a given article in order to determine their proximity and thus reveal prevalent themes and patterns within the field of study. These graphic representations of bibliometric data help researchers pinpoint the most productive research hotspots, foundational knowledge, institutions, countries, authors, and research frontiers for a certain topic of study. In this work, we used co-word analysis to examine the most often used keywords in the retrieved papers to discover the most active areas of nanoparticle research in disease suppression of fruits and vegetables. Each node is depicted as a labelled circle in VOSviewer’s visualisation maps. When analysing the frequency of occurrences, larger circles represent more frequent occurrences. Each circle’s colour represents the group to which it most closely corresponds. The significance and importance of a pair of nodes are represented by the thickness and length of the links between them. We limited the output to 1000 lines so that just the 1000 strongest connections between nodes would be shown.

3 Results and Discussion

3.1 Annual Output of Research Published on Different Keywords with Nanoparticles

According to the results of our search, there were a total of 6162 publications that were published between the years 2000 and 2021. There are 460 chapters included in books, 4583 research articles, and 1114 reviews. Figure 1 shows that between 2005 and 2021, there was an increment in the number of articles published/year in the area of disease suppression by nanoparticles. However, before 2005, there was not a single article published that demonstrated the use of nanoparticles in the disease suppression. The application of nanoparticles in disease suppression was not a well-established study field between the years 2005 and 2008, as indicated by the small number of articles. However, the average number of new publications after 2009 has increased rapidly, which may be attributable to the recognition among researchers that achieving a sustainable future must begin with agricultural sustainability, which can be accomplished by reducing the loss in agricultural production that is caused by diseases (Talukder et al. 2020). Furthermore, the number of annual global publications containing the keyword "nanoparticles" for the search keywords "fruit disease" and "vegetable disease" increased from 0 in 2000 to 370 and 148 in 2021, respectively, demonstrating a growing interest in fruit and vegetable diseases.

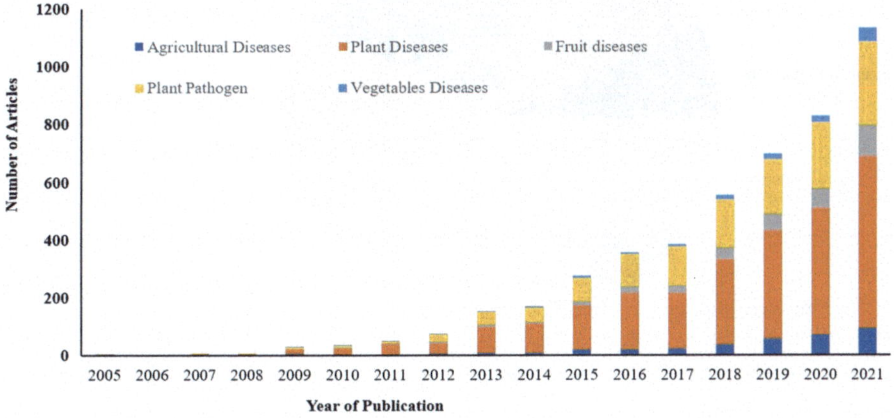

Fig. 1 Trend of nanoparticle application with different keywords based on research publication

3.2 Article Distribution Based on Journals and Subject Areas

A study was conducted on the topic of article distribution based on journals and subject areas for the keywords "nanoparticles and fruit diseases" and "nanoparticles and vegetable diseases". According to the information acquired from Scopus, the publications have been published in more than 100 different areas of study. However, for the purpose of the analysis, ten different subject groups were selected (Fig. 2a, b).

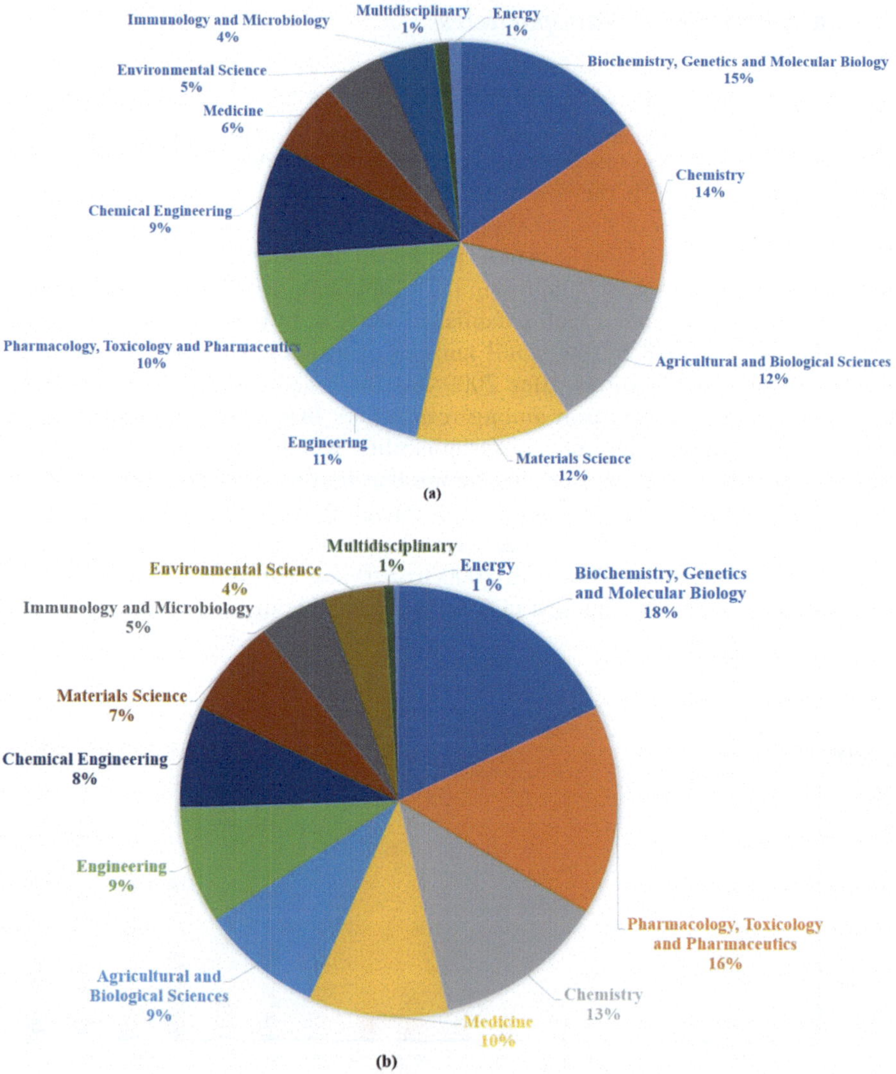

Fig. 2 The subject categories used most frequently for the role of nanoparticles in (**a**) fruit diseases and (**b**) vegetable diseases

These subject areas are nonexclusive due to their interdisciplinary nature. It is possible to draw the following conclusion after examining Fig. 2a: the publication of research on diseases of fruits that involve nanoparticles is most prevalent in the fields of biochemistry, genetics, and molecular biology (15%), followed by chemistry (14%), agricultural and biological sciences (12%), materials science (12%), and engineering (11%). A similar trend was observed, but individual percentages varied, for publications related to research on vegetable diseases, with nanoparticles, with pharmacology, toxicology, and pharmaceutics (16%) on second place (Fig. 2b). As a result, the representation of the various subjects that contributed to the use of nanoparticles in the diseases of fruits and vegetables provided a clear picture of the multidisciplinary collaboration that was concentrated among several scientific and technologically economic fields.

In addition, the contribution of published research articles on "Food Chemistry" in the eight journals that were determined to be the most productive based on their high Impact Factor (IF) and SCI Journal Rank (SJR) was recorded. The most productive journals were Environmental Research; Food Chemistry; International Journal of Biological Macromolecules; International Journal of Molecular Sciences; Journal of Agricultural and Food Chemistry; Materials Science and Engineering C; Artificial Cells, Nanomedicine, and Biotechnology; and Nanomaterials. Environmental Research is a multidisciplinary journal that publishes high-quality, original research on a wide variety of environmental topics with a focus on anthropogenic concerns with global relevance and applicability, as well as empirical evidence of environmental application. Food Chemistry is a journal that publishes articles that discuss the latest developments in the chemistry and biochemistry of foods, as well as the analytical methods and approaches that are utilised. In a similarly, the Journal of Agricultural and Food Chemistry publishes articles that discuss the chemistry and biochemistry of agricultural and food. On the other hand, nanomaterials publish reviews, regular research papers, communications, and short notes that are applicable to any area of research that incorporates nanomaterials, in terms of both their scientific basis and their practical applications. Therefore, it can be concluded that the use of nanoparticles in the suppression of disease can be considered to be an impactful and relevant issue of research within the field of scientific technology.

3.3 Co-authorship and Distribution of Countries/Regions

The country's contribution to the role of nanoparticles in vegetable and fruit diseases was estimated by using the location of at least one author's affiliation or university (Li et al. 2009; Vanga et al. 2015). As can be seen in Fig. 3, a total of 93 countries and/or regions have contributed to research on the topic of the role of nanoparticles in disease suppression, which is a subject that is of interest all over the world. Table 2 summarises the top 10 countries whose research is most focused on the role of nanoparticles in disease suppression in fruits and vegetables. India was

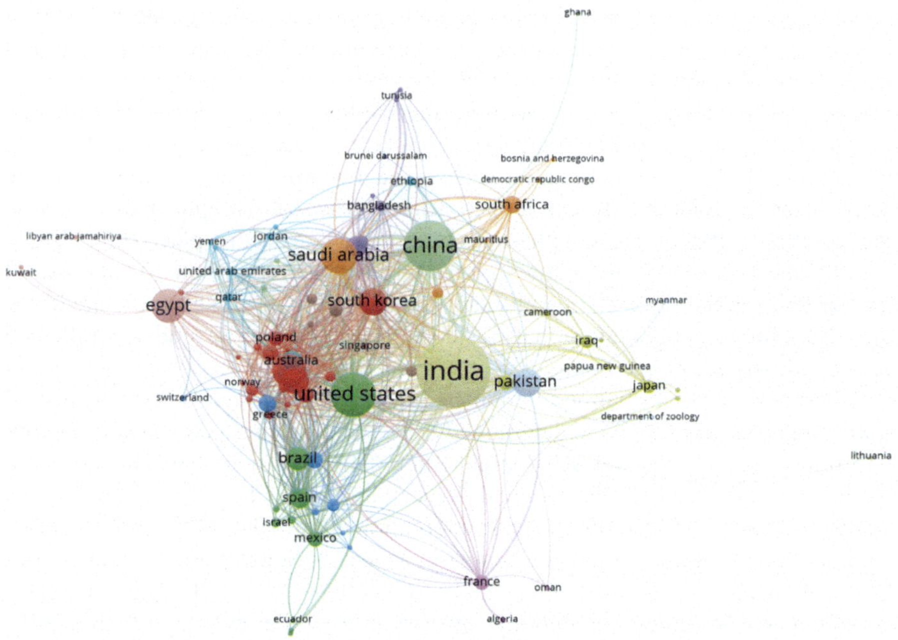

Fig. 3 Research interconnections among various countries represented by bibliographic coupling graphs

Table 2 The top 10 productive regions/countries

Search term	Rank	Publications	Region/country	Citations
Nanoparticles and fruit diseases	1	102	India	1732
	2	68	China	1996
	3	54	United States	2561
	4	26	Egypt	403
	5	25	Iran	1003
	6	25	Saudi Arabia	374
	7	17	Italy	908
	8	16	South Korea	382
	9	15	Mexico	291
	10	15	Pakistan	240
Nanoparticles and vegetable diseases	1	30	India	457
	2	23	China	528
	3	22	United States	1014
	4	13	Brazil	276
	5	13	Iran	329
	6	10	Egypt	96
	7	10	Italy	499
	8	10	Malaysia	168
	9	7	Mexico	84
	10	7	Pakistan	85

the most productive nation in terms of national research strength for the search term "nanoparticles and fruit diseases", with 102 publications and 1732 citations. This was followed by China, which had 68 publications and 1996 citations, and United States, which had 54 publications and 2561 citations. Similarly, for the search term "nanoparticles and fruit diseases", India was the most productive with 30 publications and 457 citations. This was followed by China, which had 23 publications and 528 citations, and United States, which had 22 publications and 1014 citations. It may be possible that these statistics are the result of the fact that China, India, the United States, and Brazil are the top four food-producing countries in the world. They all have sizable populations, a plentiful land area, and climate zones that are amenable to the cultivation of a wide range of crops; but the roles that food production plays in the economies of these countries couldn't be more different (Worldometer 2020).

When two different documents cite the same third document, it is known as bibliographic coupling (Kessler 1963). Figure 3 represents research interconnections among various countries represented by bibliographic coupling graphs when assessed using VOSviewer software. Co-authorship analysis of the countries/regions was performed using VOSviewer to uncover global partnerships in this area of research. Figure 3 displays the geographical co-authorship network. Out of a total of 105 countries/regions involved in the co-authorship network, 93 were placed into one of the 13 colour-coded clusters. India had the largest number of partners ($n = 62$), followed by the United States ($n = 50$) and Saudi Arabia ($n = 50$), China ($n = 46$), and other countries. When looking at the connections between different countries, it has been discovered that India, China, the United States, and Saudi Arabia are the countries that have the most connections with one another in the field of nanoparticle research and its application in the prevention of diseases in fruits and vegetables. This coupling relationship also reflected that these countries with the majority of the publications appear in the centre of the network and displayed strong research linkages with South Korea, Saudi Arabia, Egypt, Brazil, Australia, and other countries across the world. It is possible to draw the following conclusion as a result of the geographical analysis: research practices on the synthesis, characterisation, and utilisation of nanoparticles are being carried out most actively in developed countries as compared to nations that are still in the process of developing.

3.4 Keyword Co-occurrence Analysis

The primary focus of a piece of writing can often be deduced from its keyword usage, which is why it is one of the most important components of an article. In other words, we can say that keywords cover the most important aspects of publications, and high-occurrence keywords are ideally fitted for selection in the context of co-occurrence analysis. As a result, we extracted top keywords from the retrieved papers, and then we evaluated them further using the VOSviewer software. There is a correlation between a larger node size and a higher frequency of the keyword; a shorter distance between two nodes suggests a closer correlation between the specified terms, and nodes that have the same colour represent topics that are similar (Zhu et al. 2021). Figure 4a, b presents a network visualisation showing the top keywords clustered into three groups based on frequency of occurrence. VOSviewer automatically categorised all keywords with similarities into three clusters, which were visualised as cluster 1 (represented by the colour red), cluster 2 (represented by the colour green), and cluster 3 (represented by the colour blue).

In Fig. 4a, the keywords nanoparticles, metal nanoparticles, fruits, human, diseases, and non-human are located at the centre of the visualisation network map. Cluster 1, shown in red, includes 40 items related to nanoparticle synthesis and characterisation. This content identifies facile/one-pot/high-yield synthesis or green synthesis, as well as characterisation via Fourier transform infrared spectroscopy, X-ray diffraction, and other techniques for determining optical properties, size, shape, and fluorescence. Metal nanoparticles used in antibacterial research are also labelled. Cluster 2 in green also contains 40 items involving biomedical applications, especially cancer research, and applications of nanomaterials. Keywords such as drug delivery, animal experiment, cancer/tumour, in vivo/in vitro, and diagnosis are recognised in this section. The third and last cluster, denoted by the colour blue, contains 20 items and focused on the role of nanomaterials in disease prevention in fruit. This section contains a number of keywords, some of which are nanoparticles, fruits, plant diseases, plant leaves, and microbiology.

Similarly, in Fig. 4b, the keywords nanoparticles, humans, drug delivery system, diseases, and antioxidants are located at the centre of the visualisation network map. Cluster 1, shown in red, includes 46 items related to secondary metabolites of the crops. Cluster 2 in green contains 41 items focused on the action of nanomaterials in disease prevention in vegetables. This section contains a number of keywords, some of which are vegetables, plant diseases, diseases, and anti-infective agent. The third and last cluster, denoted by the colour blue, with 37 items, represents biomedical applications. Keywords such as drug release, animal experiment, cancer/tumour, controlled study, in vivo/in vitro, and drug formulations are recognised in this section.

Thus, the findings of the Scopus search and the analysis of the co-occurrence data showed that the majority of research with nanoparticles has been conducted in the fields of medicine, drug delivery, and food safety; however, their application in the prevention of diseases in vegetables and fruits had been limited.

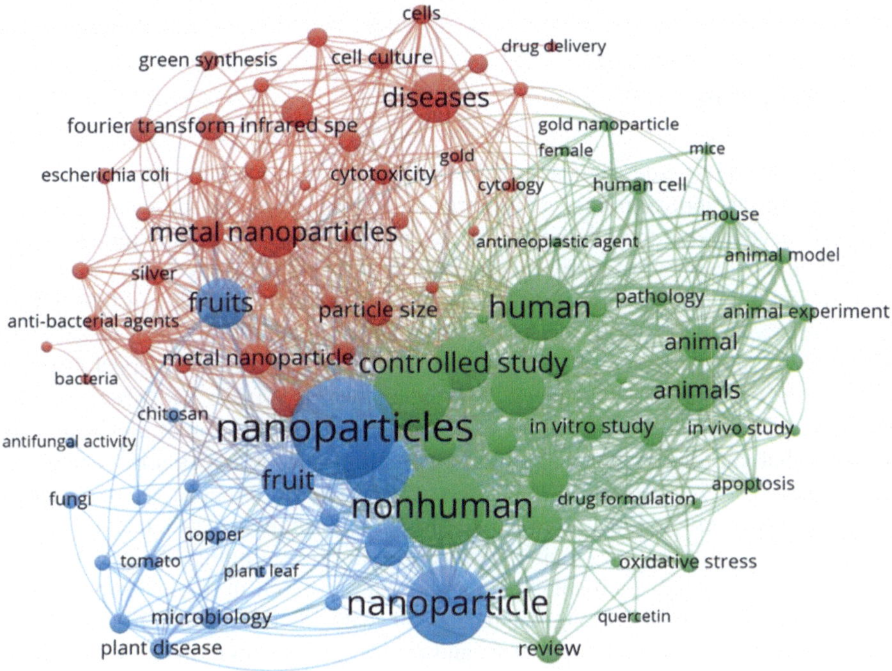

(a)

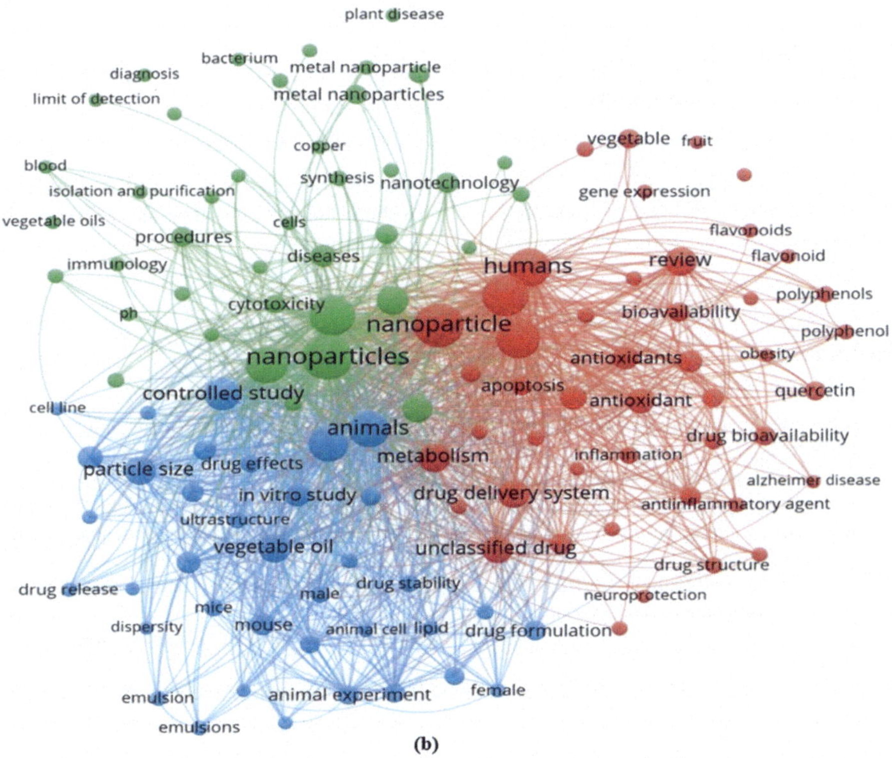

(b)

Fig. 4 (**a**) Keyword co-occurrence graphs for worldwide publication of nanoparticles and fruit diseases. (**b**) Keyword co-occurrence graphs for worldwide publication of nanoparticles and vegetable diseases

4 Conclusion

As a result of all this study, it has been determined that nanoparticles have the potential to be used in applications involving the delivery of drugs and the safety of food. It was discovered that the number of articles produced each year is rising, and significant advancements are being made in the industry. There is a greater potential than ever before for interdisciplinary research to yield results that can be used to prevent disease. In addition, it was found that nanoparticles had been researched primarily for the purposes of medicine, drug delivery, and food safety; however, their use in the prevention of diseases in vegetables and fruits had been limited. Therefore, the scientific community is now able to recognise emerging ideas and frontiers that can direct future study on the suppression of diseases in vegetables and fruits as a direct result of these discoveries. This bibliometric analysis represents the first complete look at the application of environmental nanotechnology in disease suppression in the vegetable and fruit research community. This bibliometrics produced a visualised display of nanoparticle research in disease suppression by analysing publications throughout the last 20 years, including publishing research hotspots, global collaboration patterns, and trends. Insights like this let researchers zero in on new areas of inquiry that could lead to breakthroughs in the fight against agricultural diseases using nanoparticles.

Acknowledgements The authors wish to extend their thanks to the Natural Sciences and Engineering Research Council of Canada (NSERC) for providing financial support for this investigation (Grant # RGPIN-2017-03975).

Conflicts of Interest/Competing Interests The authors declare no conflict of interest.

Funding The authors wish to extend their thanks to the Natural Sciences and Engineering Research Council of Canada (NSERC) for providing funding for this study (Grant # RGPIN-2017-03975).

References

Agarwal A, Durairajanayagam D, Tatagari S, Esteves S C, Harlev A, Henkel R, Roychoudhury S, Homa S, Puchalt N G, and Ramasamy R (2016) Bibliometrics: tracking research impact by selecting the appropriate metrics. Asian J. Androl 18: 296

Aleixandre-Tudó J L, Bolaños-Pizarro M, Aleixandre J L, and Aleixandre-Benavent R (2020) Worldwide scientific research on nanotechnology: a bibliometric analysis of tendencies, funding, and challenges. J. Agric. Food Chem. 68: 9158–9170

Bodnariuk M, and Melentiev R (2019) Bibliometric analysis of micro-nano manufacturing technologies. Nanotechnol. Precis. Eng. 2: 61–70

Elmer W, and White J C (2018) The future of nanotechnology in plant pathology. Annu. Rev. Phytopathol. 56: 111–133

Kessler M M (1963) Bibliographic coupling between scientific papers. American documentation 14: 10–25

Li L-L, Ding G, Feng N, Wang M-H, and Ho Y-S (2009) Global stem cell research trend: Bibliometric analysis as a tool for mapping of trends from 1991 to 2006. Scientometrics 80: 39–58

Mritunjay S K, and Kumar V (2015) Fresh farm produce as a source of pathogens: a review. Res. J. Environ. Toxicol. 9: 59–70

Newman M E (2004) Coauthorship networks and patterns of scientific collaboration. Proc. Natl. Acad. Sci. 101: 5200-5205

Olaimat A N, and Holley R A (2012) Factors influencing the microbial safety of fresh produce: a review. Food Microbiol. 32: 1–19

Peng C, Kuang L, Zhao J, Ross A E, Wang Z, and Ciolino J B (2022) Bibliometric and visualized analysis of ocular drug delivery from 2001 to 2020. J. control. release

Romero L, and Portillo-Salido E (2019) Trends in sigma-1 receptor research: a 25-year bibliometric analysis. Front. Pharmacol. 10: 564

Sirelkhatim A, Mahmud S, Seeni A, Kaus N H M, Ann L C, Bakhori S K M, Hasan H, and Mohamad D (2015) Review on Zinc Oxide Nanoparticles: Antibacterial Activity and Toxicity Mechanism. Nano-Micro Letters 7: 219–242.https://doi.org/10.1007/s40820-015-0040-x

Talukder B, Blay-Palmer A, and Hipel K W (2020) Towards complexity of agricultural sustainability assessment: Main issues and concerns. Environmental and Sustainability Indicators 6: 100038

van Eck N J, and Waltman L (2010) Software survey: VOSviewer, a computer program for bibliometric mapping. Scientometrics 84: 523–538. https://doi.org/10.1007/s11192-009-0146-3

Vanga S K, Singh A, Harish Vagadia B, and Raghavan V (2015) Global food allergy research trend: a bibliometric analysis. Scientometrics 105: 203–213

Worldometer (2020) Largest countries in the world (by area). Worldometers

Zhu S, Meng H, Gu Z, and Zhao Y (2021) Research trend of nanoscience and nanotechnology–A bibliometric analysis of Nano Today. Nano Today 39: 101233

Emerging Biotechnologies for Sustainable Bioenergy Production: Challenges and Outlook

Ifeanyi Michael Smarte Anekwe, Stephen Okiemute Akpasi, Atuman Samaila Joel, and Yusuf Makarfi Isa

1 Introduction

Presently, more than 50 countries have devised strategies for the bioeconomy (Svazas et al. 2019). The production of energy, electricity, heat, chemicals, and fuels, as well as other goods and services, primarily depends on fossil fuels. 81% of the entire basic energy supply comes from fossil fuels, 5% comes from nuclear energy, and 14% comes from sustainable energy sources (biomass ~70%). However, about 50% of the energy requirement for the chemical industry is utilized as raw materials for chemical products (Dwyer and Teske 2018). The use of biomass feedstock and its transformation into food, feed, and other bio-based products, including bioplastics, biofuels, and bioenergy, can serve as a substitute to fossil energy. The rising demand for biomass as an industrial feedstock will affect the production of food and other bio-based materials, ultimately resulting in an increased proportion of agriculture's contribution to gross domestic product (GDP) (Popp et al. 2021).

Given that a finite amount of land and biomass is available worldwide, ensuring global food supply while finding alternatives to fossil fuels presents an exceptional barrier (Anekwe et al. 2021). This can be attributed to the limited availability of

I. M. S. Anekwe (✉) · Y. M. Isa
School of Chemical and Metallurgical Engineering, University of the Witwatersrand, Johannesburg, South Africa

S. O. Akpasi
Department of Chemical Engineering, Durban University of Technology, Durban, South Africa

A. S. Joel
School of Chemical and Metallurgical Engineering, University of the Witwatersrand, Johannesburg, South Africa

Department Chemical Engineering, Faculty of Engineering, Abubakar Tafawa Balewa University Bauchi, Bauchi, Nigeria

I. S. Dunmade et al. (eds.), *Sustainable Engineering*, Green Energy and Technology, https://doi.org/10.1007/978-3-031-47215-2_21

arable land, which accounts for only 18% of the Earth's surface area. The primary application of biomass in today's world is for the synthesis of animal feed, followed by food, energy, fuels, and chemicals. Biomass is responsible for 13% of the total overall energy demand around the globe (other alternatives contribute extra 5% to the total energy utilization) (Popp et al. 2021). The process of "decarbonization" of the economy refers to the act of lowering the quantity of greenhouse gas (GHG) emissions generated by fossil fuel combustion. In the energy sector, sustainable resources contribute significantly in the "decarbonization" activity of the economy. Consumers and substituting fossil resources with sustainable energy are driving the rising demand for bioproducts. The market price of feedstocks, such as lignocellulose, sugars, starch, and oils, is one of the elements that can influence the viability of biochemical materials (Kircher 2019). On the other hand, biomass necessitates a more significant number of processing procedures, which results in an increased cost and more work.

The decrease in reliance on petroleum and other fossil fuels that has resulted from biotechnological innovation in the energy and chemical industries has a beneficial effect on the natural ecosystem. Both the bioeconomy and the circular economy are beginning to converge due to the incorporation of the economically beneficial components of the circular economy and the environmentally responsible practices of the bioeconomy. When developing a circular bioeconomy, one of the most important goals to achieve is the internationalization of the various bioeconomy processes, including biotechnology for bioenergy synthesis (Aguilar et al. 2018). Biotechnology is the use of biological processes, microbial species, or systems to create products that are expected to benefit human lives. In the broadest sense, this refers to engineering microorganisms to employ them in biotechnological processes. It is also possible to describe it as the collection of skills necessary for using living systems or manipulating natural processes to develop products, procedures, or conditions that are beneficial to human advancement. The development of hybrid genes and their subsequent transfer into microorganisms where the gene is not typically found, either in part or in its entirety, are presently given a greater priority in biotechnology. Table 1 shows different applications of biotechnology. This chapter discussed the application of emerging biotechnological processes in biofuel production while highlighting the limitations of this process and its potential outlook. Also, the role of biotechnology in achieving the SDGs was presented.

2 Review of Biotechnological Processes for Bioenergy Production

Due to the urgency to replace fossil fuels and develop other energy sources, bioenergy production (using biotechnology) has received much attention worldwide as a viable substitute to fossil fuels. Some biotechnological biofuel production processes include anaerobic process, algae biotechnology, microbial fermentation, and nanotechnology.

Table 1 Different applications of biotechnology

Categories	Properties	Products and processes
Bioenergy and fuel	Biotransformation of starch to sugars for bioethanol synthesis Anaerobic digestion (AD) of organic residue to CH_4 Gasoline blending with bioethanol decreases fossil energy consumption and GHG emission	Ethanol synthesized from biomass Via AD, organic waste from landfill sites is converted to biogas (~ 45% CO2 & 55% CH4) Production of liquid hydrocarbon fuel from plants, animals, and microbial oils
Commodities and biochemicals	Biocatalysis and genetic/metabolic engineering propel traditional chemical industries' advancement	Bioethanol and cellulose esters from sustainable agricultural feedstock Succinic acid and C_2H_4 glycol Genetically engineered microbes (GEM)
Specialty and life science products	Over 60% of overall biotechnological-associated sales worth for fine chemicals and pharmaceuticals accounts for 5–11%	Fermentation operation: drugs and drug precursors, biopharmaceuticals, recombinant proteins & vaccines, vitamins Enzymatic processes: selective enzymatic catalysis, extremophilic enzymes, transformation of corn starch to high-fructose corn syrup
Agricultural chemicals	Valuable products to substitute traditional agrochemicals; animal feed with enhanced nutritional value and improvement in animal production	Biopesticides: environmentally friendly, highly targeted, and do not leave toxic residues Biofertilizers and soil inoculants: contribute to pollution reduction, energy and resource consumption related to the application of traditional fertilizers
Environmental biotechnology		Application of activated sludge method in wastewater treatment Use of composting for the aerobic stabilization of solid organic waste Bioremediation using microbes and GEM

2.1 *Anaerobic Biotechnology for Bioenergy Production*

Employing a biotechnological alternative is the most economical and sustainable strategy to focus on sustainable solutions that reduce pollution, the strain on the Earth's finite natural resources, and increasing energy insecurity. Using renewable biofuels and usable by-products, anaerobic biotechnology is a sustainable waste treatment method, as shown in Fig. 1. Anaerobic biotechnology through anaerobic digestion (AD) offers significant advantages over other bioenergy sources by using readily accessible waste biomass (van Foreest 2012). In the absence of O_2, bacteria use anaerobic digestion (AD) to convert organic substrate into biogas. It is a process that produces renewable biogas energy (between 50–75% CH_4 and 25–50% CO_2)

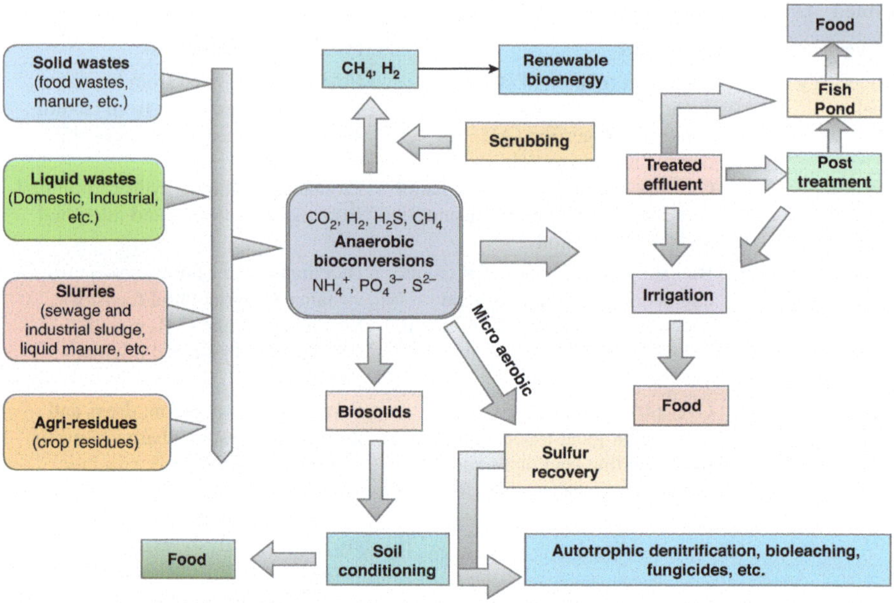

Fig. 1 Processes for recovering resources from wastes using integrated anaerobic bioconversion. (Adapted from (Khanal 2011))

while helping to reduce eutrophication, greenhouse gas emissions, dissolved oxygen loss, and other problems.

2.2 *Algae Biotechnology for Biofuel Production*

Microalgae have the prospect to transform the biofuel synthesis process without significantly affecting environmental sustainability. The use of algae biofuels has many advantages over the use of fossil fuels, including the following: algal biofuels (i) are easily accessible from widely available algal biomass sources; (ii) make a significant contribution to the reduction of greenhouse gases; (iii) are highly environmentally benign; (iv) are beneficial to the ecosystem, bioeconomy, and consumers; and (v) promote environmental sustainability. The prospect of large-scale synthesis of algae-based biofuel lies in developing technically viable solutions, such as optimizing microalgal strains for enhanced oil production. According to Chung et al. (2017), biotechnological operations could lower the cost of producing microalgal biofuel by 15–20% compared to conventional methods.

The fourth-generation biofuels are made from GEM such as yeast, fungi, and microalgae. The major biofuels from algal biomass are biodiesel, bioethanol, biogas, and biohydrogen (Hussain et al. 2021). Various initiatives have been developed to improve biofuel synthesis from microalgae (Schenk et al. 2008; Chowdhury and

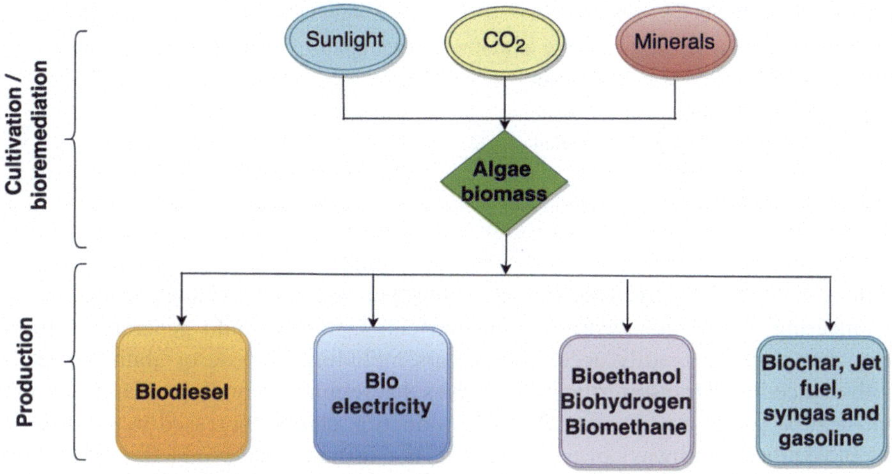

Fig. 2 Production of different biofuels from algal biomass. (Adapted from (Ahmad et al. 2022))

Loganathan 2019). Algal oil is chemically converted or transesterified into biodiesel, while algae feedstock can be burned to produce electricity. Anaerobic digestion and fermentation, in combination with other operations, can be employed for methanol and ethanol synthesis from algae using biochemical conversion technology. However, a thermochemical process can be employed to convert biomass into bio-oil and charcoal (pyrolysis), syngas and fuel gas (gasification), and bio-oil (liquefaction) (Chiaramonti et al. 2007; Anekwe et al. 2022). There may be some legislative, technological, and raw material concerns with biofuels. Figure 2 depicts the different procedures used to transform algal biomass into various biofuels.

2.3 *Microbial Fermentation*

With the help of fungi, yeast, or bacteria, sugars from lignocellulosic biomass (LCB) are converted into biochemicals (e.g., organic acids) or biofuels such as isobutanol, acetone, butanol, and others by microbial fermentation. Lipid-rich microorganisms could serve as new raw materials for biofuel production. The oils produced by oleaginous microbes may represent a promising feedstock for the transesterification process used to produce biofuels from plants (Ahmad et al. 2011). Fast-growing microbes enable the production of biofuels while using less arable land and a broad range of feedstocks, including sugarcane bagasse, which has a significantly higher yield per hectare than rapeseed. On the other hand, interest has grown in converting widely applied microorganisms including *Escherichia coli* and *Saccharomyces cerevisiae* into biofuel cell factories by incorporating an ester synthesis route that could result in direct synthesis of fatty acid ethyl esters (FAEEs) through direct esterification of ethanol with acyl residues of the fatty acid coenzyme

Table 2 Different microbial strains employed to produce biofuel (Adegboye et al. 2021)

Strain name	Type of biofuel	Titer (g/L)	Yield (g/g of sugar consumed)
Lipomyces starkeyi	Lipids	4.6	0.16
Cryptococcus curvatus	Lipids	5.8	0.20
Clostridium autoethanogenum	Ethanol	5.1	0.66
Saccharomyces cerevisiae	Isobutanol	<0.1	0.00097
Escherichia coli KO11	Ethanol	40+	0.44–0.51

A thioesters (acyl-CoA) (Ahmad et al. 2011). This approach is known as metabolic engineering. These genetically engineered cell factories could generate biofuels from inexpensive, readily accessible sugars including glucose or abundant LCB without needing a transesterification process that requires complicated isolation and purification pre-treatment step. Biofuel production can be increased by modifying the biosynthetic pathways of native strains. For instance, the production of biofuel was increased by altering the electron metabolism in *Clostridium thermocellum* (Lo et al. 2017). Several metabolically engineered microbial strains have successfully produced biofuel (Table 2).

2.4 *Nanobiotechnology for Biofuel Production: Using Nano-Immobilized Biocatalysts*

The nanobiocatalyst (NBC) is a newly developed technology that combines modern nanotechnology with biotechnology with prospects for enhancing enzymatic activities, durability, potentials, and efficiency in bioprocessing operations. In the fabrication of NBCs, enzymes are immobilized on functional nanomaterials as supports for enzymes. The bioprocessing industry is still in the early stages of utilizing NBC technologies. NBCs are not fully understood in terms of process-related characteristics. Nanocarriers have been shown to produce NBCs with a high enzyme loading and enhanced catalytic activity that are chemically and thermally stable. To create a solvent- and temperature-based nanosystem, Wang et al. (2011) synthesized NPG biocomposites that exhibit a significant enzymatic stability in thermal and organic conditions. The 35 nm pore size of the enzyme-supported composites enabled the encapsulation of numerous biomolecules of different sizes, by adsorption either inside the pores or on the outer surfaces. The NBC-driven system showed increased efficiency in converting soybean oil to biofuel with a longer process duration of up to 240 hours and maintained enzymatic operation for up to 10 subsequent batch processes. The commercialization of this technology is made possible by these outstanding laboratory-scale results. The different nano-immobilized biocatalysts utilized in the production of biofuel processes are listed in Table 3.

Table 3 Nano-immobilized biocatalysts reported in biofuel production processes

Nanobiocatalyst	Feedstock	Microorganism	Yield (%)
Alkyl grafted core–shell Fe_3O_4-SiO_2	Olive oil	*Burkholderia* sp. C20	95.74
Poly-acrylonitrile nanofibrous membrane	Soybean oil	*Pseudomonas cepacia*	90
Alkyl-celite	Sunflower oil	*Burkholderia*	85
Poly-acrylonitrile fibers	Rapeseed oil	*Pseudomonas cepacia*	80
Fe_3O_4 sub-microspheres	Waste vegetable oil	*Candida* sp.	80

Adapted from (Zhang et al. 2016; Fan et al. 2016)

2.5 Limitations of Biotechnological Processes

The growth of the bioenergy industry has been influenced by several causes, such as general concern about global warming, dependence on oil and other nonrenewable resources, and the potential depletion of fossil fuels (Jönsson and Martín 2016). Although biofuels have become one of the most important substitutes for fossil fuels, many limitations still need to be overcome before they can be considered economically viable and attractive. Against this backdrop, biotechnology has emerged as a powerful platform to address concerns related to industrial processes (Chen 2014). However, industrial biofuel production using biotechnological processes still faces several hurdles despite recent biotechnological advances. These problems include high energy demand, biological contamination, the need for microorganisms capable of fermenting pentose and hexose sugars, high concentration of enzyme mixtures to degrade the feedstock, expensive treatment of effluents containing hazardous substances, and increased capital investment in equipment and machines (Chen 2014). The high cost of algal biofuel production means that the large-scale application of this biotechnology is far from a reality. These obstacles prevent biofuel production from becoming a competitive economy, especially in contrast to fossil fuel production in times of falling oil prices. The merits and demerits of different biotechnologies for bioenergy production are presented in Table 4.

3 The Role of Biotechnology in Achieving Sustainable Development Goals (SDGs)

Bioenergy from biotechnological processes is a significant category of renewable energy. In light of this, achieving the United Nations Sustainable Development Goals (SDGs) from the perspective of climate change and energy security is essential. Integrated assessment modelling reveals a significant risk of being unable to fulfil long-term climate objectives without bioenergy, as reported by the IPCC 5th

Table 4 Merits and demerits of various biotechnologies for bioenergy synthesis

Biotechnological processes for bioenergy production	Limitations	Advantages
Algae-based biofuel production	May create variations in the quality of the biofuels reaching the market	Extremely productive from an overall production standpoint
Anaerobic-based biofuel production	Loss of feedstock, which causes a decrease in potential product yields	An energy-efficient and ecologically benign approach for producing bioenergy
Microbial fermentation-based biofuel production	The method for optimizing various processes for maximal yield is difficult	Low cost of production, as well as reduced risks of contaminations
Nanobiotechnology-based biofuel production	High production costs	Improves the economic feasibility of the bioethanol manufacturing process

Assessment Report. According to global evaluations conducted by REN 21, the IEA, and IRENA, bioenergy is currently responsible for three-quarters of all renewable energy consumption and 50 percent of the most cost-effective alternatives for boosting the utilization of renewable energy by 2030 (Skeer 2017). Bioenergy production is a component of the broader bioeconomy, including manufacturing, agriculture, and forestry. Biotechnological processes of bioenergy production can contribute significantly in the realization of the SDGs and the fulfilment of the Paris Agreement on Climate Change, thereby contributing to the advancement of climate goals, food security, improved land use, and sustainable energy for all (Mbow et al. 2017). The biotechnological processes will facilitate the attainment and actualization of the following SDGs.

3.1 Achieving SDGs 7 and 13 Through Biotechnology

The biotechnological process of biofuel production will foster the attainment of Sustainable Goal 7, which guarantees access to affordable, dependable, renewable, and efficient energy. At the same time, SDG 13 is set to pursue immediate action to tackle climate change and its effects. Biomass energy can provide sustainable energy in sectors of the economy (SDG 7). In the field of power generation, bioenergy has the potential to offer flexibility, which can help stabilize the proliferation of other resources like wind and solar that are intermittent and seasonal. In the industrial sector, biomass can produce a broad range of valuable bio-based chemicals and materials, providing an adequate supply of high-temperature process heat. The construction industry uses biomass as the fuel source for highly effective district heating systems, furnaces, and cook stoves. In the transportation sector, the utilization of liquid and gaseous biofuels, as well as electrification and increased energy efficiency in vehicles, can assist in attaining a significant decrease in the utilization of fossil fuels. In addition, biofuels are the only viable substitute for fossil fuels used in aircraft, maritime shipping, and heavy freight transportation (Skeer 2017). In most cases, the use of bioenergy improves access to regional energy sources. It decreases dependency on fossil fuels responsible for GHG, reducing climate variability and its impacts on other social and environmental concerns, which is in line with SDG 13. It has the potential to revitalize both the agricultural and forestry industries and encourage greater utilization of renewable resources as feedstocks for various industrial operations (Mbow et al. 2017).

3.2 Achieving SDGs 2 and 15 Through Biotechnology

Energy is considered sustainable when it is "obtained from resources that can sustain present activities and satisfies the demands of the present without jeopardising future generations' potential to fulfil their own needs." This energy should not compete with the global food resources and supply. The advancement in biofuel generation from the first generation to the third and fourth generation has ensured the transition from edible foods to food waste, municipal solid waste, lignocellulose biomass, and microbes (such as algae) as a sustainable biomass for biofuel synthesis. Since food security is not threatened, biomass energy will promote the actualization of SDG 2: "to put an end to hunger, ensure adequate food supply, improve nutrition, and encourage responsible agriculture." Transforming waste into bioenergy or products with a more excellent value eliminates the necessity for landfills. It can significantly minimize the accompanying methane emissions, a greenhouse gas with a much higher warming potential than CO_2. To facilitate the expansion of bioenergy in support of actualizing the SDGs, it is necessary to set rules and procedures that encourage best practices. These should consider that factors vary between regions, guarantee biodiversity protection, and promote the development of different ecosystem services in landscapes. To do this, unified land management and the participation of individual farmers, landowners, policymakers, and various other local and national stakeholders are required (Skeer 2017; Mbow et al. 2017). This will aid in maintaining, regenerating, and enhancing the sustainable utilization of terrestrial ecosystems, sustainably preserving forests, tackling desertification, halting and reversing land degradation, and halting the loss of biodiversity (SDG 15).

4 Challenges and Prospects of Biotechnology for Bioenergy Production

Bioenergy production is faced with challenges for commercial deployment in the broad energy mix. This section will look at the challenges and prospects of bioenergy production.

4.1 Challenges of Biomass Energy Production

The bottleneck that has continuously hindered biorefinery's full-scale deployment from replacing fossil-based products is mainly the feedstock and conversion technologies, which are related to low yield and poor cost competitiveness. Davison et al. (2015) reported on the main factors affecting the commercialization of the bioenergy industry, including low feedstock yields, spatially dispersed feedstock sources, logistics, land-water use, rate and titer through conversion technologies and product use.

4.1.1 Barrier to the Improvement of Feedstock and Conversion Processes

Improving feedstock yield for bioenergy production has been the goal of plant biotechnology research. This entails the overall growth of the plant in terms of height and mass and/or the plant content required for the feed, food, and fiber market. The challenge in improving the feedstock is to accurately delineate the genetic foundation of the plant trait. The degree of complexity of the feature determines the factors that must be controlled, making it more difficult to identify specific targets for transformation (Davison et al. 2015).

Difficulty in controlling metabolic flux and insufficient information on microbial activity are the two major obstacles to the further development of biotechnology in converting more feedstock streams and novel products into finished products. Only advances in synthetic biology will make it possible to achieve continuous monitoring of metabolism by coupling environmental sensing with gene expression, translational and posttranslational control, and allosteric regulation (An et al. 2021; Davison et al. 2015). Another aspect that requires attention to overcome the challenges of conversion of feedstock is the discovery of unique enzymatic activities or pathways, which is vital to be harnessed for biotechnology application. Among the discovered unique enzymatic activities is fatty aldehyde decarbonylase which showed a good performance in converting fatty acid derivatives to produce hydrocarbon. Wang et al. (2022) reported the use of alkane hydroxylase system, which is capable of converting alkanes to corresponding fatty acids. Davison et al. (2015) reported that insufficient genetic tools impede the design of industrial biocatalysts from newly isolated microbes, which demonstrate desirable properties. Therefore, careful trial-and-error genetic tool developments are required to genetically engineer any novel microorganism.

4.1.2 Logistics and Environmental Issues

Logistics such as landwater use, harvesting, densification, transportation of less dense biomass, storage and preprocessing is quite a challenge when it comes to controlling production, conversion and the supply chain of genetically improved energy crops (Balan 2014; Davison et al. 2015). In dealing with such challenges, biotechnological improvement approaches need to be deployed such as high biomass density, enhanced storability, and reduced variability and ash content. The primary goal of feedstock biotechnology researcher is to produce biomass crop with a higher yield, since about one-third of biofuel synthesis cost is coming from biomass and biomass cost is directly related to the yield. Therefore, increasing the yield will lead to improved logistic efficiency and reduce costs (Amoah et al. 2019; Avagyan and Singh 2019; Balan 2014; Davison et al. 2015; Tursi 2019).

The commercial deployment of biorefinery will lead to other pollutions due to the assembling of several technologies that will impact the environment. Among the pollutions are air pollution which is caused by emission of particulate during biomass harvesting and size reduction, noise pollution from moving machines during pre-treatment processes, and the release of pre-treatment chemicals to the environment after processing. The assessment of net environmental impact of processing stages is done using life cycle analysis (LCA). The accuracy of the LCA study depends on how good data were collected. The focus on many companies nowadays is on LCA to enhance adjustment on processes or areas that need attention to reduce emission (Balan 2014). Therefore, emission estimate by LCA will have impact on the costs of establishing biorefinery.

4.1.3 Cost and Commercialization: Economy of Scale and Cost

Biotechnological applications in the energy industry have not yet been implemented commercially; therefore, it remains to be seen whether future discoveries can overcome current limitations. Cost is the biggest barrier to implementing any technology. Scale-up and cost of commercialization require attention for an effective application of biotechnology in biofuel production. These concerns were studied by many researchers using several cost models to understand the capital cost requirement (Aui et al. 2021; Sahoo et al. 2022; Wooley et al. 1999; Wright and Brown 2007; Wu et al. 2021). Argo et al. (2013) reported on a study of different plant capacities from 0–15 ton/day plotted against total production costs ($\$m^{-3}$), and it was found that the best capacity range should be between 2000 and 5000 ton/day for a biorefinery to be economically viable (850–875 $\$m^{-3}$). However, the biorefinery that operates at a capacity above 5000 ton/day recorded a marginal increase in the cost of production which reduces biofuel competitiveness due to the high selling price (Argo et al. 2013). Aui et al. (2021) stated that input capacity and operational cost influence the economic benefit of bioethanol plant. According to their findings, the minimum selling price for ethanol fuel ranges from $0.90 to $6.00/gallon (average price = $2.65/gallon (0.583$/liters)), and this pricing is equivalent to retail rates for gasoline in the United States. Argo et al. (2013) established a correlation between the selling price of ethanol and the size of the plant. The result of this study showed that the minimum selling price of ethanol decreased from $3.75/gallon to $2.25/gallon ($0.825–$0.495/liters) as the plant capacity increases from 500 to 10,000 dry ton/day. The price of biofuels (ethanol and biodiesel) has a strong relation to the

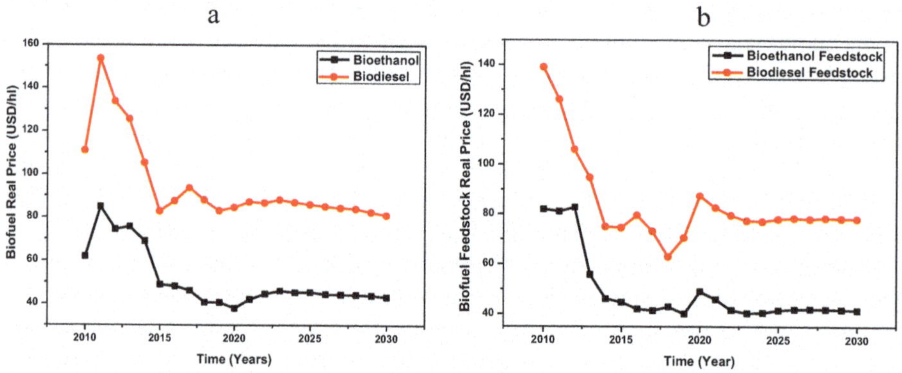

Fig. 3 Real projection for (**a**) biofuel price and (**b**) feedstock price (OECD/FAO 2021). (Note – Biofuel prices: Ethanol (USA); Biodiesel (Germany). Feedstock prices: Bioethanol (world market price for vegetable oil); Biodiesel (weighted average prices for raw sugar and maize)

prices of biofuel feedstock. According to *OECD-FAO Agricultural Outlook 2021–2030*, nominal biodiesel prices are projected to increase at 1.1% p.a, and ethanol price will increase at 1.8% p.a (OECD/FAO 2021). Figure 3 showed the real price projection for biofuels.

4.2 Outlook for the Bioenergy Industry

Bioconversion technologies have been one of the bottlenecks to realizing biomass bioenergy production (Amoah et al. 2019). Despite lots of research in producing several new products, only a few commercial applications were done, and this is due to low yields. Yield, titer, and rate are the critical parameters in the biotransformation operations. The yield of bioenergy production can be enhanced through the use of synthetic biology and metabolomics, which serve as the primary drivers for either expanding biomass streams or developing new products that are commercially viable (Davison et al. 2015).

The challenge of competition for land and water between energy crops and food production needs to be addressed using the biotechnology concept. One of the approaches in addressing the challenges of the competition is by improving biomass yield without destroying the soil fertility, thereby minimizing the land area requirement for producing bioenergy feedstock. Other approaches involve modifying plants that can tolerate severe environmental conditions such as drought and salt, which are currently unproductive and underused. Third-generation biofuels were reported to address the challenge of land competition as reported by Biofuels.org.uk (2010); algae have been used to produce up to 9000 gallons (40.91 liters) of biofuel per acre, which is 10 times the yield of the land-based feedstocks used for second-generation biofuels. The approaches designed to address the genetic complexity challenges of biomass involve understanding the pathway and genome-wide-omics which refers to large data biological analyses involving genomics, transcriptomics, proteomics, and metabolomics. The general procedure in improving feedstock is by manipulating one gene at a time, but this often had both desirable and undesirable effects on the plants. To address the challenges associated with the improvement of biomass feedstock, these routes were suggested by Davison et al. (2015): (i) plant models with shorter life cycles, (ii) investment by industries, (iii) field trials supported by greenhouse assessment, and (iv) sophisticated statistical evaluation approaches that can manage datasets that are both complicated and multifaceted.

Net-zero target by mid-century is achievable if the energy transition to renewable energy is encouraged by all the signatories at COP26 conference in Glasgow, United Kingdom. Energy from biomass as a sustainable energy resource has the potential to meet up with the net-zero targets. Ma et al. (2022) reported that algae-based technological carbon dioxide sequestration is promising for net-zero emissions. The study showed environmental benefits of using microalgae as a bioenergy source for sustainable, environmentally benign, and low-cost bioproducts and biofuels through microalgae-based CO_2 biosequestration, flue gas cultivation, and CO_2 valorization/biofuel production. Bioenergy from lignocellulosic biomass contributes to achieving net-zero emission targets by replacing fossil-based fuel, which is a threat to environmental conditions (Kostas et al. 2021; Ma et al. 2022).

5 Conclusion

There has been a rise in the demand for energy all over the world, which has led to higher fuel costs and a decline in fossil resources. Concerns about global warming have also contributed to the need to find alternate energy synthesis methods. One viable substitute for fossil fuels is the use of bioenergy which is a sub-sector of bioeconomy. Biomass is a promising sustainable feedstock for synthesizing solid, liquid, and gaseous biofuels. The difficulties involved with commercializing biotechnological processes and biofuels derived from biomass can be addressed through the integration of processes and the fine-tuning of various operational parameters that impact production. The fluctuating biomass supply throughout the year is another critical cause for concern. More intensive and systematic monitoring of biotechnology's contributions to the SDGs, promotion of knowledge and skills for the use of bio-based materials in the synthesis of biofuels and consumer products, demonstration projects for biorefineries integrating the synthesis of energy and high-value feedstocks, and research to upscale current and validate novel sustainable processes are just some examples of actions to support the biotechnological process of biofuel production and bioeconomy. The approach to biomass production should shift from a focus on a single end use to an integrated production method that ensures greater resource efficiency and promotes sustainable synthesis and application for biofuel production. The results of these systems need to be used with great prudence, to reduce waste and emissions and increase productivity as much as possible while preserving a robust resource base for future generations.Funding InformationThis work is based on research funded partly by the National Research Foundation of South Africa (NRF) (BRIC190321424123).

References

ADEGBOYE, M. F., OJUEDERIE, O. B., TALIA, P. M. & BABALOLA, O. O. 2021. Bioprospecting of microbial strains for biofuel production: metabolic engineering, applications, and challenges. *Biotechnology for biofuels,* 14**,** 1–21.

AGUILAR, A., WOHLGEMUTH, R. & TWARDOWSKI, T. 2018. Perspectives on bioeconomy. Elsevier.

AHMAD, A., BANAT, F., ALSAFAR, H. & HASAN, S. W. 2022. Algae biotechnology for industrial wastewater treatment, bioenergy production, and high-value bioproducts. *Science of The Total Environment,* 806**,** 150585.

AHMAD, A. L., YASIN, N. M., DEREK, C. & LIM, J. 2011. Microalgae as a sustainable energy source for biodiesel production: a review. *Renewable and sustainable energy reviews,* 15**,** 584–593.

AMOAH, J., KAHAR, P., OGINO, C., KONDO, A., 2019. Bioenergy and Biorefinery: Feedstock, Biotechnological Conversion, and Products. Biotechnol J. https://doi.org/10.1002/biot.201800494

AN, Y., LIU, Yu, LIU, YIJING, LU, M., KANG, X., MANSFIELD, S.D., ZENG, W., ZHANG, J., 2021. Opportunities and barriers for biofuel and bioenergy production from poplar. GCB Bioenergy. https://doi.org/10.1111/gcbb.12829

ANEKWE, I. M. S., ARMAH, E. K. & TETTEH, E. K. 2022. Bioenergy Production. *Biomass.* IntechOpen.

ANEKWE, I. M. S., KHOTSENG, L. & ISA, Y. M. 2021. The Place of Biofuel in Sustainable Living; Prospects and Challenges. *Reference Module in Earth Systems and Environmental Sciences.* Elsevier.

ARGO, A.M., TAN, E.C., INMAN, D., LANGHOLTZ, M.H., EATON, L.M., JACOBSON, J.J., WRIGHT, C.T., MUTH, D.J., WU, M.M., CHIU, Y., GRAHAM, R.L., 2013. Investigation of biochemical biorefinery sizing and environmental sustainability impacts for conventional bale system and advanced uniform biomass logistics designs. Biofuels, Bioproducts and Biorefining 7, 282–302. https://doi.org/10.1002/bbb.1391

AUI, A., WANG, Y., MBA-WRIGHT, M., 2021. Evaluating the economic feasibility of cellulosic ethanol: A meta-analysis of techno-economic analysis studies. Renewable and Sustainable Energy Reviews 145. https://doi.org/10.1016/j.rser.2021.111098

AVAGYAN, A.B., SINGH, B., 2019. Biodiesel: Feedstocks, Technologies, Economics and Barriers, Biodiesel: Feedstocks, Technologies, Economics and Barriers. Springer Singapore. https://doi.org/10.1007/978-981-13-5746-6

BALAN, V., 2014. Current Challenges in Commercially Producing Biofuels from Lignocellulosic Biomass. ISRN Biotechnol 2014, 1–31. https://doi.org/10.1155/2014/463074

BIOFUELS.ORG.UK, 2010, Third Generation Biofuels. https://biofuel.org.uk/third-generation-biofuels.html (accessed on 26/11/2022).

CHEN, H. 2014. Applications of lignocellulose biotechnology in bioenergy. *Biotechnology of Lignocellulose.* Springer.

CHIARAMONTI, D., OASMAA, A. & SOLANTAUSTA, Y. 2007. Power generation using fast pyrolysis liquids from biomass. Renewable and sustainable energy reviews, 11, 1056–1086.

CHOWDHURY, H. & LOGANATHAN, B. 2019. Third-generation biofuels from microalgae: a review. Current Opinion in Green and Sustainable Chemistry, 20, 39–44.

CHUNG, Y.-S., LEE, J.-W. & CHUNG, C.-H. 2017. Molecular challenges in microalgae towards cost-effective production of quality biodiesel. *Renewable and Sustainable Energy Reviews,* 74**,** 139–144.

DAVISON, B.H., BRANDT, C.C., GUSS, A.M., KALLURI, U.C., PALUMBO, A. V., STOUDER, R.L., WEBB, E.G., 2015. The impact of biotechnological advances on the future of US bioenergy. Biofuels, Bioproducts and Biorefining 9, 454–467. https://doi.org/10.1002/bbb.1549

DWYER, S. & TESKE, S. 2018. Renewables 2018 global status report. *Renewables 2018 Global Status Report.*

FAN, Y., WU, G., SU, F., LI, K., XU, L., HAN, X. & YAN, Y. 2016. Lipase oriented-immobilized on dendrimer-coated magnetic multi-walled carbon nanotubes toward catalysing biodiesel production from waste vegetable oil. *Fuel,* 178**,** 172–178.

HUSSAIN, F., SHAH, S. Z., AHMAD, H., ABUBSHAIT, S. A., ABUBSHAIT, H. A., LAREF, A., MANIKANDAN, A., KUSUMA, H. S. & IQBAL, M. 2021. Microalgae an ecofriendly and sustainable wastewater treatment option: Biomass application in biofuel and bio-fertiliser production. A review. Renewable and Sustainable Energy Reviews, 137, 110603.

JÖNSSON, L. J. & MARTÍN, C. 2016. Pre-treatment of lignocellulose: formation of inhibitory by-products and strategies for minimising their effects. *Bioresource technology,* 199**,** 103–112.

KHANAL, S. K. 2011. *Anaerobic biotechnology for bioenergy production: principles and applications*, John Wiley & Sons.

KIRCHER, M. 2019. Bioeconomy: Markets, implications, and investment opportunities. *Economies,* 7, 73.

KOSTAS, E.T., ADAMS, J.M.M., RUIZ, H.A., DURÁN-JIMÉNEZ, G., LYE, G.J., 2021. Macroalgal biorefinery concepts for the circular bioeconomy: A review on biotechnological developments and future perspectives. Renewable and Sustainable Energy Reviews. https://doi.org/10.1016/j.rser.2021.111553

LO, J., OLSON, D. G., MURPHY, S. J.-L., TIAN, L., HON, S., LANAHAN, A., GUSS, A. M. & LYND, L. R. 2017. Engineering electron metabolism to increase ethanol production in *Clostridium thermocellum. Metabolic engineering, 39*, 71–79.

MA, Z., CHEAH, W.Y., NG, I.S., CHANG, J.S., ZHAO, M., SHOW, P.L., 2022. Microalgae-based biotechnological sequestration of carbon dioxide for net zero emissions. Trends Biotechnol 40, 1439–1453. https://doi.org/10.1016/J.TIBTECH.2022.09.002

MBOW, H.-O. P., REISINGER, A., CANADELL, J. & O'BRIEN, P. 2017. Special Report on climate change, desertification, land degradation, sustainable land management, food security, and greenhouse gas fluxes in terrestrial ecosystems (SR2). *Ginevra, IPCC,* 650.

OECD/FAO 2021, "OECD-FAO Agricultural Outlook", OECD Agriculture statistics (database), https://doi.org/10.1787/agr-outl-data-en. (Accessed on 25/03/2022)

POPP, J., KOVÁCS, S., OLÁH, J., DIVÉKI, Z. & BALÁZS, E. 2021. Bioeconomy: Biomass and biomass-based energy supply and demand. *New Biotechnology,* 60**,** 76–84.

SAHOO, G., WANI, A.M., SWAMY, S.L., SHARMA, A., 2022. Environmental Impact and Economic Benefits of Biofuel Production. pp. 349–378. https://doi.org/10.1007/978-981-16-8090-8_16

SCHENK, P. M., THOMAS-HALL, S. R., STEPHENS, E., MARX, U. C., MUSSGNUG, J. H., POSTEN, C., KRUSE, O. & HANKAMER, B. 2008. Second generation biofuels: high-efficiency microalgae for biodiesel production. Bioenergy research, 1, 20–43

SKEER, J. 2017. Bioenergy for Sustainable Development.

SVAZAS, M., NAVICKAS, V., KRAJNAKOVA, E. & NAKONIECZNY, J. 2019. Sustainable supply chain of the biomass cluster as a factor for preservation and enhancement of forests. *Journal of International Studies,* 12**,** 309–321.

TURSI, A., 2019. A review on biomass: Importance, chemistry, classification, and conversion. Biofuel Research Journal. https://doi.org/10.18331/BRJ2019.6.2.3

VAN FOREEST, F. 2012. *Perspectives for biogas in Europe*, Oxford Institute for Energy Studies.

WANG, S., LI, G., LIAO, Z., LIU, T., MA, T., 2022. A novel alkane monooxygenase (*alkB*) clade revealed by massive genomic survey and its dissemination association with IS elements. PeerJ 10, e14147. https://doi.org/10.7717/peerj.14147

WANG, X., LIU, X., YAN, X., ZHAO, P., DING, Y. & XU, P. 2011. Enzyme-nanoporous gold biocomposite: excellent biocatalyst with improved biocatalytic performance and stability. *PLoS One,* 6, e24207.

WOOLEY, R., RUTH, M., SHEEHAN, J., IBSEN, K., MAJDESKI, H., GALVEZ, A., 1999. Lignocellulosic Biomass to Ethanol Process Design and Economics Utilizing Co-Current Dilute Acid Prehydrolysis and Enzymatic Hydrolysis Current and Futuristic Scenarios. Golden, CO (United States). https://doi.org/10.2172/12150

WRIGHT, M., BROWN, R.C., 2007. Establishing the optimal sizes of different kinds of biorefineries. Biofuels, Bioproducts and Biorefining 1, 191–200. https://doi.org/10.1002/bbb.25

WU, B., WANG, Y.W., DAI, Y.H., SONG, C., ZHU, Q.L., QIN, H., TAN, F.R., CHEN, H.C., DAI, L.C., HU, G.Q., HE, M.X., 2021. Current status and future prospective of bio-ethanol industry in China. Renewable and Sustainable Energy Reviews. https://doi.org/10.1016/j.rser.2021.111079

ZHANG, Q., ZHENG, Z., LIU, C., LIU, C. & TAN, T. 2016. Biodiesel production using lipase immobilised on epoxychloropropane-modified Fe3O4 sub-microspheres. *Colloids and Surfaces B: Biointerfaces,* 140, 446–451.

Current Advances and Potentials of Nanotechnology for Biofuel Production

Ifeanyi Michael Smarte Anekwe, Emmanuel Kweinor Tetteh, Stephen Okiemute Akpasi, and Yusuf Makarfi Isa

1 Introduction

The world is currently experiencing a severe energy resource crisis due to the exhaustion of petroleum-based fuels. Owing to the growing global population, increased energy consumption, and limited supplies of fossil fuels, humanity is now searching for alternate fuel sources. The indiscriminate utilization of petroleum-based fuels is the cause of various present catastrophes, including global warming, ozone layer depletion, biosphere damage, and environmental degradation (Adeniyi et al. 2018). Thus, the scientific world is constantly working to develop sustainable energy alternatives to replace fossil-based fuels. The lack of fossil fuels and their adverse impacts on the ecosystem have caused the focus to move to biofuels. Among the wide varieties of biofuel, biogas, biohydrogen, bioethanol, and biodiesel can all be considered potential substitute forms of renewable energy (Saravanan et al. 2018; Sharma and Kumar 2018). Biofuels are produced from microbes and are safe for the ecosystem. The reduction of greenhouse gas emissions, sustainability, and renewability are advantages of biofuels. The most significant biofuels are biodiesel, biohydrogen, biogas, and bioethanol. They are eco-friendly and have a far higher energy content than fossil fuels. The yield of these fuels is strongly reliant on operational conditions. Additionally, their processing has several challenges, such as increased production costs, infrastructure difficulties, and a lack of innovative

I. M. S. Anekwe (✉) · Y. M. Isa
School of Chemical and Metallurgical Engineering, University of the Witwatersrand, Johannesburg, South Africa

E. K. Tetteh · S. O. Akpasi
Green Engineering Research Group, Department of Chemical Engineering, Faculty of Engineering and the Built Environment, Durban University of Technology, Durban, South Africa

I. S. Dunmade et al. (eds.), *Sustainable Engineering*, Green Energy and Technology, https://doi.org/10.1007/978-3-031-47215-2_22

technologies (Guin et al. 2022; Anekwe et al. 2021). As a result, numerous important concerns related to the industrial synthesis of biofuels need to be addressed.

The present focus of research is on improving biofuel synthesis using nanotechnology (Rai and Da Silva 2017; Palaniappan 2017). Nanotechnology can provide a workable answer to every problem associated with changing the feedstock used to produce biofuels. The nanoparticles are very efficient as nanocatalysts due to their small size and other special features (Rai and Da Silva 2017). The recent years have seen a significant expansion of the nanotechnology sector and numerous achievements in the biofuel industry. Nanomaterials have enormous promise for commercialization and cost reduction in the generation of biofuels. Integrating nanotechnology with the bioenergy industry significantly impacts the processing and improvement of the quality of biofuels. The global state of nanotechnology's contribution to enhancing biofuel synthesis in this context is an intriguing area to research. Different nanomaterials are attracting much interest as potential building blocks for synthesizing high, cost-effective biofuels. Nanotechnology is used to alter the characteristics of the feed materials, which ultimately increases both the production rate and biofuel quality. Nanomaterials are significantly vital to the improvement of biofuel synthesis as a result of the extraordinary structural, optical, mechanical, chemical, electronic, and magnetic characteristics they possess. Because of both their diminutive size and their large specific surface area, they have a considerable catalytic impact. The use of nanofibers, nanotubes, metallic nanoparticles, and nanocatalysts is considered for synthesizing many biofuels, including biohydrogen, biodiesel, bioethanol, and biogas, as well as for improving their quality (Guin et al. 2022). This chapter discussed a summary of various nanotechnology techniques, their application in biofuel synthesis, the role of nanotechnology in achieving the Sustainable Development Goals (SDGs), potential challenges, and prospects.

2 Classification of Nanoparticles for Biofuel Production

In general, nanoparticles are characterized as having one or more dimensions with diameters between 1 and 100 nanometers (nm). A nanoparticle can be grouped based on the materials employed in its fabrication, the origin of the materials, and its shape or dimension (Fig. 1).

2.1 Carbon Nanotubes (CNTs)

CNTs are carbon allotropes created by cylindrically rolling graphene sheets. These nanotubes are basically applied in the construction of biosensors and microbial fuel cells because of their capability for carrying electron transfer kinetics and redox reactions (Liu et al. 2012). CNTs can be grouped into two types: single-walled

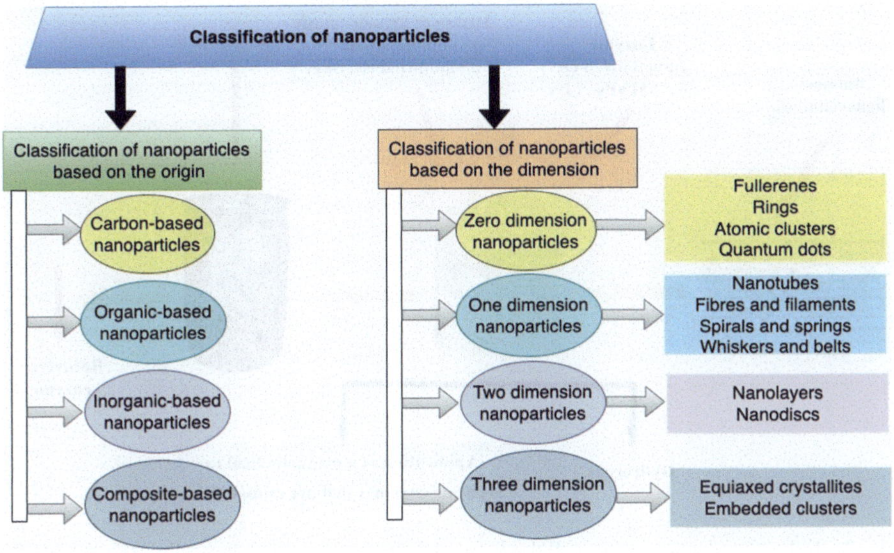

Fig. 1 Primary classification of nanoparticles

carbon nanotubes (SWCNTs), which have one layer of carbon, and multi-walled carbon nanotubes (MWCNTs), which have several layers of graphene (Dresselhaus et al. 2000). CNTs are utilized as a catalyst in the synthesis of biofuels and offer distinguishing qualities like stability, a large surface area, and low toxicity. CNTs have become a viable nanomaterial due to their affordability and renewable precursors. According to a study, growing *Enterobacter aerogenes* immobilized on top of functionalized multi-walled carbon nanotubes (MWCNT-COOH) increased the rate of hydrogen synthesis (2.72 L/L/h), glucose degradation efficiency (96.20%), and H_2 yield (2.2 mol/mol glucose) compared to the free cells (Boshagh et al. 2019).

2.2 *Magnetic Nanoparticles (MNPs)*

The biofuel industries typically use enzymes like cellulases and lipases (Verma et al. 2013). According to studies on MNPs, they are essential for immobilizing enzymes to produce biofuel. After becoming attached to a support matrix coated with a particular nanomaterial, enzymes can be recycled, and this procedure is excellent for hydrolyzing LCB (Puri et al. 2013). The immobilization of enzymes used for LCB hydrolysis can be enhanced by modifying specific enzyme features (Singh and Chandel 2018). MNPs such as $CaSO_4/Fe_2O_3$-SiO_2 have been applied in biodiesel synthesis using *Jatropha curcas* as feedstock (Teo et al. 2019). This NP has a volume of 0.55 cm^3/g, with a large surface area of 391 m^2/g, and a pore size of 90 nm. In ideal circumstances, the biodiesel generation rate from crude *Jatropha* is

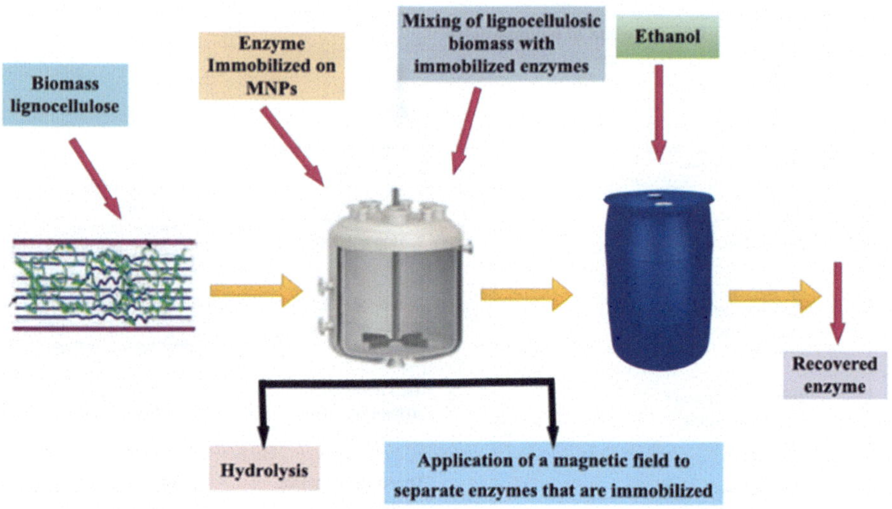

Fig. 2 Biofuel production using cellulase, an enzyme that breaks down cellulose, and MNPs (magnetic nanoparticles)

94%; however, after four cycles, it dropped to 85% and gradually reduced because the nanoparticles became inactive with extended time-on-stream. The cause of the inactivation of NPs was the subject of further research. The results showed that the pores were closed due to the reaction medium elements deposited after the fourth, seventh, and ninth cycles (Teo et al. 2019). Previous studies have shown that utilizing MNPs to immobilize cellulase on MNPs, in addition to lipid extraction, has the potential to effectively hydrolyze the cell wall of microalgae (Fig. 2) (Duraiarasan et al. 2016; Teo et al. 2019). These findings demonstrated that adding a small amount of NPs stimulated the development of microorganisms and the activities of essential enzymes, increasing the production of biogas.

2.3 Metallic Nanoparticles

Although metallic nanoparticles have not received much attention, numerous studies have been done to assess their efficacy in producing biofuels. Owing to their appreciable surface area and small size, metallic NPs can attract a variety of enzymes, including oxidoreductase, which improves electron transport (Vincent et al. 2007). Many catalytic nanoparticles have been developed for increased ion exchange and O_2 reduction rate performance. It has been suggested that metallic nanoparticles can be employed in an organized pattern to improve electrocatalytic performance and synthesize a biofuel cell with a reasonable electron transfer rate and a high loading capacity when used in a layer-by-layer construction with appropriate polymers and enzymes (Kwon et al. 2018). To make hybrid nanocatalysts, metallic NPs of Au, Pt, and Pt0.75-Tin0.25 were incorporated into multi-walled NCTs with acid functionalization. With a poly(amidoamine) PAMAM dendrimer structure, Au nanoparticles were encapsulated in a new way. In biofuel cells fabricated with Au, Pt, and Pt0.75-Tin0.25 supported by MWCNTs, Au-NPs showed superior electrical conductivity, biocompatibility, and more excellent catalytic performance than Pt-NPs (Neto et al. 2014).

2.4 Metal Oxide Nanoparticles

For applications to be successful and for solution phase approaches to give the synthesis product significant control, metal oxide NPs must first be synthesized. Sol-gel techniques, which halt the reaction before gelation, like precipitation techniques, are commonly used to synthesize metal oxide NPs. The preparation, nucleation, and aging mechanisms of NPs reveal their characteristics. In electronic materials, natural remediation, catalysis, and sensors, metal oxide NPs are well known for their applications. The process of transforming vegetable oil into biofuel has been done using metal oxides. Oil can be transformed into organic liquid products with the aid of the metal oxide catalysts such as NiO, CO_3O_4, V_2O_5, ZnO, MoO_3, and KOH (Yigezu and Muthukumar 2014). Metal oxides are utilized as a support system because of their solid catalytic activity but lower selectivity. CaO and Al_2O_3 nanocatalysts have been used in biodiesel production. The optimum yield for biodiesel synthesis from the transesterification of *Jatropha curcas* oil catalyzed by CaO-Al2O3 nanoparticles was reported to be 82.3% at a 5:1 ratio of methanol to oil (Hashmi et al. 2016). It has been demonstrated that the esterification and transesterification processes can be performed simultaneously by the metal oxide catalyst ZrO_2 utilizing a mixture of free fatty acids and soybean oil as input. High stability, hardness, and primary and acidic qualities have all been attributed to ZrO_2. Both simultaneous processes produced 68–69% of fatty acid methyl ester (FAME). In addition, various NP catalysts, like MeO-SBA-15, ZnO-SBA-15, La_2O_3-SBA-15, etc., have been utilized to boost biofuel synthesis from spent cooking oils (Cao et al. 2019).

2.5 Nanoparticles in Heterogeneous Catalysis

Due to their lack of excessive water requirements and ease of separation from the process mixture, heterogeneous catalysts have replaced homogeneous catalysts as the preferred kind of catalyst in biofuel generation (Ferreira et al. 2019). They are simple to separate, and the resulting products are normally noncorrosive and eco-friendly, have a high specificity, and have long lifespans. In several research, lignocellulosic biomass was converted to biofuel employing NPs as heterogeneous catalysts (Akia et al. 2014). Utilizing hybrid support composed of partly reduced graphene oxide (PRGO) nanosheets and metal-organic framework (MOF) crystals, dispersed metal nanoparticle catalysts improved the catalytic performance and specificity of heterogeneous catalysis. The hydrodeoxygenation of vanillin, a typical raw material in lignin-derived bio-oil, was achieved under moderate process parameters using a 3D hierarchical nanocomposite termed Pd/PRGO/Ce-MOF, which is composed of a Ce-based MOF wrapped in thin PRGO nanosheets containing palladium nanocatalysts. The findings indicate that the PRGO/Ce-MOF hybrid scaffold offers good support for Pd-NPs for converting vanillin into 2-methoxy-4-methyl phenol. This vital high-value phenol product may be employed immediately in the petrochemical and pharmaceutical sectors (Ibrahim et al. 2016).

3 Applications of Nanoparticles in Bioenergy Production

As an environmentally friendly and renewable fuel source, bioenergy production has drawn the attention of nanotechnology applications for developing efficient catalysts or nanoparticles (NPs) (Manikandan et al. 2022; Arya et al. 2021). These NPs (Fig. 3) have a larger surface area, making it easier to immobilize enzymes and increase the catalytic activity of reactions that generate biofuels. There have been reports that using NPs speeds up the production process and higher yielding of biofuels (Manikandan et al. 2022; Khan et al. 2022; Arya et al. 2021). More so, the rate at which fossil fuels are used is much higher than the rate at which they are replaced. So, this section focused on how NPs can be used to make biohydrogen, bioethanol, biodiesel, and biogas production easier. This may alleviate the problem of the ongoing depletion and high price of fuel derived from petroleum.

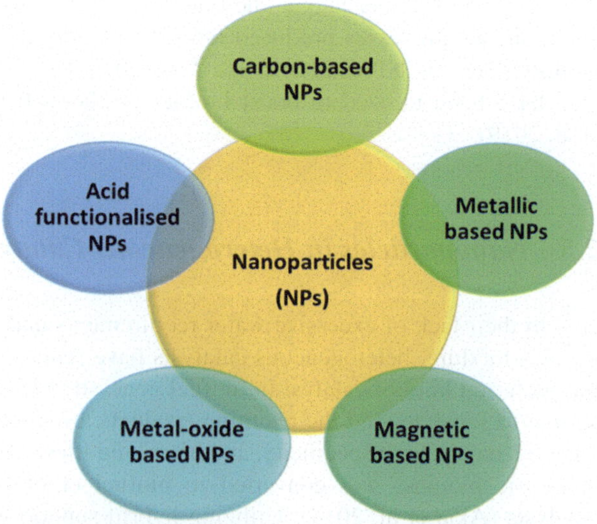

Fig. 3 Types of nanoparticles (NPs) used for bioenergy production. (Adapted from Sharma et al. (2022))

3.1 Biohydrogen Production

Biological hydrogen can be made in two different fermentation processes: (i) photo-fermentation and (ii) dark fermentation (Sharma et al. 2022; Chung et al. 2022). In photo fermentation, microbes like cyanobacteria and green algae are exposed to light and water. During dark fermentation, it is important to have anaerobic bacteria break down substrates or biomass into biohydrogen (Sharma et al. 2022; Manikandan et al. 2022). Even though dark fermentation has been used to make biohydrogen for a long time, it is a slow process that produces little hydrogen. Using NPs is an excellent way to boost hydrolysis, liquification, and enzyme activity during fermentation. This is because NPs have their own physical and chemical characteristics, and they can be employed to increase the hydrogen yield in the dark fermentation process. In the past few years, scientists have successfully studied several nanoparticles made of metals (Ag, Au, Cu, Fe, and Ni) and metal oxides (Fe_2O_3, Fe_3O_4, TiO_2). Table 1 shows how different NPs can be used to synthesize biohydrogen.

Table 1 Effect of NPs on biohydrogen synthesis

Substrate	Sources	Nanoparticles	Operating parameters	Biohydrogen yield
Acetate	Anaerobic sludge	2.5 mg/L Au	$T = 35$ °C, pH 7.2	>65%H_2
Wastewater from industries	Anaerobic sludge	2.5 mg/L Ni	$T = 55$ °C, pH 7	67% H_2
Synthetic wastewater	Anaerobic culture	5 mg/L Au	$T = 35$ °C, pH 7.2	56% H_2
Water hyacinth	Mixed culture	57 mg/L Fe	$T = 35$ °C	85.5% H_2
Glucose	Anaerobic sludge	391 mg/L Fe	$T = 37$ °C, pH 5.5	55% H_2
Glucose	*Enterobacter cloacae*	100 mg/L Fe	$T = 37$ °C, pH 7	75%H_2
Glucose	Anaerobic sludge	2.5 mg/L Ni	$T = 37$ °C, pH 5.5	55%H_2
Water hyacinth	Mixed culture	57 mg/L	$T = 35$ °C, duration 4 days	85.5%H_2
Glucose	*Enterobacter cloacae DH-89*	60 mg/L Fe	$T = 37$ °C, pH 7	95%H_2
Industrial wastewater	Anaerobic digestion	5.44 mg/L Ni	$T = 55$ °C, pH 7, rotation 180 rpm	67%H_2
Glucose	*Enterobacter cloacae*	125 mg/L Fe_2O_3	$T = 37$ °C, pH 7, duration 24 h	85%H_2
Glucose	Anaerobic sludge	1.92 mg/L Fe2O3	$T = 60$ °C, pH 5.5, rotation 150 rpm	51%H_2
Wastewater	Mixed culture	200 mg/L Fe2O3	$T = 37$ °C, pH 6, rotation 200 rpm	62.14%H_2
Sugarcane bagasse	Anaerobic sludge	1.54 mg/L Fe3O4	$T = 30$ °C, pH 5	65%H_2
Malate	*R. sphaeroides* NMBL-02	60 mg/L TiO2	$T = 32$ °C, pH 8	63.27%H_2

Adapted from Sharma et al. (2022) and Liu et al. (2021b)

3.2 Bioethanol Production

Bioethanol is seen as a viable and eco-friendly biofuel. Bioethanol has good chemical characteristic, including a high octane number and a high dissipation enthalpy. Bioethanol is made from edible and nonedible vegetable oils, waste materials, algal biomass, and bacterial biomass (Sharma et al. 2022; Chung et al. 2022). Microalgae have been a good resource for bioethanol production due to their ready availability and how much they are produced (Sharma et al. 2022; Manikandan et al. 2022). It is also reported that genetically modified microorganisms make more bioethanol than typical microorganisms (Manikandan et al. 2022; Khan et al. 2022; Arya et al. 2021). However, NPs play a significant role in the production of bioethanol. In the real world, it has been reported that using agricultural waste and sugarcane leaves along with MnO_2 nanoparticles increases the production of bioethanol (Sharma et al. 2022). Table 2 shows some of the natural products and NPs used to make bioethanol.

3.3 Biodiesel Production

Biodiesel offers several exciting prospective applications because it releases less contaminants, is environmentally friendly, and is derived from edible and nonedible oils. The transformation of oils into biodiesel is accomplished by a process known as transesterification (Sharma et al. 2022; Arya et al. 2021). The procedure employs both heterogeneous and homogeneous catalysts in the reaction. Also, nanomaterials

Table 2 Effect of NPs on bioethanol production

Substrate/ feedstock	Sources	Nanoparticles	Operating conditions	Bioethanol yield
Potato peel waste	*Saccharification fermentation*	200 mg/L NiO	*T* = 37 °C, rotation 120 rpm, duration 24 h	69.6%
Rice straw	*Fusarium oxysporum*	20 mg/L ZnO	*T* = 25 °C, pH 6, rotation 100 rpm, duration 72 h	75%
Potato peels	*Saccharomyces cerevisiae (BY4743)*	22 g/L Fe3O4	*T* = 30 °C, pH 6, rotation 120 rpm, duration 72 h	51%
Corn starch	*Saccharomyces cerevisiae (BY4743)*	264 mg/L Fe3O4	*T* = 60 °C, pH 4, duration 30 days	62%
Potato peel waste	*Saccharomyces cerevisiae (BY4743)*	0.25 g/L NiO 1.6 g/L Fe3O4	*T* = 37 °C, rotation 120 rpm, duration 24 h	59.96% 51%
Rice straw	*Fusarium oxysporum*	200 mg/L ZnO	*T* = 25 °C, pH 6, rotation 200 rpm, duration 72 h	66%

Adapted from Manikandan et al. (2022) and Sharma et al. (2022)

have demonstrated some promising effects in biodiesel production by increasing the catalytic performance during transesterification (Manikandan et al. 2022). The enormous surface area and nanoscale size of nanostructures make them ideal for immobilization support. It is established that the functionalization of nanoconjugates can boost biodiesel yield and quality (Giannakopoulou et al. 2020). For instance, nanoconjugates of iron and silica, such as Fe_3O_4/SiO_2, are utilized to produce biodiesel from cooking and algae oils (Manikandan et al. 2022; Chung et al. 2022). The utilization of NPs in the manufacture of biodiesel is detailed in Table 3.

3.4 Biogas Production

The production of biogas involves four distinct stages: (i) hydrolysis, which breaks down organic materials into simple monomeric or dimeric units; (ii) acidogenesis, which uses the product of hydrolysis for fermentation; (iii) acetogenesis, which results in the synthesis of acetate along with hydrogen and carbon dioxide; and (iv) methanogenesis, which is the last stage and produces CH_4 from acetate, H_2, and CO_2 (Anekwe et al. 2022; Khan et al. 2022; Heikal et al. 2022). Since nanotechnology

Table 3 Effect of NPs on biodiesel

Substrate/ feedstock	Sources	Nanoparticles	Operating conditions	Biodiesel yield
Potato peel waste	Saccharification fermentation	NiO	$T = 37$ °C, rotation 120 rpm, duration 24 h	59.96%
Microalgal oil	Methanol	Fe_3O_4/ ZnMg(Al)O	$T = 65$ °C, duration 3 h	94%
Chlorella vulgaris	Methanol/sulfuric acid (85:15 v/v)	SiO_2	$T = 70$ °C, duration 40 min	89%
Waste cooking oil	Methanol	CaO	$T = 65$ °C, duration 6 h	98.95%
Castor oil	Methanol	Ni/ZnO (2:8)	$T = 55$ °C, duration 60 min	95.2%
Soybean oil	Methanol	Cu	$T = 180$ °C, duration 60 min	96.2%
Bombax ceiba oil	Methanol	CaO	$T = 65$ °C, duration 70 min	90–96%
Sunflower	Methanol	Calcite/Au	$T = 65$ °C, duration 6 h	95.7%
Sunflower	Methanol	$MgO/MgAl_2O_4$	$T = 110$ °C, duration 3 h	96.2%
Jatropha oil	Methanol	Mg/Al	$T = 45$ °C, duration 1.5 h	95.2%
Palm oil	Methanol	TiO_2-ZnO	$T = 80$ °C, duration 5 h	92.2%
Rice brain oil	Methanol	CaO	$T = 65$ °C, duration 120 min	93%

Adapted from Sharma et al. (2022) and Manikandan et al. (2022)

Table 4 Effect of NPs on biogas production

Substrate/feedstock	Nanoparticles	Operating conditions	Biogas yield
Granular sludge	ZnO	*T* = 30 °C, pH 7.2, rotation 120 rpm, duration 24 h	50%
Waste activated sludge	ZnO	*T* = 37 °C, pH 6, rotation 120 rpm, duration 14 h	53.7%
Granular sludge	Cu	*T* = 30 °C, pH 7.2, rotation 120 rpm	50%
Manure slurry	Ni	*T* = 30 °C, pH 7.2, rotation 20 rpm	61%
Domestic sludge	nZVI	*T* = 37 °C, rotation 20 rpm, duration of 14 days	88%
Activated sludge from municipality	CuO	*T* = 35 °C, duration 30 days	84%
Sludge	CeO_2	*T* = 30 °C, duration 40 days	11%
Wastewater sludge	Fe_3O_4	*T* = 37 °C, duration 60 days	80%
Wastewater sludge	Co/SiO_2	*T* = 55 °C, duration 30 days	48%
Wastewater sludge	Ni/SiO_2	*T* = 55 °C, duration 30 days	71%
Raw manure	Ni	*T* = 37 °C, duration 40 days	78%

Adapted from Sharma et al. (2022) and Heikal et al. (2022)

has a bio-stimulating influence on the methanogenic stage, it plays an essential part in the synthesis of biogas and CH_4 (Heikal et al. 2022; Arya et al. 2021; Giannakopoulou et al. 2020). Several studies have shown that trace element-based NPs (Co, Fe, Fe_3O_4, ZnO, TiO_2, CeO_2, and Ni) at different concentrations and large particle size can decrease the length of the lag phase and the time it takes to reach the transformation peak (Khan et al. 2022; Heikal et al. 2022; Tetteh and Rathilal 2021; Tetteh et al. 2021). It has been demonstrated that nano-zero-valent iron (nZVI) can influence anaerobic digestion by causing an increase in the synthesis of both biogas and CH_4 (Molla et al. 2022). In addition, nZVI has been shown to inhibit dichlorination while simultaneously promoting methanogenesis in the AD process. Biogas synthesis has used a variety of NPs in their various forms (Table 4).

4 Achieving the SDGs Through the Application of Nanotechnology

One of the most severe concerns confronting the twenty-first century is global climate change. An understanding has progressively evolved over the past two decades that the primary causes of global climate change are rising emissions of greenhouse gases like CO_2 from burning fossil fuels (such as coal and petroleum). Meeting rising energy demand while considerably lowering CO_2 emissions would necessitate the implementation of technologies that will foster the synthesis of clean and sustainable energy (Ogbuagu and Akubue 2015). Nanotechnology offers unparalleled prospects for the advancement of clean and sustainable energy systems. It is

Table 5 Examples of how the application of nanotechnology can contribute to attaining the SDGs

Sustainable Development Goal	SDG Objective	Nanotechnology's contribution to SDG	Ref.
SDG 1	No poverty	Nanotechnology-based biofuel production can boost local economies in low-income nations in Africa and South America and create new jobs	Ahiahonu et al. (2021)
SDG 2	Zero hunger	Provide support to human nutrition, nutraceuticals, functional foods, animal feed supplements, and dietary supplements	Galanakis (2019)
SDG 7	Affordable and clean energy	Microalgal biofuels have been recognized as a substitute for renewable energy sources	Jacob et al. (2021), Oliveira et al. (2021)
SDG 13	Climate action	Reduces emissions of particulate matter and the most toxic compounds found in car and truck exhaust, including benzene, toluene, and xylene	Niven (2005)
SDG 15	Life on land	Human livelihoods in vulnerable areas, such as deserts, can be improved through nanotechnology and microalgal biomass	de Oliveira et al. (2018), He et al. (2018)

anticipated that nanotechnology and materials will also aid in the growth of a sustainable society and achieving the SDGs by assuring water purification, lowering CO_2 emissions, and boosting material circulation using recycling technologies (Pokrajac et al. 2021). Table 5 shows the potential contributions of nanotechnology towards the actualization of SDGs.

5 Challenges and Prospects of Biofuel Production Using Nanotechnology

5.1 Challenges of the Application of Nanotechnology in Biofuel Synthesis

The importance of nanotechnology has been a significant factor in the most recent advancements in manufacturing biofuels. To replace the petroleum-based oils available commercially, there are still a lot of obstacles to overcome and room for development. It is challenging to use nanotechnology in industrial production since nanocatalyst-based biofuel synthesis is still in its early stages. The deployment of nanotechnology to biofuel production will help improve the marketability of biofuel production, both economically and environmentally. However, there is a significant increase in operational expenses associated with pre-treatment techniques for LCBs (Manikandan et al. 2022; Khan et al. 2022), and nanomaterials are subject to risk assessment due to the minute physicochemical changes at the nanoscale.

Nanotechnology still has to work on converting new fundamental discoveries in the scientific literature that laypeople can understand into practical technical biofuel production applications to overcome significant obstacles.

To some extent, nanotechnologies have produced manufactured ignorance, making it highly dependent on scientists and specialists to be aware of global concerns. For instance, nuclear and industry experts possess a dual position: they both produce and evaluate risks, using their clout based on relationships between risk concepts. In the same way, researchers investigating and utilizing nanomaterials are responsible for determining whether they pose any risks. Similar inquiries have also been made regarding nanomaterials' effectiveness and environmental friendliness in manufacturing biofuels, such as nanoparticles. The manufacturing sector that uses nanoparticles is unconcerned by this case because it makes several levels of risk a reality and boosts that sector. In the synthesis of biofuels, the difficulties related with the protection of nanomaterials must be considered, as many of these nanomaterials have the potential to cause human disease and significant environmental problems. Therefore, a safety review is crucial for the safe usage of nanoparticles.

5.2 Prospects and Future Work Consideration for Effective Biofuel Production Using Nanotechnology

Biofuel is the fuel of the future for sectors that rely on petroleum since it is safer, greener, and more sustainable and can be produced in infinite quantities (Kumar et al. 2021; Azadbakht et al. 2021; Institute, 2012). However, the restricted availability of petroleum-derived fuels coupled with an increase in demand has led to a price hike for fossil fuels (Popp et al. 2014). This has necessitated researchers to consider an effective way of developing biofuels as a viable alternative. In essence, the production of biofuels can be improved by wide varieties of nanomaterials in different ways. This includes enhancing the catalytic production of biohydrogen and the biological and chemical digestion and improving the stability of the enzymatic activities (Kurniadi et al. 2021; Kumar et al. 2021; Jung et al. 2021). The influence of nanoparticles on the process is uttered by the distinct catalytic activity, which is based on the structure, shape, and size of the NPs (Heikal et al. 2022; El Ghandoor et al. 2012; Lu et al. 2007). To achieve cost-effective biofuel production, the nanoparticle synthesis technique must also be considered. Co-precipitation, sol-gel, sputtering deposition, spray pyrolysis, electrophoretic deposition, chemical vapor deposition, plasma-enhanced, and hydrothermal are among the synthesis techniques (Lamichhane et al. 2022). The production of biofuel may become economically viable through the synthesis of low-cost nanomaterials. This needs to be determined by producing nanomaterials with distinctive catalytic properties using commercially feasible nanoparticles (Nakhlband et al. 2022; Heikal et al. 2022). Also, biofuel depends on biomass availability, which is easy to get from wood, plants, organic residue, agricultural residue, municipal solid waste, etc. Being oil-rich, carbon-neutral, and capable of rapid growth, algal biomass has the potential to be exploited to synthesize biodiesel (Armah et al. 2022; Liu et al. 2021a). Additionally, there is ongoing research to increase biofuel generation using nano-catalysts and natural resources and the deployment of biological nanoparticle synthesis methods. Food crops, including sugarcane, maize, and others, have historically been utilized for commercial biofuel synthesis (Institute 2012; Festel 2008). The production of nonedible biofuel is significantly less than the production of edible biofuel. The prospect of nanotechnology to accelerate biofuel production from edible and nonedible sources has a great potential for green energy and environmental sustainability.

6 Conclusion

The present study makes it abundantly clear that the application of NPs in biofuel production has led to a significant improvement in biofuel quality. This improvement is attributed to the exceptional physicochemical features of NPs, which include high reactivity, large surface-to-volume ratio, good dispersibility, and high selectivity. To increase the amount of biofuel that can be synthesized from several feedstocks, various NPs, such as metals, metal oxides, and magnetic and carbonaceous materials, have been successfully used. Nanoparticles are used in the pre-treatment stage in addition to the synthesis process to increase the digestibility of the feedstock, which results in an increase in the amount of biofuel that can be produced. However, some of the existing technical limitations must be overcome to successfully commercialize this technology. These limitations include the production and use of NPs that are not harmful to microbes, the application of inexpensive and environmentally friendly NPs, and the adaptation of biological NP fabrication techniques rather than chemical methods that require stringent process variables. The advancement of a sustainable future and the achievement of the SDGs will be facilitated by nanotechnology and materials that ensure a reduction in carbon footprints and an increase in material circulation through the utilization of recycling technologies.Funding InformationThis work is based on research funded partly by the National Research Foundation of South Africa (NRF) (BRIC190321424123).

References

ADENIYI, O. M., AZIMOV, U. & BURLUKA, A. 2018. Algae biofuel: current status and future applications. *Renewable and sustainable energy reviews,* 90**,** 316–335.

AHIAHONU, E. K., ANKU, W. W., ROOPNARAIN, A., GREEN, E., GOVENDER, P. P. & SEREPA-DLAMINI, M. H. 2021. Bioprospecting wild South African microalgae as a potential third-generation biofuel feedstock, biological carbon-capture agent and for nutraceutical applications. *Biomass Conversion and Biorefinery***,** 1–16.

AKIA, M., YAZDANI, F., MOTAEE, E., HAN, D. & ARANDIYAN, H. 2014. A review on conversion of biomass to biofuel by nanocatalysts. *Biofuel Research Journal,* 1, 16–25.

ANEKWE, I. M. S., ARMAH, E. K. & TETTEH, E. K. 2022. Bioenergy Production. *Biomass.* IntechOpen.

ANEKWE, I. M. S., KHOTSENG, L. & ISA, Y. M. 2021. The Place of Biofuel in Sustainable Living; Prospects and Challenges. *Reference Module in Earth Systems and Environmental Sciences.* Elsevier.

ARMAH, E. K., CHETTY, M., RATHILAL, S., ASANTE-SACKEY, D. & TETTEH, E. K. 2022. Chapter 11 - Lignin: value addition is key to profitable biomass biorefinery. *In:* SAHAY, S. (ed.) *Handbook of Biofuels.* Academic Press.

ARYA, I., POONA, A., DIKSHIT, P. K., PANDIT, S., KUMAR, J., SINGH, H. N., JHA, N. K., RUDAYNI, H. A., CHAUDHARY, A. A. & KUMAR, S. 2021. Current trends and future prospects of nanotechnology in biofuel production. *Catalysts,* 11, 1308.

AZADBAKHT, M., ARDEBILI, S. M. S. & RAHMANI, M. 2021. Potential for the production of biofuels from agricultural waste, livestock, and slaughterhouse waste in Golestan province, Iran. *Biomass Conversion and Biorefinery*, 1–11.

BOSHAGH, F., ROSTAMI, K. & MOAZAMI, N. 2019. Biohydrogen production by immobilised Enterobacter aerogenes on functionalised multi-walled carbon nanotube. *International Journal of Hydrogen Energy,* 44, 14395–14405.

CAO, X., LI, L., SHITAO, Y., LIU, S., HAILONG, Y., QIONG, W. & RAGAUSKAS, A. J. 2019. Catalytic conversion of waste cooking oils for the production of liquid hydrocarbon biofuels using in-situ coating metal oxide on SBA-15 as heterogeneous catalyst. *Journal of Analytical and Applied Pyrolysis,* 138, 137–144.

CHUNG, K.-H., LAM, S. S., PARK, Y.-K., KIM, S.-J. & JUNG, S.-C. 2022. Application of liquid-phase plasma for the production of hydrogen and carbon from the plasma-induced cracking of liquid hydrocarbons. *Fuel,* 328, 125297.

DE OLIVEIRA, C. Y. B., DA SILVA ALMEIDA, A. J. G., DE VIVEIROS, M., DE LIMA SANTOS, I. & DE MACEDO DANTAS, D. M. 2018. Prospecting the first culture collection of algae of the pernambuco semiarid, Brazil. *Revista Brasileira de Engenharia de Pesca,* 11, 27–34.

DRESSELHAUS, M. S., DRESSELHAUS, G., EKLUND, P. & RAO, A. 2000. Carbon nanotubes. *The physics of fullerene-based and fullerene-related materials.* Springer.

DURAIARASAN, S., RAZACK, S. A., MANICKAM, A., MUNUSAMY, A., SYED, M. B., ALI, M. Y., AHMED, G. M. & MOHIUDDIN, M. S. 2016. Direct conversion of lipids from marine microalga C. salina to biodiesel with immobilised enzymes using magnetic nanoparticle. *Journal of environmental chemical engineering,* 4, 1393–1398.

EL GHANDOOR, H., ZIDAN, H., KHALIL, M. M. & ISMAIL, M. 2012. Synthesis and some physical properties of magnetite (Fe3O4) nanoparticles. *Int. J. Electrochem. Sci,* 7, 5734–5745.

FERREIRA, R. S. B., DOS PASSOS, R. M., SAMPAIO, K. A. & BATISTA, E. 2019. Heterogeneous catalysts for biodiesel production: a review. *Food Public Health,* 9, 125–137.

FESTEL, G. W. 2008. Biofuels–economic aspects. *Chemical Engineering & Technology: Industrial Chemistry-Plant Equipment-Process Engineering-Biotechnology,* 31, 715–720.

GALANAKIS, C. 2019. *Proteins: Sustainable source, processing and applications*, Academic Press.

GIANNAKOPOULOU, A., GKANTZOU, E., POLYDERA, A. & STAMATIS, H. 2020. Multienzymatic nanoassemblies: recent progress and applications. *Trends in biotechnology,* 38, 202–216.

GUIN, M., KUNDU, T. & SINGH, R. 2022. Applications of Nanotechnology in Biofuel Production. *Bio-Clean Energy Technologies Volume 2.* Springer.

HASHMI, S., GOHAR, S., MAHMOOD, T., NAWAZ, U. & FAROOQI, H. 2016. Biodiesel production by using CaO-Al2O3 Nano catalyst. *International Journal of Engineering Research & Science,* 2, 43–49.

HE, Q., YANG, H. & HU, C. 2018. Effects of temperature and its combination with high light intensity on lipid production of Monoraphidium dybowskii Y2 from semi-arid desert areas. *Bioresource technology,* 265, 407–414.

HEIKAL, G., Shakroum, M., Vranayova, Z. & Abdo, A. 2022. Impact of Nanoparticles on Biogas Production from Anaerobic Digestion of Sewage Sludge. *Journal of Ecological Engineering,* 23, 222–240.

IBRAHIM, A., LIN, A., ZHANG, F., ABOUZEID, K. & EL-SHALL, M. 2016. Palladium nanoparticles supported on hybrid MOF-PRGO for catalytic hydrodeoxygenation of vanillin as a model for biofuel upgrade reactions. *ChemCatChem,* 9, 469–480.

INSTITUTE, W. 2012. *Biofuels for transport: global potential and implications for sustainable energy and agriculture,* Earthscan.

JACOB, A., ASHOK, B., ALAGUMALAI, A., CHYUAN, O. H. & LE, P. T. K. 2021. Critical review on third generation micro algae biodiesel production and its feasibility as future bioenergy for IC engine applications. *Energy Conversion and Management,* 228, 113655.

JUNG, S., SHETTI, N. P., REDDY, K. R., NADAGOUDA, M. N., PARK, Y.-K., AMINABHAVI, T. M. & KWON, E. E. 2021. Synthesis of different biofuels from livestock waste materials and their potential as sustainable feedstocks–A review. *Energy Conversion and Management,* 236, 114038.

KHAN, S. Z., ZAIDI, A. A., NASEER, M. N. & ALMOHAMADI, H. 2022. Nanomaterials for biogas augmentation towards renewable and sustainable energy production: A critical review. *Frontiers in Bioengineering and Biotechnology*, 1470.

KUMAR, Y., YOGESHWAR, P., BAJPAI, S., JAISWAL, P., YADAV, S., PATHAK, D. P., SONKER, M. & TIWARY, S. K. 2021. Nanomaterials: stimulants for biofuels and renewables, yield and energy optimisation. *Materials Advances,* 2, 5318–5343.

KURNIADI, A. P., AIMON, H. & AMAR, S. 2021. Determinants of Biofuels Production and Consumption, Green Economic Growth and Environmental Degradation in 6 Asia Pacific Countries: A Simultaneous Panel Model Approach. *International Journal of Energy Economics and Policy,* 11, 460–471.

KWON, C. H., KO, Y., SHIN, D., KWON, M., PARK, J., BAE, W. K., LEE, S. W. & CHO, J. 2018. High-power hybrid biofuel cells using layer-by-layer assembled glucose oxidase-coated metallic cotton fibers. *Nature communications,* 9, 1–11.

LAMICHHANE, N., SHARIFABAD, M. E., HODGSON, B., MERCER, T. & SEN, T. 2022. Superparamagnetic iron oxide nanoparticles (SPIONs) as therapeutic and diagnostic agents. *Nanoparticle Therapeutics.* Elsevier.

LIU, L., WANG, J., WANG, F. & YANG, X. 2021a. The impact of the planting of forest biomass energy plants under the embedded Internet of Things technology on the biodiversity of the local environmental ecology. *Environmental Technology & Innovation,* 24, 101894.

LIU, Y., LIU, J., HE, H., YANG, S., WANG, Y., HU, J., JIN, H., CUI, T., YANG, G. & SUN, Y. 2021b. A review of enhancement of biohydrogen productions by chemical addition using a supervised machine learning method. *Energies,* 14, 5916.

LIU, Z., LV, F., ZHENG, H., ZHANG, C., WEI, F. & XING, X.-H. 2012. Enhanced hydrogen production in a UASB reactor by retaining microbial consortium onto carbon nanotubes (CNTs). *international journal of hydrogen energy,* 37, 10619–10626.

LU, A. H., SALABAS, E. E. L. & SCHÜTH, F. 2007. Magnetic nanoparticles: synthesis, protection, functionalisation, and application. *Angewandte Chemie International Edition,* 46, 1222–1244.

MANIKANDAN, S., SUBBAIYA, R., BIRUNTHA, M., KRISHNAN, R. Y., MUTHUSAMY, G. & KARMEGAM, N. 2022. Recent development patterns, utilisation and prospective of biofuel production: Emerging nanotechnological intervention for environmental sustainability–A review. *Fuel,* 314, 122757.

MOLLA, A., KIM, A. Y., WOO, J. C., CHO, H. S. & YOUK, J. H. 2022. Study on Preparation Methodology of Zero-Valent Iron Decorated on Graphene Oxide for Highly Efficient Sonocatalytic Dye Degradation. *Journal of Environmental Chemical Engineering*, 107214.

NAKHLBAND, A., KHOLAFAZAD-KORDASHT, H., RAHIMI, M., MOKHTARZADEH, A. & SOLEYMANI, J. 2022. Applications of magnetic materials in the fabrication of microfluidic-based sensing systems: Recent advances. *Microchemical Journal,* 173, 107042.

Neto, S. A., Almeida, T., Palma, L., MINTEER, S. & DE ANDRADE, A. 2014. Hybrid nanocatalysts containing enzymes and metallic nanoparticles for ethanol/O2 biofuel cell. *Journal of Power Sources,* 259, 25–32.

NIVEN, R. K. 2005. Ethanol in gasoline: environmental impacts and sustainability review article. *Renewable and sustainable energy reviews,* 9, 535–555.

OGBUAGU, J. & AKUBUE, P. 2015. ACHIEVING SUSTAINABLE DEVELOPMENT GOALS THROUGH NANOTECHNOLOGY IN POLYMER AND TEXTILE. *European Journal of Engineering and Technology Vol,* 3.

OLIVEIRA, C. Y. B., D'ALESSANDRO, E. B., ANTONIOSI FILHO, N. R., LOPES, R. G. & DERNER, R. B. 2021. Synergistic effect of growth conditions and organic carbon sources for improving biomass production and biodiesel quality by the microalga Choricystis minor var. minor. *Science of the Total Environment,* 759, 143476.

PALANIAPPAN, K. 2017. An overview of applications of nanotechnology in biofuel production. *World Appl Sci J,* 35, 1305–1311.

POKRAJAC, L., ABBAS, A., CHRZANOWSKI, W., DIAS, G. M., EGGLETON, B. J., MAGUIRE, S., MAINE, E., MALLOY, T., NATHWANI, J., NAZAR, L., SIPS, A., SONE, J. I., VAN DEN BERG, A., WEISS, P. S. & MITRA, S. 2021. Nanotechnology for a Sustainable Future: Addressing Global Challenges with the International Network4Sustainable Nanotechnology. *ACS Nano,* 15, 18608–18623.

POPP, J., LAKNER, Z., HARANGI-RÁKOS, M. & FÁRI, M. 2014. The effect of bioenergy expansion: Food, energy, and environment. *Renewable and Sustainable Energy Reviews,* 32, 559–578.

PURI, M., BARROW, C. J. & VERMA, M. L. 2013. Enzyme immobilisation on nanomaterials for biofuel production. *Trends Biotechnol,* 31, 215–216.

RAI, M. & DA SILVA, S. S. 2017. *Nanotechnology for bioenergy and biofuel production*, Springer.

SARAVANAN, A. P., MATHIMANI, T., DEVIRAM, G., RAJENDRAN, K. & PUGAZHENDHI, A. 2018. Biofuel policy in India: a review of policy barriers in sustainable marketing of biofuel. *Journal of cleaner production,* 193, 734–747.

SHARMA, M. & KUMAR, A. 2018. Promising biomass materials for biofuels in India's context. *Materials Letters,* 220, 175–177.

SHARMA, P., BANO, A., SINGH, S. P., ATKINSON, J. D., LAM, S. S., IQBAL, H. M. & TONG, Y. W. 2022. Nanomaterials as highly efficient photocatalysts used for bioenergy and biohydrogen production from waste toward a sustainable environment. *Fuel,* 329, 125408.

SINGH, O. V. & CHANDEL, A. K. 2018. *Sustainable biotechnology-enzymatic resources of renewable energy*, Springer.

TEO, S. H., ISLAM, A., CHAN, E. S., CHOONG, S. T., ALHARTHI, N. H., TAUFIQ-YAP, Y. H. & AWUAL, M. R. 2019. Efficient biodiesel production from Jatropha curcus using CaSO4/Fe2O3-SiO2 core-shell magnetic nanoparticles. *Journal of cleaner production,* 208, 816–826.

TETTEH, E. K., AMO-DUODU, G. & RATHILAL, S. 2021. Synergistic effects of magnetic nanomaterials on post-digestate for biogas production. *Molecules,* 26, 6434.

Tetteh, E. K. & Rathilal, S. 2021. Application of biomagnetic nanoparticles for biostimulation of biogas production from wastewater treatment. *Materials Today: Proceedings,* 45, 5214–5220.

VERMA, M. L., NAEBE, M., BARROW, C. J. & PURI, M. 2013. Enzyme immobilisation on amino-functionalised multi-walled carbon nanotubes: structural and biocatalytic characterisation. *PloS one,* 8, e73642.

VINCENT, K. A., LI, X., BLANFORD, C. F., Belsey, N. A., WEINER, J. H. & ARMSTRONG, F. A. 2007. Enzymatic catalysis on conducting graphite particles. *Nature chemical biology,* 3, 761–762.

YIGEZU, Z. D. & MUTHUKUMAR, K. 2014. Catalytic cracking of vegetable oil with metal oxides for biofuel production. *Energy Conversion and Management,* 84, 326–333.

Part V
Artificial Intelligence and Socio-Economic Impact in Sustainable Engineering

Perspective and Attitudes of Engineering Students Towards the Fourth Industrial Revolution: A Qualitative Study

Jaya Bharti and Hitaishi Singh

1 Introduction

Industrial revolutions in human civilization have a long history. It is typically characterized by the use of machines with various degrees of automation in the production process yielding multiple number of identical products replacing completely the handmade products. The industrial revolutions took place in phases, each phase marked by certain significant features. Whereas the First Industrial Revolution took place with the power of water and steam, the Second one was based on the use of electrical energy. Subsequently, the Third Industrial Revolution which is still in its intermediate stage in many parts of the world was focussed on electronic and information technology. Simultaneously, the Fourth Industrial Revolution, commonly known as "Industry 4.0", has also taken place almost a decade ago in many developed countries and is in the process of taking place in many developing and underdeveloped countries. Here the industrial procedures are carried out by combining IT with manufacturing sector. From 2010 onwards it was initiated by countries like America and Germany. However, the concept of "Fourth Industrial Revolution" was coined by the founder of the World Economic Forum Klaus Schwab in 2016. It is also termed as "Digital Revolution". There is a big difference between the earlier industrial revolutions and the Fourth Industrial Revolution in terms of shrinkage in the gap between digital, physical, and biological aspects. The technology is changing faster than ever and is fundamentally changing our ways of living. Since the technology is changing rapidly, therefore, the industry is also changing at a faster

J. Bharti (✉)
Department of Psychology, A.N.D.N.N.M.M. (CSJM University) Kanpur, Kanpur, UP, India
e-mail: jayabharti_kn18@csjmu.ac.in

H. Singh
Department of Home Science, A.N.D.N.N.M.M. (CSJM University) Kanpur,
Kanpur, UP, India

I. S. Dunmade et al. (eds.), *Sustainable Engineering*, Green Energy
and Technology, https://doi.org/10.1007/978-3-031-47215-2_23

pace, and now people have started talking about "Industry 5.0". However, "Industry 4.0" has emerged as a global force and is being considered as the base for the future industrial revolution. It is primarily based on applications such as the Internet of Things (IoT), seamless internet connectivity, high-speed communication technologies, and 3D printing, which envisages greater digitization and greater interconnectivity of products, value chains, business models, etc.

The Fourth Industrial Revolution and the Indian Scenario: Some Facts

- One of the initiatives of the Government of India in this direction – "Digital India campaign" – has taken information technology to the remote villages of the country to a greater extent. In the near future, communication density, internet coverage, and the number of mobile internet users are expected to increase significantly.
- There has been a 93% growth in the telecom sector, and around 500 million Indians now have mobile phones.
- India has the highest mobile data consumption in the world, and India is one of the countries where data is available at the lowest cost. India's digital infrastructure and its interfaces including Aadhaar, UPI, e-NAM, and GeM have played a major role in this context.
- The work of connecting all 2.5 lakh gram panchayats with optic fibre will be completed soon. In 2014, only 59 gram panchayats were connected with optic fibre, while at present this number has reached more than 1 lakh.
- A few months back, a national strategy has been formulated to create a strong infrastructure for research in artificial intelligence, and a new centre has been created which is expected to strengthen researches in the area of artificial intelligence.
- This centre can play an important role in areas like traffic and smart mobility. The Prime Minister said that "Solution for India, Solution for the World" is our goal in view of the progress in these areas.
- The expansion of "Industry 4.0" and artificial intelligence will lead to better improvements in the health sector and lower healthcare costs.
- Likewise, it will help the farmers, and it will be very helpful for the agriculture sector. This will help the government achieve its goal of "smart agriculture".
- Government initiatives like "Skill India Mission", "Start-up India", and "Atal Innovation Campaign" are preparing the youth for new and emerging technologies.

The whole scenario of the current industrial revolution is majorly based on technically sound human resources. Another important issue is clarity in the understanding of the concept of Industry 4.0 and its requirements among current and future professionals. One also needs to know how ready our current generation is especially technical professionals to face the challenges posed by IR 4.0.

It is also pertinent that challenges are not limited to the financial investment for the acquisition of new technologies for IR4.0, but also getting qualified professionals that are able to handle the complexities of future production systems for all levels of operations is going to be a task. In the words of Caballero and Walker (2010), "it is very important to see how much graduates are getting ready to demonstrate their potential in terms of job performance under the changing nature of work". "It is crucial to adapt engineering education in order to prepare them for fulfilling the requirements of Industry 4.0" (Coskun et al. 2019).

Due to the necessities of cross-functional roles caused by the Fourth Industrial Revolution, an entirely different skill and knowledge set that combines information technology and production processes will be required by the engineering students. Therefore, "the department of engineering of various universities and institutions has to play a vital role in fulfilling this need by determining and bringing change in the patterns of delivery of education. Simultaneously, content also needs to be upgraded and changed" (Cevik et al., September 2018). Aminul Islam (2022) also tried "to investigate answer to the research question of readiness of students to perform modern tasks requirements in the era of competitive business arising due to fourth Industrial revolution and suggested training and development opportunities to upgrade and update their skills".

Further, the Fourth Industrial Revolution has also imposed an extremely speedy re-evolution of human resources especially those who are in engineering field, and it is expected that these iMillennials and the new paradigm both should adopt each other mutually. Cotet et al. (January, 2020) conducted a study aimed at identifying the relevant characteristics of technological background of contemporary iMillennial generation and mapped the abilities of these youngsters against the need that aroused due to the evolution of new technologies that will replace the simple systems with intelligent system; hence, the need for creative employees will increase. Researchers of this group a new technology-based tool for educational transformation in the form of a constellation of soft skills – abilities and personal qualities – to match the characteristic features of iMillennials and the necessary competency of Industry 4.0 ecosystem and call for not only a paradigm shift in the curriculum skill set but also revamping of the skill mapping system that can bring change in values, attitude, and behaviour so that they can cope with the changed environment.

The individuals trained today in various technological institutes across the globe with special reference to India are going to be the skilled professionals in the future and will become the primary workforce of the Fourth and Fifth Industrial Revolutions. Various studies have been conducted in this field worldwide. In the study conducted by Ismail et al. 2020 aimed at identifying the knowledge, attitudes, interests, and students' readiness to face the challenges of Industrial Revolution 4.0 found that the knowledge of students on IR 4.0 was weak; however, their readiness to deal with IR 4.0 was observed to be high.

Another research study carried out by Abd Rahman Ahmad et al. in July 2019 on another crucial dimension of IR 4.0 was "about factors that influences students' readiness for Industrial revolution 4.0″. The results of the study show that, "students from all backgrounds, be it technical or non- technical education, were equally but partially ready to face the challenges and required some form of assistance to be able to do so".

This will be largely based on their basic understanding of the term, the concept of IR 4.0, and its future requirements by the students of all categories, technical or nontechnical. But since the IR 4.0 is converting the whole industry into highly sophisticated, digital, and artificial intelligence technology-based industry, therefore obtaining data on perspective and attitudes of engineering students towards the Fourth Industrial Revolution becomes very significant to decide on the future course of action. Therefore, the present study was undertaken.

2 Methodology

"Industry 4.0" was the theme of the World Economic Forum's annual meeting held in 2016, after which the idea of the Fourth Industrial Revolution became increasingly popular. The purpose of the present research in this chapter was to assess or explore the effect of Industry 4.0 on the perception, attitude, and viewpoint of engineering students across courses (Degree & Diploma). Data were collected using self-developed interview schedule. The sample consisted of 100 persons (50 Engineering Degree students and 50 Engineering Diploma students). Their age range varied from 18 years to 23 years. Ex post facto research with exploratory orientation was used. Incidental sampling was used to collect the data.

The interviews were recorded after taking participants' prior consent. The language of the schedule was both Hindi and English. There were 12 questions to understand the perception, attitude, and viewpoint of engineering (Degree & Diploma) students towards Industry 4.0. The responses were recorded in verbatim.

The questions that were included in the interview schedule were designed very thoughtfully, so that it could extract the meaning from the depth and make more statements in fewer words without losing its meaning. The questions used in the schedule were open ended, so that it could meet the target of the current study. With reference to results, each question is being discussed in detail about the various responses given by the interviewee. The researcher tried to explore Industry 4.0 in three parts of cognition, i.e. perceptions, attitudes, and viewpoints.

2.1 Cognition

According to psychologists, "cognition" means "understanding" or "knowledge". In the process of teaching, the main focus of learning is the cognitive domain. In this area, learning is related to those mental activities in which information is received from the environment. In this way, many activities take place in this area, which start from receiving information and continue to the brain of the learner. The various forms of cognition are often closely intertwined. Human cognition is a complex combination of a person's "Perception, Attitude and Viewpoint" at psychological level. In short, cognition can be described as one's understanding or knowledge: How do the circumstances in which you live affect you? How does your thinking affect your thought processes and actions? How do you think about the environment around you?

2.1.1 Cognition: Perception

Perception is a complex cognitive process. As our primary mode of contact with the world, "perception" is the causal and informational basis for our higher cognitive functions. It guides our thinking and acting on the world. Stimulation of the sense organs is essential for perception. Given that perception is the input of cognition, the two systems must be able to "talk" to each other. At the very least, the information supplied by perception must be sufficient to be "picked up" by our various cognitive systems. The first part of the interview schedule deals with the "Perception of Engineering (Degree & Diploma) students towards Industry 4.0".

2.1.2 Cognition: Attitude

The cognitive component of attitudes is the way people think about and interpret their attitudes. This component includes the thoughts, feelings, beliefs, and behaviours that are related to how individuals think about their attitudes. People often have a hard time understanding what they think about their attitude. Attitudes build on prior experience and help people make sense of their environment. Thus, attitudes play a central role in life and make up a large part of our daily thoughts, feelings, and behavioural processes. In the second part, an attempt has been made to discuss the "Attitude of Engineering (Degree & Diploma) students towards Industry 4.0".

2.1.3 Cognition: Viewpoint

Cognition is the study of how people think and process information. Human cognition seeks to understand how the mind thinks and how various factors affect motivation, problem-solving, decision-making, etc. Thus, the third part deals with the "Viewpoint of Engineering (Degree & Diploma) students towards Industry 4.0".

2.2 *Narratives with Data Presentation*

A narrative is a way of presenting connected events to tell a real story of a person's life. It is a way for understanding others' viewpoint. It could be a narrative essay, a biography, or a novel. A narrative connects events separated by concept, idea, or plot. Common types of narratives usually have a beginning, a middle, and an end.

In this chapter, perspective, attitudes, and views of engineering students have been tried to comprehend by asking some questions based on identified themes that focused on the Fourth Industrial Revolution and its related issues.

3 Results and Discussion

Theme 1: Semantics after listening to the word "Fourth Industrial Revolution or Industry 4.0"

Question: What comes to your mind after listening to the word "Fourth Industrial Revolution or Industry 4.0"?

Categories	Engineering degree students (50)	Engineering diploma students (50)
Terminology/founder-Industry 4.0, also known as the Smart Factory	40 (80%)	5 (10%)
Physical components complemented by sensors, actuators, and embedded software	10 (20%)	1 (2%)
Connecting resources, services, and humans in real time	–	4 (8%)
Contemporary automation, data exchange, and manufacturing technologies	–	40 (80%)

The first question tried to explore the conceptualization of the word "Fourth Industrial Revolution or Industry 4.0" in the minds of respondents. For the majority of Engineering Degree students (80%), after listening to the word "Fourth Industrial Revolution or Industry 4.0", the first thing that comes to their mind is something that could be "a Terminology/founder-Industry 4.0, also known as the Smart Factory", etc. On the other hand, a majority (80%) of Engineering Diploma students considered the term as "contemporary automation, data exchange, and manufacturing technologies". Empirical evidence show that the term Industry 4.0 was shaped by the German researchers Henning Kagermann, Wolf-Dieter Lukas, and Wolfgang Wahlster as a paradigm shift for maintaining the future competitiveness of the German economy (Kagermann and Lukas 2011).

Narrative of Engineering Degree Students "The Fourth Industrial Revolution (4IR) is a term coined by Klaus Schwab, founder and executive chairman of the World Economic Forum (WEF) in 2016. The incorporation of ADP technologies into industrial production processes has led to the concept of Industry 4.0, also known as the Smart Factory. It can also be deciphered as - One who learns by doing, continually adapting and optimizing its processes accordingly".

Narrative of Engineering Diploma Students "Industry 4.0 or Fourth Industrial Revolution is a collective term that encompasses many contemporary automation, data exchange and manufacturing technologies".

Theme 2: Starting point of these types of revolution
Question: What is the starting point of these types of revolution?

Categories	Engineering Degree students (50)	Engineering Diploma students (50)
As a continuous revolution initiated from the 18th, 19th, and twentieth century	50 (100%)	–
Computerization of manufacturing	–	50 (100%)

When asked about the starting point of these types of revolutions in the world, the Engineering Degree students responded that it is "A continuous revolution initiated from the 18th, 19th, and 20th century". On the other hand, Engineering Diploma students were of the view that "Computerization of manufacturing" is the starting point of these types of revolution.

Narrative of Engineering Degree Students "The first industrial revolution was triggered by the invention of the steam engine in the eighteenth century, the second was driven by widespread electrification in the nineteenth century, and the third, mainly the product of advances in computing in the 1960s. Although 4IR is also a product of technological development".

Narrative of Engineering Diploma Students "The term Industry 4.0 originated from a project of a high-tech strategy by the German government, which encouraged the computerization of manufacturing".

Theme 3: Need of this revolution
Question: What is the need of this revolution?

Categories	Engineering degree students (50)	Engineering diploma students (50)
New technology and equipment to improve manufacturing efficiency but also revolutionize the way of operating and expanding the businesses and further creating the right ecosystem	50 (100%)	–
Nanotechnology, renewable energy to quantum computing with a wider scope and need	–	50 (100%)

With reference to the need of this revolution, Engineering Degree students responded that this will not only lead to the development of "New technology and equipment to improve manufacturing efficiency but also revolutionize the way of operating and expanding the businesses and further creating the right ecosystem". All the Engineering Diploma students responded in this regard that "Nanotechnology, renewable energy to quantum computing with a wider scope and need" is the main need of this revolution.

Narrative of Engineering Degree Students "Industry 4.0 isn't just about investing in new technology and equipment to improve manufacturing efficiency – it's about revolutionizing the way your entire business operates and grows. Industry 4.0 refers to a new phase in the industrial revolution that focuses heavily on interconnectivity, automation, machine learning, and real-time data. Industry 4.0, including IoT and smart manufacturing, combines physical production and operations with smart digital technologies, machine learning and big data, to build a more holistic and better-connected business environment for companies focused on manufacturing and supply chain management and creating right eco-system".

Narrative of Engineering Diploma Students "Need of Fourth Industrial Revolution is not just about smart and connected machines and systems. Its scope and need are much wider. There are waves of simultaneous breakthroughs in fields ranging from gene sequencing to nanotechnology, renewable energy to quantum computing. It is the fusion of these technologies and their interaction across physical, digital and biological domains that makes the Fourth Industrial Revolution fundamentally different from previous revolutions".

Theme 4: Understanding about Industry 4.0
Question: What is your understanding about Industry 4.0?

Categories	Engineering degree students (50)	Engineering diploma students (50)
Revolutionizing the way companies manufacture, improve, and distribute their products	40 (80%)	10 (20%)
Potential to change the present and future of human life in terms of industrial change for India but also a social change	10 (20%)	40 (80%)

To get a clearer insight on the understanding of engineering students about Industry 4.0, another question was asked, and in response quite enriched responses were received from both the groups. Whereas 80% of the Engineering Degree students told that Industry 4.0 is "Revolutionizing the way companies manufacture, improve, and distribute their products", 80% of the Engineering Diploma students were of opinion that it has a "Potential to change the present and future of human life in terms of industrial change as well as social change for India".

Narrative of Engineering Degree Students "Industry 4.0 is revolutionizing the way companies manufacture, improve and distribute their products. Manufacturers are integrating new technologies, including the Internet of Things (IoT), cloud computing and analytics, and AI and machine learning, into their production facilities and throughout their operations".

Narrative of Engineering Diploma Students "'Industry 4.0' has the potential to change the present and future of human life. This is not only an industrial change for India but also a social change. 'Industry 4.0' has the potential to bring about an irreversible creative change in India. This will give the necessary speed to the works in India and will help in improving the work. Emerging areas such as artificial intelligence, machine-learning, Internet of Things, blockchain and big data can take India to new heights of development and improve the lives of citizens".

Theme 5: Differences between the current and previous industrial revolutions.
Question: What are the differences between the current and previous industrial revolutions?

Categories	Engineering degree students (50)	Engineering diploma students (50)
Industrial revolutions were brought to increase productivity per person, but in Industry 4.0 automation is all about eliminating the human involvement	50 (100%)	–
As compared to previous industrial revolutions, Industry 4.0 optimizes the computerization of Industry 3.0 making smart factory a reality	–	50 (100%)

When the participants were asked about the differences between the past and present industrial revolutions, 70% of the Engineering Degree students responded "Industrial revolution was brought to increase productivity per person; automation is all about eliminating the human involvement", while counterparts reported that "Industry 4.0 optimizes the computerization of Industry 3.0 making smart factory a reality".

Narrative of Engineering Degree Students "The difference between the Industrial Revolution of the eighteenth century and the automation revolution of today is that the Industrial Revolution was the mechanized process of human labor that allowed humans to achieve greater productivity per person; Humans were still needed and necessary. This allowed industry to hire thousands of people and build factories; Hiring more people benefited the business to the point of achieving maximum output per person. The essential difference here is that even as productivity increased, it was up to people to produce more. With the automation revolution of the twenty-first century, the goal is no longer about mechanizing the process of manual labor, the goal is now about automating the entire process to eliminate human involvement, so that productivity is no longer measured in output per person. Go The end goal is to eliminate people from the entire process, with lines running 24/7 and no change except for a few people keeping the machines running. The more reliable the machines, the less mechanics they require until the place eventually goes dark. So, in short, the goal of the Industrial Revolution was to increase productivity per person; Automation is all about eliminating the person".

Narrative of Engineering Diploma Students "When computers were introduced in Industry 3.0, it was disruptive thanks to the inclusion of an entirely new technology. Now, and in the future as Industry 4.0 unfolds, computers are connected and communicate with each other so that decisions can ultimately be made without human involvement. The combination of Cyber-Physical Systems, the Internet of Things and the Internet of Systems is what makes Industry 4.0 possible and the Smart Factory a reality. As a result of supporting smart machines that get smarter as they gain access to more data, our factories will become more efficient and productive and less wasteful. Ultimately, it is the network of these machines that are digitally connected to each other and create and share information that results in the real power of Industry 4.0".

Theme 6: Important pillar of Industry 4.0
Question: Which one is the most important pillar of Industry 4.0?

Categories	Engineering degree students (50)	Engineering diploma students (50)
Industrial Internet of Things (IIoT)	–	50 (100%)
Autonomous robots	5 (10%)	–
Simulation/digital twins	5 (10%)	–
Additive manufacturing/3D printing	40 (80%)	–

When participants were probed about their opinion about the *most important pillar of Industry 4.0*, all the Engineering Diploma students respond "Industrial Internet of Things (IIoT)" as the main pillar out of the nine pillars. On the contrary, 80% of the Engineering Degree students considered "Additive manufacturing/3D printing followed by autonomous robots and simultaneous/digital twins as the next important pillars of industry 4.0".

Narrative of Engineering Degree Students "According to me out of nine pillars, Additive manufacturing, or 3D printing, is a main key technology driving Industry 4.0. 3D printing was initially used as a rapid prototyping tool, but now serves a wide range of applications from mass customization to distributed manufacturing. With 3D printing, for example, parts and products can be stored as design files in a virtual inventory and printed on demand at the point of need- reducing both transport distance and cost".

Narrative of Engineering Diploma Students "The Internet of Things (IoT)—more specifically, the Industrial Internet of Things—is so central to Industry 4.0 that the two terms are often used interchangeably. Most physical things in Industry 4.0 – tools, robots, machinery, equipment, products – use sensors and RFID tags to provide real-time data about their status, performance or location. This technology lets companies run smoother supply chains, rapidly design and modify products, prevent equipment downtime, stay on top of consumer preferences, track products and inventory, and much more".

Responses under this theme show that both categories of students not only were very clear about which pillar they consider as most important but also gave reasons for their responses.

Theme 7: Benefits of Industry 4.0
Question: What are the main benefits of Industry 4.0?

Categories	Engineering degree students (50)	Engineering diploma students (50)
Computers and automation togetherness	30 (60%)	–
Intelligent assets and technicians can monitor asset performance in real time and anticipate and prevent downtime	20 (40%)	–
Fast emerging, dynamic, and continuous flow	–	70 (35%)
Empowered people in terms of technologies like AI and access to live sensor data	–	30 (15%)

While responding on the benefits of Industry 4.0, 60% of the Engineering Degree respondents quoted "computers and automation togetherness" as the main benefit. However, "Fast emerging, dynamic and continuous flow" was mentioned as the main benefit by 70% of the total Engineering Diploma students.

Narrative of Engineering Degree Students "In this era of business competition, the whole world has entered into 'Industry 4.0'. Under which computers and automation have come together in a new avatar. Robot-assisted remote computer systems equipped with machine learning algorithms can be guided by vast-in-put with minimal need for human operations".

Narrative of Engineering Diploma Students "A more in-depth look at the concept of 'Industry 4.0' reveals that it is a fast emerging, dynamic and continuous flow with strong potential to change the way industrial units work. Is. To an extent, 'Industry 4.0' is the second industrial revolution. The only difference is that while the first industrial revolution in the world was limited to Europe, mainly England, the current industrial revolution in the form of Industry 4.0 is spread all over the world. Its presence and spread is not limited to any country".

One worth noting and very interesting benefit that was mentioned by Engineering Diploma students was that the previous industrial revolutions were restricted mainly to Europe where Industry 4.0 has worldwide coverage.

Theme 8: Industry 4.0 – main challenges and risks
Question: What is the main challenge of Industry 4.0 according to you?

Categories	Engineering degree students (50)	Engineering diploma students (50)
Data and IT security as well as increasing better connectivity and a greater number of connections	50 (100%)	–
Managing culture change, i.e. change of work culture	–	50 (100%)

In Industry 4.0 technologies are developing rapidly, helping businesses become "smarter" and more efficient. To remain competitive, companies must adapt to, or at least accept, these changes. Using digital transformation tools is not always straightforward. Adopting Industry 4.0 can present some challenges and risks for many businesses.

While exploring about the views of respondents towards the challenges and risks of Industry 4.0, the Engineering Degree students responded that "Data and IT security as well as increasing better connectivity and a greater number of connections" is the biggest challenge and risk of Industry 4.0, whereas the Engineering Diploma students mentioned "Managing Culture Change".

Narrative of Engineering Degree Students "IT security can pose substantial risks in an Industry 4.0 setting. Online integration of processes, systems and potentially left room for security breaches and data leaks are to name few. IT security is not limited to cyber-attacks. Other major threats include network misconfiguration, and software or device failures that could potentially disrupt business operations and production. IT infrastructure will also need to be up to the task of handling the additional connectivity required by digital transformation".

Narrative of Engineering Diploma Students "Managing culture change is a critical part of Industry 4.0 success, as well as a potential bottleneck if not executed well. If the workers are not prepared for the changes, they may be reluctant, resistant or unable to adapt. Preparing and getting them on board for technology changes is critical to the success of digital project. Top leadership can play an important role in bringing about the cultural change required for digital transformation".

Theme 9: Ways that need to be developed for efficient employment
Question: Which ways need to be developed for more efficient employment in Industry 4.0?

Categories	Engineering degree students (50)	Engineering diploma students (50)
Apprenticeship/pre-training/probation period	50 (100%)	10 (20%)
Skill development and upgradation	–	40 (80%)

On asking about the *ways that need to be developed for more efficient employment in Industry 4.0,* 100% of Engineering Degree student respondents opted "Apprenticeship/pre-training/probation period", while 80% of Engineering Diploma students opted "Skill development" as the major way.

Narrative of Engineering Degree Students "The skill gap is not a new problem in the engineering and manufacturing sector. The Manufacturer Report states that 71% of manufacturers believe that apprenticeships are increasingly becoming a realistic alternative to higher education. The incoming talent can and should learn from the highly skilled and experienced employees already embedded in the workforce. By encouraging individuals to start careers in this field early, companies can train and develop employees as experts – and more importantly, keep them up to date with new technology that's on the horizon. Engaging with adolescents and young adults, engaging them in businesses through apprenticeship and work placement programs are of extraordinary benefit, especially within Industry 4.0".

Narrative of Engineering Diploma Students "Machine operators and technicians play a vital role in most manufacturing and engineering occupations. Along with recruiting into these roles, there is also a need to upskill the people already working in the organizations. If the UK is to be a driving force for Industry 4.0, every manufacturer needs to engage in skills development, understand the skills needed in the factories of tomorrow, and invest in developing these skills today. Enhancing the skills of machine operators to diagnose faults and repair machines at source will mean increased productivity. It seems like a simple equation".

Theme 10: Rational of the progression of Industry 5.0
Question: What is the rationale following the progression of Industry 5.0?

Categories	Engineering degree students (50)	Engineering diploma students (50)
Humans and machines are working together	45 (90%)	10 (20%)
People and machines complimenting their activities, not people being replaced by machines and *beyond the focus of Industry 5.0 lies Society 5.0*	5 (10%)	40 (80%)

In response to the question, 90% of the Engineering degree respondents gave their rationale following the progression of Industry 5.0 as "Humans and machines are working together", while out of the total Engineering Diploma students, 80%

shared their perspective about Industry 5.0 as “People and machines complimenting their activities, not people being replaced by machines”.

Narrative of Engineering Degree Students “The revolution of Industry 5.0 means that humans and machines are working together, improving the efficiency of industrial production. Human workers and universal robots are increasing the productivity of the manufacturing industry. The future direction of Industry 5.0 is the creation of industrial robots. The advancement of artificial intelligence and cognitive computing technologies is taking the manufacturing world at a higher pace and increasing business efficiency. Apart from the benefits in manufacturing business, Industry 5.0 will also benefit in sustainability as it aims to develop a sustainable system that runs on renewable energy”.

Narrative of Engineering Diploma Students “Industry 5.0 marks a sea change in perspective; The core of Society 5.0 focuses on people as the fundamental pivot of the production sector. Both the production and marketing sectors agree that beyond the focus of Industry 5.0 lies Society 5.0. The products or services to be offered in Society 5.0 will be customized as per the needs of the customer. The intention is to reach a fusion between technological advances and humans, with the main objective being people and machines complimenting their activities, not people being replaced by machines. The use of robots and robots to assist with repetitive, dangerous and unsafe tasks is a fundamental change. Furthermore, human work would be intellectual production, which means being able to be active in this society model would be essential”.

Theme 11: Regarding general population’s perspective on the deadly effects of Industry 4.0
Question: What is a deadly effect of Industry 4.0 on general population?

Categories	Engineering degree students (50)	Engineering diploma students (50)
Sense of privacy, our concepts of ownership, our purchasing patterns etc.	30 (60%)	–
Causing pollution to our planet and its ill effects on human health	20 (40%)	–
Biotech and AI revolutions	–	30 (60%)
Will lead to rethinking our moral and ethical boundaries	–	20 (40%)

Out of the total Engineering Degree students, 60% opted “Sense of privacy, our concepts of ownership, our purchasing patterns, etc.” as the deadly effect of Industry 4.0. A serious concern towards “Causing pollution to our planet and its ill effects on human health” due to improper waste disposal was also shown by these students (40%). On the other side, 60% of Engineering Diploma students responded “Biotech and AI revolutions” as its deadly effects. “Raised income inequality” was also expressed as a major concern by the respondents (40%) of this category.

Narrative of Engineering Degree Students "The Fourth Industrial Revolution will not only change what we do, but also who we are. Our sense of privacy, our concepts of ownership, our purchasing patterns, the time we devote to work and leisure, and how we enhance our jobs, develop our abilities, meet people, and nurture relationships. do, all of them will be affected by it. It is already affecting our health and contributing to the 'quantified' self, and it may be causing human evolution sooner than we believe. The list goes on and on as we are only limited by our imagination".

Narrative of Engineering Diploma Students "Today, the biotech and AI revolutions are redefining what it means to be human by pushing back current limits on lifespan, health, intelligence and capabilities, forcing us to rethink our moral and ethical boundaries. Privacy is one of the most important personal challenges presented by modern information technology. We instinctively understand why this is so important, yet recording and sharing information about ourselves is an important component of new connections. In the coming years, the debate about fundamental concerns such as the impact on our inner lives of losing control of our data will only increase".

Theme12: Regarding applicability
Question: Explain the applicability of Industry 4.0 in current scenario.

Categories	Engineering degree students (50)	Engineering diploma students (50)
Additive manufacturing	30 (60%)	10 (20%)
Autonomous robots	20 (40%)	40 (80%)

Most of the Engineering Degree respondents reported (60% & 40%) "Additive manufacturing", while 80% of Engineering Diploma students reported "Autonomous robots" as the main applicability of Industry 4.0 in current scenario.

Narrative of Engineering Degree Students "Additive manufacturing is a term that refers to the process of making something by adding material as needed, rather than by removing material as is done in traditional types of manufacturing processes such as milling, drilling, boring, etc. 3D printing is the most famous example of additive manufacturing. Companies can now manufacture small batches of bespoke products instead of prototyping individual components. Advantages include the ability to quickly create complex, lightweight designs".

Narrative of Engineering Diploma Students "Autonomous robots can communicate with each other and work alongside people in a secure environment. Over time these robots will be less expensive and have a wider range of capabilities. An autonomous robot, also known as an auto-robot or auto-bot, is a robot that has a high degree of autonomy in its behavior or activities without any external influence".

4 Implications and Suggestions

Industry 4.0 refers to the collection, use, and analysis of actionable data and information as a means of realizing smart industries and industrial innovation and ecosystems of big data, people, processes, services, systems, and collaboration in an IoT-enabled world. Our engineering students are going to be the future workforce for making the vision of Industry 4.0 a reality for India as a developed and technologically advanced nation. And, if they know the concept well, understand the requirements of Industry 4.0, and are well aware of its threats and challenges, it gives an edge to the country over other countries. It also shows that as a nation, we are well prepared for a sustainable and bright future. Very optimistic, enriched, and creative responses were received by both groups of respondents. The results of narratives also reflect that the engineering students have a clearer understanding of the beginning of industrial revolutions as well as the current and future industrial revolutions in terms of their processes and functioning especially in terms of development and use of technology, impacts, risks, and challenges. On the whole the responses obtained from both groups portrayed a very positive picture. According to the respondents, Industry 4.0 is a comprehensive vision with well-defined framework and reference design, defined primarily by the integration of physical industrial assets clubbed with digital technology. Therefore, the results of the current study will help our workforce to be better prepared and upgraded to handle the challenges and job requirements of Industry 4.0.

Further, the results of this research are quite pertinent and meaningful for framing policies related to several aspects raised by the respondents such as IT security issues, social issues, adaptation to changed work culture, skill upgradation, etc.

The findings have raised future research needs through this study. Certain key questions found from this study on which more probe and data set will be required are as follows: What could be the solution to our sense of privacy, our concepts of ownership, IT security, and social issues occurring due to the advent of new technology, especially robots replacing humans in terms of change in the time we devote to work and leisure, meet people, build relationships and nurture them, and develop our abilities? How can we grow our jobs and develop which and what amount of extremely highly specialized skills? Strong laws for handling moral and ethical issues will be the calling for the future. In a nutshell the way we look at the world and the way we live and think will also change drastically for which a great deal of research and development work will be needed for a sustainable future.

Sustainable development that has become part of every international and international agenda and conventions is directly linked with various developments occurring under Industrial Revolution 4.0. Therefore, "important factors that needs consideration are all types of threats occurring out of IR 4.0 in future, especially that are pertaining to accidents causing due to operations, automation, widening of income and social inequality gaps and finally environmental threats for instances possibility of any nuclear disaster. These all need to be given utmost importance for sustainable future and political intervention was found to be indispensable" (Zervoudi, January, 2020).

An exhaustive review of research articles in Scopus journals on the topic revealed that the concept of Industry 4.0 has been one of the most trending and debated topics of current times not only among policy planners, administrators, corporates, and managers but also among scientists, academicians, researchers, etc. The research found that the main focus of the studies done on Industry 4.0 so far was focussed on developed nations and very little concern has been given to the developing countries. It suggests for a "much wider and more mature scope for researches on various dimensions related to the industrial revolution especially focussed on developing and underdeveloped countries" (Elnadi and Abdullah, January, 2023).

In a nutshell, it is proposed that every country needs to devise its own strategies and plans for all levels covering every concerned domain to meet the challenges of the Fourth Industrial Revolution and further the upcoming Fifth Industrial Revolution too. Yang Fengwei and Sai Gu (2021) say that "the governmental policies, technical advancement and market, along with innovations have never been so intertwined before as currently under IR 4.0" and, thus, go a step ahead. They further advocate "for designing and developing updated national-level strategies and plans that are tailor-made to the requirements of that particular country and then driving all efforts on those lines".

References

Abd Rahman Ahmad, Pannirchelvi, S., Ng Kim Soon, Hairul Rizad Md. Sapry, Siti Sarah Omar. (July 2019). Factor Influence the Students' Readiness on Industrial Revolution 4.0. International Journal of Recent Technology and Engineering (IJRTE). Published by Blue Eye Intelligence Engineering & Sciences Publication. 8 (2S). 461–468.

Caballero, C. L. & Walker, A. (2010). Work Readiness in graduate and Selection: A review of Current Assessment Methods. Journal of Teaching and Learning for Graduate Employability, 1, 13–25.

Cevik S., Alp Ustundag, Cigdem Kadaifci & Basar Oztaysi. (September 2018). The Changing Role of Engineering Education in Industry 4.0 Era. chapter in book "Industry 4.0: Managing the Digital Transformation" Part of Springer Series in Advanced Manufacturing book Series (SSAM), 137–15. https://doi.org/10.1007/978-3-319-57870-5_8. https://www.researchgate.net/publication/319863634_The_Changing_Role_of_Engineering_Education_in_Industry_40_Era

Coskun S., Yasanur K., & Eray, G. (2019). Adapting Engineering Education to Industry 4.0 Vision. Technologies (special issue on Technologies for Industry), 7 (1), 10. https://www.mdpi.com/2227-7080/7/1/10. https://doi.org/10.3390/technologies7010010

Cotet, G. B., Nicoleta Luminita Carutasu & Florina Chiscop. (January, 2020). Industry 4.0 Diagnosis from an iMillennial Educational Perspective. Special issue on Education for the Profession in Times of Change, Education Science, 10 (1), 21. https://doi.org/10.3390/educsci10010021, htps://www.mdpi.com/2227-7102/10/1/21

Elnadi., Moustafa., Yasser Oman Abdullah. (January, 2023). Industry 4.0: Critical Investigations and Synthesis of Key Findings. Management Review Quarterly, Springer nature Switzerland. JEL Classification L6.03, 1–34. htts://link.springer.com/article/10.1007/s11301-022-00314-4

Islam, Aminul. (2022). Industry 4.0: Skill set for Employability. Social Sciences and Humanities Open, Science Direct, Elsevier, 6 (1), 1–9. https://www.sciencedirect.com/science/article/pii/S2590291122000341

Ismail, A., Wan Hassan. W. A. S., Ahmed, F., Affan, Z., Harun, M. I. (August 2020). Student' Readiness in Facing Industrial Revolution 4.0 among Students of Technical Teacher's Education. International Journal of Scientific & Technology Research, 9(8), 300–305.

Kagermann, W.-D. Lukas, Industrie 4.0, VDI Nachrichten (2011).

Yang Fengwei & Sai Gu. (2021). Industry 4.0, A Revolution that Requires Technology and National Strategies. Complex and Intelligent Systems, Volume 7, 1311–1325. https://link.springer.com/article/10.1007/s40747-020-00267-9

Zervoudi, E. K., (January, 2020). Fourth Industrial Revolution: Opportunities, Challenges and Proposed Policies. Chapter in an edited volume 'Industrial Robotics' edited by Antoni Grau and Zhuping Wang. https://www.intechopen.com/chapters/70877. https://doi.org/10.5772/intechopen.90412

Facial Emotion Recognition (FER) with Deep Learning Algorithm for Sustainable Development

Abhishek Kumar, Manoj Sindhwani, and Shippu Sachdeva

1 Introduction

Facial emotion recognition (FER) is a crucial component of human-machine interaction and has numerous applications in sustainable development. FER involves detecting and analyzing human emotions using facial expressions. With the increasing adoption of computer vision and machine learning techniques, FER has become a more accurate and efficient process (Yang et al. 2017). One such application is in behavior understanding. FER can be used to analyze the behavior of people in public places, such as airports or train stations. By analyzing their emotions, we can understand their behaviors and patterns, which can help in improving public safety and security. FER also has a potential in detecting disorders such as depression or anxiety. It can help in identifying people who may be experiencing mental health issues and provide timely intervention and support. This can significantly improve the quality of life for individuals and contribute to the achievement of progress (Savoiu and Wong 2017). Facial emotion recognition (FER) has a significant potential to contribute to the achievement of Sustainable Development Goals (SDGs). Some of the ways FER can contribute to SDGs are as follows (Cowie et al. 2001; Lebobics 1999):

1. **Improving mental health**: FER can be used to detect emotional changes in individuals and identify potential mental health issues. By detecting these changes, timely interventions can be made, leading to better mental health outcomes and adding to SDG 3 – Good Health and Well-being.
2. **Enhancing public safety**: FER can be used in public places to detect the emotions and behavior patterns of people. By analyzing this data, public safety

A. Kumar (✉) · M. Sindhwani · S. Sachdeva
School of Electronics and Electrical Engineering, Lovely Professional University, Phagwara, Punjab, India

I. S. Dunmade et al. (eds.), *Sustainable Engineering*, Green Energy and Technology, https://doi.org/10.1007/978-3-031-47215-2_24

measures can be improved, contributing to SDG 11 – Sustainable Cities and Communities.

3. **Promoting inclusivity**: FER can be used to enable communication accessibility for individuals with hearing or speech impairments. By recognizing and interpreting facial expressions, FER can translate sign language into spoken language or text, contributing to SDG 10 – Reduced Inequalities.
4. **Improving education**: FER can be used in the education sector to detect and understand the emotions of students. By analyzing this data, educators can personalize their teaching approach and provide support to students who may be experiencing emotional challenges, contributing to SDG 4 – Quality Education.

Facial emotion recognition has a significant potential to add up to the attainment of Sustainable Development Goals. By analyzing and detecting emotions, FER can improve mental health outcomes, enhance public safety, promote inclusivity, and improve education. As technology continues to evolve, FER can make a significant contribution to creating a more sustainable and inclusive world. FER with deep learning algorithms has numerous applications in sustainable development. It can be used in behavior understanding, mental health, and communication accessibility. Deep learning algorithms are particularly useful in FER as they can learn complex representations of facial features and emotions. These algorithms can learn to recognize patterns in large datasets and can improve them over time with more data. Using deep learning algorithms, FER can detect and recognize emotions with a high accuracy of greater than 70%, making it a useful tool for various applications. As technology continues to improve, FER has the potential to make a significant contribution to creating a more sustainable and inclusive world.

Facial expressions have long been used to communicate feelings (emotions), such as anger, disgust, happiness, fear, neutrality, sadness, and surprise. Humans often strive to convey their feelings through their facial expressions. Facial emotion recognition provides a simple and commercially effective way to gauge the interest of an audience, or consumer's retailers can use this technology to assess customer interest, while healthcare practitioners can provide better services by determining the emotional state of patients during therapy.

Humans are naturally adept at reading other people's emotions, and even babies as young as 14 months old can distinguish between joyful and sad expressions. With the facial expression recognition process, a computer can also interact with humans, enabling improved communication and interaction. The facial expression recognition process typically comprises three steps, including image acquisition, feature extraction, and classification. By analyzing the facial features, the system can recognize and classify the emotions expressed on the face.

In this work, we used a face comparison system for facial emotion recognition so that it not only reads the emotion but also compares the human with others in the database and shares them with that person. A deep learning neural network (DNN) enhances the capacity to make inferences (understanding) about emotional states. Figure 1 present the Monalisa illusion images posses different emotions, 83% happy, 9%disgusted, 6% fearful, and 2% angry presented in Fig. 1 available at

Fig. 1 The *Mona Lisa* illusion

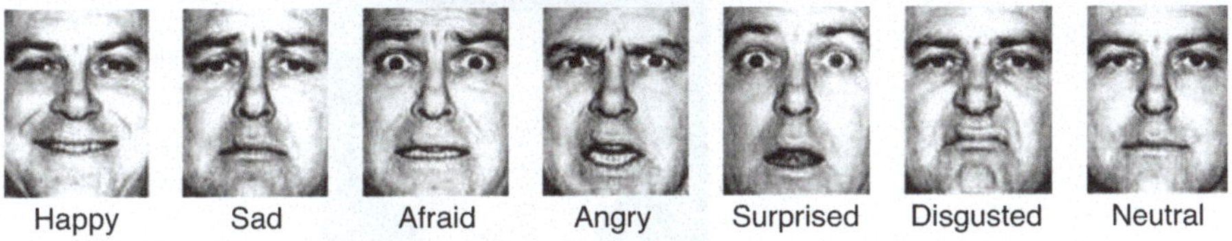

Fig. 2 Facial expression. (Scherer 2019)

(Sebe 2005). Studies on machine learning and prediction based on extracted facial features can identify the kind person in a crowd and allow for preventative action based on the emotion observed.

2 Types of Facial Expression

This work is dedicated to creating a facial emotion recognition system (Ekman 1983) that can accept human facial images with some expression as input and detect and distinguish the emotion in the image into seven basic emotions as shown below, as well as create an alert system to broadcast the emotion in Fig. 2.

2.1 *Anger*

Anger is like a fingerprint presented in Fig. 3, unique to each individual. It is a complex emotion that can be triggered by a variety of factors, such as interference, injustice, physical or psychological harm, betrayal, abandonment, rejection, and violation of the law or societal norms. While anger is a fundamental human emotion, it can be hazardous and can lead to violence in extreme cases. As a result,

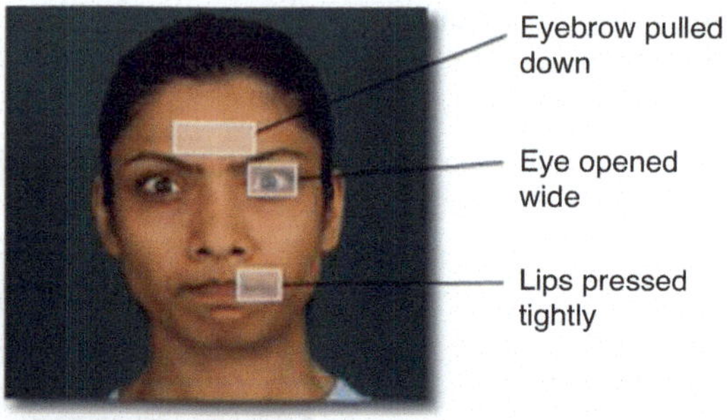

Fig. 3 Anger face. (Ekman 2020)

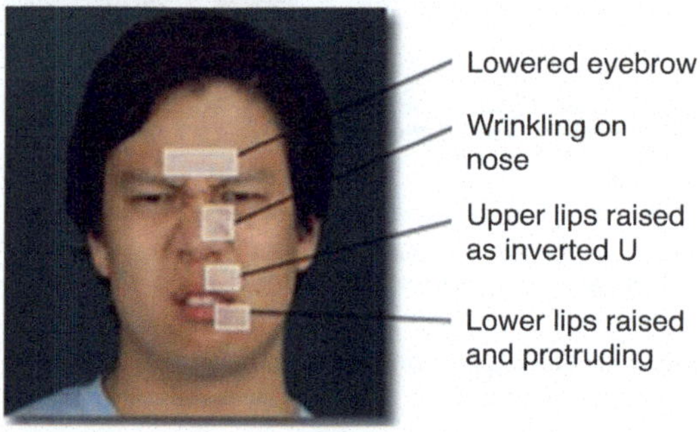

Fig. 4 Disgust face. (Ekman 2020)

people often seek therapy to manage their anger. Although everyone gets angry about similar things, each of us has our own unique set of triggers that can make us feel angry (Valk et al. 2015).

2.2 *Disgust*

The disgust expression shown in Fig. 4 is a complex human emotion that is characterized by a strong aversion or revulsion towards something offensive, disagreeable, or harmful (Wegrzyn et al. 2017). It can be prompted by a variety of stimuli that we see, smell, touch, hear, or taste, as well as by people's actions or appearances. Disgust can also be triggered by thoughts, making it a particularly powerful emotion. While some disgust triggers are culturally and individually influenced by food, others are more universal, such as the sight of expelled bodily products like feces, vomit, urine, mucus, and blood. Other triggers include something rotten, diseased, or dying, injuries, surgeries, and exposure to bodily insides. Physical examination of a person, animal, or object can also elicit disgust. What triggers disgust varies from person to person, and some people may be more sensitive to certain stimuli than others, making it a unique and individual experience.

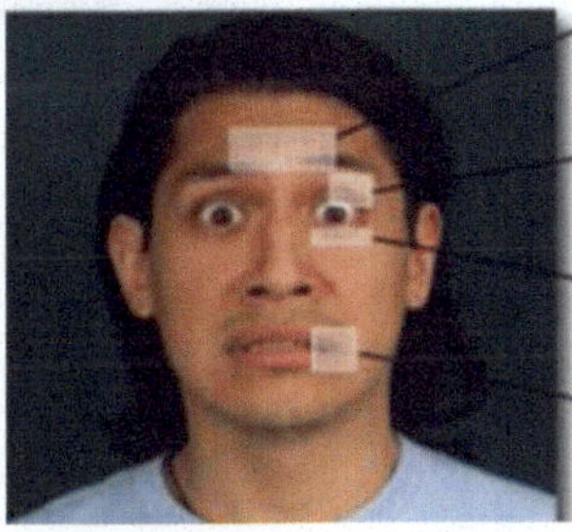

Fig. 5 Fear face. (Ekman 2020)

2.3 *Fear*

Fear is a universal and fundamental human emotion that is triggered by the perception of a genuine or imagined threat of physical, emotional, or psychological harm.

While fear is often seen as a negative and depressing emotion, it serves an important function in keeping us safe by mobilizing us to face danger presented in Fig. 5. People can be scared of a variety of things, and we can learn to be afraid of anything. A common symptom of fear is the upper eyelid lifted high and the lips being tensed and stretched, which is often referred to as the fear grin. Frightening experiences can be categorized based on three factors: threat intensity, time to effect, and action that can be taken to mitigate the threat. Some typical fear triggers include darkness or loss of vision of surroundings, social interaction or rejection, dangerous animals, and death. However, what triggers fear varies from person to person, making it a unique and individual experience.

2.4 *Happiness*

Happiness is, for many, the most desirable of the seven universal emotions (Sel et al. 2015). Happiness is one of the most basic human emotions, and it usually stems from a sense of belonging or sensory pleasure. The words "happiness" and "enjoyment" are interchangeable. The term "happiness" refers to a person's general feeling of well-being. Typical happiness triggers for Fig. 6 are as below:

- Touch, taste, smell, sight, and sound pleasures
- Witnessing and engaging in acts of human generosity, kindness, and compassion
- Relieving suffering from extreme circumstances
- When we or our loved ones accomplish something or observe someone else do something
- When we come across something beautiful, surprising, or astonishing
- Feeling connected to ourselves, other people, places, animals, and nature

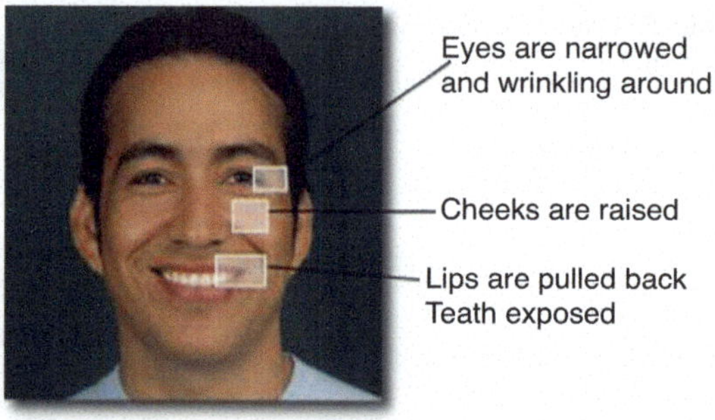

Fig. 6 Happiness face. (Ekman 2020)

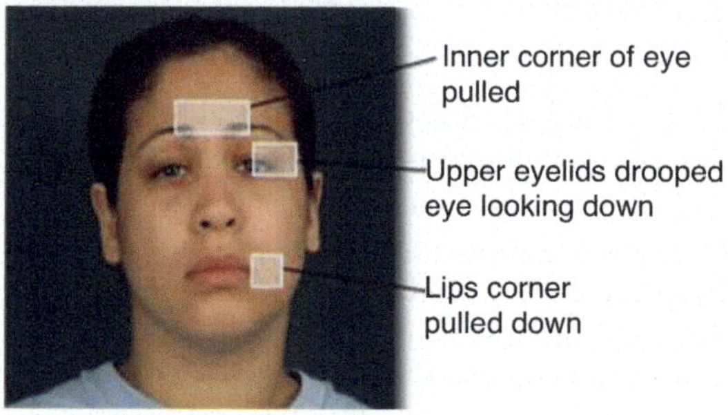

Fig. 7 Sadness face. (Ekman 2020)

2.5 Sadness

Sadness is one of the most common universal feelings felt by people all over the world when they lose someone or something they care about. The cause of grief differs from person to person depending on their level of attachment. Although sadness is commonly thought of as a negative emotion, it plays a crucial role in expressing a need for assistance or consolation. The loss of a valued person or object is the most prevalent cause of sadness, though this varies considerably between people depending on the person's worth. Sadness can be induced by a variety of emotions presented in Fig. 7. Sadness triggers are as follows and can be clubbed with other emotions: rejection, endings, and goodbyes, death of a loved one, loss of some aspect of identity, and being disappointed by an unexpected outcome.

2.6 Surprise

One of the most basic common emotions is surprise, which occurs when we are confronted with sudden or unexpected actions. The purpose of surprise is to draw our attention (Zhao et al. 2017). The surprise emotion in Fig. 8 has a short duration, lasting only a few seconds. Other emotions can last a long time, whereas surprise has a set and limited lifespan. The amazement wears off after a few seconds when we image out what's going on. Then, depending on what surprised us, surprise can

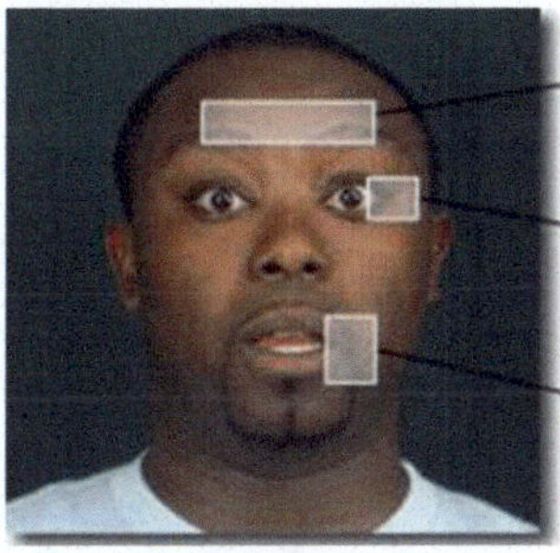

Fig. 8 Surprise face. (Ekman 2020)

transform into amusement, relief, anxiety, rage, and so on. If we judge that the unexpected incident was of no relevance to us, it may be followed by no emotion. The typical surprise triggers are loud noise and sudden movements.

3 Implementation of Facial Emotion Recognition (FER) System

The presented FER system detects human emotion via convolutional neural networks (CNNs), convolution layers, Haar cascade classifier, and so on which are used to implement the facial emotion recognition system. The integration of computer vision (CV) and machine learning (ML) achieved a high level of accuracy of 76.05% in emotion identification and accessibility based on the available dataset (Kim et al. 2017). Hybrid recurrent neural network (RNN)-CNN framework has been applied to recognize seven emotions, i.e., anger, fear, happiness, disgust, sadness, surprise, and neutral (Fan et al. 2016), using the dataset of EmotiW. A dataset of MMICASMEII recognized six emotions with spatial image characteristics learned using a short-term memory model (Ko 2018). CNN temporal appearance and temporal geometry feature analyze seven emotions using dataset CK + MMI (Jeong et al. 2020) achieving an accuracy value of 55.87%. The extraction of image features is relevant to the key feature of the facial point CNN to predict labels for the facial expressions.

3.1 *Convolutional Neural Networks (CNNs)*

CNN is constructed with neurons that possess weights and biases and can learn. Each neuron accepts inputs, performs a dot product, and is optionally followed by nonlinearity. Deep convolutional neural networks (DCNNs) are unidirectional feed-forward neural networks in which information flows from inputs to outputs. CNNs are motivated by biological processes in the same way with those artificial neural networks (ANN) (Mollahosseini et al., 2016). The architecture of the cortical region

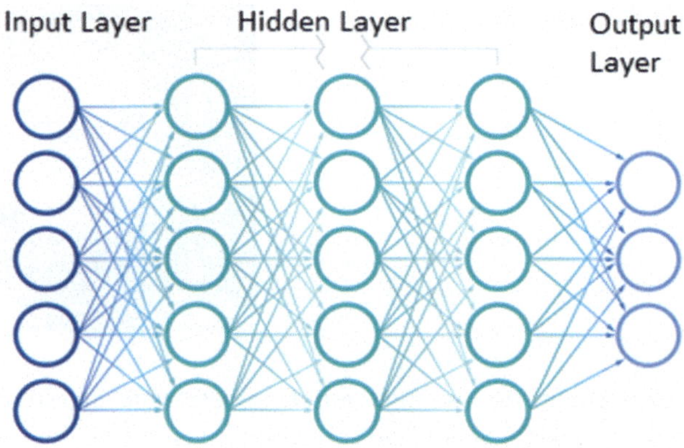

Fig. 9 Deep convolution neural network architecture. (Tu et al. 2017)

in the mind is motivated by another layer of basic and sophisticated cells. CNN architectures (Khan et al., 2020) are available in multiple shapes and sizes, but they all have convolutional layers and pooling layers that are segregated into modules. Following these modules, typical feed-forward neural networks with one or more fully connected layers are used (Fig. 9).

Modules are frequently piled on top of one another to produce a deep model. For a toy image classification job, it illustrates standard convolutional neural network (CNN) design. The network receives an image as input, which is then processed through various steps of convolution and pooling. As a result, these operations' representations feed one or more completely connected layers. Finally, the class label is output by the last layer. Because it is the most often used base circuit in the literature, various architectural modifications have been presented in recent years to boost image classification with lower computation costs.

3.2 *Convolutional Layers*

Convolutional layers are a crucial component of convolutional neural networks (CNNs), which are commonly used in artificial vision tasks like image classification and object detection. These layers are designed to extract parameters from input images and learn feature representations that are useful for solving the task at hand. The process of convolution involves sliding a filter (also known as a kernel) over the input image and computing dot products between the filter and each overlapping patch of the image. The resulting parameter/feature maps highlight different aspects of the input image, such as edges, textures, and shapes (Khan 2022).

Pooling layers are often used in conjunction with convolutional layers to minimize the spatial size of the attribute maps and make the network more computationally efficient. Pooling involves dividing the feature map into non-overlapping regions and computing a summary statistic (such as the maximum or average value) for each region. This minimizes the number of attributes in the network and helps prevent overfitting.

Fully connected layers are typically used at the end of a CNN to perform high-level reasoning functions and make predictions. These layers take the flattened output of the convolutional and pooling layers and pass it through a series of densely connected neurons, where each neuron is joined to every neuron in the preceding layer. The softmax function is often used at the output layer to compute the probability distribution over the different classes. This allows the network to make predictions and estimate the likelihood of each class.

Training a CNN involves adjusting the weights of the neurons to reduction into the difference between the predicted output and the actual output. Backpropagation is a popular algorithm for calculating the gradients of the objective function concerning the weights, which is used to update the weights in the opposite direction of the gradient. This cycle is repeated multiple times until the network achieves a satisfactory level of performance. One common issue with training CNNs is overfitting or underfitting, where the network is unable to generalize well to new data. Regularization techniques such as dropout and weight decay can be used to prevent overfitting and improve the network's ability to generalize to new data.

3.3 Haar Cascade Classifier

Haar cascade classifier is a machine learning-based object detection algorithm used to identify objects in images or videos. It uses Haar-like features and a cascade classifier to detect objects by analyzing the intensity variations in the image. Haar cascade is commonly used in face detection and has been implemented in the OpenCV library. Haar cascade classifier is an ML object detection algorithm that aids in the recognition of objects in images and videos. The algorithm is broken down into four steps:

- Haar feature calculation
- Creating integral images
- Implementing cascading classifiers
- Using AdaBoost

It's vital to note that training the Haar classifier, which is similar to other machine learning models, necessitates a higher number of positive (+ve) photos of faces and negative (−ve) images of non-faces (Padilla et al. 2012).

3.3.1 Haar Feature Calculation

Haar features are similar to Kernels and are often used to detect edges. In human faces, the eye region is generally darker than the upper cheek region, and the nose region is brighter than the eye region (Shetty et al. 2021). The placement and size of these matchable traits will aid us in detecting a face presented in Fig. 10.

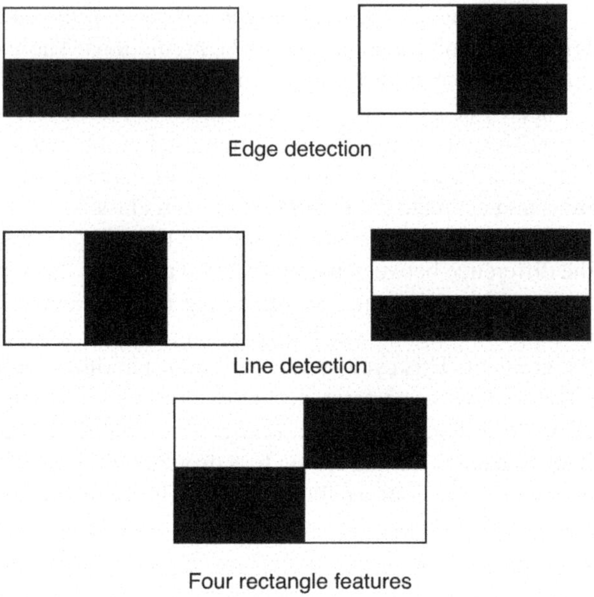

Fig. 10 Haar feature. (Mita et al. 2005)

Here are various Haar traits that can be used to determine whether or not there is a face. The black region is +1, while the white region is −1, according to the Haar feature. An image is displayed in a 24 × 24 window. The features in the Haar cascade are computed using black and white rectangular areas, where the value of each feature is determined by subtracting the sum of pixel intensities in the white region from the sum of pixel intensities in the corresponding black region. We must find the summation of pixels highlighted in white and black shape for each attribute computation. For a 24 × 24 window, there will be 160,000+ Haar features. They devised integral pictures to tackle the problem. It also reduces the sum of pixel computation and determines how great the number of pixels can be to 4 -a pixel process.

3.3.2 Integral Images

The primary concept behind an integral picture is to compute the area. So that we don't have to add up all of the pixel values, we can only take the corner significant values and execute a basic calculation. The addition of the pixels left and above of x,y makes up the integral image at x,y: $(x, y) = \sum x' \leq x,' \leq y i(x',y')$. The integrated picture for a given image input will be generated by adding all of the above and left pixels together (Fig. 11).

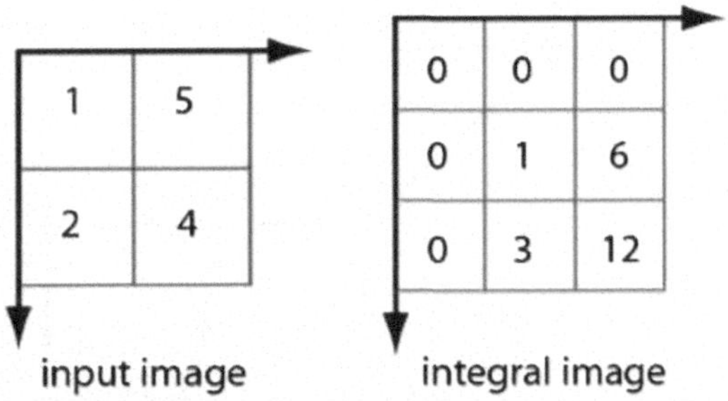

Fig. 11 Histograms of oriented gradients. (Déniz et al. 2011)

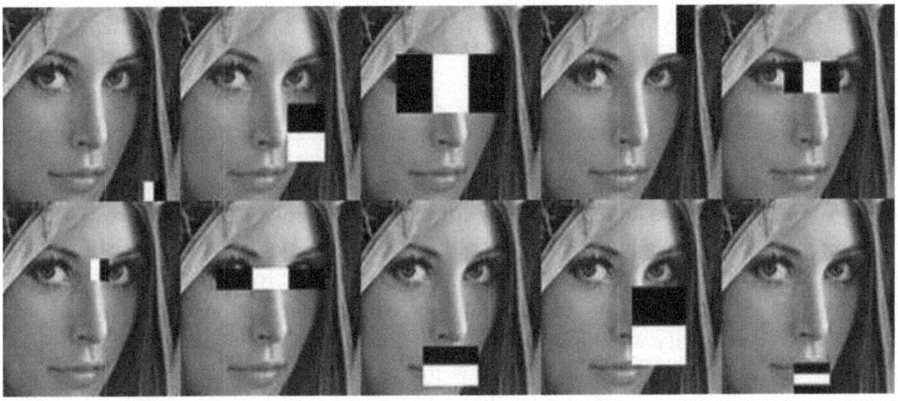

Fig. 12 Cascade method for image processing. (Al-Zaydi et al. 2016)

3.3.3 AdaBoost

AdaBoost is used to remove Haar's superfluous attributes. A classifier can be created by combining a small number of these features. The most difficult part is locating these characteristics, choosing the features, and training the classifier (Fig. 12).

Another feature utilized to identify the nose bridge does not apply to the upper portion of the lips because they have a more or less continuous attribute. As a result, we should have little trouble eradicating it. Out of 160,000+ features, AdaBoost can help you find the ones that matter. After locating all of the features, a weighted value is applied, to determine whether or not a specific window is a face.

$F(x)$ is referred to as a strong classifier, while $f(x)$ is referred to as a weak classifier. Weak classifiers always return a binary result, such as 0 or 1. If the feature is identified, the value will be 1, else it will be 0. A strong classifier is usually made up of 2500 classifiers. Now, the chosen features are regarded to be acceptable if they perform random guessing, which implies they must identify more than half of the situations.

$$F(x) = a_1 f_1(x) + a_2 f_2(x) + a_3 f_3(x) + a_4 f_4(x) + a_5 f_5(x) + \dots \quad (1)$$

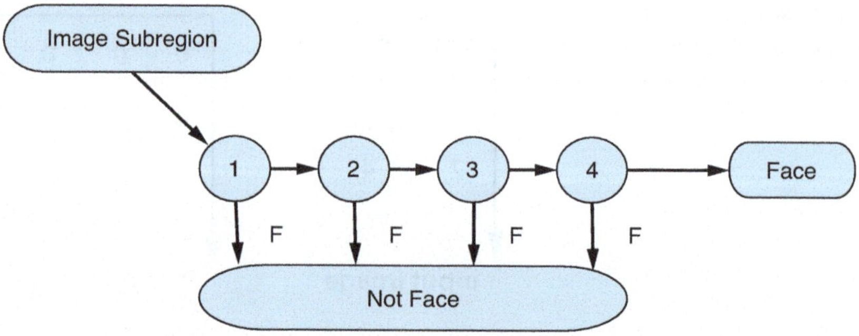

Fig. 13 Cascade of simple features. (Viola et al. 2001)

3.3.4 Cascading

There is an input image of 640 × 480 resolutions, which must move a 24 × 24 window over the image and analyze 2500 features for a window. Linearly using all 2500 features, it determines whether there is a threshold and, if so, whether it is a face or not. The first 10 features are classified into 1 classifier, the following 20–30 features into another classifier, and the remaining 100 features into yet another classifier. As a result, the complexity will be increased. The advantage of instead of going through the 2500 features of a 24 × 24 window is that we can eliminate non-face from the first step. Assume we have a picture. It could be a face if the image passes past the first step, which stores ten classifiers. After then, the image will be checked for the second time. If the image fails to pass the first stage, we may easily remove it. The smallest and most efficient classifier is cascading. We may simply cover non-face parts with cascading (Fig. 13).

3.4 Haar Cascade Classifier in OpenCV

To train a classifier using this technique, a significant number of positive images containing faces and negative images without faces are required. The Haar features shown in the graphic below are employed for this. Our convolutional kernel is similar to those. Feature value is evaluated by diminishing (subtracting) the sum of pixels in the white color rectangle from the count of pixels in the black rectangle. Many features are now computed using all possible sizes and positions of every kernel. Consider how much computing power is required. Even in a 24 × 24 window, there are about 160,000 features to be found. We should find the total of pixels under black and white rectangles. They devised integral pictures to tackle this problem. Integral pictures simplify the calculation of pixel summation in a four-pixel process, regardless of how great the count of pixels is. It facilitates the completion of tasks in a short amount of time. However, the traits we calculated are irrelevant.

Figure 12 is shown below, with the header row displaying two good features: the first is focused on the region of the eyes, and in the second the eyes are darker than the bridge of the nose. The characteristics that work well for identifying faces, such as the ones around the eyes, may not work well for the other parts of the face, such as the cheeks. AdaBoost algorithm applies each feature to all training images and calculates the optimal threshold to classify them as positive (faces) or negative (non-faces). The features with the lowest error rates are selected, indicating that they are the most effective in distinguishing between face and non-face images. However, the process is not straightforward. At the start, each image is assigned an equal weight, and the weights of misclassified images are increased after each classification. The process is repeated to calculate new error rates and weights until the desired error rate is achieved or until the desired number of features is obtained. The weighted sum of these weak classifiers is the final result, which becomes a strong classifier when combined with other features. According to the report, even 200 characteristics yield 95% accuracy in detection. Around 6000 characteristics were included in their final arrangement. Consider reducing the number of features from 160,000 to 6000, which is a significant improvement. Now you take an image, apply 6000 features to each 24 × 24 window, and check whether it is a face or not a little inconvenient and time-consuming. Because non-facial regions make up the majority of the image area, it's a good idea to have a straightforward way of determining whether or not a window is a face region, using the concept of a cascade of classifiers to accomplish rather than applying all 6000 characteristics.

The total numbers of features are segregated into the separate level of classifiers and applied one at a time. Discard a window if it fails at the initial stage. We don't think about any remaining ones. Once successful, move on to the next stage and repeat. In this technique, a face region is defined as a window that passes through all stages. Through the detection process, nearly 6000 features were detected in 38 stages, and the first 5 stages were found to have 1, 10, 25, 25, and 50 features, respectively. The two features in the above graphic were derived from AdaBoost as the best two characteristics. According to the authors, 10 features out of 6000+ are assessed on average in every sub-window. So that's a basic intuitional description of how Viola-Jones face detection works. A picture is fed directly into the network, which is then processed through multiple steps of convolution pooling. Then, one or multiple completely connected layers are sustained by representations from these operations. Finally, the class label is output by the last fully connected layer.

4 Implementation of Face Comparison

Face comparison can be implemented using various techniques such as feature-based comparison, template matching, or deep learning-based methods. It involves comparing the facial features of two or more faces to determine their similarity or dissimilarity. This can be useful in various applications such as face recognition, identity verification, and security systems.

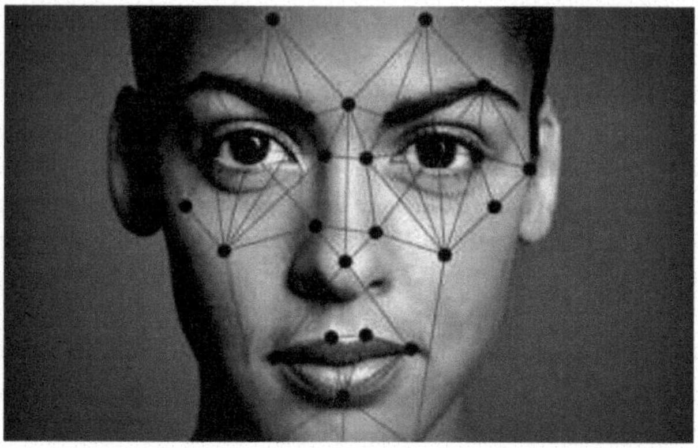

Fig. 14 Face comparison. (Wu and Ji 2019)

4.1 Face Detection

Figure 14 shows the landmark identification for feature extraction. The first challenge we completed was to recognize the face in the provided image. The exact coordinates/location of the face may then be determined, and the face can be extracted for further processing. Once the face is identified and isolated using the Haar cascade classifier, the next step is to extract features from it. Landmark identification is one such technique used for feature extraction. Landmarks are specific points on the face, namely, the corners of the eyes or the tip of the nose, that can be detected using a pre-trained model or manually annotated data. The coordinates of these landmarks are then used to create a feature vector that can be used for various applications, namely, face recognition, emotion detection, or facial expression analysis. Accurate landmark identification is crucial for achieving high-quality feature extraction, and it is a challenging task that requires careful consideration of various factors such as lighting conditions, head pose, and occlusions.

4.2 Feature Extraction

Face detection is followed by feature extraction, which extracts the cropped-out face from the given image so that certain traits can be extracted. We'll show you how to extract these aspects of the face using face embeddings. An NN inputs a picture of a person's face and produces a vector that reflects the face's most traits as output. When we first begin training the NN, it learns to place comparable vectors for faces that appear to be similar. In the case of many photographs of faces taken over some time, for example, it is evident that some characteristics of the face will change. As a result of this criterion, the vectors linked with facial features are similar or comparable in vector space. We've learned how this network works up to this point; now we'll explore how to apply it to our data. We transmit all of the photos in

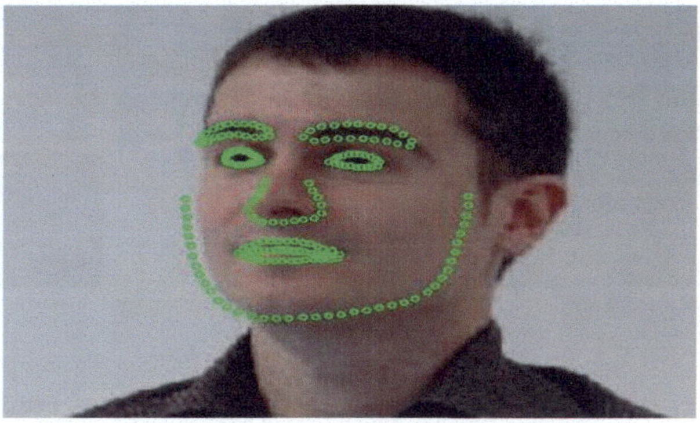

Fig. 15 Face landmark detection. (Efraty et al. 2011)

our data to this pre-trained network, which extracts the appropriate embeddings and saves them in a file for the next step (Fig. 15).

4.3 Comparing Faces

After we've saved face embeddings in our data in a file, the following step is to identify a new image that isn't already in our database. As a result, the first step is to use the same network we used before to figure out the face embedding for the image and then compare it to the rest of the embeddings in the data. If the created embedding is similar to any other embedding, the face will be recognized.

5 Experimental Work

In this study, a unique approach was taken to train a model using supervised learning. Three different datasets were used to create the model, which was then evaluated using an unbiased test dataset. The Kaggle Facial Expression Recognition Challenge (FER2013) was used for training, which consisted of grayscale face images with a resolution of 48 × 48 pixels. The training dataset included 28,709 samples, while the test dataset had 3589 samples. The faces in the dataset were registered if they were centered and occupied a certain portion of the image. The model was trained to categorize the faces based on seven different emotions, namely, anger, disgust, fear, happiness, sadness, surprise, and neutrality (Fig. 16).

The test images have been used by our team members shown in Fig. 17. The presented model has been implemented with PyCharm. The library package required is OpenCV (Open-Source Computer Vision Library) an open-source computer vision and machine learning software library. OpenCV provides many functions for facial detection and facial recognition. It comes with a trainer and a detector – Pandas for analyzing, cleaning, exploring, and manipulating data.

Fig. 16 FER2013 images. (Open source dataset, https://www.kaggle.com/datasets/msambare/fer2013)

Fig. 17 Test samples

Algorithm

- Step 1: Gather photos for the dataset.
- Step 2: Preprocess the photos in the dataset.
- Step 3: Read photographs from the database, collect their names, and locate their faces.
- Step 4: Collect image encoding from the database, and store it in a list.
- Step 5: Detect faces in each image.
- Step 6: Create a grayscale image of the clipped face.
- Step 7: Convert each image to a NumPy array (1, 48, 48) to enable it to be applied to the input layer.
- Step 8: Apply the NumPy array to the 2D convolution layer.

- Step 9: Use convolution to generate feature maps.
- Step 10: Apply MaxPooling2D pooling with (2,2) windows to the feature maps to retain the maximum pixel value.
- Step 11: Conduct neural network forward and backward propagation on the pixel values during training.
- Step 12: Use the softmax function to represent the probability composition of each emotion class.
- Step 13: Locate the face locations and encodings of the input images, and compare them to the encodings in the database.
- Step 14: If the comparison fails, display the input image's emotion and "NO MATCH FOUND," and terminate the program.

Figure 18 shows the execution results of the happy face; it compares the test image with the available image in the database and replies with the label. The extracted feature identifies the facial expression "mouth open" and predicts it's a happy face.

The work presented in this study accurately identifies human facial expressions, which can have numerous benefits for human privacy and security. The ability to store facial images or recognize real-time facial expressions can help prevent unwanted situations by providing early warnings and alerts with 95% accuracy. For example, in public spaces, facial expression recognition technology can detect individuals who display expressions of anger or aggression, allowing for early intervention to avoid potential conflicts. This technology can also be used in security settings, such as airports, to identify individuals who display suspicious behaviors or emotions. Ultimately, the accurate identification of facial expressions can improve safety and security for individuals and communities, aligning with the United Nations' Sustainable Development Goals of promoting peace, justice, and strong institutions (Jain et al. 2021). Face detection technology can potentially contribute to the achievement of several SDGs, including:

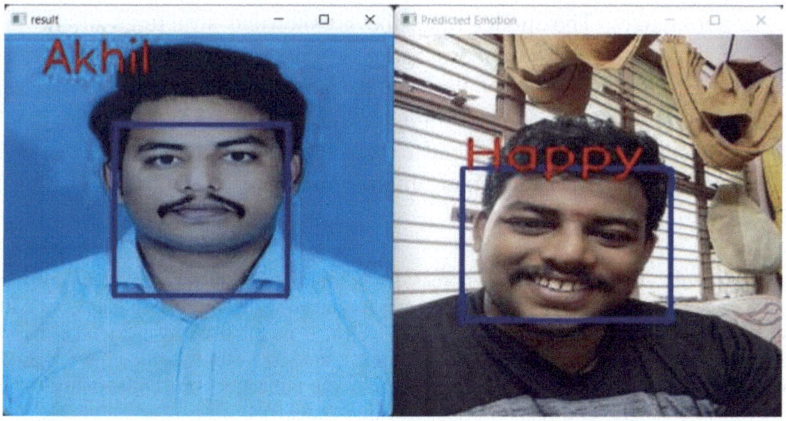

Fig. 18 Happy face recognition

- SDG 1: No Poverty – Face detection technology can help identify individuals living in poverty who may require targeted assistance.
- SDG 2: Zero Hunger – Face detection technology can assist in monitoring food security programs and identifying areas with high levels of food insecurity.
- SDG 3: Good Health and Well-being – Face detection technology can be used in medical applications, such as identifying skin cancer and other diseases.
- SDG 4: Quality Education – Face detection technology can be used to monitor student attendance and engagement in educational settings.
- SDG 5: Gender Equality – Face detection technology can be used to monitor and prevent gender-based violence and discrimination.
- SDG 8: Decent Work and Economic Growth – Face detection technology can be used in various industries, such as security, retail, and entertainment, to enhance efficiency and productivity.
- SDG 11: Sustainable Cities and Communities – Face detection technology can be used in smart city applications to monitor and improve public safety and transportation.

Overall, the potential uses of face detection technology can contribute to the achievement of various SDGs, particularly in areas related to health, education, and safety. However, it is essential to ensure that the technology is developed and used ethically, with respect for privacy and human rights.

6 Conclusion

Automated facial emotion recognition (FER) systems have revolutionized the way we interact with machines, as facial expressions provide crucial information about a person's emotions. FER systems have a wide range of applications, including detecting disorders, understanding behavior, and automating sign language. This work presents a robust FER model that maps both behavioral and physiological biometric parameters. The model links the geometrical structures of the face to diverse expressions such as happiness, sadness, fear, anger, neutrality, surprise, and disgust. The behavioral aspect of the system captures the attitude behind various expressions. By comparing detected images with available images in the database, the system accurately identifies facial expressions. Overall, this FER model has significant implications for enhancing human-machine interaction and improving quality of life.

Acknowledgments We would like to express our gratitude to all those who have contributed to the successful completion of this chapter. First and foremost, we would like to thank our editor for the valuable guidance and feedback throughout the writing process. We also extend our appreciation to the reviewers for their constructive comments and suggestions, which have significantly improved the quality of the content. We are grateful to our colleagues and collaborators who have provided valuable insights and support in our research endeavors.

References

Savoiu, A., & Wong, J. (2017). Recognizing facial expressions using deep learning. Recognizing Facial Expressions Using Deep Learning.

Cowie, R., Douglas-Cowie, E., Tsapatsoulis, N., Votsis, G., Kollias, S., Fellenz, W., & Taylor, J. G. (2001). Emotion recognition in human-computer interaction. IEEE Signal processing magazine, 18(1), 32–80

Lebovics, H. (1999). Mona Lisa's escort: André Malraux and the reinvention of French culture. Cornell University Press.

Tian, Y., Kanade, T., & Cohn, J. F. (2011). Facial expression recognition. In Handbook of face recognition (pp. 487–519). Springer, London.

Scherer, K. R., Ellgring, H., Dieckmann, A., Unfried, M., & Mortillaro, M. (2019). Dynamic facial expression of emotion and observer inference. Frontiers in psychology, 10, 508

Ekman, P. (2020). Universal Emotions: What are Emotions.

Ekman, P., Levenson, R. W., & Friesen, W. V. (1983). Autonomic nervous system activity distinguishes among emotions. Science, 221(4616), 1208–1210.

Valk, J. M. D., Wijnen, J. G., & Kret, M. E. (2015). Anger fosters action. Fast responses in a motor task involving approach movements toward angry faces and bodies. Frontiers in psychology, 6, 1240.

Wegrzyn, M., Vogt, M., Kireclioglu, B., Schneider, J., & Kissler, J. (2017). Mapping the emotional face. How individual face parts contribute to successful emotion recognition. PloS one, 12(5), e0177239

Steimer, T. (2022). The biology of fear-and anxiety-related behaviors. Dialogues in clinical neuroscience.

Sel, A., Calvo-Merino, B., Tuettenberg, S., & Forster, B. (2015). When you smile, the world smiles at you: ERP evidence for self-expression effects on face processing. Social cognitive and affective neuroscience, 10(10), 1316–1322.

Zhao, K., Zhao, J., Zhang, M., Cui, Q., & Fu, X. (2017). Neural responses to rapid facial expressions of fear and surprise. Frontiers in psychology, 8, 761.

Kim, D. H., Baddar, W. J., Jang, J., & Ro, Y. M. (2017). Multi-objective based spatio-temporal feature representation learning robust to expression intensity variations for facial expression recognition. IEEE Transactions on Affective Computing, 10(2), 223–236.

Fan, Y., Lu, X., Li, D., & Liu, Y. (2016, October). Video-based emotion recognition using CNN-RNN and C3D hybrid networks. In Proceedings of the 18th ACM international conference on multimodal interaction, pp. 445–450.

Ko, B. C. (2018). A brief review of facial emotion recognition based on visual information. Sensors, 18(2), 401.

Jeong, D., Kim, B. G., & Dong, S. Y. (2020). Deep joint spatiotemporal network (DJSTN) for efficient facial expression recognition. Sensors, 20(7), 1936.

Mollahosseini, A., Chan, D., & Mahoor, M. H. (2016, March). Going deeper in facial expression recognition using deep neural networks. In 2016 IEEE Winter conference on applications of computer vision (WACV) (pp. 1–10). IEEE.

Khan, A., Sohail, A., Zahoora, U., & Qureshi, A. S. (2020). A survey of the recent architectures of deep convolutional neural networks. Artificial intelligence review, 53(8), 5455–5516.

Tu, F., Yin, S., Ouyang, P., Tang, S., Liu, L., & Wei, S. (2017). Deep convolutional neural network architecture with reconfigurable computation patterns. IEEE Transactions on Very Large Scale Integration (VLSI) Systems, 25(8), 2220–2233.

Padilla, R., Costa Filho, C. F. F., & Costa, M. G. F. (2012). Evaluation of haar cascade classifiers designed for face detection. World Academy of Science, Engineering and Technology, 64, 362–365.

Shetty, A. B., & Rebeiro, J. (2021). Facial recognition using Haar cascade and LBP classifiers. Global Transitions Proceedings, 2(2), 330–335.

Mita, T., Kaneko, T., & Hori, O. (2005, October). Joint haar-like features for face detection. In Tenth IEEE International Conference on Computer Vision (ICCV'05) Volume 1 (Vol. 2, pp. 1619–1626). IEEE.

Déniz, O., Bueno, G., Salido, J., & De la Torre, F. (2011). Face recognition using histograms of oriented gradients. Pattern recognition letters, 32(12), 1598–1603.

Al-Zaydi, Z., Ndzi, D., & Sanders, D. (2016, March). Cascade method for image processing based people detection and counting. In Proc. International Conference on Image Processing, Production and Computer Science, pp. 30–36

Viola, P., & Jones, M. (2001, December). Rapid object detection using a boosted cascade of simple features. In Proceedings of the 2001 IEEE computer society conference on computer vision and pattern recognition. CVPR 2001 (Vol. 1, pp. I-I). Ieee.

Efraty, B., Huang, C., Shah, S. K., & Kakadiaris, I. A. (2011, October). Facial landmark detection in uncontrolled conditions. In 2011 International Joint Conference on Biometrics (IJCB), pp. 1–8. IEEE.

Wu, Y., & Ji, Q. (2019). Facial landmark detection: A literature survey. International Journal of Computer Vision, 127(2), 115–142.

Valstar, M. F., Mehu, M., Jiang, B., Pantic, M., & Scherer, K. (2012). Meta-analysis of the first facial expression recognition challenge. IEEE Transactions on Systems, Man, and Cybernetics, Part B (Cybernetics), 42(4), 966–979.

Sebe, N. (2005). Software Decodes Mona Lisa's Enigmatic Smile. New Scientist, 17

Yang, B., Cao, J., Ni, R., & Zhang, Y. (2017). Facial expression recognition using weighted mixture deep neural network based on double-channel facial images. IEEE access, 6, 4630–4640.

Khan, A. R. (2022). Facial emotion recognition using conventional machine learning and deep learning methods: current achievements, analysis and remaining challenges. Information, 13(6), 268.

Jain, A., Sah, H. R., & Kothari, A. (2021). Study for emotion recognition of different age groups students during online class. 8th International Conference on Computing for Sustainable Global Development (INDIACom),pp. 621–625 IEEE.

AI-Integrated Solar Energy Systems for Sustainable Energy in Africa

McRonald Oyedapo, Philip Olufemi Babalola, and Sunday O. Oyedepo

1 Introduction

One of the most important factors that have impeded the growth of African economies is the absence of a reliable power supply. In sub-Saharan Africa, for example, there live about 17 out of 20 of the 1.3 billion people in third-world nations who have no means of electrical power (Kaygusuz 2012), with a proportion of electrification estimated in metropolitan cities as 64% and in pastoral areas as 13% (Scarlat et al. 2015). According to (Karekezi 2002a; Wolde-Rufael 2005) sub-Saharan Africa has the majority of the least electrified nations on the globe. Developing economies like most African economies are, in many ways, well aligned to enjoy renewable technological inventions. Usually, they are loaded with innately occurring deposits of renewables, especially sunlight; are usually void of any meaningful power systems heritage; are especially probable to undergo surges in power demands; and are often marked out by quite immense and scattered pastoral and suburban natives with little or zero contact with contemporary sources of power (Arndt et al. 2019). For instance, the African energy sector is divided into three: North Africa, which is characterized by heavy dependence on petroleum; South Africa, which is characterized by dependence on coal power; and lastly, sub-Saharan Africa, whose main source of power is biomass energy (Karekezi 2002b). According to (Karekezi and Ranja 1997a; Levodo et al. 2015), Africa is known to have great renewable energy sources including 4–6 kW/m^2 of solar radiation average per day, almost 10 gigawatts worth of geothermal potential, over 1 gigawatt of hydropower capacity, superabundance of biomass, and humongous levels of solar and wind resources. However,

M. Oyedapo
Department of Mechanical Engineering, Obafemi Awolowo University, Ile-Ife, Nigeria

P. O. Babalola (✉) · S. O. Oyedepo
Department of Mechanical Engineering, Covenant University, Ota, Nigeria
e-mail: phillip.babalola@covenantuniversity.edu.ng

I. S. Dunmade et al. (eds.), *Sustainable Engineering*, Green Energy and Technology, https://doi.org/10.1007/978-3-031-47215-2_25

most of these are left untapped due to narrow policy interest and regrettably low investment levels. This shows that the main source of power supply in African economies is largely the burning of fossil fuels. However, with environmental concerns being raised globally as a result of the ozone layer removal consequent of greenhouse gas emissions rising from the constant need for energy supply and its attendant disasters, there is a need to diversify into the consumption of renewables as power supply especially to drive the industry in African countries, since renewable energy can be deployed to end electricity problems on a global scale (and consequently that of Africa) and the resources far exceed global electricity demand (Ellabban et al. 2014). Other barriers to the deployment of renewables in African economies include technical and financial barriers (Karekezi and Ranja 1997b), which have made the much-needed revolution of these economies seem like a mirage. However, prospects for exploitation, renewable energy development, and dissemination abound especially with the aid of AI (artificial intelligence) in these emerging economies.

AI describes complex efforts dedicated to designing smart or intelligent contrivances, whereby this smartness or intelligence is the inherent ability of an object to carry out its duties appropriately and with foreknowledge in its environment (Nilsson 2010). AI techniques are paramount to the designing, analysis, and forecasting of the efficiency and control of renewable energy and its systems (Belu 2012). Besides this, AI and big data can effectively produce accurate power generation predictions making it possible and easy to integrate more renewable energy into energy grids (MIT 2014). By affording precise predictions of its demand and supply, AI techniques can increase the efficiency of systems operation, especially decentralized systems with bidirectional electricity flow, thereby increasing the complexities of power systems and ensuring that power grids always operate at optimal load and can furthermore optimize customer energy consumptions (IRENA 2019). In fact, AI can help forecast renewable energy generation and its demand with high precision such that it can be used as a tool to achieve economic load dispatch optimization and improved demand-side management and efficiency. Moreover, it allows for the effective creation of policies and gives insights into human motivations as regards renewable energy adoption, and consumer behavior can be tweaked to achieve an optimal energy system (Jucikas 2017). Furthermore, AI can help facilitate optimized energy storage operation by operating novel battery technologies in the most efficient way for the maximum integration of renewable electricity.

2 Applications of AI Techniques in Solar Energy

According to Kalogirou (Kalogirou 1996), AI can be broadly divided into two, namely, expert systems (ES) and artificial neural networks (ANNs). The former aids computers in making decisions through the interpretation of data (by preestablished rule systems) and selection amidst alternatives. On the other hand, ANN is an information-processing system modeled to the operations of the brain (Belu 2012). ANNs are not pre-programmed for specific tasks but are rather trained to identify patterns in data sets which are later used for predictive or classification purposes of new data sets.

AI has been widely employed in solar power systems in the past. Meteorological data which includes radiation of the sun, atmospheric temperature and its humidity, velocity of the wind, the clearness index, and the sunlight lifespan all serve as vital data needed for the smooth running of these systems. However, although Africa basks in an abundance of sunlight, these data may be unavailable consequent to the elevated expenses and intricacies of the tools required for their report, which most communities in African countries may not be able to afford. Furthermore, according to (Belu 2012), it's necessary that global sunlight irradiance incident on inclined surfaces be estimated since the inclination of most solar power systems is dictated by their installation location as well as the usage. Yet, these data are unavailable and at best can only be predicted and estimated. Due to these reasons, it becomes compelling, therefore, that models be formulated for the accurate prediction and estimation of these data (Fig. 1).

Elizondo et al. (1994) were part of the first pioneers of using the neural network to make forecasts of day-to-day sunlight radiation through the aid of easily accessed weather information and various atmospheric input variables as far back as 1994. This data was gotten from four sites in the southeastern parts of the USA. These sites were chosen due to the presence of long-term day-to-day weather data like solar radiation. Their data comprised of 23 years of ample weather information collection which were divided into 11 and 12 years for the training information collection and the testing information collection respectively. The inputs for their model include the daily highest and lowest air temperature and precipitation values, coupled with day-to-day computed data for the length of day and the sunlight radiation from a bright horizon. The model's accuracy was compared with the autonomous data collection by employing a root mean square (RMS) error ranging from 2.92 to 3.64 MJ/m^2 and a determination coefficient ranging from 0.52 to 0.74 for each year. The limitation of this approach however is its inability to be used to forecast

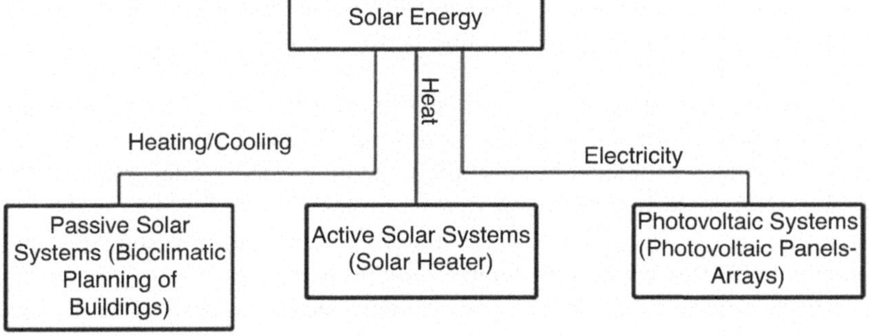

Fig. 1 Global solar irradiance applications. (Behrang et al. 2010)

day-to-day sunlight radiation when the readings of the daily highest and lowest values of the air temperature and precipitation become unavailable.

On the other hand, (Williams and Zazueta 1994) suggested employing feed-forward neural networks to evaluate everyday sunshine radiation whereby they employed for inputs various meteorological parameters like temperature, precipitation, radiation from a bright horizon, length of day, and the days of a year.

Moreover, (Williams and Zazueta 1996) employed input values longitude, latitude, height, and sunlight period after collecting their data from 41 stations, 31 of which were used for the network training and 10 for its testing. These measurements were obtained within the years 1998 and 2002 from Saudi Arabia, in Abha City. The results which were about 16.4% tell the feasibility of their approach for solar radiation spatial modeling.

Furthermore, (Mohandes et al. 1998) also employed ANNs to forecast sunlight radiation in a novel approach vis-a-vis establishing a model for evaluating the global radiation (GR) relationship with meteorological values, as well as forecasting this global sunlight radiation for sites uncovered while training the model which was then juxtaposed with the real values. The network inputs include site, months, average temperature, average wind velocity, average vapor pressure, average pressure, average relative humidity, and the average period of sunshine.

The ANN model forecasts sunlight radiation with an accuracy of 93% and an average absolute error of 7.3%, which depicts a considerable improvement to previous works done, as the results depict this new model as being superior to invisible data while also showcasing its capability to give accurate evaluations. A radial basis function (RBF) neural network was employed in forecasting data for daily solar radiation in Algeria (Alawi and Hinai 1998) (Fig. 2).

Applications of ANNs also include modeling diverse parts of the solar steam generator, for example, to calculate the intercept factor (proportion of receiver power absorption to incident power at concentrator aperture) using a disparity lesser than 0.4% when juxtaposed with the exceedingly intricate Energy DEPosition (EDEP) computer code estimation (Kemmoku et al. 1999). ANN has also been deployed to imbibe the profiles of radiation of angle measurements that experimentations might be done from and to forecast other angular measurements, the normal incidence angle inclusive (Kemmoku et al. 1999), which is hard to do due to the collector size. These predictions when juxtaposed alongside the real experiment readings give a satisfactory 3.2% disparity (Figs. 3, 4 and 5).

Furthermore, Kalogirou et al. (2005) report that the ANN has been employed for modeling the system start-up, as the system performance is drained through the energy dissipated during the morning starting-up. Analytic methods prove ineffective in solving such problems because the generator works under transient conditions. The ANN could forecast temperature profiles at various points in the system to an adequate 3.9%. The estimate of the infused energy in the heat-up duration may be computed effortlessly from the temperature profiles.

Kalogirou reported that an ANN was conditioned with respect to 30 specific scenarios of solar water heating systems, with the areas of collector ranging from 1.81 m^2 to 4.38 m^2. Data inputs included the area of the collector, coefficient of heat

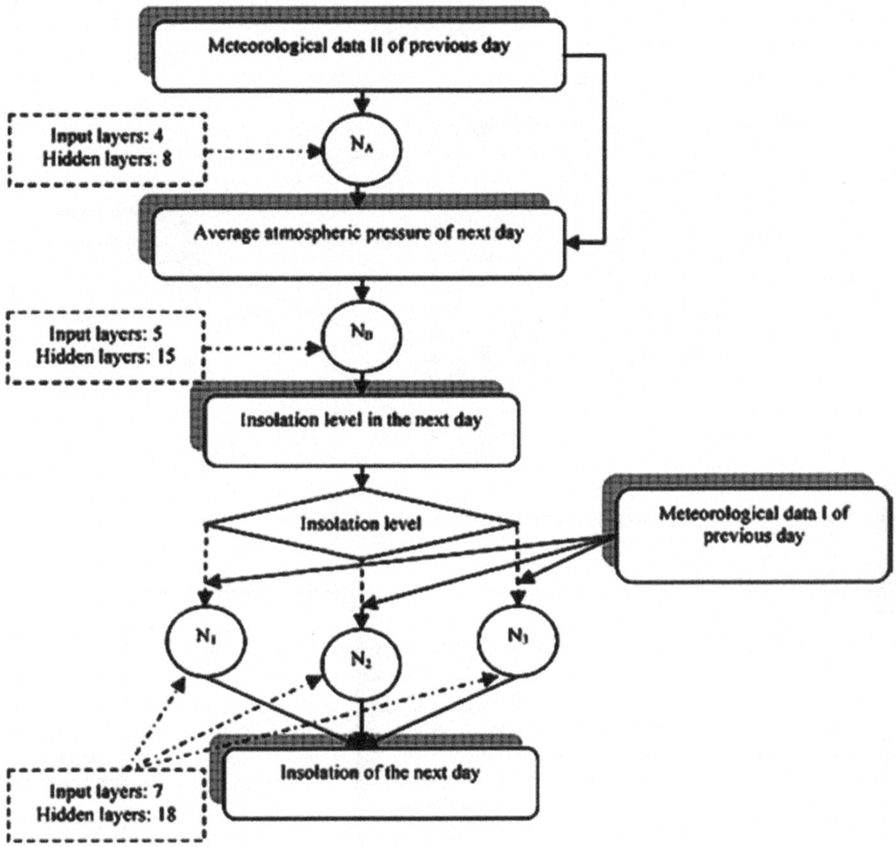

Fig. 2 Insolation forecast flowchart (while employing a multistage neural network). (Guessoum et al. 1988)

loss, storage tank capacity, tank type, and system type, as well as ten measurements from the sum total of the day-to-day sunlight radiation experiments, average atmospheric air temperature, and temperature of water in the tank early in the morning. The system's extracted energy and the water's temperature increase while in the tank were the network outputs (Table 1).

Forecast accuracy was probed by unknown data. Forecasts within 7.1% and 9.7% were achieved respectively (Kalogirou et al. 1996b) (Fig. 6).

These same systems were further tested and modeled according to the standard ISO 9459-2 outlined procedures in three regions in Greece so as to evaluate their performance in the long term. Data extracted from these for 27 systems were used to train and test the network, while data for the remaining three were employed for experimental validation. When unfamiliar data were employed in probing the forecasting accuracy, statistical R2 values equivalent to 0.9913 were realized for the first network and 0.9733 and 0.9940 for the second network (Table 2).

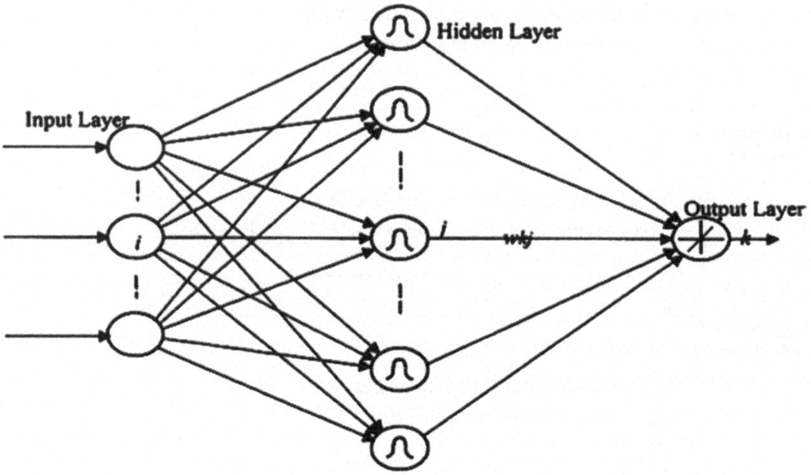

Fig. 3 A RBF neural network. (Kalogirou et al. 1996a)

Fig. 4 MLP architecture for clearness indexes forecasting. (Mohandes et al. 2003)

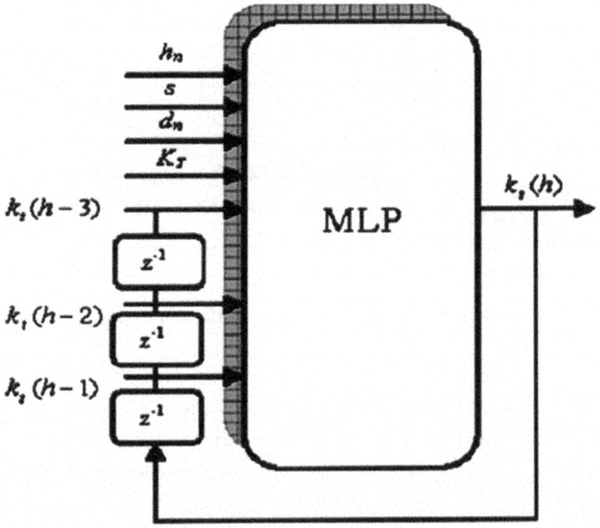

3 Techno-economic Feasibility of AI-Enabled Solar Systems in Africa

The economic competitiveness of solar PV systems, for example, in several African countries showed that even after including externality costs, the economics of PV systems applications (excluding the introduction of AI) are improbable to make for unsubsidized, extensive assimilation of this technology in the nearest future outside of remarkable technological breakthroughs (Khare et al. 2016). For example, the

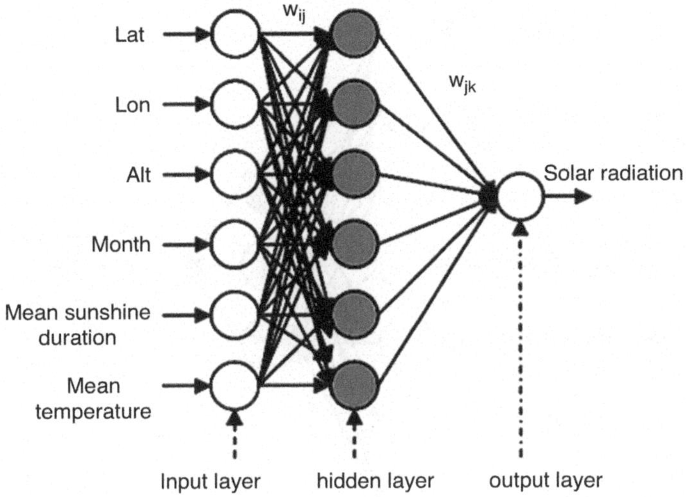

Fig. 5 Lone unseen-layer artificial neural network of sunlight radiation forecasting. (Hontoria et al. 2005)

Table 1 Diverse solar energy resources in Nigerian societies

Class of resources	Yield		Production	Household usage
Natural units	Unit of energy	Unit of energy (BTOE)		
Solar radiation	3.5 KWh/m^2/day to 7.0 KWh/m^2/day (4.2 million MWh/day using 0.1% of Nigeria's land area)	5.2 (40 years and 0.1% of Nigeria's land area)	Approximately 6 MWh/day solar photovoltaic	Approximately 6 MWh/day solar photovoltaic

Source: Alam et al. (2009)

economic viability of a self-contained PV system as juxtaposed to most common conventional alternative systems, i.e., diesel-powered systems, has been analyzed using sensitivity analysis for energy demand (Drennen et al. 1996). The analysis reveals that PV-powered systems are actually the lowest-cost alternative with daily energy demands reaching 15 kWh even under hostile economic conditions and reaching 68 kWh/day under friendly economic conditions (Levodo et al. 2015). The analysis's most important results, however, are the computation of the gross energy requirement (GER) of 1494 MJ/panel (0.65 m^2 surface) and of the global warming potential (GWP) of 80 kg of CO_2 panel. These perceived benefits of renewable energy make it necessary for AI techniques to be incorporated into it. This, however, shall require further research and development programs as the continent presently doesn't have the technological know-how to achieve this. The techno-economic assessment of these systems includes a high utilization rate of solar power generation; optimal load satisfaction; accurate prediction of power generation that makes

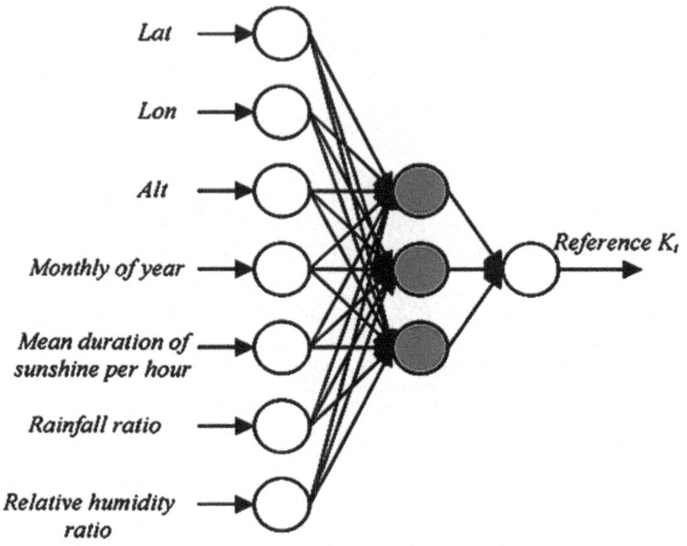

Fig. 6 ANN model employed in estimating beam solar radiation. (Kalogirou et al. 1999)

Table 2 Reviews of machine learning methods as applied to renewables

Literature	Years	Energy sources
Mellit et al.	2020	Solar
Bermejo et al.	2019	solar, hyHydropower and wind
Wang et al.	2019	Solar and wind
Mosavi et al.	2019	Solar, hydropower, and wind
Ahmed and Khalid.	2019	Solar and wind
Zendehboudi et al.	2018	Solar and wind
Das et al.	2018	Solar
Voyant et al.	2017	Solar
Pérez-Ortiz et al.	2016	Solar, wave, and wind
Khare et al.	2016	Solar and wind

Source: Ojosu (1990), Mellit et al. (2020), Bermejo et al. (2019), Wang et al. (2019), Mosavi et al. (2019), Ahmed and Khalid (2019), Zendehboudi et al. (2018), Das et al. (2018), Voyant et al. (2017), Pérez-Ortiz et al. (2016)

it possible to add more renewable energy to the energy grid (MIT 2014); maximum energy efficiency with minimum maintenance; and reliable power supply. It is an established fact that renewable energy sources are able to close the gap between power generation and demand (Kolhe et al. 2002) in a sustainable way such as not to compromise future energy and the environment (Babalola et al. 2019) (Figs. 7, 8 and 9).

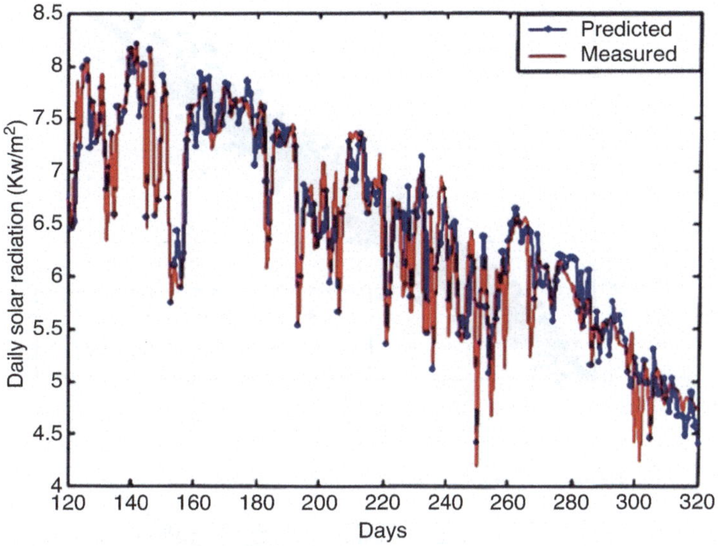

Fig. 7 Predicted sample of daily radiation. (Oyedepo and Babalola 2017)

4 Fulfillment of the Affordable and Clean Energy SDG: Challenges and Solution

The most critical issue hindering the expansion of the renewable energy industry in Africa is the inefficiency of solar cells which has consequently led to their extremely high cost as which translates to the purchase and use of more solar cells to meet the target energy demand. This challenge has prevented the industry from helping Africa fulfill its SDG of affordable and clean energy. However, new ways to make solar energy more affordable have now been discovered. One is to improve solar panel efficiency. If efficiency is increased, this will translate to the installation of fewer panels, employment of lesser labor, use of lesser land, and purchase of lesser hardware. Tandem solar cells manufactured from multiple light-absorbing materials (like silicon and the novel metal halide perovskite) can help increase the theoretical efficiency of solar cells to about 46%. Also, scalable distributed energy systems designed with battery cubes that comprise tandem cells can be used to power a whole village. Furthermore, AI can help achieve optimized energy consumption and storage by triggering electrical appliances to be switched off when electricity becomes costly or triggering solar batteries to store electricity when power becomes affordable or solar energy becomes abundant. All these will make solar energy more accessible and affordable and help Africa achieve its SDG.

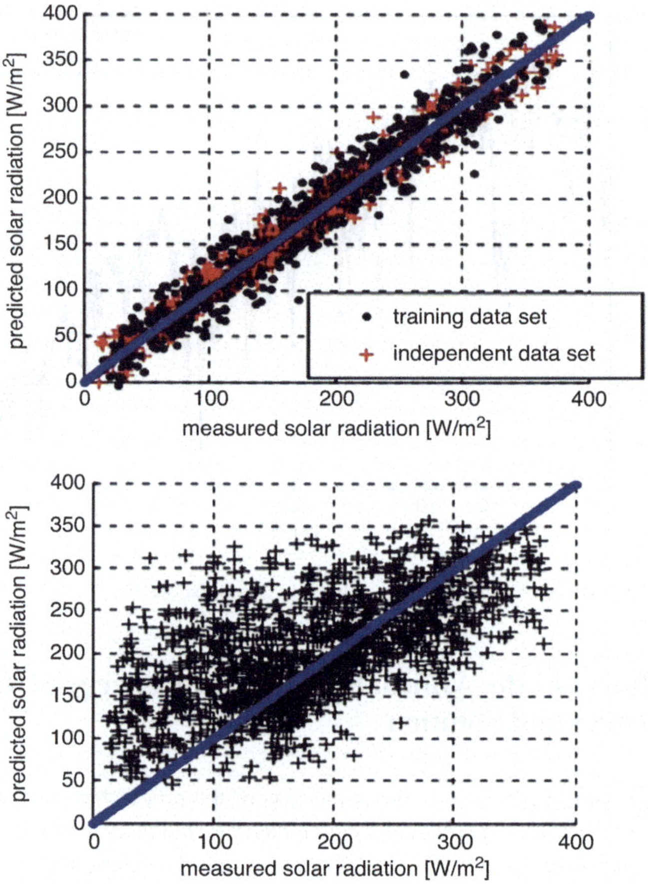

Fig. 8 Daily average values of the predicted versus the global solar radiation ground measured on horizontal surface employing the ARPS model (left panel) and the ANN-based model (right panel). (Iqdour and Zeronal 2006)

5 Recommendations

Possible future research endeavors for AI models in renewable energy forecasts might include other kinds of renewable energy forecasts, like wave energy, tidal energy, hydraulic power, biomass, and geothermal energy. Secondly, machine learning techniques and hybrid models could turn out to be auspicious ways in renewable energy predictions, and research could be carried out in those areas. Thirdly, data preprocessing methods and AI model analysis in renewable energy forecasts are probably another future research area, since data preprocessing techniques do influence forecasting performances of AI models, and not much research has been carried out yet along that line. Finally, considering that parameter selection influences the performance of AI models in renewable energy forecasts, employing new

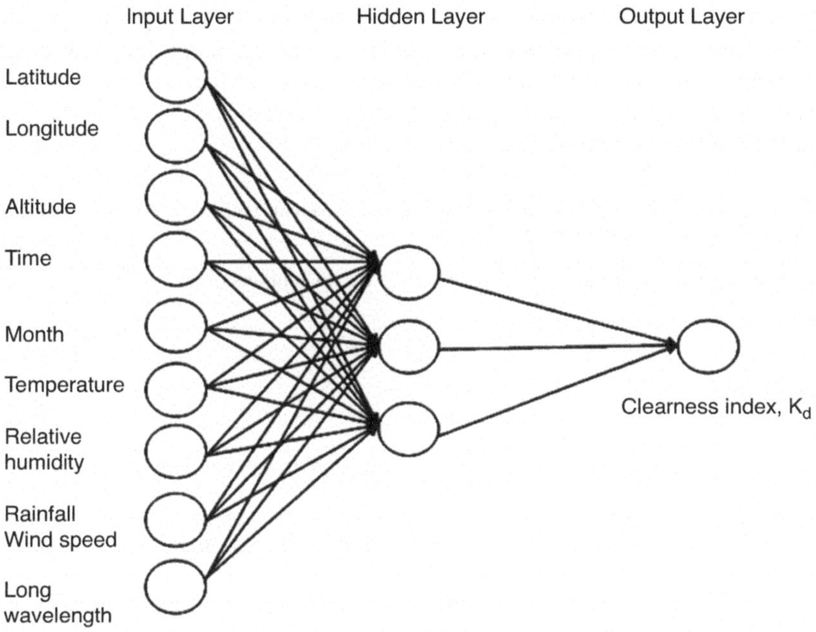

Fig. 9 ANN architecture for diffuse solar radiation forecasts. (Kalogirou et al. 1999)

metaheuristics, for example, a coronavirus optimization algorithm (Kratzenberg et al. 2008), for AI parameter selection to improve the performances of AI models in renewable energy forecasts is an encouraging opportunity for research in the nearest future.

6 Conclusions

The absence of stable power, increasing greenhouse gas emission, and consequent global warming have made it necessary for Africa to diversify its energy production and generation into renewable energy. It becomes necessary therefore that there be procedures and systems designed for accurate forecasting of renewable energy power, and much work has been done in this line, especially since the complexity of several environmental conditions in renewable energy generation systems led to the unsuitability of using confined mathematical forms to describe them. This has led to AI techniques like expert systems, ANNs, and machine learning models finding applications in a variety of fields for modeling and forecasting in renewable energy systems. Some of those applications have been presented here, but they are not exhaustive. If Africa will achieve an AI-enabled renewable energy future, it must employ a transdisciplinary approach in its research. This must be developed from an indigenous perspective with regard to the knowledge, values, and beliefs of the natives while giving cognizance to how these cultures and value systems are evolving. Furthermore, the importance of a visionary government and favorable policies in bringing this dream to life cannot be overemphasized. There must also be ample relevant data available and protection and effective collaborations with other stakeholders inside and outside Africa.

Acknowledgments We appreciate the support of the Covenant University Centre for Research, Innovation and Discovery (CUCRID).

Conflicts of Interest The authors declare that there are no conflicts of interest.

References

Kaygusuz, K., 2012. Energy for sustainable development: A case of developing countries. Renew. Sustainable Energy Rev. 16(2),1116–1126.

Scarlat, N., Motola, V., Dallemand, J. F., Monforti-Ferrario, F. and Mofor, L., 2015. Evaluation of energy potential of municipal solid waste from African urban areas. Renew. Sustainable Energy Rev. 50, 1269–1286.

Karekezi, S., 2002a. Poverty and energy in Africa—a brief review. Energy Policy. 30(11–12), 915–919.

Wolde-Rufael, Y., 2005. Energy demand and economic growth: the African experience. J. Policy Model. 27(8), 891–903.

Arndt, C., Arent, D., Hartley, F., Merven, B., and Mondal, H. A., 2019. 'Faster Than You Think: Renewable Energy and African Countries' vol. 11:149–168 (Volume publication date October 2019) First published as a Review in Advance on June 10, 2019

Karekezi, S., 2002b. 'Renewables in Africa – Meeting the Energy Needs of the Poor', Energy Policy, Vol. 30 Nos. 11–12, Special Issue – Africa: Improving Modern Energy Services for the Poor. Oxford: Elsevier Science Limited

Karekezi, S., and Ranja, T., 1997a. Renewable Energy Technologies in Africa. ZED Books and AFREPREN. Oxford UK.

Levodo, J., Ford, A., and Chaer, I., 2015. 'Implementation-of-Renewable-Energy-Technologies--in-Rural-Africa-j' https://doi.org/10.5281/zenodo.1100104

Ellabban, O., Abu-Rub, H. and Blaabjerg, F., 2014. Renewable energy resources: Current status, future prospects and their enabling technology. Renew. Sustainable Energy Rev. 39, 748–764.

Karekezi, S., and Ranja, T., 1997b. Renewable Energy Technologies in Africa. ZED Books and AFREPREN. Oxford UK.

Nilsson, Nils J., 2010. The Quest for Artificial Intelligence: A History of Ideas and Achievements (Cambridge, UK: Cambridge University Press, 2010)

Belu, R., 2012. Artificial intelligence techniques for solar energy and photovoltaic applications, pg 376–382

MIT., 2014. Smart wind and solar power. MIT Technology Review, www.technologyreview.com/s/526541/smartwind-and-solar-power

IRENA (2019), Innovation landscape brief: Artificial intelligence and big data, International Renewable Energy Agency, Abu Dhabi.

Jucikas, T., 2017. "Artificial intelligence and the future of energy", Medium, https://medium.com/wepower/artificial-intelligence-and-the-future-ofenergy-105ac6053de4

Kalogirou S.A., 1996. Proc. of Eurosun'96, Freinberg, Germany, pp. 470–475.

Behrang, M. A., Assareh, E., Ghanbarzadeh, A., and Noghrehabadi, A. R., 2010. The potential of different artificial neural network (ANN) techniques in daily global solar radiation modeling based on meteorological data. Solar Energy, 84(8), 1468–1480. https://doi.org/10.1016/j.solener.2010.05.009

Elizondo, D., Hoogenboom, G., and Mcclendon, R. W., 1994. Development of a neural network model to predict daily solar radiation. Agricultural and Forest Meteorology, 71(1–2), 115–132. https://doi.org/10.1016/0168-1923(94)90103-1

Williams, D. B., and Zazueta, F. S., 1994. Solar radiation estimation via neural network. 6th International Conference on Computers in Agriculture, Cancun, Mexico, (pp. 140–146).

Williams, D. B., and Zazueta, F. S. (1996). Solar radiation estimation via neural network. 6th International Conference on Radiation, Computers in Agriculture, Cancun, Mexico (pp. 1143–1149).

Mohandes, M., Rehman, S., and Halawani, T. O., 1998. Estimation of global solar radiation using artificial neural networks. Renewable Energy, 14(1–4), 179–184. https://doi.org/10.1016/S09601481(98)00065-2

Alawi, S. M., and Hinai, H. A., 1998. An ANNbased approach for predicting global radiation in locations with no direct measurement instrumentation. Renewable Energy, 14(1-4), 199–204. https://doi.org/10.1016/S0960-1481(98)00068-8

Guessoum, A., Boubkeur, S., and Maafi, A., 1988. A global irradiation model using radial basis function neural network. World Renewable Energy Congress (WREC 1988), UK, (pp. 332–336).

Kemmoku, Y., Orita, S., Nakagawa, S., and Sakakibara, T., 1999. Daily insolation forecasting using a multi-stage neural network. Solar Energy, 66(3), 193–199. https://doi.org/10.1016/S0038-092X(99)00017-1

Kalogirou S.A., Neocleous C.C. and Schizas C.N., 1996a. Proc. of Int. Conf. EANN'96, London, UK, pp. 5-8

Mohandes, M., Balghonaim, A., Kassas, M., Rehman, S., and Halawani, T. O., 2003. Use of radial basis functions for estimating monthly mean daily solar radiation. Solar Energy, 68(2), 161–168. https://doi.org/10.1016/S0038-092X(99)0007

Hontoria, L., Aguilera, J., and Zuria, P., 2005. A new approach for sizing stand-alone photovoltaic systems based on neural networks. Solar Energy, 78, 313–319. https://doi.org/10.1016/j.solener.2004.08.018

Adnan, S., Arcaklýog¢lub E., Ozalpa, M., and Agclarc, N. C., 2005. Forecasting based on neural network approach of solar potential in Turkey. Renewable Energy, 30(7), 1075–1090. https://doi.org/10.1016/j.renene.2004.09.02

Kalogirou S.A., Neocleous C.C. and Schizas C.N., 1996b. Proc. of Int. Conf. EANN'96, London, UK, pp. 1–4

Kalogirou S.A., Panteliou S. and Dentsoras A., 1999. Solar Energy, 65, 335

Alam, S., Kaushik, S. C., and Garg, S. N., 2009. Assessment of diffuse solar energy under general sky condition using artificial neural network. Applied Energy, 86(3), 554–564. https://doi.org/10.1016/j.apenergy.2008.09.004

Ojosu, J., 1990. The iso-radiation map for Nigeria. Solar Wind Technol. 7, 563–75.

Mellit, A., Massi Pavan, A., Ogliari, E., Leva, S., and Lughi, V., 2020. Advanced methods for photovoltaic output power forecasting: A review. Appl. Sci. 2020, 10, 487.

Bermejo, J.F., Fernández, J.F.G., Polo, F.O., and Márquez, A.C., 2019. A review of the use of artificial neural network models for energy and reliability prediction. a study of the solar PV, hydraulic and wind energy sources. Appl. Sci. 2019, 9, 1844.

Wang, H.Z., Lei, Z.X., and Zhang, X., 2019. A review of deep learning for renewable energy forecasting. Energy Convers. Manag. 2019, 198, 111799.

Mosavi, A., Salimi, M., Ardabili, S.F., Rabczuk, T., Shamshirband, S., and Varkonyi-Koczy, A.R., 2019. State of the art of machine learning models in energy systems, a systematic review. Energies 2019, 12, 1301.

Ahmed, A., and Khalid, M., 2019. A review on the selected applications of forecasting models in renewable power systems. Renew. Sustain. Energy Rev. 2019, 100, 9– 21.

Zendehboudi, A., Baseer, M.A., and Saidur, R., 2018. Application of support vector machine models for forecasting solar and wind energy resources: A review. J. Clean. Prod. 2018, 199, 272–285.

Das, U.K., Tey, K.S., Seyedmahmoudian, M., Mekhilef, S., Idris, M.Y.I., Deventer, W.V., Horan, B., and Stojcevski, A., 2018. Forecasting of photovoltaic power generation and model optimization: A review. Renew. Sustain. Energy Rev. 2018, 81, 912–928.

Voyant, C., Notton, G., Kalogirou, S., Nivet, M.L., Paoli, C., Motte, F., and Fouilloy, A., 2017. Machine learning methods for solar radiation forecasting: A review. Renew. Energy 2017, 105, 569–582.

Pérez-Ortiz, M., Jiménez-Fernández, S., Gutiérrez, P.A., Alexandre, E., Hervás-Martínez, C., and Salcedo-Sanz, S., 2016. A review of classification problems and algorithms in renewable energy applications. Energies 2016, 9, 607.

Khare, V., Nema, S., and Baredar, P., 2016. Solar–wind hybrid renewable energy system: A review. Renew. Sustain. Energy Rev 2016, 58, 23–33.

Drennen, T. E., Erickson J. D., and Chapman, D., 1996. Solar Power and Climate Change Policy in developing Countries," Energy Policy, Vol. 24, No. 1, 1996, pp. 9–16. https://doi.org/10.1016/0301-4215(95)00117-4

Kolhe, M., Kolhea S., and Joshi, J. C., 2002. Economic Viability of Stand Alone Solar Photovoltaic System in Comparison with Diesel-Powered System for India, Energy Economics, Vol. 24, No. 2, 2002, pp. 155165. https://doi.org/10.4236/sgre.2012.34037

Babalola, P.O., Bolu, C.A., Inegbenebor, A.O., Oyedepo, S.O., Kilanko, O. and Adeyemi, G.A., 2019. Application of solar photovoltaic system to power air blower and mixing mechanism in a tilting furnace, World Review of Science, Technology and Sustainable Development, Vol. 15, No. 2, pp. 160–169

Oyedepo, S.O and Babalola, P.O., 2017. Renewable Energy Technologies in Nigeria: Challenges and Opportunities for Sustainable Development. In S.A. Kale (Eds.), Renewable Energy Systems, pp. 226, Nova Science Publisher, New York

Iqdour, R., and Zeronal, A., 2006. The multi layered perceptrons neural networks for the prediction of daily solar radiation. International Journal of Signal Process, 3(1), 24–29.

Kratzenberg, M. G., Colle, S., and Beyer, H. G., 2008. Solar radiation prediction based on the combination of a numerical weather prediction model and a time series prediction model. EURPSUN 2008, Portugal Lisbon, (pp. 1–12)

A Procedure Model for Developing Gerontechnological Solutions to Achieve Demographic Sustainability in Aging Society

Rongbo Hu, Thomas Bock, Yuan Lu, and Thomas Linner

1 Introduction

Population aging is arguably one of the greatest challenges facing human society today. In addition to being a serious challenge in wealthy nations, it poses a rigorous threat to emerging economies. As of today, there are more than 8 billion individuals living on this planet, of which 9.6% (771 million) are over 65 years old. Globally, the population aged 65 years or over is the fastest-growing age group, whose proportion increased from 6.9% in 2000 to 10% in 2022 and is projected to reach 16% by 2050 (United Nations 2022).

The World Health Organization (WHO) refers societies where older population aged 65 years and over takes up 7%, 14%, and 21% of the total population as "aging society," "aged society," and "super-aged society," respectively. The general trend is that the higher the income of a country, the more serious the aging problem. Specifically, the situation of population aging varies drastically in different countries. As a result, the challenge of population aging is especially prominent in the most developed countries. In Germany, for example, the current percentage of

R. Hu (✉) · T. Bock
Chair of Building Realization and Robotics, Technical University of Munich, Munich, Germany
e-mail: rongbo.hu@tum.de; thomas.bock@tum.de

Y. Lu
Department of Industrial Design, Eindhoven University of Technology, Eindhoven, Netherlands
e-mail: Y.Lu@tue.nl

T. Linner
Faculty of Civil Engineering, Ostbayerische Technische Hochschule Regensburg, Regensburg, Germany
e-mail: thomas.linner@oth-regensburg.de

I. S. Dunmade et al. (eds.), *Sustainable Engineering*, Green Energy and Technology, https://doi.org/10.1007/978-3-031-47215-2_26

population aged 65 and over has exceeded 21.7% (i.e., 18.2 million), which already met the standard of "super-aged society." Japan is another prominent example of "super-aged society" that faces an even more serious aging situation than Germany. Due to a high life expectancy and a low fertility rate, the percentage of the Japanese population aged 65 and over reached 28.4% in 2020 and will continue to increase to 37.7% in 2050 (United Nations 2019). As the population's income increases, population aging also becomes a rigorous challenge in emerging economies such as China and Africa. In China, the population is currently experiencing a rapid, accelerated aging process due to several reasons such as improved medical standards, economic pressures, and decades-long family planning policies. By the end of 2020, Chinese older adults aged 65 or over reached 191 million, accounting for 13.5% of the total population, which was nearly equal to the total population of Germany and Japan combined (National Bureau of Statistics 2021). Moreover, it is anticipated to spend the shortest period (i.e., 33 years) to transfer from "aging society" to "super-aged society" compared to any major developed countries in the world, which gives it much less preparation time to brace for social impacts caused by population aging (Chen et al. 2019). Even in Africa, although the median age of the continent's population is only 19.7 years as of 2020, the number is projected to be 34.9 by 2100. Accordingly, Africa's percentage of population aged 65 years or over will increase to 13.9% by 2100 (United Nations 2019), which will essentially meet the definition of aged society as well.

Population aging has many consequences for society, such as increased chance of getting chronic noncommunicable diseases among older adults, growing healthcare expenditures especially on long-term care (de Meijer et al. 2013), labor shortages, and slowing economic growth (Bloom et al. 2011). Therefore, novel approaches are needed to address these consequences. This chapter mainly focuses on proposing a methodological framework as well as technological solution to address the population aging phenomenon and its consequences, eventually contributing to demographic sustainability.

2 Methods

There are many types of methods for continuous product improvement which is also known as "kaizen," meaning change for the better in Japanese (Imai 1986), such as Kanban (Powell 2018), Six Sigma (Antony 2006), Look-Ask-Model-Discuss-Act (LAMDA) (Tortorella et al. 2015), and Plan-Do-Check-Act (PDCA), among which the PDCA method is arguably one of the most known and applied one worldwide (Realyvásquez-Vargas et al. 2018).

The PDCA Cycle (also referred to as the Deming Cycle) is an iterative process management model for the continuous development and improvement of products, which is widely popular for its originality, simplicity, and practicality. It was first introduced to the Japanese by W. E. Deming in the 1950s in order to improve product quality and satisfy customers, which was further developed by the Japanese into a systematic tool for continuous product development called the PDCA model (Imai 1986), which was also adopted by the Western industry in the 1980s (Nilsson-Witell et al. 2005). The model consists of four phases: Plan, Do, Check, and Act. The phases are worded slightly differently in different literature but have the same purpose which is to continuously improve (Lodgaard et al. 2013). Generally speaking, the Plan phase includes identifying and analyzing the problem; the Do phase includes the development and implementation of the solution; the Check phase includes the evaluation of the result to check whether the goal is achieved; and the Act phase includes activities such as standardization and knowledge transfer (Johnson 2002).

The PDCA model was also adopted in the development process of automated building technology systems. For example, mainly based on the consultancy project investigating the potential of implementing robotics and automation in the context of large-scale housing development for Hong Kong as well as several other research projects on developing construction robots, Linner et al. proposed a PDCA-based technology management system for the development of single-task construction robots (STCRs) based on the analysis of several exemplary research and development (R&D) projects in that field (Linner et al. 2022). The Ambient Rehabilitation Kit reported in this chapter can also be considered a special type of robotic system due to their common purpose of supplementing human labor with automation technology. Therefore, it is reasonable to apply this methodology in the development process of ambient health-promoting system as well.

As shown in Fig. 1, the basic structure of this methodology is based on the PDCA model. Flexible starting point and continuous improvement are important for developing gerontechnology products because it is difficult to immediately specify the requirements, system architecture, work process, and business strategy. Therefore, an iterative and cyclic approach is appropriate. The primary goal of the proposed procedure model is the continuous evolution of the system architecture and the technical readiness levels (TRLs) of its subsystems (Layer 1). The core elements around the center are the four primary development phases based on the V-Model, which are requirements engineering, development, performance evaluation, and implementation (Layer 2). Guided by the Deming Cycle (Layer 3), each phase consists of various modular steps belonging to this phase such as specific activities and tasks (Layer 4). The non-inclusive activities and tasks listed in Layer 4 essentially constitute a modular toolkit to complete the system development goal in several iterations, some of which can be repeated in different cycles to fit different contexts and regions.

3 Results

The results of this chapter are based on the case study of a large European research project entitled REACH (Bock 2017), which implemented this procedure model in its development cycle.

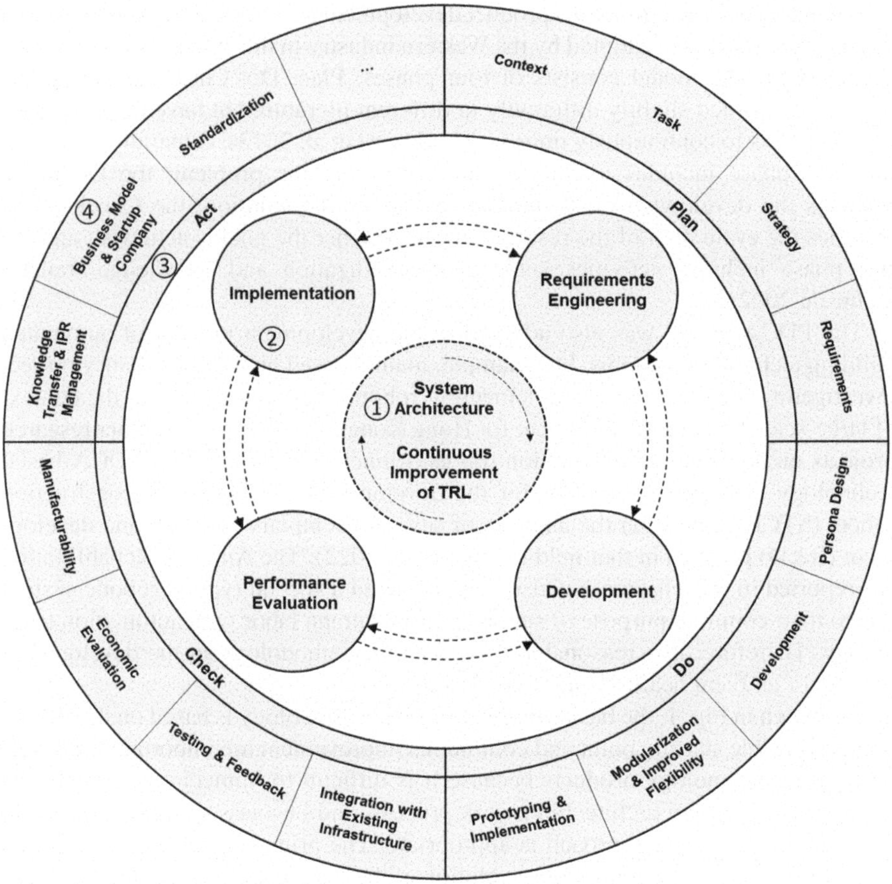

Fig. 1 The procedure model for developing gerontechnology systems. (Adapted from Linner et al. 2019)

3.1 State-of-the-Art Analysis and Shortcomings (Plan)

The original ideas of assistive home environment for aging society can date back to the 1980s where co-author T. Bock developed the life support system (LSS) for older adults, which was sponsored by the Japan Science Society. The concept depicted visionary functions such as assisted dressing, smart home control, automated body turnover, telemedicine, and emergency evacuation (see Fig. 2).

In recent years, due to the pressure from social challenges such as aging societies, limited resources, and a continuously increasing demand for productivity and efficiency, the topics of assistive smart furniture and smart homes have attracted more attention.

It has been demonstrated that smart furniture to a limited extent can integrate sensory and intervention functionality into the living environment. Notable

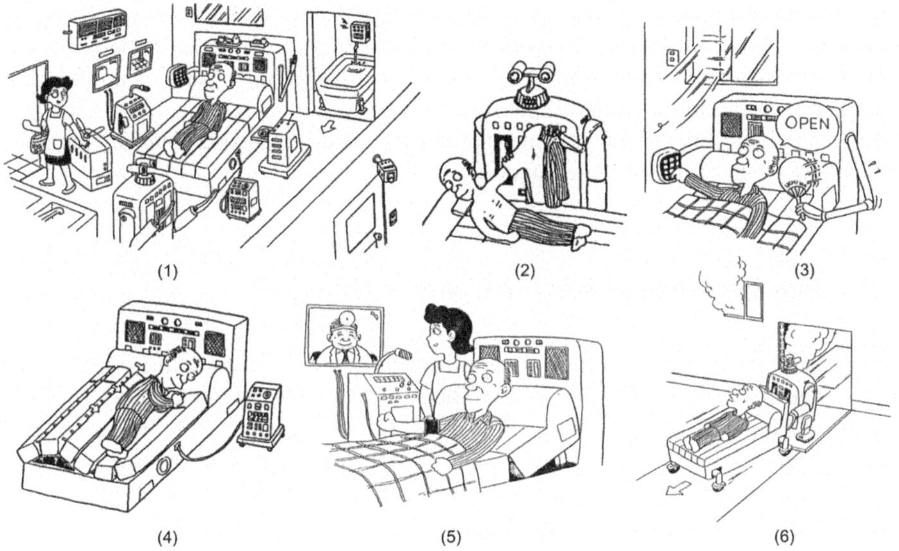

Fig. 2 The first concept of life support systems for aging society in Japan. (Image: T. Bock)

examples are smart chair (Erdt et al. 2012), smart bathroom (Manoel et al. 2017), smart kitchen (Beetz et al. 2008), and smart bed (Spillman et al. 2004). An obvious shortcoming of these furniture schemes is that the majority are stand-alone devices that are not integrated into a larger system or are incapable of monitoring the entire area where they are positioned. An additional weakness is the lack of modularity and the limited amount of functionality the systems provide. In the meantime, researchers have also proposed several intelligent sensing solutions, including the Aware Home (Kidd et al. 1999) and the RoboticRoom (Sato et al. 2004), both of which are ambitious concepts of smart environments using intelligent robots and sensor networks. However, they lack the capacity to be integrated into a regular home or care environment due to the high cost, the high complexity, and the need for fundamental changes to the existing living environment. Recently, researchers have been focused on transforming the built environment into smart assistive environment for older adults. However, many of these attempts employed intrusive sensory and motivation functions into the built environment to support the monitoring and care to improve older adults' health and independence (Sakamura 1996; Intille 2002; Sato et al. 2004; Pyo et al. 2014).

In order to address the shortcomings of the state-of-the-art gerontechnological systems, a series of research projects, including but not limited to GEWOS (Linner and Schulz 2015), PASSAge (Bock et al. 2015), USA^2 (Linner et al. 2016), LISA Habitec (Güttler et al. 2017), and BaltSe@nior (Langosch et al. 2019), were carried out by the Chair of Building Realization and Robotics at the Technical University of Munich (TUM), tackling various aspects of gerontechnology such as exercise, mobility, working, activities of daily living (ADLs), and fall detection respectively. However, there are various shortcomings still existing in the state-of-the-art assistive technology as well as latest research projects, including a lack of modularity, compatibility, flexibility, rehabilitation functionality, fall prevention, financial feasibility, and scalability outside Europe. Therefore, a comprehensive solution that goes beyond the state of the art is demanded.

3.2 *Touchpoints and Engine Concept (Plan)*

The research team proposed the Touchpoints and Engine concept as the comprehensive system architecture for the project. Based on this concept, the product-service-system architecture of REACH is divided into six manageable research and development clusters: four "Touchpoint" clusters that represent tangible connections between the REACH system and users such as older adults and caregivers; one "Engine" cluster, which is a cloud-based digital platform serving as the brain of the project; and one "Interface" cluster, which comprises a set of means that allow the Touchpoints and other products and services to interact with the Engine (see Fig. 3). Each cluster is associated with a dedicated and independent development team coming from the project consortium members. The work presented in this chapter is mainly related to Touchpoint 2 – Active Environment.

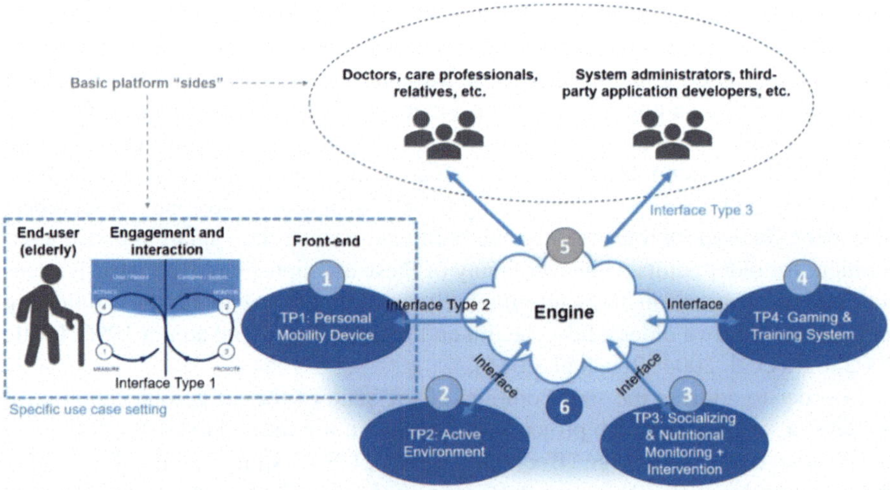

Fig. 3 "Touchpoints and Engine" concept in the REACH project

3.3 Determining Users' Requirements (Plan)

In order to achieve a successful development process, it is necessary to conduct requirements analysis before the design process starts. A description of the target user including the user environment and an analysis of the entities associated with the user were performed. Various methods were applied to describe and analyze these key components of the REACH system, including describing use cases, defining personas, creating experience maps, and analyzing stakeholders (Schäpers et al. 2017). Subsequently, the major hypotheses of four Touchpoints were concluded based on these analyses. For example, the hypothesis of Touchpoint 2 is formulated as follows: the REACH system based on smart furniture elements with sensing systems enables the patients to reduce the duration of their hospitalization, reduce decline after discharge, reduce risk of readmission, and be able to perform their ADLs with reduced support from professional caregivers.

Based on the result of the requirements engineering, the extracted raw requirements were further formalized and assigned to specific requirements categories (e.g., health outcome requirement, regulatory requirement, stakeholder requirement, etc.). All outcomes of the activities in the requirements engineering (e.g., stakeholder analysis, co-creation workshop, analysis of best practice, business strategy, motivation analysis, data management, ethics study, intellectual property management, etc.) were systematically summarized, and the key points were extracted and translated into raw requirements.

3.4 Ambient Rehabilitation Kit: An Unobtrusive Interdisciplinary Approach to Achieve Demographic Sustainability (Do)

The relevant requirements of the system can be materialized by a series of Personalized Intelligent Interior Units (PI^2Us). The PI^2U is a special type of smart furniture that seamlessly integrates the required concepts and functionality into the different use case settings. This chapter focuses on a series of PI^2Us developed by the project team in Touchpoint 2 (i.e., the "Active Environment" work group of the project). The relevant PI^2Us include SilverArc, MiniArc, SilverBed, iStander, and ActivLife. The term "Silver" implies the color of older adults' hair, which is in line with the booming silver economy.

The SilverArc is a multifunctional device developed for the use in a large kitchen or dining space (e.g., a community kitchen). The dimensions can be easily adapted due to the telescopic design. It offers an interactive projection area in the kitchen, where recipes and training programs can be displayed. It also has a foldaway projection area where an elderly-friendly graphical user interface (GUI) can be displayed. The MiniArc can be considered as a flexible and smaller variant of the SilverArc, which is meant to assist in the training and moving of older adults who are

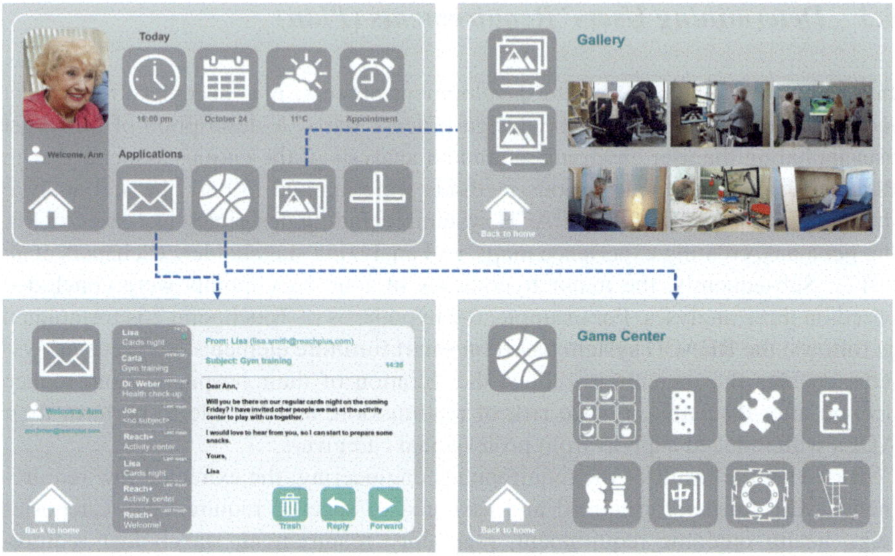

Fig. 4 Design of the GUI's home screen and second screens

hospitalized or reside in smaller apartments. This prototype was fitted with wheels and is thus mobile. The philosophy of inclusive design was also considered so that a user in wheelchair can easily push the wheelchair in between the wheels (i.e., 895 mm). An ultrashort projector can project the GUI on its foldaway table or on a separate table as needed. Meanwhile, a motion-sensing camera (i.e., Microsoft Kinect) is integrated to detect the user's gestures, enabling the interactive gesture control and gaming function. The major features of the GUI for both SilverArc and MiniArc include calendar, weather, appointment reminder, email, game center, and photo gallery (see Fig. 4). In addition, an infrared camera can detect user's respiratory rate using machine learning.

SilverBed is a carpentry-based modular bed incorporating Sara Combilizer that assists older adults to move autonomously to a sitting, standing, or supine position in a comfortable and safe manner. Physical exercise is offered in combination with entertainment, motivating its users to become more active. More importantly, health functions can be also integrated such as vital signs and skin pressure monitoring.

ActivLife is equipped with a mechanism to assist the user to stand up and to perform motor exercises of the ankles, knees, and hip joints. It also allows the user to maintain a safe, upright standing position and perform physical-mental serious gaming and balance exercises through the motion-sensing TV component.

Furthermore, based on the PI[2]Us, a modularized smart home solution, namely, Ambient Rehabilitation Kit, was proposed, integrating the smart furniture and key technologies in REACH to create a complete interior living and care environment (also known as the smart infill system in open building) for older adults in different

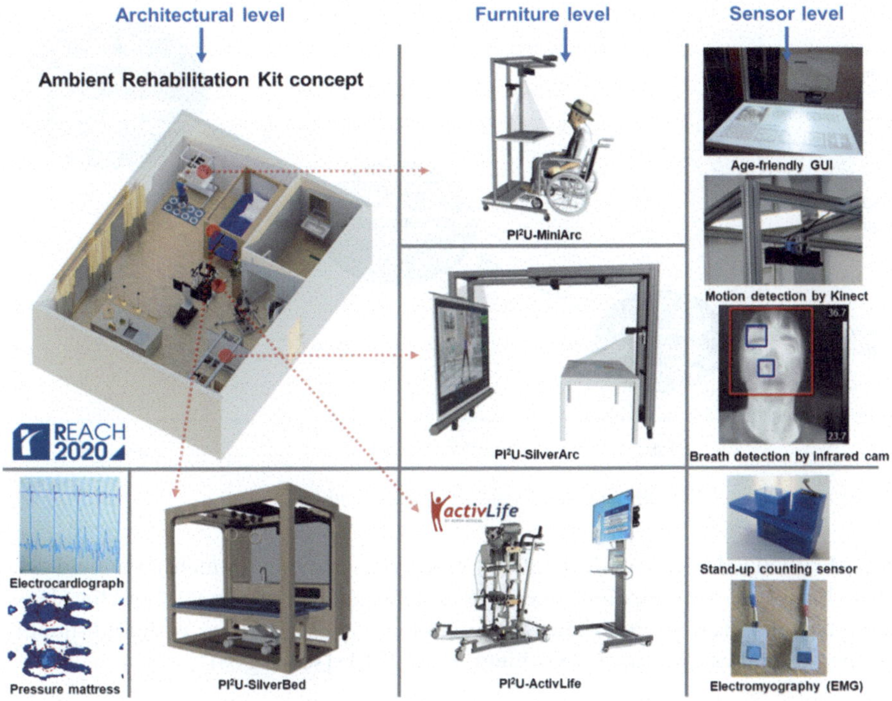

Fig. 5 Ambient Rehabilitation Kit concept

living environments such as home, hospital, and community in a flexible and adaptable manner (see Fig. 5).

Thereafter, all the data collected via a variety of sensors integrated into the PI²Us (e.g., electrocardiography, thermal camera for breath detection, body pressure mapping system, stand-up counter, etc.) will be transferred, exchanged, and stored securely via the CARP platform, which is a set of open-source software components and frameworks developed by project partner Technical University of Denmark (DTU). The platform enables the design and development of mobile health applications for digital phenotyping research (https://carp.cachet.dk/). The PI²Us, the distributed intelligent home concept, and the CARP platform were tested and validated in several exhibitions and tests across Europe later on (Hu et al. 2020).

3.5 *Strategy for Testing and Exhibition (Check)*

After the development of the various PI²U prototypes, a series of testing and exhibition activities were carried out to verify the functionality, operability, and usability of the PI²Us. These activities followed a “lab-field-showroom” approach that is also an important component of the Ambient Rehabilitation Kit where the lab

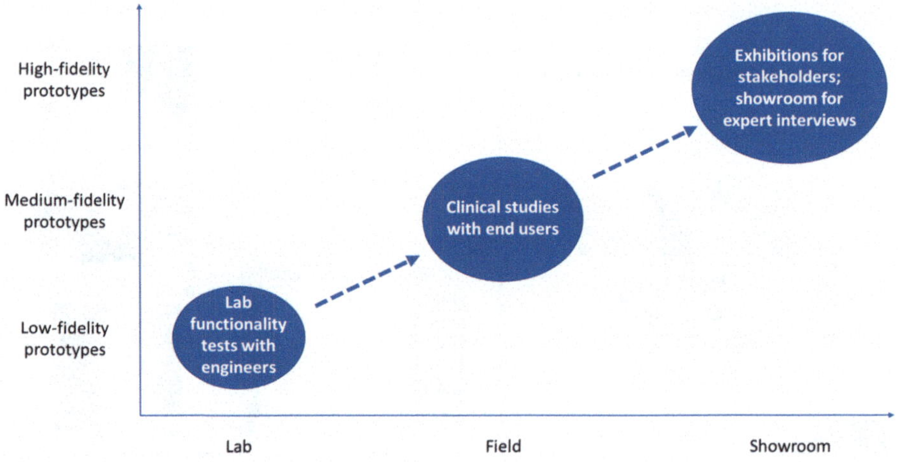

Fig. 6 "Lab-field-showroom" strategy for testing and exhibitions

decontextualized (i.e., with low-fidelity prototypes and environment), the field contextualized (i.e., with medium-fidelity prototypes and environment), and the showroom enabled users to experience the tangible technologies (i.e., with high-fidelity prototypes and environment) (Koskinen et al. 2013) (see Fig. 6).

3.5.1 Main Sensing, Monitoring, and Analysis Functionality Testing by Engineers ("Lab")

As mentioned in the previous sections, a series of PI^2Us are developed in Touchpoint 2. The PI^2U prototypes are manufactured, deployed, and tested for main sensing, monitoring, and analysis activities in the laboratory (see Fig. 7). The qualitative functionality testing of these activities is presented as follows, which clearly demonstrates how various sensors and functions can be integrated into the PI^2Us due to their modularity and customizability.

3.5.2 Clinical Studies Using PI^2Us in Touchpoint 2 ("Field")

After the prototypes of PI^2Us were built, the evaluation was conducted in various care facilities with partners. In the REACH project, more than 30 testing activities including tests, studies, and clinical trials were conducted. All trials in the project were in conformity with ethical principles set by the Declaration of Helsinki. Data were collected and processed following the legal requirements for data protection. The testing activities demonstrated that the PI^2Us are safe and useful for the rehabilitation and independence of older adults (Steinböck et al. 2019; Lu et al. 2020; Randriambelonoro et al. 2020).

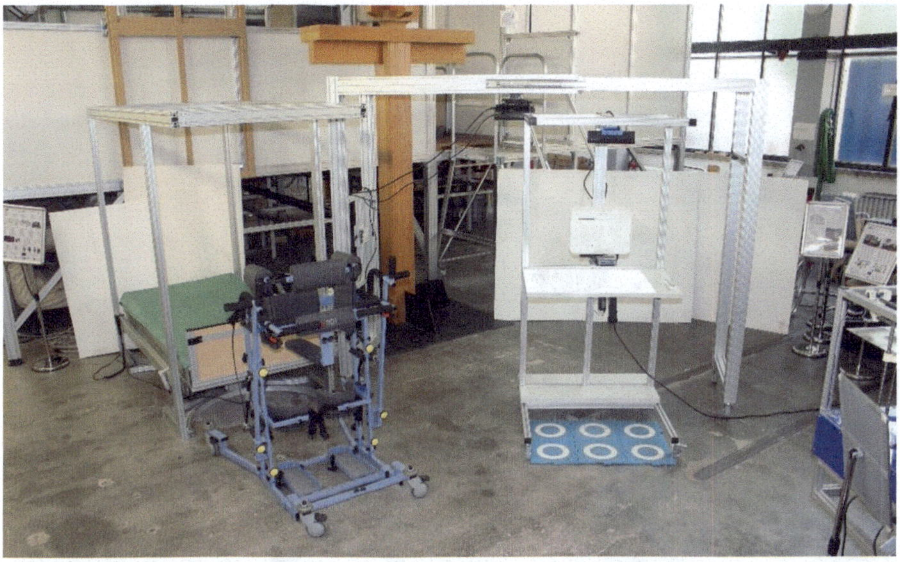

Fig. 7 The prototypes of various PI²Us deployed and tested in the laboratory

3.5.3 Exhibitions and Expert Interviews ("Showroom")

During the REACH project, the developed smart furniture was exhibited in several exhibitions and trade fairs (see Fig. 8), which is an important and effective approach to disseminate the products due to its personal communication nature (Kellezi 2014). In addition, an expert interview was conducted in the simulated home environment at TUM, which is another effective qualitative research method to improve the future iteration of the abovementioned PI²U prototypes (Döringer 2021). As a result, all five experts from the related fields (e.g., engineering, fashion, healthcare, business, education) all had highly favorable opinion towards the prototypes in the showroom, which established a strong endorsement to the Ambient Rehabilitation Kit (see Fig. 9). In addition, the experts suggested the following improvements to be considered in the next round of iteration:

(1) The volume and weight of the prototypes shall be further reduced to better fit the home environment.
(2) Classic functions that are familiar to older adults such as a radio module as well as information about its programs shall be integrated into the GUI of the MiniArc and SilverArc.
(3) The algorithm of the GUI of MiniArc and SilverArc needs to be further optimized to enhance the user experience of gesture maneuver.
(4) The ergonomics of iStander and SilverBed can be further improved.
(5) The costs of the PI²Us shall be considered and optimized to improve affordability.

Fig. 8 Aerial view of the REACH exhibition venue that more than 200 people attended

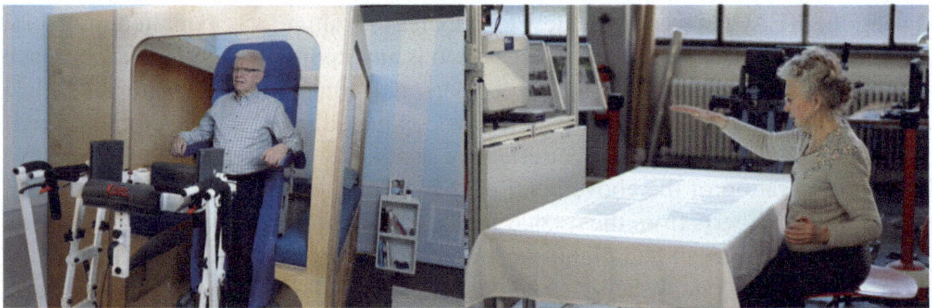

Fig. 9 Testing process with the experts in simulated home environment ("showroom") at TUM

3.6 A Framework for Economic Evaluation of Robotic and Automated Systems (Check)

No matter how advanced and efficient an innovation is, if it cannot achieve profitability, it will not be successful due to the lack of investment motivation. Therefore, this section aims at proposing a simple framework for the economic evaluation of robotic and automated systems for the healthcare industry. It is widely known that cost-benefit analysis (CBA) is considered to be the most important problem-solving tools in the decision-making process (Munger 2000), yet there is a lack of research on the CBA of smart home solutions for aging society. Therefore, based on the methods proposed in this chapter, the following section focuses on using the simple framework to evaluate economically smart furniture solutions compared to traditional care methods in various care environments.

The CBA followed the analytical framework described in Fig. 10 (Hu et al. 2021). In the calculation process of the CBA, all relevant factors that affect the main stakeholder need to be considered. Normally, the cash flow analyses for CBA range from at least three (small-scale projects) to more than ten years (e.g., large-scale public projects) (Li and Madanu 2009; Jones et al. 2014; Wang et al. 2014). Due to the public welfare attribute of the Ambient Rehabilitation Kit, it is appropriate to set

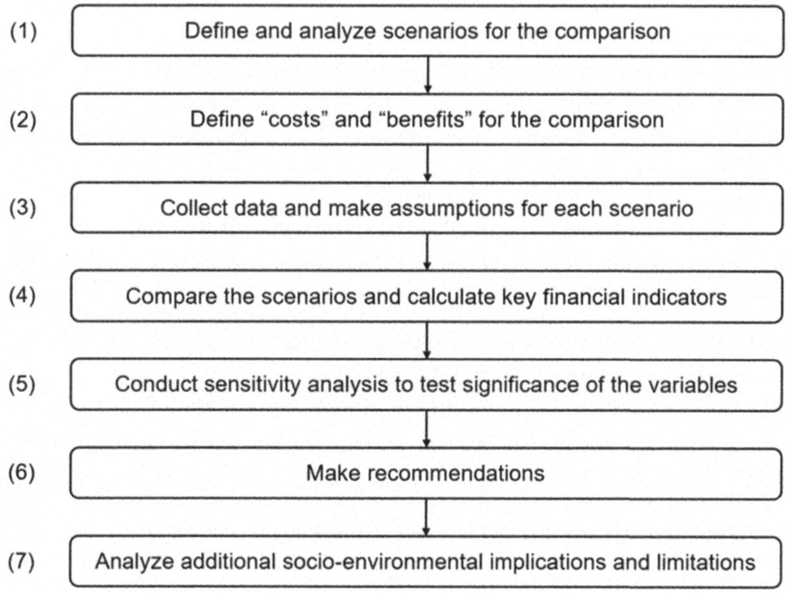

Fig. 10 The analytical framework of cost-benefit analysis (CBA) applied in this research

AAL Service Financial Evaluation									
Cash outflows	Year 1	Year 2	Year 3	Year 4	Year 5	Year 6	Year 7	Total (€)	Remarks
Central - hardware costs								0	
Central - software costs								0	
Central – network costs								0	
Service implementation								0	
Per patient costs - hardware								0	
Per patient costs - software								0	
Per patient costs - network								0	
Per patient costs - install								0	
Per patient costs - support								0	
Per patient costs - discharge								0	
Other - maintenance								0	
Total outflow	0	0	0	0	0	0	0	0	
Savings - in-patient								0	
Savings - emergency dep.								0	
Savings - home visits								0	
Savings - primary care								0	
Savings - long term care								0	
Other								0	
Cost savings	0	0	0	0	0	0	0	0	
Net annual cashflow	0	0	0	0	0	0	0	0	
Net cumulative cashflow	0	0	0	0	0	0	0	0	

Fig. 11 Cash flow analysis table for the Ambient Rehabilitation Kit compared to conventional caregiving method

the calculation period to seven years. As a result, a comparison table between conventional care methods and the Ambient Rehabilitation Kit solutions is designed which takes every factor during the care process into consideration in a seven-year period (see Fig. 11).

In order to define the scenarios for the comparison, a workshop was organized among the key partners within the REACH project and beyond, including medical device manufacturers, professional caregivers, and business consultants. The workshop participants co-created three suitable comparison scenarios for the Ambient Rehabilitation Kit (e.g., home and community prevention, hospital rehabilitation, care home solution). The data inputs for this CBA were acquired via market search and interviews with professional care providers.

Based on the results of the cash flow analysis table of the three scenarios, key financial indicators for each scenario can be calculated based on the following equations:

- BCR = (|present value of benefits|)/(|present value of costs|)
- ROI = (total cost savings – total outflows)/(|total outflows|)
- PBP = n + (|net accumulative cash flow of year n|)/(net annual cash flow of year $n + 1$), where n represents the number of the final year with negative net accumulative cash flow
- IIV = (initial hardware cost) + (initial deployment cost)
- NPV = (net annual cash flow)/(1 + cost of money)$^{\text{Years in the future}}$

Later on, sensitivity analysis (i.e., what-if analysis) was also performed, which concluded that caregivers' pay rate is the most important factor to consider.

As a result, the Ambient Rehabilitation Kit overall can be recommended to the key beneficiaries in various scenarios (e.g., home and community prevention, hospital rehabilitation, care home solution) in Germany.[1] Furthermore, other implications of the CBA are worth noting here.

- Given that the assumptions for CBA made here are relatively conservative, as the production number increases and manufacturing cost lowers over time, it is expected that the cost can be reduced further.
- This CBA only calculated mostly the direct benefits of the Ambient Rehabilitation Kit. More importantly, other social benefits of the proposed system that were not directly calculated here are also worth mentioning, such as improved quality of life and liberated productivity from both formal and informal caregivers.
- When transferring the developed technology to other markets outside Germany, the corresponding CBA based on the local market and situation will be necessary.
- Furthermore, researchers and entrepreneurs beyond the REACH project can also benefit from the proposed methodology to conduct their CBA of smart furniture and smart home solutions for various scenarios in aging society. Due to the lack of research on the CBA of smart furniture solutions for aging society, the results of this CBA will provide first-of-its-kind insights for the wise use and efficient allocation of smart healthcare resources in aging society.

3.7 *Towards an Elderly-Oriented Smart Furniture Solution in China: A Survey and an Action Plan (Act)*

Population aging is one of the major challenges not only facing the developed countries but also threatening many emerging economies. Specifically, China's upcoming aging society caused by various factors will pose an imminent threat to its future development. Therefore, a variety of measures must be taken to achieve its demographic sustainability. Smart home technology has gained substantial popularity over the past decade. It also started to show its prominence in the fields of aging in place and home care for older adults. Smart furniture can be considered as a novel subcategory of smart home technology. However, gerontechnology research in China is still lagging behind compared to developed countries such as the USA, Germany, and Japan. Specifically, according to a study in 2015, China was still considered as an academic laggard in gerontechnology compared to leaders such as the USA and the UK, although it began to catch up in most recent years (Huang et al. 2015). At the same time, there is also a lack of research on the adoption of gerontechnology among Chinese older adults. Therefore, to conduct a survey to investigate their adoption, attitude, and preference for smart home and smart furniture technology will be helpful for the future implementation of these technologies.

[1] The detailed data analysis of the CBA can be found in Chapter 4 of the thesis "Ambient Rehabilitation Kit: Developing Personalized Intelligent Interior Units to Achieve Demographic Sustainability in Aging Society" by co-author R. Hu (https://mediatum.ub.tum.de/680889?show_id=1638907).

In the following sections, the opinion survey will be introduced, and its implications will be analyzed. Finally, a project action plan for implementing elderly-oriented smart furniture technology in China will be illustrated.

3.7.1 Survey Methods

The opinion survey to investigate the attitudes and opinions of Chinese older adults towards using smart furniture technologies based on the PI²Us as an example is introduced. The survey was conducted via two main channels (e.g., mainly via WeChat app on smartphones and email links and in-person questionnaires as a supplement for users who do not use any smartphones or tablets). With over one billion active users only in China, WeChat provided an excellent platform to distribute the questionnaires. The survey consisted of 11 close-ended questions (see Table 1) related to the current situation of the participants and their opinion towards the smart furniture and relevant technologies for older adults, of which questions 8–11 were measured by Likert scale (Likert 1932). In particular, photos with descriptions of older adults using the PI²Us were shown as an example of smart furniture to give the survey participants an intuitive impression of the appearance and functions of the PI²Us. The questions were designed with principles of simple language, common concepts, manageable tasks, and widespread information (Converse and Presser 1986). The survey was kept short as much as possible, which can be easily finished by older adults in 3–5 minutes. The survey was pretested with several older adults before formerly sending out in order to optimize the understandability and order of the questions.

Once individuals' data is involved, protecting their data privacy will be critical. Therefore, the good practice for protecting individual privacy in the survey is reported as follows. First and foremost, the survey was conducted in a fully anonymous manner, meaning that information such as names, birth dates, addresses, and

Table 1 List of survey questions for the older adults

No.	Question
1	What year were you born?
2	What is your gender?
3	What area are you currently living?
4	What is your highest level of education (including enrollment)?
5	Where is your main place of residence?
6	Which of the following smart digital products have you used?
7	Which of the following smart home devices have you used?
8	What do you think is the ease of use of current technology products for older adults?
9	How interested are you in using smart furniture with health functions? (examples from the REACH project are given.)
10	How important are the following attributes to you for using smart furniture?
11	What do you think of the prospects of China's elderly-oriented smart furniture market?

resident identity card numbers were not collected. This approach excludes any possibility to identify any individual survey participant. Furthermore, in order to participate in the online survey, all the participants needed to give consent to provide their basic demographic information such as age, gender, province of residence, and level of education. In addition, the data collected from older Chinese citizens was stored in the server of Tencent within the territory of the People's Republic of China during the research, which was in line with the newly established Data Security Law of the People's Republic of China. After the analysis was completed, the data collected in this survey was securely deleted in the user account, which according to Tencent Questionnaire's user account policy means permanent erasure in the company's server. These measures are also in line with Guide on Good Data Protection Practice in Research (European University Institute 2019).

3.7.2 Survey Results

The survey lasted 45 days from January 7, 2021, to February 20, 2021. In total, 1313 questionnaires were sent out, and 403 responses were collected, of which 384 were valid, leading to an effective return rate of 29.2% (19 responses were removed due to reasons such as incomplete data). The average completion time for each participant was 4 min and 17 s, which well met the expectation for the questionnaire design. The vast majority of the older survey participants completed the survey without issues. Necessary guidance or explanation was provided to the participants if needed. In order to better perform the data analysis, the study categorized the survey subjects as the "young old" (60–69), the "middle old" (70–79), and the "very old" (80+) (Forman et al. 1992), because the current retirement age for Chinese workers is 60 for male employees and 55 for female employees. The analyses of the survey data including general analysis and cross analysis are shown as follows.

3.7.2.1 General Analysis

In general, 384 older adults from 26 out of the total 34 provincial-level administrative divisions of China provided valid questionnaires during the survey. As of the beginning of 2021, there are approximately 260 million Chinese older adults aged 60 and over (People's Daily Online 2021). Therefore, it can be calculated with a 95% confidence level that the survey can represent the Chinese older population with a margin of error (MOE) of ±5%, which is acceptable for categorical data in social research (Bartlett et al. 2001). The average age of the survey participants was 68.64 years old. Among these participants, 39.1% were male and 60.9% were female. The education level of the survey participants was relatively balanced, among which 49.5% had college degree or above and 50.5% had high school education or below. Regarding the places of residence, 75.2% of the survey participants lived at home either alone or with spouse, and 20.8% of them lived with their children. Only 0.8% of them were living in retirement homes or nursing homes. This phenomenon is likely because (1) over 90% of Chinese citizens own their homes (Kharas and Dooley 2020), and (2) in Chinese culture, aging in place is a common practice, and older adults tend to rely on family members for primary care in later life due to the cultural norm of filial piety (Bai et al. 2020), although living in retirement homes has started to pick up momentum in recent years.

Regarding the user adoption rate of personal smart devices (e.g., smartphones, tablets, PCs/laptops, wearables), 93.5% of the participants used smartphones. On the contrary, only 8.9% of them used wearables. A possible explanation could be the inadequate functionality and frequent need for charging for current wearables, which also suggests that the ambient sensing solution integrated in smart furniture could provide a good alternative to wearables. In this survey, 5.5% of the participants did not use any of these devices. Although admittedly it is likely that older adults who did not use any personal smart devices were under-sampled because the majority of the questionnaires were completed via WeChat app, it is fair to say the adoption rate of personal smart devices among Chinese older adults is satisfactory.

In terms of user adoption of smart home devices (e.g., smart speaker, smart TV, robot vacuum, pet robot, smart door lock, smart appliance, smart furniture, etc.), three quarters of the participants had experience with at least one of them, with smart TV having the highest adoption rate of 49.5%. On the contrary, smart furniture had the second lowest user adoption rate of 3.1%, only higher than companion robots. This is mainly because smart furniture is relatively a new field without many mature applications on the market. On the other hand, however, it also indicates a substantial market potential.

Regarding the usability of today's technology products for older adults, only 45.3% of the participants thought that they were easy or very easy to use. Therefore, improving the usability for older adults is highly important for developing new or improving current technology products.

When the interest in using elderly-oriented smart furniture was asked, examples of four PI^{2}U prototypes from the REACH project were given. As a result, 60.9% of the participants are interested or very interested in using elderly-oriented smart furniture. This indicates substantial interest and market opportunities for smart furniture among Chinese older adults.

In terms of the importance of various attributes in smart furniture, the survey participants valued the safety of the products the most (i.e., 85.4% of the participants find it important or highly important), followed by usability, quality, privacy protection, affordability, multifunctionality, and the aesthetics the least (i.e., 48.2%). This result indicates that when developing elderly-oriented smart furniture products for Chinese older adults, more attention shall be paid to aspects such as safety, ease of use, quality, privacy protection, and affordability.

Furthermore, most of the participants (i.e., 73.7%) reckoned that there will be a substantial market potential for elderly-oriented smart furniture in China, which further verifies the inference above.

3.7.2.2 Cross Analysis

This section focuses on analyzing the correlation between participants' demographics (e.g., age, gender, education level) and attitude towards smart home and smart furniture technology. The statistics were tested by Pearson's chi-squared test ($n \geq 40$) to evaluate the statistical significance between any two groups (Cochran 1952; Gravetter and Wallnau 2013).

Regarding the correlation between age and adoption rate, difficulty, interest, and expectations, there is a sharp decline in smart home technology adoption rate when the participants are older. In the "young old" group, 84.0% of the participants had experience in at least one smart home product, while in the "very old" group, only 35.6% had experience in using any smart home technology (Fig. 12-i). The

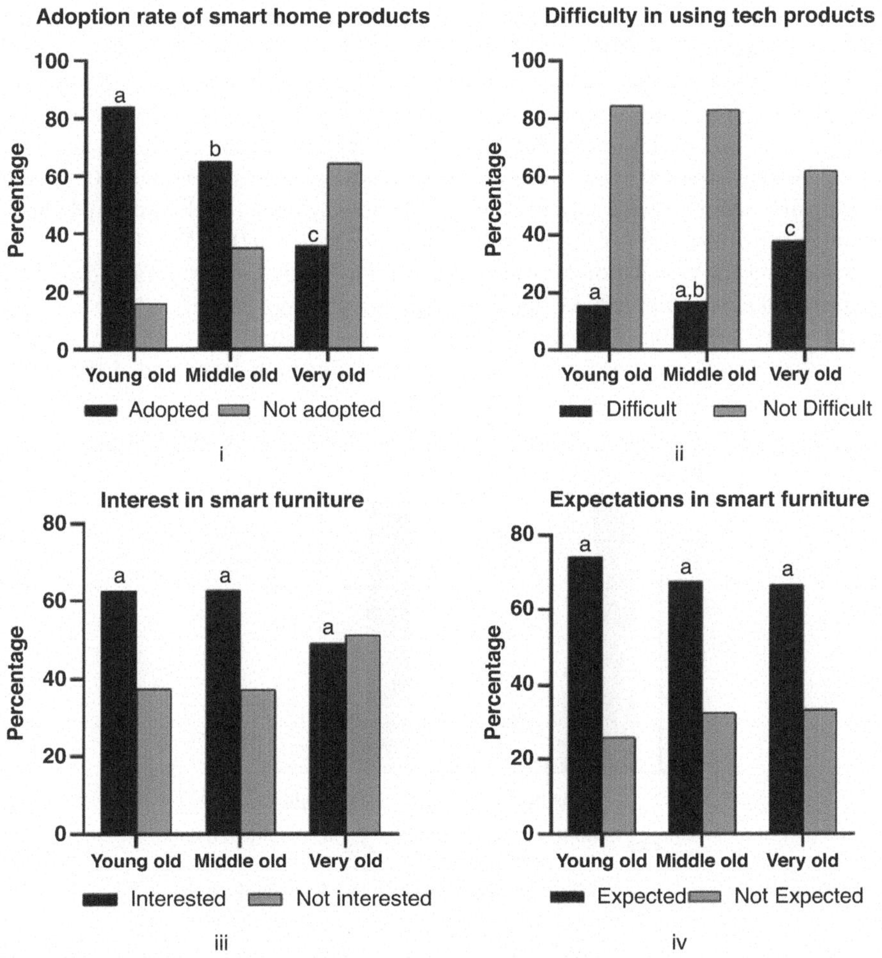

Fig. 12 Correlation between age and smart home technology adoption rate (the difference between any two groups sharing the same letters is not statistically significant)

differences between any two age groups are extremely significant ($p < 0.01$). Regarding the correlation between age and difficulty in using technology products (Fig. 12-ii), only around 15% of the older adults in both "young old" and "middle old" groups found it difficult or very difficult to use technology products ($p > 0.05$). However, the percentage is significantly higher in the "very old" group compared to the first two age groups ($p < 0.01$). Regarding the correlation between age and interest in using elderly-oriented smart furniture (Fig. 12-iii), more than 60% of older adults in the "young old" and "middle old" groups were interested or very interested in using the smart furniture developed in the REACH project. The percentage dropped slightly in the "very old" group, but no statistical significance can be observed between the "very old" group and other two age groups regarding user interest ($p > 0.05$). Therefore, the overall interest in using smart furniture is strong among Chinese older adults. Regarding the correlation between age and expectations in elderly-oriented smart furniture technology (Fig. 12-iv), all three groups of older adults expressed high expectations for its future market potential with no significant difference ($p > 0.05$).

Regarding the correlation between age and adoption rate, difficulty, interest, and expectations, it is impossible to observe statistically significant differences in the adoption rate of smart home technology, difficulty, and interest as well as expectation in elderly-oriented smart furniture between different genders ($p > 0.05$). Male participants seem to have slightly more difficulty in using technology products, although the difference is not statistically significant ($p > 0.05$) (Fig. 13).

Regarding the correlation between education level and adoption rate, difficulty, interest, and expectations, older adults with primary school education have a significantly lower adoption than those with any other education levels ($p < 0.05$) (Fig. 14-i). Regarding the correlation between education level and difficulty in

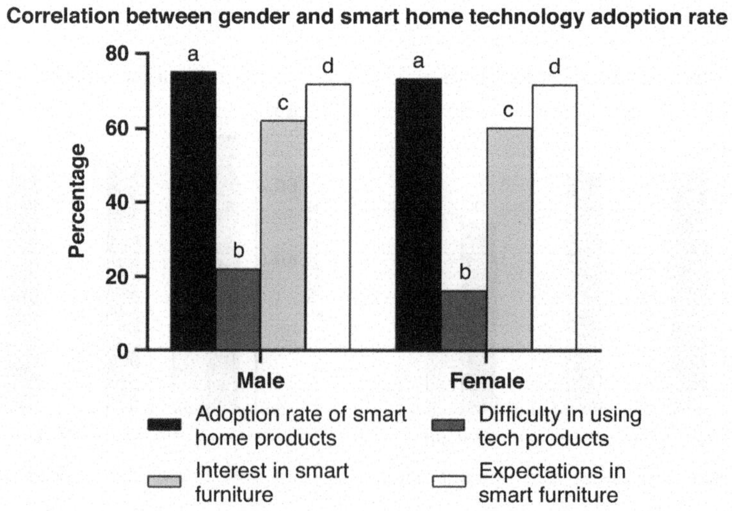

Fig. 13 Correlation between gender and smart home technology adoption rate (the difference between any two groups sharing the same letters is not statistically significant)

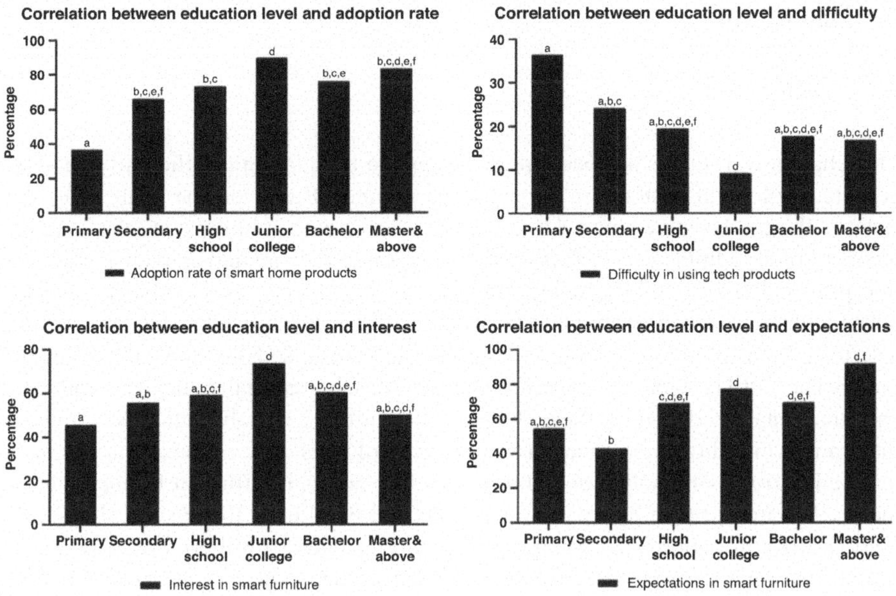

Fig. 14 Correlation between gender and smart home technology adoption rate (the difference between any two groups sharing the same letters is not statistically significant)

using technology products (Fig. 14-ii), older adults with education level lower than high school find it significantly more difficult than those with junior college education ($p \leq 0.01$). Meanwhile, participants with junior college education have a significantly higher interest in using elderly-oriented smart furniture than those with high school education or below ($p < 0.05$) (Fig. 14-iii). Finally, the majority of most education groups (expect for secondary school) have high expectations for elderly-oriented smart furniture (Fig. 14-iv). In particular, older adults with junior college education or above have significantly higher expectations than those with secondary school education ($p \leq 0.01$).

The survey results show that there is a substantial amount of interest and optimism towards elderly-oriented smart furniture among Chinese older adults. Although living in retirement homes and nursing homes has started to pick up momentum, the focused application scenario for developing smart furniture technology in China shall be mainly homes due to cultural considerations. In the process of developing localized elderly-oriented smart furniture products for China, aspects such as safety, ease of use, quality, privacy protection, and affordability shall be prioritized. The digital literacy among Chinese older adults is decent, but there is a clear digital gap among older adults aged 80 or over and with a lower education level. As a result, it is important to close the digital gap especially for older adults over 80 years old and with a lower education level using measures such as improving safety, increasing ease of use, improving quality, ensuring privacy protection, and bringing down the costs. The method of this survey is highly adaptable and scalable and thus can be easily adopted by researchers in other regions.

4 Discussion

4.1 Summary

This chapter introduces a procedure model as the foundation to efficiently develop gerontechnological systems in a systematic manner. Based on the analysis of the state-of-the-art assistive smart furniture and smart home solutions and the requirements of older adults, a service system transforming clinical and care environments into personalized modular sensing, prevention, and intervention systems was proposed, encouraging older adults to become healthier through activities. As a result, a series of special smart furniture integrating advanced sensing and assistive technology (i.e., PI²Us) were developed which seamlessly materialize the care concepts and functionality. Based on these devices, the Ambient Rehabilitation Kit concept integrating the abovementioned assistive technologies was proposed to create a unique interior living and care solution for older adults in different living environments. Accordingly, an approach for testing and exhibiting the system (i.e., "lab-field-showroom" approach) was proposed and executed. Furthermore, a practical framework for the economic evaluation of the gerontechnological solutions compared to traditional care methods was also proposed. In addition, the prospects for implementing the Ambient Rehabilitation Kit in the Chinese market are discussed, and a research and development action plan is suggested based on the findings of an opinion survey in order to tackle the severe challenge that rapid population aging poses to the social sustainability of emerging economies.

4.2 Limitations and Future Work

The work reported in this chapter acted as a springboard for the further iterations and enhancements of the Ambient Rehabilitation Kit. Nevertheless, there are a few areas that this development cycle has not yet fully addressed. The primary goals of the follow-up phase of this research will be to (1) improve the usability of the prototypes and grow the PI²U family, (2) create an organization to carry out a sustainable business model for the abovementioned innovations, (3) manage intellectual property rights for the key outcomes, and (4) seek standardization of the methodology and components developed in this research in order to push them to the market in a sustainable way.

4.2.1 Optimizing and Expanding the PI²U Family (Do)

In the future design iterations, the work priority will be focused on enhancing the user experience of the PI²Us together with end users. In particular, attributes of the PI²Us such as usability, quality, privacy protection, affordability, multifunctionality, and aesthetics need to be improved. In order to better appeal to older adults, persuasive design principles and behavior change techniques need to be applied in this process (Valk et al. 2017). Furthermore, PI^2Us with additional functions and forms such as mobile robot, exoskeleton suit, and unobtrusive wearables will be developed to expand the PI^2U family in order to appeal to a broader user group.

4.2.2 Business Strategy and Scalability (Act)

Innovative products can never be successful if they have no access to the global market, even if appreciated by experts. Establishing the appropriate business strategy is the key to success of an innovation. Therefore, key members of the project have already initiated a start-up (i.e., CREDO GmbH) which will serve beyond the project as an integrator of the developed products and services and a solution provider to well-defined market segments based on suitable business models. Currently, the start-up is cooperating with a large Chinese furniture manufacturer to develop localized and customized elderly-oriented smart furniture technology in order to tackle the challenges imposed by rapid aging society in China and beyond.

4.2.3 Intellectual Property Protection (Act)

In the first cycle of the development, an innovative Ambient Rehabilitation Kit comprising a series of original smart furniture devices was developed. Thus, suitable legal protection will be the key to continue the success of the products developed so far. In the next phase of the R&D process, the PI^2Us (e.g., SilverArc, MiniArc, SilverBed, etc.) including their installation process and user interfaces shall be protected through industrial designs and utility patents in respective countries in order to better push the developed systems into the market and sustain the motivation of innovation.

4.2.4 Towards Standardization (Act)

The standardization activities play a crucial role in catalyzing not only the competitiveness and social impact of innovations but also the economic growth of society. Standardization in Germany only, for example, accounts for at least 0.9% of the annual economic growth (Blind et al. 2011). There are three levels of standardization bodies that can facilitate the standardization of innovations, which are national (e.g., DIN in Germany, ANSI in USA, SAC in China, etc.), continental (e.g., CEN in Europe), international (e.g., ISO, ITU, IEC, etc.). Based on the experience and know-how gained in this research, the technical committee ISO/TC 314 – Ageing societies has been established to develop standards and guidance to tackle global challenges imposed by aging society. Furthermore, in order to better suit the needs and preference of older adults, standards for developing Ambient Rehabilitation

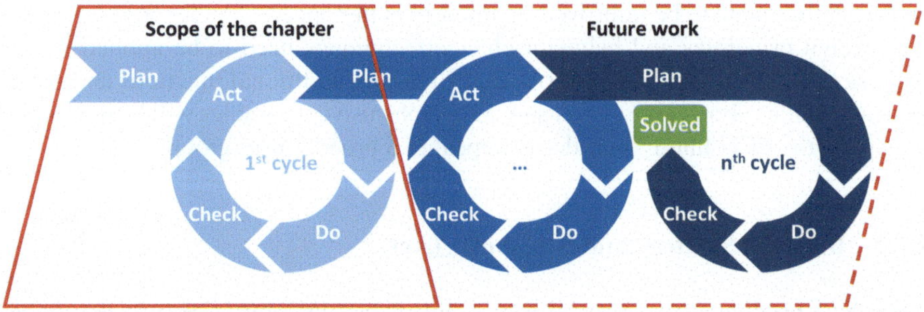

Fig. 15 Perpetual development cycle to adapt to the ongoing situation. (Figure redrawn and adapted from Christoph Roser at AllAboutLean.com, licensed under CC BY-SA 4.0, https://creativecommons.org/licenses/by-sa/4.0/)

Kits can be further formulated with different standardization bodies in other parts of the world to facilitate product localization and market optimization.

These ongoing or upcoming tasks comprise the crucial procedures of the future iterations of the proposed procedure model. The cycle will be repeated until the ideal solution is achieved (see Fig. 15). The case study reported in this chapter provides a valuable toolbox for researchers and developers to develop assistive technology for older adults and beyond in the future.

Acknowledgments This chapter is partly based on co-author R. Hu's doctoral thesis titled "Ambient Rehabilitation Kit: Developing Personalized Intelligent Interior Units to Achieve Demographic Sustainability in Aging Society" (Hu 2022). The research has received funding from the European Union's Horizon 2020 research and innovation program under grant agreement No. 690425.

References

Antony, J. (2006) 'Six sigma for service processes', *Business Process Management Journal*, 12(2), pp. 234–248. Available at: https://doi.org/10.1108/14637150610657558.

Bai, X., Lai, D.W.L. and Liu, C. (2020) 'Personal care expectations: Photovoices of Chinese ageing adults in Hong Kong', *Health & Social Care in the Community*, 28(3), pp. 1071–1081. Available at: https://doi.org/10.1111/hsc.12940.

Bartlett, J., Kotrlik, J. and Higgins, C. (2001) 'Organizational research: Determining appropriate sample size in survey research appropriate sample size in survey research', *Information technology, learning, and performance journal*, 19(1), p. 43. Available at: http://citeseerx.ist.psu.edu/viewdoc/download?doi=10.1.1.486.8295&rep=rep1&type=pdf

Beetz, M. et al. (2008) 'The Assistive Kitchen – A demonstration scenario for cognitive technical systems', in *RO-MAN 2008 – The 17th IEEE International Symposium on Robot and Human Interactive Communication*, pp. 1–8. Available at: https://doi.org/10.1109/ROMAN.2008.4600634.

Blind, K., Jungmittag, A. and Mangelsdorf, A. (2011) *The Economic Benefits of Standardization*. Available at: https://www.din.de/blob/89552/68849fab0eeeaafb56c5a3ffee9959c5/economic-benefits-of-standardization-en-data.pdf

Bloom, D.E., Canning, D. and Fink, G. (2011) 'Implications of Population Aging for Economic Growth', *National Bureau of Economic Research Working Paper Series*, No. 16705. Available at: https://doi.org/10.3386/w16705.

Bock, T. et al. (2015) 'The Development of Intra-House Mobility, Logistics and Transfer Solutions in PASSAge', *Journal of Robotics and Mechatronics*, 27(1), p. 108. Available at: https://doi.org/10.20965/jrm.2015.p0108.

Bock, T. (2017) 'REACH: Responsive Engagement of the Elderly promoting Activity and Customized Healthcare', *Gerontechnology*, 16(3), pp. 125–128. Available at: https://doi.org/10.4017/gt.2017.16.3.001.00.

Chen, R. et al. (2019) 'China has faster pace than Japan in population aging in next 25 years', *BioScience Trends*, 13(4), pp. 287–291. Available at: https://doi.org/10.5582/bst.2019.01213.

Cochran, W.G. (1952) 'The Chi-Squared Test of Goodness of Fit', *The Annals of Mathematical Statistics*, 23(3), pp. 315–345. Available at: https://doi.org/10.1214/aoms/1177729380.

Converse, J.M. and Presser, S. (1986) *Survey Questions: Handcrafting the Standardized Questionnaire*. Newbury Park, CA: SAGE Publications, Inc.

Döringer, S. (2021) '"The problem-centred expert interview". Combining qualitative interviewing approaches for investigating implicit expert knowledge', *International Journal of Social Research Methodology*, 24(3), pp. 265–278. Available at: https://doi.org/10.1080/13645579.2020.1766777.

Erdt, S. et al. (2012) 'Systematic development of a complex personal health assistance system explained by the example of GEWOS activity furniture', in *5th German Ambient Assisted Living Congress*. Berlin. Available at: https://mediatum.ub.tum.de/doc/1484187/1484187.pdf

European University Institute (2019) *Guide on Good Data Protection Practice in Research*. Available at: https://www.eui.eu/documents/servicesadmin/deanofstudies/researchethics/guide-data-protection-research.pdf

Forman, D.E. et al. (1992) 'PTCA in the Elderly: The "Young-Old" versus the "Old-Old"', *Journal of the American Geriatrics Society*, 40(1), pp. 19–22. Available at: https://doi.org/10.1111/j.1532-5415.1992.tb01823.x.

Gravetter, F.J. and Wallnau, L.-A.B. (2013) *Essentials of Statistics for The Behavioral Sciences*. Cengage Learning, Inc.

Güttler, J. et al. (2017) 'Development and Evaluation of Assistive Terminals for the Improvement of Functional Performance of the Elderly in a Variety of Life Centers', in *Proceedings of the first international symposium on applied abstraction and integrated design (AAID)*. Yokohama, Japan. Available at: https://mediatum.ub.tum.de/doc/1484225/1484225.pdf

Hu, R. et al. (2020) 'Developing a Smart Home Solution Based on Personalized Intelligent Interior Units to Promote Activity and Customized Healthcare for Aging Society', *Journal of Population Ageing*, 13(2), pp. 257–280. Available at: https://doi.org/10.1007/s12062-020-09267-6.

Hu, R. et al. (2021) 'A Simple Framework for the Cost–Benefit Analysis of Single-Task Construction Robots Based on a Case Study of a Cable-Driven Facade Installation Robot', *Buildings*, 11(1), p. 8. Available at: https://doi.org/10.3390/buildings11010008.

Hu, R. (2022). Ambient Rehabilitation Kit: Developing Personalized Intelligent Interior Units to Achieve Demographic Sustainability in Aging Society [Technical University of Munich]. https://mediatum.ub.tum.de/680889?show_id=1638907

Huang, L. et al. (2015) 'International Comparison of Research on Gerontechnology (in Chinese)', *Journal of Intelligence*, 34(10), pp. 22–27. Available at: https://doi.org/10.3969/j.issn.1002-1965.2015.10.004.

Imai, M. (1986) *Kaizen: The Key to Japan's Competitive Success*. New York City, NY: McGraw-Hill Education Ltd.

Intille, S.S. (2002) 'Designing a home of the future', *IEEE Pervasive Computing* [Preprint]. Available at: https://doi.org/10.1109/MPRV.2002.1012340.

Johnson, C.N. (2002) 'The benefits of PDCA: Use this cycle for continual process improvement', *Quality Progress*, 35(5), p. 120. Available at: https://search.proquest.com/openview/6fb24b731a9c0c8bafd90096fd751e76/1?pq-origsite=gscholar&cbl=34671

Jones, H., Moura, F. and Domingos, T. (2014) 'Transport Infrastructure Project Evaluation Using Cost-benefit Analysis', *Procedia - Social and Behavioral Sciences*, 111, pp. 400–409. Available at: https://doi.org/10.1016/j.sbspro.2014.01.073.

Kellezi, J. (2014) 'Trade Shows: A Strategic Marketing Tool for Global Vompetition', *Procedia Economics and Finance*, 9, pp. 466–471. Available at: https://doi.org/10.1016/S2212-5671(14)00048-3.

Kharas, H. and Dooley, M. (2020) *China's influence on the global middle class*. Available at: https://www.brookings.edu/wp-content/uploads/2020/10/FP_20201012_china_middle_class_kharas_dooley.pdf

Kidd, C.D. et al. (1999) 'The Aware Home: A Living Laboratory for Ubiquitous Computing', in *Proceedings of the International Workshop on Cooperative Buildings*, pp. 191–198. Available at: https://doi.org/10.1007/10705432_17.

Koskinen, I. et al. (2013) 'Design Research Through Practice: From the Lab, Field, and Showroom', *IEEE Transactions on Professional Communication*, 56(3), pp. 262–263. Available at: https://doi.org/10.1109/TPC.2013.2274109.

Langosch, K., Güttler, J. and Bock, T. (2019) 'The BaltSe@nioR System – addressing senior's needs with Smart Furniture', in *BaltSe@nioR – Challenges and innovative solutions in product development for seniors´ home living*. Hamburg, Germany.

Li, Z. and Madanu, S. (2009) 'Highway Project Level Life-Cycle Benefit/Cost Analysis under Certainty, Risk, and Uncertainty: Methodology with Case Study', *Journal of Transportation Engineering*, 135(8), pp. 516–526. Available at: https://doi.org/10.1061/(ASCE)TE.1943-5436.0000012.

Likert, R. (1932) 'A technique for the measurement of attitudes', *Archives of Psychology*, p. 55. Available at: https://doi.org/2731047

Linner, T. et al. (2016) 'Development and Evaluation of an Assistive Workstation for Cloud Manufacturing in an Aging Society', in R. Wichert and H. Klausing (eds) *Ambient Assisted Living: Advanced Technologies and Societal Change*. Cham: Springer International Publishing, pp. 71–82. Available at: https://doi.org/10.1007/978-3-319-26345-8_7.

Linner, T., Pan, W., Hu, R., Zhao, C., Iturralde, K., Taghavi, M., Trummer, J., Schlandt, M., & Bock, T. (2019). A technology management system for the development of single-task construction robots. Construction Innovation, 20(1). https://doi.org/10.1108/CI-06-2019-0053

Linner, T. et al. (2022) 'A Procedure Model for the Development of Construction Robots', in S.H. Ghaffar et al. (eds) *Innovation in Construction – A Practical Guide to Transforming the Construction Industry*. Cham: Springer International Publishing, pp. 321–352. Available at: https://doi.org/10.1007/978-3-030-95798-8_14.

Linner, T. and Schulz, T. (2015) 'GEWOS—A Mechatronic Personal Health & Fitness Assistance System', *Journal of Robotics and Mechatronics*, 27(1), pp. 105–106. Available at: https://doi.org/10.20965/jrm.2015.p0105.

Lodgaard, E., Gamme, I. and Aasland, K.E. (2013) 'Success Factors for PDCA as Continuous Improvement Method in Product Development', in C. Emmanouilidis, M. Taisch, and D. Kiritsis (eds) *Advances in Production Management Systems. Competitive Manufacturing for Innovative Products and Services*. Berlin, Heidelberg: Springer Berlin Heidelberg, pp. 645–652.

Lu, Y. et al. (2020) 'Promoting Fall Prevention among Community Dwelling Older Adults through ActivLife: a Physical and Social Activation', *Journal of Population Ageing*, 13(2), pp. 223–237. Available at: https://doi.org/10.1007/s12062-020-09261-y.

Manoel, F. et al. (2017) 'Managing Natural Resources in a Smart Bathroom Using a Ubiquitous Multi-Agent System', in *Proceedings of 11th Workshop-School on Agents,* Environments and Applications. São Paulo, Brazil, pp. 101–112. Available at: https://www.researchgate.net/publication/316741200_Managing_Natural_Resources_in_a_Smart_Bathroom_Using_an_Ubiquitous_Multi-Agent_System de Meijer, C. et al. (2013) 'The effect of population aging on health expenditure growth: a critical review', *European Journal of Ageing*, 10(4), pp. 353–361. Available at: https://doi.org/10.1007/s10433-013-0280-x.

Munger, M.C. (2000) *Analyzing Policy: Choices, Conflicts, and Practices*. W.W. Norton (New Institutionalism in Americ).

National Bureau of Statistics (2021) *Communiqué of the Seventh National Census (No. 5)*. Available at: http://www.stats.gov.cn/tjsj/zxfb/202105/t20210510_1817181.html

Nilsson-Witell, L., Antoni, M. and Dahlgaard, J.J. (2005) 'Continuous improvement in product development', *International Journal of Quality & Reliability Management*, 22(8), pp. 753–768. Available at: https://doi.org/10.1108/02656710510617210

People's Daily Online (2021) *National Bureau of Statistics: 18.7% of the population aged 60 and above, the aging process has accelerated significantly (in Chinese), People's Daily Online*. Available at: http://finance.people.com.cn/n1/2021/0511/c1004-32100026.html (Accessed: 26 May 2021).

Powell, D.J. (2018) 'Kanban for Lean Production in High Mix, Low Volume Environments', *IFAC-PapersOnLine*, 51(11), pp. 140–143. Available at: https://doi.org/10.1016/j.ifacol.2018.08.248.

Pyo, Y. et al. (2014) 'Floor sensing system using laser reflectivity for localizing everyday objects and robot', *Sensors (Switzerland)* [Preprint]. Available at: https://doi.org/10.3390/s140407524.

Randriambelonoro, M. et al. (2020) 'Hospital-to-Home Transition for Older Patients: Using Serious Games to Improve the Motivation for Rehabilitation – a Qualitative Study', *Journal of Population Ageing*, 13(2), pp. 187–205. Available at: https://doi.org/10.1007/s12062-020-09274-7.

Realyvásquez-Vargas, A. et al. (2018) 'Applying the Plan-Do-Check-Act (PDCA) Cycle to Reduce the Defects in the Manufacturing Industry. A Case Study', *Applied Sciences*, 8(11). Available at: https://doi.org/10.3390/app8112181.

Sakamura, K. (1996) 'The TRON Project', *Information and Software Technology* [Preprint]. Available at: https://doi.org/10.1016/0950-5849(95)01083-1.

Sato, T., Harada, T. and Mori, T. (2004) 'Environment-type robot system "RoboticRoom" featured by behavior media, behavior contents, and behavior adaptation', *IEEE/ASME Transactions on Mechatronics*, 9(3), pp. 529–534. Available at: https://doi.org/10.1109/TMECH.2004.834650.

Schäpers, B. et al. (2017) 'Determining design requirements for active aging: Personas, experience maps, and stakeholders', *Gerontechnology*, 16(3), pp. 139–150. Available at: https://doi.org/10.4017/gt.2017.16.3.003.00.

Spillman, W.B. et al. (2004) 'A "smart" bed for non-intrusive monitoring of patient physiological factors', *Measurement Science and Technology*, 15(8). Available at: https://doi.org/10.1088/0957-0233/15/8/032.

Steinböck, M. et al. (2019) 'Biomechanical evaluation of a transfer device for sit-to-stand training: effects on sagittal lower-extremity joint kinematics', *Neurologie & Rehabilitation* [Preprint], (S3). Available at: https://www.researchgate.net/publication/348415811_Biomechanical_evaluation_of_a_transfer_device_for_sit-to-stand_training_effects_on_sagittal_lower-extremity_joint_kinematics

Tortorella, G.L., Viana, S. and Fettermann, D. (2015) 'Learning cycles and focus groups', *The Learning Organization*, 22(4), pp. 229–240. Available at: https://doi.org/10.1108/TLO-02-2015-0008.

United Nations (2019) *World Population Prospects: the 2019 Revision*. Available at: https://population.un.org/wpp/Publications/

United Nations (2022) *World Population Prospects 2022*. New York, NY, USA. Available at: https://population.un.org/wpp/Publications/

Valk, C. et al. (2017) 'Towards personalized persuasive strategies for active ageing', *Gerontechnology* [Preprint]. Available at: https://doi.org/10.4017/gt.2017.16.3.005.00.

Wang, R. et al. (2014) 'A Cost-benefit Analysis of Commuter Train Improvement in the Dhaka Metropolitan Area, Bangladesh', *Procedia – Social and Behavioral Sciences*, 138, pp. 819–829. Available at: https://doi.org/10.1016/j.sbspro.2014.07.231.

Eradication of Global Hunger at UN Initiative: Holacracy Process Enriched by Human Will and Virtue

Radhakrishnan Nair and Justin Joy

1 Prolegomena

Enough food is produced today to feed everyone on the planet, but hunger is on the rise in some parts of the world, and some 821 million people are considered to be "chronically undernourished". What steps are being taken to ensure that everyone, worldwide, receives sufficient food? (Nations 2019).

The researchers take note of the significance of revelation given in the above cited Newsletter resolved to investigate into the vital issue affecting the life of human beings on earth. A few facts vexed them, a few defied them, and a few challenged them to find a way out. The researchers reread them, examined them, and scanned them: (a) "enough food is produced today to feed everyone on the planet", (b) "Hunger is on the rise in some parts of the world", and (c) "some 828 million people are considered to be 'chronically undernourished'". The ideas cited and stated are contradictory and upsetting.

The researchers proposed to investigate into the problems of hunger which the world has been confronted with. The impossible answers till date to the perennial problems were both at the tip of their tongue and at the tip of their pen. But they failed to get expressed. It looked unconvincing. But on closer look, persistent thoughts helped the answers get unravelled, and they appeared fully convincing. The answer is derived from the following question: Can we contain and conquer the fatal hunger in the world? We produce food for all living humans in the globe. But the required food produced does not reach the starving mouths.

R. Nair
School of Arts and Humanities, Christ University, Bangalore, India
e-mail: radhakrishnan.nair@christuniversity.in

J. Joy (✉)
School of Business and Management, Christ University, Bangalore, India

I. S. Dunmade et al. (eds.), *Sustainable Engineering*, Green Energy and Technology, https://doi.org/10.1007/978-3-031-47215-2_27

The researchers took this as a challenge and as a mandate to be taken up at the instance of the UN. The first part of the work lingers around the UN and their resolution and findings. The UN shall deem it their primary task. The UN shall institute a GLOBAL HUNGER INFORMATION DATA Centre that can provide data live to all concerned. The centre will be the nerve centre for available real information to solve the global hunger problem. The third part of the research work handles a process devised and proposed by the researchers. The function of the process will be performed on the strength of the goodwill and ethics of the people concerned. The goodwill is the cause for initiating a philosophy to persuade people to a managerial process. The process is a novel managerial theory, holacracy. The issue has to be globally considered by the UN. Using the managerial philosophy holacracy, human beings applying the love and goodwill for the fellow beings, with data provided by the Global Data Centre of Food Availability, Excess, and Need, the UN will operate the process with all the nations of the world.

The researchers are shocked to learn the significance of these data. They resolve to investigate the paradox of the above data materials. They feel the presence of the fruit, forbidden to eat, so is the availability of food forbidden to humans to taste. It is here the world can note the ultimate success of the investigation. The researchers ring out the old theories of management to ring in a new theory to be confident to overcome the incapability of human beings in fetching food. The researchers finally assert that the world can wipe off poverty and hunger from the globe if every human being in the globe is responsible for the well-being of the rest of humans by keeping a website with hunger food digital data open and visible to the world to get subjected to the food management process called holacracy to be implemented by all countries and all humans inhabited in them, denouncing the obsolete hierarchy, implemented through a process with multilevel management using the good will of the humans.

Timely response action from the real-time data centre reports retrieved is propelled by the noble values of holacracy and the latest digital transformation technologies that address a variety of concerns against the successful implementation of the UN goals. This needs to get subjected to the food management process called holacracy. It is to be implemented by all countries and all humans inhabited in them. Holacracy denounces the obsolete hierarchy, implemented through a process with multilevel management using the goodwill of humans.

The researchers finally assert that the world can wipe off poverty and hunger from the globe if every human being on the globe is responsible for the well-being of the rest of humans.

This data nerve centre, followed by the execution of actionable plans with its contents expounded here, will help wipe off food shortage. Human beings all over will be able to solve the issue of hunger and, thereupon, poverty.

2 Context

At the outset, the researchers examine the UN's 2030 Agenda for Sustainable Development and its 17 Goals (SDGs) as shown in Fig. 1.

However, the research focuses on only 2 of the 17: hunger and poverty. The greatest UN concerns are linked to FOOD, as the UN continues to chant the eradication of hunger and poverty. It's not enough; the researchers feel the need for a total change to occur in man with transformation in the mood of the world by devising and proposing a new approach to holacracy in food production, distribution, and management "The World Is One Family".

The researchers remind the world of the Indian dictum Vasudhaiva Kutumbakam – "The World Is One Family" – as the base: it may appear utopian, but using holacracy the researchers see the goal yonder. The hypothesis comprises producing

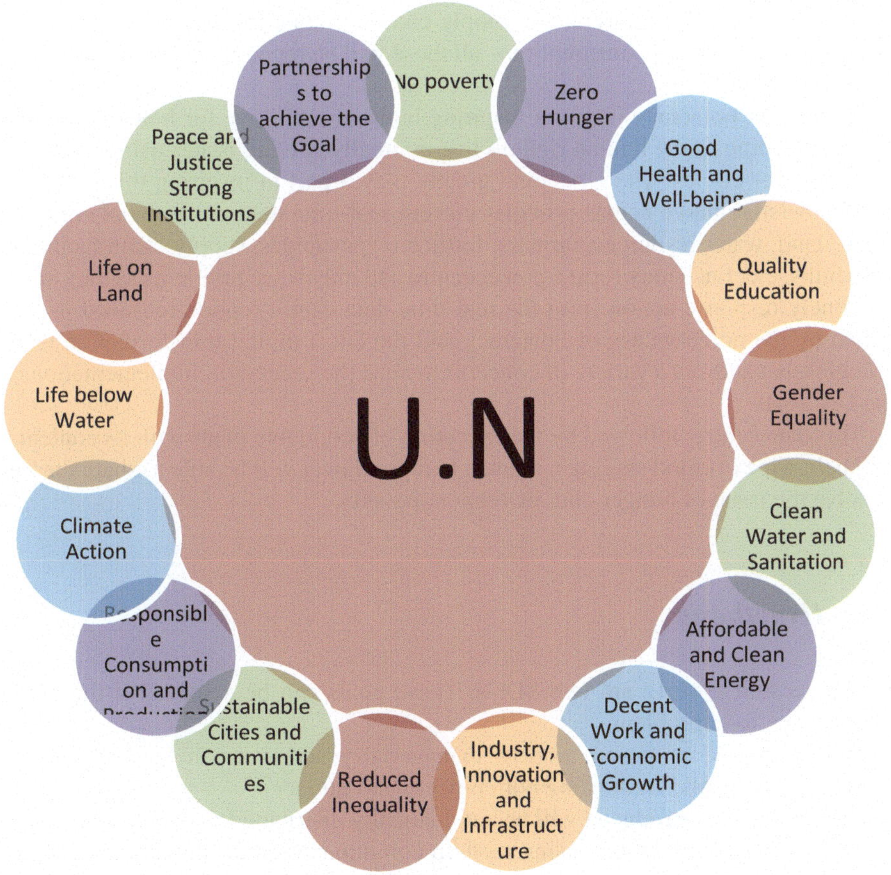

Fig. 1 UN 17 Sustainable Development Goals

food for all and using food together by all, treating the world as one. The mood change will generate the thought that food is not for nations, for farmers, for industrialists, for businesses, or for import or export. Food is for all members of the family, for all human beings on earth. The research here wishes that the UN could inspire nations with the dictum, "The World is one Family", followed by an exhortation: "human values" will not sound good when human beings starve.

The researchers propose how the UN could opt for holacracy as the viable process to achieve the goal. Holacracy is used not as a model, idea, or theory, but as a practice, as a new social technology. Holacratic organizations are governance structures that have the potential to wipe off all obstacles that stand in the way.

The implementation begins with the UN functioning as a nerve centre for data. It will be instituted with 193 member nations as constituents, and they contribute data to the centre as equi-important, mutually dependent, mutually complementary, and mutually supportive units. The centre will exhibit the global food scenario on a common website. The UN will make the moral and ethical exhortation to all human beings generating a SPARK, partly emotional, partly ethical, and moral. Data on the availability of food, requirements, supply chain options, fulfilment of needs, and ensuring of supply and consumption – all these will be reported on the website. The site is open to all.

Embracing holacracy involves inspiring humans with love for humanity as the panacea leading to the human right to live, mutually solving issues using interconnected power circles. The UN could greatly play a part in inspiring this embrace. The knowledge nerve centre provides current real-time data on the availability of arable land, water resources, farmers, fertilizers, pesticides, supply chain facilities, crop mills, etc. Each may form a power centre and may solve problems as they arise.

Timely response action from the real-time data centre reports retrieved is propelled by the noble values of holacracy and the latest digital transformation technologies that address a variety of concerns against the successful implementation of the UN goals.

This data centre, followed by the execution of actionable plans with its contents, will help wipe off food shortage. Human beings all over will be able to share the toil to solve the issue of hunger and, thereupon, poverty.

2.1 Hypothesis

- Hunger and poverty are two SDGs. These goals can be achieved if the globe solves its food problem.
- Holacracy, the proposed modern managerial practice can help the globe to achieve the goals; eradicating both poverty and hunger.
- The excess food produced will be automatically distributed to enable all humans to share food with others, using modern communication technology and supply chain facilities dexterously.

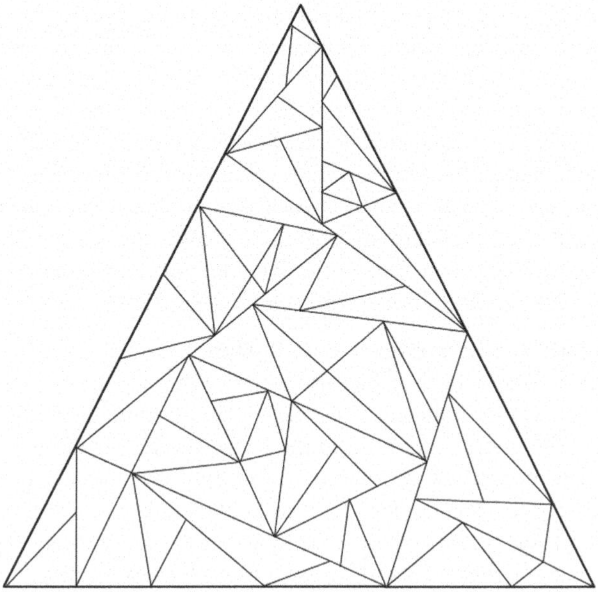

Fig. 2 Composite constitution of the UN

- Above all, these goals can be fully attained only if the humans on earth have ethical values and love for other humans (Fig. 2).

(1) The UN comprises 193 countries.
(2) All countries differ in area, geography, and socio-economic characteristics.
(3) All countries have a presence of common human race.
(4) Countries differ in production of their food.
(5) The total production of food may suffice their requirements.
(6) The human will alone can solve the world food problem by better food distribution.
(7) The world comprises one family spread in different countries.

2.2 *Chapter and the Crux of the Matter*

This chapter of course could be a potential breakthrough in among the changes mooted by the UN. At the outset, the researchers examine the UN's 2030 Agenda for Sustainable Development and its 17 Goals (SDGs). Hunger and poverty are two SDGs. These goals can be together achieved if the globe solves its food problem. Employing the modern managerial process called holacracy, the global goals of eradicating poverty and hunger can be achieved. The excess food produced will be automatically distributed to enable all humans to share food with others, using dexterously the modern communication technology and supply chain facilities. Above all, these goals can be fully attained only if the humans on earth have ethical values and love for other humans. However, the research here focuses only on two SDGs, as the greatest UN and human concerns are linked to food and as the UN continues to chant eradication of hunger and poverty. It's not enough; the researchers feel that there is a need for total transformation in human minds towards fellow beings.

The research chapter proposes four sections of activities: (a) will and virtue can only change the world, not the resources; (b) the total transformation from the present world to a hunger free world; (c) the practices of meticulous holacracy management modus operandi will help solve the issue; and (d) the exemplary new management concerns will efface the current business tools.

The research modes will opt for four tracks:

Mode 1: Change the Mood of Man
"The World Is One Family", is a twine old thought widely prevalent in India. It demands in propriety to consider that all humans are brother and sister.

The researchers remind the world the Indian dictum Vasudhaiva Kutumbakam – "The World Is One Family" – as the base: it may appear utopian, but using holacracy the researchers see the goal, yonder. The hypothesis comprises producing food for all and using food together by all, treating the world as one. The mood change will generate the thought that food is not for nations, for farmers, for industrialists, for businesses, or for import or export. Food is for all members of the family, for all human beings on earth. The research here wishes that the UN could inspire nations with "The World Is One Family", followed by an exhortation: "human values" will not sound good when some human beings starve.

Mode 2: Ring in the Centre of "Food Data of the Total Globe"
The implementation of corrective activities begins with the UN functioning as a nerve centre for data. It will be instituted with 193 member nations as constituents, and they will contribute data to the centre as equi-important, mutually dependent, mutually complementary, and mutually supportive units. The centre will exhibit the global food scenario in a common website. The UN will make the moral and ethical exhortation to all the human beings generating a SPARK, partly emotional, partly ethical and moral. Data on the availability of food, requirements, supply chain options, fulfilment of needs, and ensuring supply and consumption – all these will be reported on the website. The site is open to all. The knowledge nerve centre provides current real-time data on the availability of arable land, water resources, farmers, fertilizers, pesticides, supply chain facilities, crop mills, etc. Each may form a power centre and may solve the problems as they arise. This data centre followed by the execution of actionable plans with its contents will help wipe off food shortage. Human beings all over will be able to share the toil to solve the issue of hunger and thereupon poverty.

Mode 3: Ring in the Managerial Policy
The researchers propose how the UN could opt for holacracy as the viable process to achieve the goal. Holacracy is used not as a model, idea, or theory, but as a practice, as a new social technology. Holacracy organizations are governance structures that have the potential to wipe off all obstacles that stand in the way. Embracing

holacracy involves inspiring human beings with love for humanity as the panacea leading to the human right to live, mutually solving issues using interconnected power circles. The UN could greatly play a part to inspire this embrace. Timely response action from the real-time data centre reports retrieved is propelled by the noble values of holacracy and the latest digital transformation technologies that address a variety of concerns against the successful implementation of the UN goals.

Mode 4: Bring in the Data Response "World Family Forming" Process
Embracing holacracy involves inspiring human beings with love and concern for Humanity as the panacea leading to human right to live, mutually solving issues using interconnected power circles. The UN could greatly play a part to inspire this embrace. The knowledge nerve centre provides current real-time data on the availability of arable land, water resources, farmers, fertilizers, pesticides, supply chain facilities, crop mills, etc. Each may form a power centre and may solve the problems as they arise.

Timely response action from the real-time data centre reports retrieved is propelled by the noble values of holacracy and the latest digital transformation technologies that address a variety of concerns against the successful implementation of the UN goals.

This data centre followed by the execution of actionable plans with its contents will help wipe off food shortage. Human beings all over will be able to share the toil to solve the issue of hunger and thereupon poverty. If the data from one country provides information that the people of a part of their country do not have enough food, the reader shall respond with action of sending the excess food it has, and then the file is closed. This is done as a natural response. This is not the result of a hierarchical process (Fig. 3).

The data centre will carry data as provided by the concerned UN agencies with regard to the following fields as in Fig. 4.

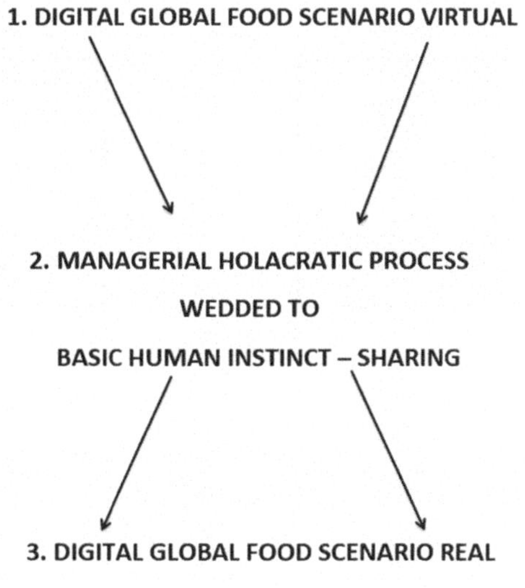

Fig. 3 The UN's initiative for food self-reliance – Data Reserve Bank

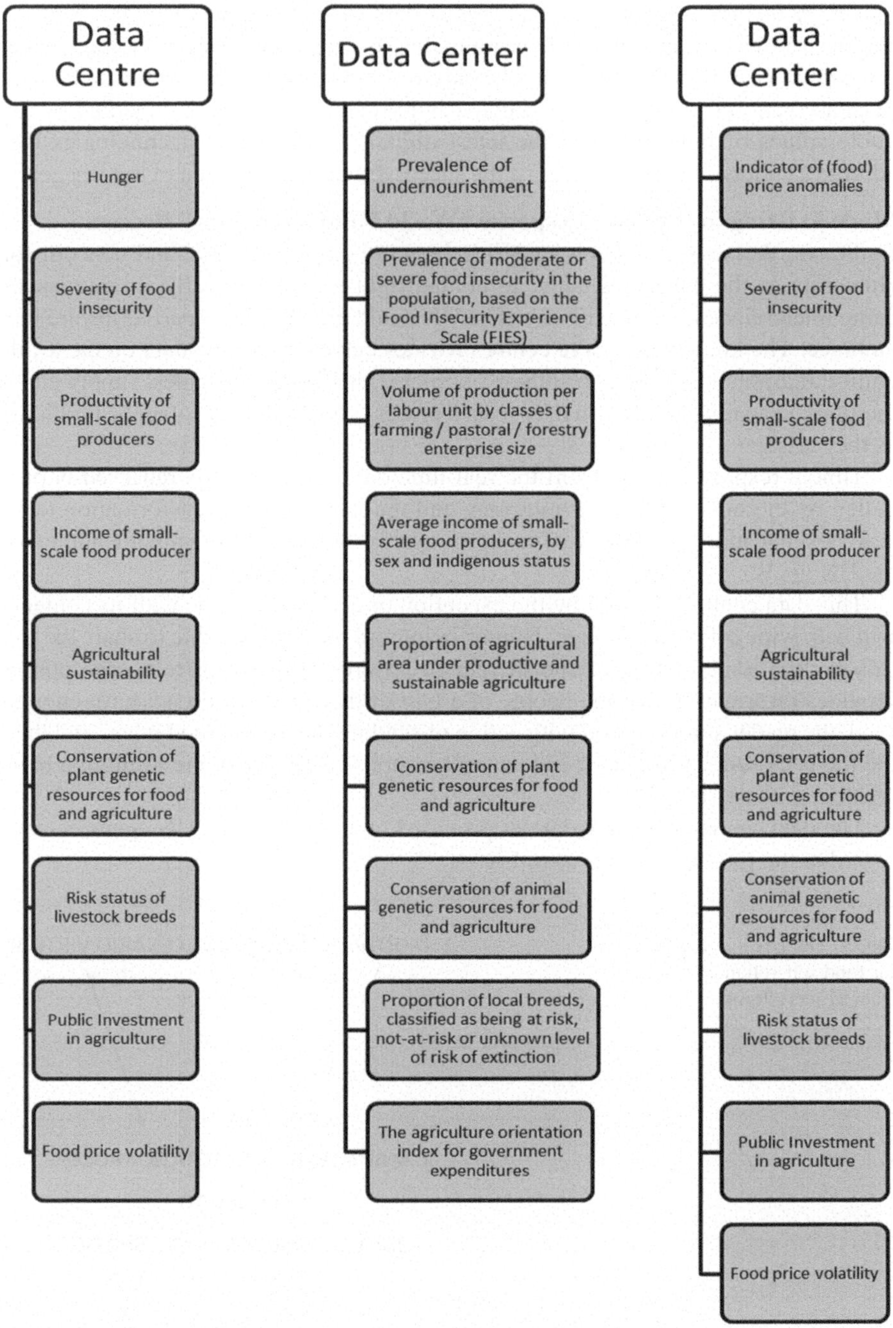

Fig. 4 Fields in the data centre

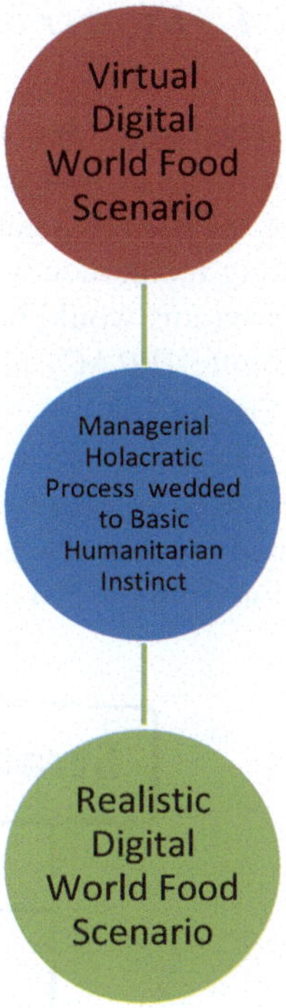

Fig. 5 Virtual to realistic process

Every human being may use the digital data center to access relavant data and information. This knowledge sharing will lead to the achievement of the goal of moving the food from the virtual to the realistic process, as shown in Fig. 5.

2.3 *Wiping Out Hunger and Poverty*

Of all the UN's 2030 Agenda for Sustainable Development and its 17 Goals (SDGs) as in Fig. 1, hunger is the greatest threat to mankind. Hunger destroys many human lives every year. Chronic hunger is the cause of death of more than 828 million people in the world. It causes premature death, children being the main victims. Above 20.5 million babies are annually born with low birth weight as their mothers didn't get proper food. "Starvation only accounts for about 8% of the hungry people in the world", says Pedro Sanchez, co-chair of the UN Millennium Project's Hunger Task Force. "The other 92% suffer from hunger silently, and they die in droves because of mal-nutrition-related diseases" (GlobalGiving 2021).

2.4 *Hunger in the United States*

According to Joel Berg, the CEO of Hunger Free America, it would cost $25 billion to eliminate hunger in the United States. Unlike hunger caused by war, crop failures, or inflation, hunger in America is primarily due to a lack of funds to purchase food. Berg recommends that a combination of increased wages and improved safety net programs would be necessary to address this issue. The Food Research and Action Centre (FRAC) also provides key strategies to end hunger in the United States (Fig. 6).

Even in the United States, the essential strategies to end hunger in America are reduced to chants. It is in this context that the researchers have undertaken their study with a few constructive suggestions. The researchers endorse the first deviation from the last of the suggestions. It is not the political will that must be built, but

Fig. 6 Strategies to end hunger in America

a strong humanitarian will, the will reinforced and rejuvenated by human values like providing welfare for our brothers and sisters in our "One Family".

2.5 UN Agenda

The UN General Assembly in 2015 adopted the 2030 Agenda for Sustainable Development. This document includes 17 Sustainable Development Goals (SDGs). Building on the principle of "leaving no one behind", "the new Agenda emphasizes a holistic approach to achieving sustainable development for all (Oommen et al. 2022)"; we have examined "the well thought out seventeen" for the world. The greatest concerns on hunger and poverty will continue threatening the world.

2.6 Food Is the Most Significant Human Right

As the UN continues chanting "the well thought out seventeen", the researchers resolve to change the mood of the world by devising a new approach. The researchers exhort to think on the Indian dictum: Vasudhaiva Kutumbakam – "The World Is One Family". This appears to be a utopian idea. But using a management concept, the researchers suggest achieving one of the major goals.

2.7 Provide Food to All

The research takes up food as the greatest of the targeted goals. The hypothesis comprises producing food and using food together, treating the world as one. For that, we need to change the mood of the world. We can use a panacea of instituting holacracy, a novel management concept for practice. We propose a global application of holacracy in food production and its management. The main objective, mood change, will relate to the concept: food is not for nations, for farmers, for industrialists, for businesses, or for import or export. Food is for all members of the family, for all human beings on earth (Serrini 2018; Skripak et al. 2018; Thompson and McHugh 1990; Ulieru 2005; Viðarsson 2017; Wilber 1982).

2.8 Holonic Systems

The metaphor of the holonic (holacracy) systems has been the subject of research for more than 20 years, exploited with different semantics in the fields of organization, management, and even computer science (Ravarini and Martinez 2019).

Among organizational studies, scholars have been introducing different terms over time: such as holonic enterprise (Thompson and McHugh 1990), virtual organization (Mowshowitz 2002), self-organization (Ulieru 2005), etc. These concepts all share a common objective of presenting and elucidating an organizational model characterized by weak social connections and constantly evolving structures, control mechanisms, and power dynamics. This area of research can be attributed to the ongoing discourse surrounding centralized versus decentralized governance, where the holonic system serves as a metaphor endorsing the notion of decentralized governance.

2.9 Holacracy Concept

Holacracy was introduced by Brian Robertson and Tom Thomison in 2007 as a system of organizational governance where decisions are made by teams and authority is distributed throughout the organization rather than concentrated in top-level management. The term "holacracy" was first used by Arthur Koestler in his book *The Ghost in the Machine* (Koestler and Wagmuth 1968) to describe a hierarchical system of self-organizing holons. This concept has since evolved into a management approach that emphasizes distributed authority, self-management, and fluid roles within an organization. Holacracy is a social terminology and governance system that has gained recognition in the modern workplace (Kamp 2014; Krasulja et al. 2016; Kristensen and Shafiee 2019; Kurki et al. 2016; Kurki and Wilenius 2016).

2.10 Self-Management Organizations

Self-managing organizations (SMOs) are characterized by a formal and systematic decentralization of power throughout the organization, according to Elman (2018). This approach contrasts with the hierarchical reporting relationship that characterizes traditional managerial structures. The decentralization of power is a key mechanism of control in SMOs, which also feature well-defined decision-making processes that enable employees to exercise self-rule. In a study by Davis, Eisenhardt, and Bingham (2009), it was found that SMOs perform well in dynamic and unpredictable environments. These environments tend to have low levels of structure, which can facilitate the implementation of tasks. Small organizations with little structure lack guidance to make suitable behaviours efficiently. Yet, larger organizations with strict structures are compelled and lack inflection (Baden-Fuller and Volberda 1997).

2.11 Holacracy and Hierarchy

Holacracy is a unique approach to organizational structure that replaces the traditional hierarchical pyramid with a system of circles and sub-circles, as outlined by holacracy (Robertson 2015). Each circle has specific roles that connect and coordinate the work of the sub-circles, and each person in a holacratic organization can take on multiple roles and belong to multiple circles. Unlike traditional job descriptions, roles in holacracy are dynamic and can change in response to internal or external needs (Mahlo and Spencer 2022; Mella 2009; Mosamim and Ningrum 2020; Mowshowitz 2002).

2.12 Organizational Structure

In holacracy, the traditional structure of an organization is replaced with a more organic, self-organizing system of circles and sub-circles, as described by holacracy (Robertson 2015). Each circle is connected to its sub-circles through a double-link, which facilitates two-way communication and fast feedback loops. In this system, each circle is responsible for determining the roles needed to achieve its goals and assigning members to fill those roles. This empowers each circle to rule itself and make decisions in alignment with its purpose (Baden-Fuller and Volberda 1997; Campbell et al. 2003; Davis et al. 2009; Elman 2018).

2.13 Organizational Control

According to Kristensen and Shafiee (2019), holacracy is effective in increasing organizational speed by improving the methods of controlling activities. With holacracy, decisions can be made quickly and incrementally with the maximum available information. This allows organizations to adjust their course continuously as new information emerges. Moreover, holacracy encourages individual action based on sound judgement when the expected decisions and actions are not specific. This approach helps individuals take ownership of their impact and provides the organization with valuable knowledge from experimentation (Hollenbeck et al. 2011; Iqbal and Kureshi 2016).

3 Managing Food Distribution Through Holacracy

3.1 Production, Distribution, and Consumption

The researchers propose uploading of digital representation of the world scenario of all details of food production, food distribution, and food consumption. All countries in the UN shall provide all information with them to the digital globe food website voluntarily. There is no advice, instruction, or direction behind the voluntary disclosure. The whole world shall see the statistics and respond to the needed information faithfully as humans. This is the prior requirement of holacratic managerial endeavour. What research suggests is that all conferences convened by the UN shall focus mainly on materials required for the data reserve. With reference to collecting information and uploading to the UN sites, the research suggests the following.

3.2 International Conference on Agricultural Statistics (ICAS)

The UN has taken decision to convene an International Conference on Agricultural Statistics (ICAS) with focus on fostering and leveraging best practices and research in response to the changing needs and opportunities for agricultural statistics. ICAS brings together experts from around the world to share research and operational accomplishments and to explore the latest methodological innovations by countries and development partners in May 2023 in Washington, DC.

3.3 Proposal for Similar UN Gestures

The UN may take initiative to conduct international conferences on data analysis of food statistics, food requirements, food distribution and statistics. The researchers propose to the UN to organize similar international conferences on Data Analysis of Food Statistics, on Food Requirements and Food Availability Statistics, and on Holacracy Process in Food Production, Distribution, and Utilization of Agricultural Statistics.

Enough food is produced today to feed everyone on the planet, but hunger is on the rise in some parts of the world, and some 828 million people are considered to be "chronically undernourished". What steps are being taken to ensure that everyone, worldwide, receives sufficient food? (Nations 2019).

The researchers conclude the chapter with the contented note: the UN will be a facilitator of exact information of the food position to all the countries, and those who can respond will promptly respond to current situation, and the daily status will state in the world website the absence of the words *hunger* and *poverty*.

3.4 Potential Themes for Sessions and Papers

The suggestions of themes for the UN conference agenda are given below. These conferences could bring in more insights and awareness and could benefit from focusing on the following themes:

- Food security diets and nutrition.
- Rural development, poverty, and social issues.
- Climate change and environmental issues.
- Sustainable agriculture production and consumption.
- Markets, prices, and value chains.
- Shocks, risks, and resilience.
- Natural resource use.
- Data collection/data quality.
- Alternative data sources/data integration/data interoperability.
- Innovative approaches to data analysis.
- Data access/data dissemination/data use for policymaking.
- Capacity building in agricultural statistics.
- Training session for young (or new) statisticians, particularly from developing countries, which is organized by the USDA.

The outcome of all these conference proceedings as shown in Fig. 7 will be displayed in the global data centre and can be used by the users for supply chain use.

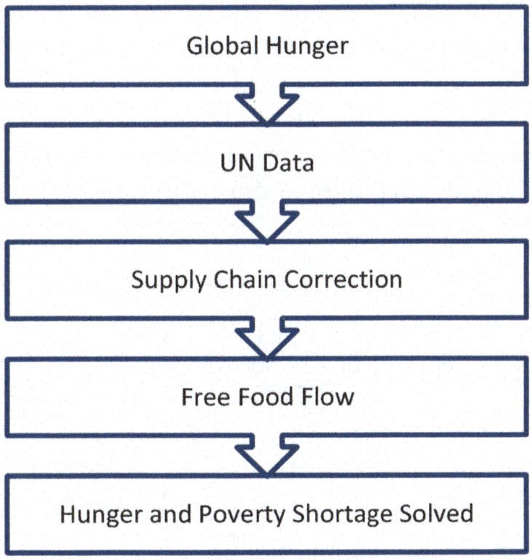

Fig. 7 Outcomes of Conference Initiatives

4 Our Recommendations

Our recommendations to address hunger and poverty could be realizable through a multi-faceted approach. We suggest the following:

- The website of UN food status is open to all in the world.
- The website will show the virtual and digital food scenario as in (Fig. 8).
- Convene an International Conference on Agricultural Statistics (ICAS).
- Convene an International Conference on Data Analysis of Food Statistics (ICAS) for generation of statistics.
- Convene an International Conference on Food Requirements and Food Availability Statistics (ICAS) for generation of statistics.
- Convene an International Conference on Holacracy Process in Food Production, Distribution, and Utilization Agricultural Statistics (ICAS) for generation of statistics.
- Convene a Holacracy Implementation Conference.

4.1 Findings

The research uses modern technology to access the food status quo using the digitally latest data visibility of the global scenario pertaining to food production, distribution, and consumption to solve issues of poverty and hunger besides reaching other EP goals partly. At the same time, the recommendations put forth here by the researcher's final consolidation in the chapter legitimize the strong presence of the largest organization – the UN – to take care of mankind. The UN invokes the greatest ethical virtue in humans in the world at large, vindicating the fruitful culmination of UN's human welfare initiative of giving life to all. The whole world after the huge success can be contented to try holacracy for the rest of management undertakings of the UN and will continue similar gestures for all humans (Viðarsson 2017; Wilber 1982).

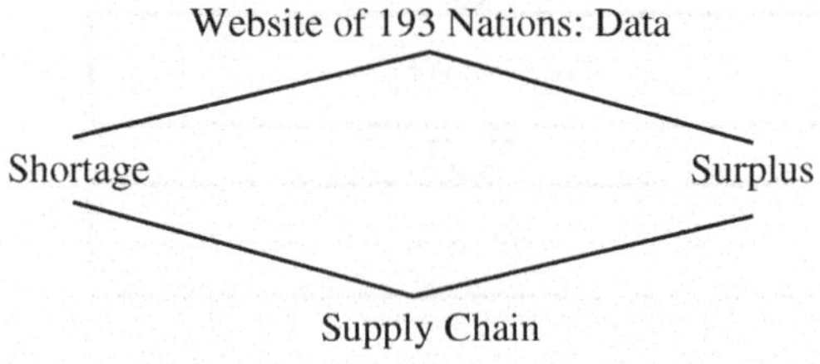

Fig. 8 Elements of Digital Food Scenario in Website

- Website of 193 Nations: Data
- Shortage
- Surplus
- Supply Chain

5 Conclusion

The research takes up food as the greatest of the targeted goals. The hypothesis comprises producing food and using food together, treating the world as one. For that we need to change the mood of the world. We can use a panacea of instituting holacracy, an untried management concept for practice. We propose global application of holacracy in food production and its management. The main objective, mood change, will relate to the concept: food is not for nations, for farmers, for industrialists, for businesses, or for import or export. Food is for all members of the family, for all human beings on earth.

References

Baden-Fuller, C., & Volberda, H. W. (1997). Strategic renewal: How large complex organizations prepare for the future. *International Studies of Management & Organization*, *27*(2), 95–120.

Campbell, D. J., Dardis, G., & Campbell, K. M. (2003). Enhancing incremental influence: A focused approach to leadership development. *Journal of Leadership & Organizational Studies*, *10*(1), 29–44.

Davis, J. P., Eisenhardt, K. M., & Bingham, C. B. (2009). Optimal structure, market dynamism, and the strategy of simple rules. *Administrative Science Quarterly*, *54*(3), 413–452.

Elman, B. (2018). *The coordination mechanisms of self-managing organizations: An explorative case-study of three pioneers*.

GlobalGiving. (2021). *How Much Would It Cost to End World Hunger?* https://www.globalgiving.org/learn/how-much-would-it-cost-to-end-world-hunger/

Hollenbeck, J. R., Ellis, A. P. J., Humphrey, S. E., Garza, A. S., & Ilgen, D. R. (2011). Asymmetry in structural adaptation: The differential impact of centralizing versus decentralizing team decision-making structures. *Organizational Behavior and Human Decision Processes*, *114*(1), 64–74.

Iqbal, S., & Kureshi, N. (2016). Dissatisfaction With Performance Management System In A Bank; Possible Reasons And Remedies. *Journal of Strategy and Performance Management*, *4*(1), 4.

Kamp, P. V. D. (2014). Holacracy – a radical approach to organizational design. *Elements of The Software Development Process: Influences on Project Success and Failure* (Eds. H. Dekkers, W. Leeuwis, I. Plantevin) University of Amsterdam, 13–26.

Koestler, A., & Wagmuth, W. (1968). *Das Gespenst in der Maschine*. Molden Wien.

Krasulja, N., Radojević, I., & Janjušić, D. (2016). Holacracy-the new management system. *Proceedings of the International Scientific Conference, Njs, Serbia, 13*.

Kristensen, S. S., & Shafiee, S. (2019). Rethinking organization design to enforce organizational agility. *11th Symposium on Competence-Based Strategic Management, Stuttgart*, 1–13.

Kurki, S., Pura, M., & Wilenius, M. (2016). *RE-Acting The Future: New Ways to Work: The Case of Reaktor*.

Kurki, S., & Wilenius, M. (2016). Trust makes this organisation unique. *European Journal of Futures Research*, *4*(1), 1–12.
Mahlo, K. E., & Spencer, P. (2022). *The Use of digital project management solutions by project businesses: a case study of a selected platinum mine in the Limpopo province*. Journal of Construction Project Management and Innovation.
Mella, P. (2009). The holonic revolution. *Holons, Holarchies and Holonic*.
Mosamim, P., & Ningrum, S. (2020). Holacracy and Hierarchy Concepts: Which One is More Effective in an Organizational Leadership and Management System? *Malaysian Journal of Social Sciences and Humanities (MJSSH)*, *5*(12), 257–271.
Mowshowitz, A. (2002). *Virtual organization: Toward a theory of societal transformation stimulated by information technology*. Greenwood Publishing Group.
Nations, U. (2019). *UN News – UN News audio and video coverage from around the world*. https://news.un.org/en/story/2019/10/1048452
Oommen, K. P., Kurian, S., & Joy, J. (2022). Carbon Dioxide Neutralization Across the Global Supply Chain. *ECS Transactions*, *107*(1), 12131.
Ravarini, A., & Martinez, M. (2019). Lost in holacracy? The possible role of e-HRM In dealing with the deconstruction of hierarchy. In *HRM 4.0 for human-centered organizations*. Emerald Publishing Limited.
Robertson, B. J. (2015). *Holacracy: The revolutionary management system that abolishes hierarchy*. Penguin UK.
Serrini, L. (2018). *Rethinking organizations: the holacracy practice: an empirical analysis to assess the performability of the practice for small-medium and large-sized companies: ARCA and Zappos cases*.
Skripak, S. J., Cortes, A., Walz, A. R., Parsons, R., & Walton, G. (2018). *Fundamentals of business*. VT Publishing.
Thompson, P., & McHugh, D. (1990). *Work organisations: a critical introduction*. Springer.
Ulieru, M. (2005). Adaptive information infrastructures for the e-society. *Engineering Self-Organising Systems: Methodologies and Applications 2*, 32–51.
Viðarsson, E. R. (2017). *Learning Holacracy Fundamentals Through Play*. OCAD University.
Wilber, K. (1982). *The holographic paradigm and other paradoxes: Exploring the leading edge of science*.

Scope of System of Systems (SoS) in Industrial Technology and Examples

Nitaigour Premchand Mahalik

1 Introduction

Sustainability plays an important role in technology systems in industry. Sustainability means optimal design, cost-performance trade-off, energy, space and time savings, scalability, and interoperability (open systems). One of the designs that focuses on advanced industrial systems is system integration (SI) (Tabim et al. 2021) that can facilitate sustainability. System integration is key to SoS (system of systems) because it inherits a collection of task-oriented subsystems that together perform optimal and global operations. "A system of systems (SoS) is the collection of multiple, independent systems in context as part of a larger, more complex system. A system is a group of interacting, interrelated and interdependent components that form a complex and unified whole" (https://www.techtarget.com/searchapparchitecture/definition/system-of-systems-SoS; Ray 1989; Ashiku and Dagli 2019). Immediate examples where the SoS can be applied are business, education, government, healthcare, media, engineering, transportation, agriculture, safety, and so on. The SoS paradigm is configurable in that it can also be used for related applications. The SoS design approach requires knowledge base from several disciplines including sociology; physics; chemistry; biology; biochemistry; instrumentation engineering; mechanical engineering; manufacturing engineering; mechatronics engineering; electrical, electronics, and communication engineering; production engineering; and all under the banner of engineering technology. In the sequel, keeping in view sustainability, this chapter presents topics on industrial technology (IT) and systems of systems (SoS). First it presents an introduction to Industry 4.0 standards and systems, an important aspect of SoS, and then some application areas

N. P. Mahalik (✉)
Department of Industrial Technology, Jordan College of Agricultural Sciences and Technology, California State University, Fresno, CA, USA
e-mail: nmahalik@csufresno.edu

I. S. Dunmade et al. (eds.), *Sustainable Engineering*, Green Energy and Technology, https://doi.org/10.1007/978-3-031-47215-2_28

of IT as far as developments are concerned. The chapter includes (i) introduction to Industry 4.0, (ii) safety and privacy standards in industrial technology systems, (iii) agricultural safety and standards, (iv) advanced food processing and packaging, (v) space and ocean technology systems, (vi) smart home with energy implications, (vii) hypertube technology in transportation, (viii) MEMS and NEMS technology, (ix) entertainment technology, (x) business opportunity on social media and networking, and (xi) SoS concluding remarks.

2 Background and Introduction to Industry 4.0

Industry 4.0 paradigm has become the norm recently (https://www.i-scoop.eu/industry-4-0/). Open and sustainable technology systems back the paradigm. In essence, Industry 4.0 is the new age industrial revolution that brings digital and physical technology together which helps in smart factories and supply chains. The estimated revenue generation due to companies adapting to Industry 4.0 can reach one trillion US dollars by the year 2030. Industry 4.0 uses cyber-physical systems (CPS), and their applications are found in industrial robotics, sensors, IoT, and cloud computing as the implications increase productivity. Since manufacturing companies improve and distribute products with new technology using AI, the Internet of Things (IoT), analytics, cloud computing, and Industry 4.0 are there to connect with the factory floor and send real-time data to the cloud created for smart factories. There are multiple challenges for Industry 4.0 because the strategic implication intends to collect vast data across a wide scope of activity. As an example, almost 50% of an industrial asset can lead to maintaining a data collection system. And the integration of digital and physical systems enables cyber threats that affect data privacy even though the secondary data deals with critical hardware that assures management issues with AI automation. As claimed by the stakeholders, Industry 4.0 is the information-intensive transformation of manufacturing and related industries in a connected environment of big data, people, processes, services, systems, and IoT-enabled industrial assets with the generation, leverage, and utilization of actionable data and information as a way and means to realize smart industry and ecosystems of industrial innovation and collaboration. Industry 4.0-based smart system is capable of coordinating production for global distribution. A coordination might involve two billion components from the built-to-order process. In such an environment, it was estimated that the predictive information can lead to reducing machine downtime by 30–50% while improving life span by 20–40%. This in turn increases time to market by 30–50%. Some of the companies that have adapted Industry 4.0 are IBM, Qualcomm, Intel, and Microsoft, to name a few.

3 Safety and Standards in Industrial Technology Systems

The Occupational Safety and Health Act (OSHA) of 1970 is a US labor law which governs the Federal law of occupational health and safety in the private as well the public sector in the USA. OSHA has the authority to both set and enforce workplace safety standards. Religious institutes come under the act with necessary clauses. Within the paradigm, a Process Safety Management (PSM) program is another aspect of hazardous materials at onsite factories. PSM standards such as "29 CFR 1910.110" explain elements associated with a management program. PSM is for risk management that aligns with environmental health and safety, emergency preparedness, business continuity, and public safety. PSM programs are usually reviewed and adjusted to improve safe environments every 3–5 years. PSM is meant for sustainability. Legislation ensures that a product or event within an industrial environment is safe and not dangerous. OSHA legislations can be found in Title 29 of the Code of Federal Regulations (CFR) and are divided into separate standards for common industries including construction. For example, enforcement of 29 CFR 1910 that is regulated by OSHA is applicable to all companies, and it is about inspections and citations.

The hazards that the industries must deal with are chemical hazards, compressed gas, lifting, noise, fire, and so on. A chemical affects the human body in a variety of ways causing harm to either a single cell, a group of cells, an organ system, or the entire body. Chemicals may also find a way to enter the body via the skin, inhalation, and ingestion. Both employers and employees must have appropriate procedural knowledge or training facilities when working in an environment where pressurized or poisonous gasses are involved. It is equally important that they should know what to do in case of a gas leakage or an emergency. In this respect, guidelines must be available for safe cylinder handling, for instance. Also, PPE (personal protective equipment) such as safety glasses plays an important role when handling compressed gasses. Protective gloves and safety shoes are designed using appropriate standardized materials. Note that selection of materials to develop PPE items that are needed to follow OSHA safety regulations in the industry is itself a SoS approach. An OSHA safety regulation could be a frequent inspection of the area around the valve for corrosion damage or mechanical strain using PPE such as gloves and glasses. The most demanding factor is to check documentation periodically. There are strict procedures for transporting the compressed gas as well. As such a vehicle that has intrinsic safety (IS-78)-compliant wiring in the engines is preferred. Repeated lifting causes back injury pain that can be acute or chronic occurring between the neck and spine. The lifting procedure entails knowledge on ergonomics as well as anatomy. As regards noise, anything over 90 dBA is known to be harmful. Workers must either be provided hearing protective equipment or have limited exposure time. Similarly, for electrical safety, methods such as the lock-out-tag-out device will prevent unauthorized people from accessing the machine operated by high-voltage electrical systems. Fire safety is described in OSHA standards for general industries as well as in recordkeeping, maritime, and construction industries. The standards exclusively illustrate the types of extinguishers, the fire classes for which each is used, and the limitations of each extinguisher. Above all, biological hazards involve bacteria, viruses, fungi, or other living organisms that are able to infect. The last level and the least effective form of control for safety is PPE or personal protective equipment that was already introduced above. Examples of PPE are hazmat suits, goggles, gloves, hard hats, and special protective clothing. One of the safety inspections includes the status of PPE equipment and must meet the standards. For example, all hard hats must meet ANSI Standard Z 89.1. Reporting failures in machinery and PPE equipment are ways of prevention of mistakes. The main foci on safety and standards in industry and industrial systems are (i) continuous improvement of SMS; (ii) identification of principles of safety, health, and environmental management; (iii) the prevention principles; (iv) facilitating team-based organizations; (v) incorporation change; (vi) budgeting; and (vii) total quality management approach.

Example Electrical safety and fire safety are the safety procedures that inform a person on how to act in such situations and how the latest technologies can be used to alert prior to occurrences. To apply mandate approaches to these problems in a particular way, we must not only first create awareness and standard drills and use the latest technology but also learn about the incident types and then communicate as early as possible. This is called the system-of-systems (SoS) approach to safety regulations. Awareness includes sharing of information, for example, about what a particular signboard means and places where extinguishers and pull stations are placed. Apparently, it has five aspects: self-concept, thoughts, feelings, body, and emotions. Secondly, standard drills are practices that teach a person to behave and act in such a way to avoid any stampede, panic attack, and blockage. It has four aspects such as massed, distributive, fixed, and variable. Similarly, technology is a vital part of safety such as the mass notification system (MNS) which broadcasts messages to the public during critical events. Tools such as smoke detector, sprinkler, safety fuse, and sorts of sensors must be used. It has five aspects: analyze, design, implement, execute, and evaluate. SoS procedural methods must be available in accessible form for technical as well as nontechnical audiences including the public to create awareness, where all these approaches must be practiced to prevent widespread incidents, panic, and deaths. Figure 1 shows an SoS architecture as far as electrical safety is concerned in large-scale vehicle charging stations. Sustainable safety regulations are developed to ensure safety from electrical and hazardous accidents as the supply equipment and the battery are the two main driving factors in this case. For details refer to (Wang et al. 2019).

Diversity and inclusion (D&I) in industrial technology systems is a must, meaning the enterprise software must provide insights at the individual or organizational level, in support of organizations' efforts to become more diverse and inclusive. It is now more appropriate to pronounce the D&I technology market, and it is convincing that tremendous opportunities exist for D&I technology vendors to sell their solutions to larger organizations as can be seen from the fig (Micu et al. 2021). This approach entails sociology as one of the aspects of SoS strategy, as indicated (Fig. 2).

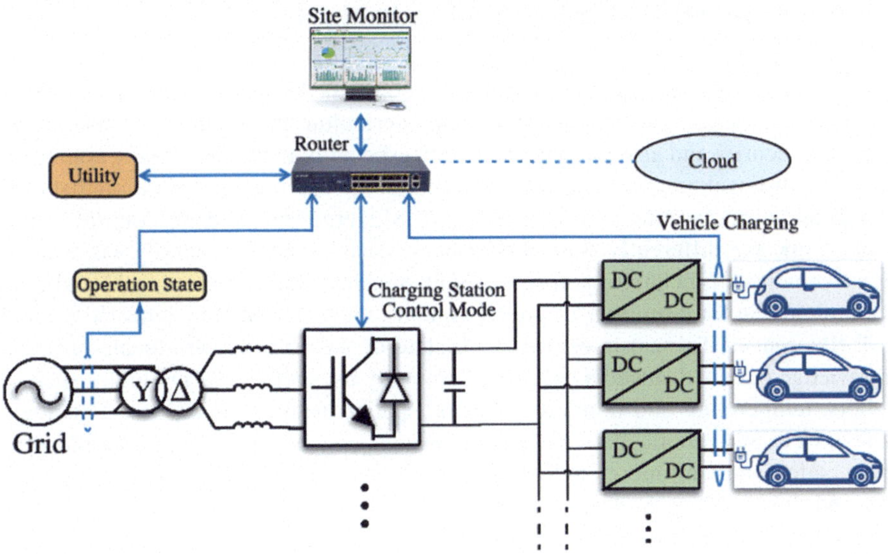

Fig. 1 Overall architecture of an EVCS. (Wang et al. 2019)

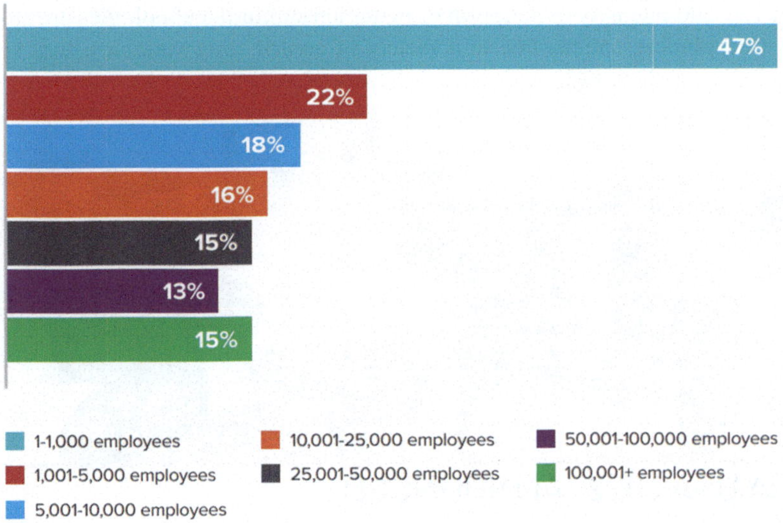

Fig. 2 Average customer size for D&I technology vendors (n = 42). (Source: RedThread D&I Technology Survey, 2018; Garr and Jackson 2019)

4 Agricultural Technology and Industry 4.0

According to analysis, it is believed that by 2050, the human population can reach virtually ten billion. The benefits of adopting agricultural technology include ensuring food security and also poverty reduction, especially in the developing countries. This entails development of agricultural technology and their contribution that now stands as "smart farming and sustainability." Modern farms and agricultural operations work far differently than those many decades past, primarily because of advancements in technology, together with sensors, devices, machines, and data technology. Smart farming uses digital systems, cyber-physical systems (SPS), and IoT (Internet of Things) integrated with state-of-the-art tools and techniques and knowledge base. Advanced agriculture habitually uses refined technologies such as robots, temperature and humidity sensors, spatial pictures, and GPS technology. The devices and robotic platform permit businesses to be a lot more profitable, efficient, safer, and a lot more environmentally friendly, in addition to managing hunger and thus holding the key to economic condition. Autonomous farming begins with collaborative farm robots. Figure 3 shows Industry 5.0-type robotic platform (displayed by Blue White Robotics at FIRA USA 2022) displayed at Fresno State campus in April 2022. Blue White Robotics revolutionizes agriculture through autonomy. It is claimed that the self-driving tractor is fully autonomous through multiple sensors, ultrahigh-precision navigation system, and robust AI models ensuring real-time situational awareness with no downtime. The entire agriculture and food sectors apply SoS architecture.

The rate and intensity of adoption of better agricultural technologies are still low in most developing nations due to a variety of contributing variables. It is vital to

Fig. 3 Robot tractors and farm equipment on display at Fresno State. (Courtesy: Photo - FOX 26 Photojournalist Olen Hogenson; Article – Stephen Hawkins on April 13th 2022)

enhance food security and boost poverty reduction. The result of the study shows that the adoption of agricultural technology in developing countries will lead to enhanced food security and poverty reduction. These low rates of adoption are prevalent even in most third-world countries. There are potential barriers that prevent farmers in developing countries from adopting and employing new agricultural technologies. A meta-analysis of the empirical literature needs to be produced in order to reflect various aspects of bottleneck, hindrance, and setbacks on adaptability.

OSHA tries to provide a safer work environment for farmworkers. It requires the farm employers to follow the safety program and the farmers to follow the safety instructions provided by manufacturers of machinery, tools, and supplies. Also, the ISO (International Standards Organization) standards cover safety aspects of agriculture. Both the organizations help to promote effective farming methods to adequate levels of safety and quality. Some of the important titles on agricultural safety and standards include standards for machinery (ISO-17989 & ISO/TC-23), irrigation standards (ISO/TC-23/SC-18), food safety standards (ISO/TS-22002-6), fertilizer standards (ISO/TC-134), environmental standards (ISO-14055), PPE standards (ISO-27065), and feeding-related standards (ISO-22000). It was observed that there are differences in international standards for agricultural products as different nations use different standardized fertilizers for the production even during germination and crops. The international standards are present to manage inconsistent markets in different countries. Asian countries cannot sell their cereal products in the EU (European Union) and the rest of the world market without implementation of proper standards. For example, Indonesia, Bangladesh, and India have shown growth of 17.34% in export in 2020–2021 given the fact that it has been declining due to noncompliance with international standards. Detailed information on the usage of standards can be found in respective countries' government sites. It is seen that most of the Asian countries suffer in the EU market as they use different standard fertilizers in agricultural productions. The global standards and safety practices for agriculture should bring the right balance to help entrepreneurs from developing countries find a place in the international market.

5 Advanced Food Processing and Packaging

Food industry is one of the biggest known industries in today's world, and the recent design and development include the system-of-systems (SoS) approach. The industrial infrastructures have been growing steadily toward Industry 5.0 which is nothing but SoS. Traditionally, the processing of food was achieved by means of manpower, which was a little expensive and had a higher rate of inaccuracies as far as measurements of grading quality and packaged quantity are concerned. However, with the help of advanced technology, the food processing and packaging industry has taken a huge leap (Akyazi et al. 2020). Accuracy is 99.99%, and the production speed has become a negative coefficient, meaning supply exceeds demand. The advanced packaging methods are better for the environment as they use biodegradable, water-soluble, and edible materials.

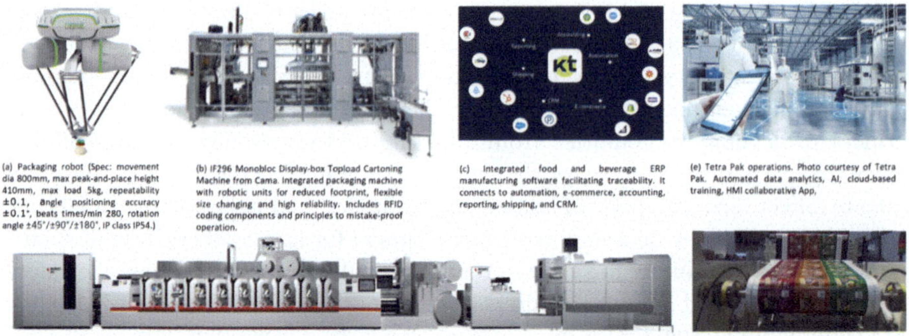

(a) Packaging robot (Spec: movement dia 800mm, max peak-and-place height 410mm, max load 5kg, repeatability ±0.1, angle positioning accuracy ±0.1°, beats times/min 280, rotation angle ±45°/±90°/±180°, IP class IP54.)

(b) IF296 Monobloc Display-box Topload Cartoning Machine from Cama. Integrated packaging machine with robotic units for reduced footprint, flexible size changing and high reliability. Includes RFID coding components and principles to mistake-proof operation.

(c) Integrated food and beverage ERP manufacturing software facilitating traceability. It connects to automation, e-commerce, accounting, reporting, shipping, and CRM.

(e) Tetra Pak operations. Photo courtesy of Tetra Pak. Automated data analytics, AI, cloud-based training, HMI collaborative App,

(f) BOBST M6 Line (Printing press for food packaging). Innovative printing technologies; Choose UV flexo as printing method; Nine-color printing and a 670mm web width; Consistency and high printing quality; 80 lines per cm/200, lines/in; Automatic job changeovers (1-minute on-the-fly without stopping the press, with up to 95% press up-time); Uses reels of carton board instead of sheets; High health safety (no solvents in the air - no odors).

Fig. 4 Industry 4.0 and Society 5.0 food processing and packaging principles and methods

One of the roles of technology is to find ways to cut down on the processing time, without compromising the nutrients of the food. This can be done with the help of science and techniques which include ultrasound, cold plasma, pulsed fields of electricity, high-pressure processing, and much more. For example, advances in enzyme research facilitate additive constituents to treat food itself as well as biological cells in food. They also try to make it healthier and lessen the processing time. Food processing and packaging heavily involves materials science, biochemistry, micro–/nano-technology, engineering and technology, optoelectronics and sensing technology, RFID, embedded and communication systems, printing technology, quality control safety, manufacturing and JIT (Just in Time), scheduling, production, automation and process control, and SCMS (supply chain management systems). The continuous SoS development chain keeps the food industry running, and the advancements try to cut off the extra time. Engineering technology (ET) concepts are adopted to build processing and packaging machineries, and optimization and sustainability are keys to the designs (Fig. 4).

6 Smart Home with Energy Implications

Considering the terminology and standards, this section summarizes smart homes with energy implications and provides suggestions to producers, distributors, and users. In essence, besides security, smart home technology (SHT) allows customers the ability to manage and control energy consumption within their homes by creating a network of connected smart devices. SHT are designed using sustainable materials and devices like thermostats, lights, and appliances that can track and optimize energy usage in real time so that users can choose to make more informed energy-efficient decisions for their homes. It is estimated that customers who choose to have an SHT-based system that includes zone heating-cooling and lighting, as well as monitoring and control of appliances, can reduce their energy consumption by 25% (Increasing Efficiency of Building Systems and Technologies, 2022). It is important to note that a number of variables may potentially affect this outcome. The most common barrier associated with implementation of successful systems is awareness about the state-of-the-art smart home products. In fact, results of the review found that the largest barriers associated with implementation of home energy management (HEM) systems include the lack of knowledge about the SHT devices, hardware installation difficulties, network connectivity issues, and user error. In reality, the SHT devices are not expensive. However, because of the lack of knowledge, it becomes expensive as its installation requires the cost of labor. It is suggested that the next generation of SHT products will require companies to prioritize devices that are more user-friendly and intuitive, error-proof, robust, and resilient. This may lead to short-term energy saving in the form of reduced energy consumption; however, the long-term energy saving implications will require users to continue to integrate more areas of their home which can only happen when the SHT devices get easier for users to use. The secondary goal of HEM and SHT is to develop a common platform for all in order to not only optimize the knowledge base but also reduce the resources such as installation, unnecessary time involved in returning the product, and applicability.

Figure 5 shows a 2000 sq. ft. residential smart home that has automated grid solar systems; security system; context-switching operation as regards energy usage; energy-efficient windows; centralized AC; brightness monitoring system; intuitive warning system; surrounding and guest monitoring system; fire and electrical safety system; front-and-backyard sprinkling system; air, floor, cleanliness, and insect monitoring; EV charging station; etc. It is estimated that the renewable fuel-based smart home designs save more energy and carbon footprint than an open-ended system. This is the reason why the government is providing rebates and

Fig. 5 A smart home that has zero electric bill and is good for 25–30 years. [Note: for the purpose of security, the image has been made such]

incentives for such SoS-based smart home architecture (Reinisch et al. 2011; Micu et al. 2021). The progress is slow; however, this is the future. Smart homes will be a norm of the future because, while we are running out of fossil fuels, the scope of SoS establishment is widening. The future smart homes entail knowledge in (i) facility planning; (ii) CAD-based building architecture design; (iii) materials science and engineering; (iv) engineering technology; (v) civil, electrical, mechanical, production, and manufacturing engineering; (vi) 3D building printing technology; (v) communication and sensor network; (vi) safety and monitoring; and (vii) entertainment technology, home robotics, and automatic irrigation.

7 Space and Ocean Technology Systems

Space technology mostly focuses on vehicles and infrastructure that work together to perform a task in the space environment. Space technology and systems also include weather forecasting, remote sensing, and satellite television, as well as long-distance communication systems. Space education and paradigm not only improves our knowledge of the physical universe through celestial observation and planetary exploration but also provides intelligence and surveillance that is critical for national defense. There are a variety of categories under the scope of space technology systems as far as research and development is concerned. The major advancements that have been made in this field are compact satellite systems with local intelligence; reliable, robust, and stable systems; and information integration and embedded decision support. Major developments are seen at NASA and MIT Lincoln facilities. Each key area presents stages of development, benefits, and future challenges. As regards CEC (Cooperative Engagement Capability), researchers have developed and operated sensor network systems that discover, track, and characterize location maps in the earth as well as space through satellites in the earth orbit to support national security and space objectives. Advanced satellite systems are used to monitor the activity of objects in space and perform remote sensing. Information integration systems are software tools that can help the military track, monitor, control, and protect these valuable satellites. Particular to a nation, the global satellite systems and network collectively perform critical civil, scientific, and national defense functions, such as navigation, communications, weather forecasting, astronomical observations, and surveillance.

Oceans have long been regarded as an immediate prospect for exploration, scientific discovery, commerce, and trade as they cover two-thirds of the earth and contain more than 90% of the world's biodiversity. Scientists, researchers, and stakeholders have been exploring the oceans as renewable sources and therefore have developed technologies to extract sustainable energy from waves, tides, and ocean thermal activity. The research and development is to critically review, study, and compare the developed systems systematically. Ocean systems have numerous applications in marine archeology, military, biotechnology, medical, energy, hydrography, fisheries and aquaculture, and many industries. All of the advancements leading to ocean data collection accounts to what is called the blue economy. The study also aims to provide trends in exploration sustainability. The latest breakthroughs and technological advancements that include Industry 4.0 concepts and the application of machine learning and artificial intelligence (MLAI), IoT, and biomimic engineering have undoubtedly opened new markets for ocean-based exploration and sustainability technologies. This is due to the fact that ocean technology systems can help us understand the ocean, which is instrumental in determining how to build future cities surrounded by oceans, lakes, and rivers with food supply systems that are required to feed an increasingly newer population. Research has shown that people would like to lead life in such an environment rather than living on Mars, for example. With rapid development in technology, our understanding about the relationship between oceans and atmosphere has improved, including how the oceans affect the climatic conditions and water cycle therein boosting the research in ocean energy. Other aspects that need to be addressed are analysis of a new generation of engineering and scientific noninvasive tools such as co-robotic systems, SONAR, advanced sensors, geographic information systems, and eDNA, which will widen in situ observational capabilities and develop the knowledge required to maintain a safe, clean, and sustainable blue economy.

8 Hypertube Technology in Transportation

Hypertube (hyperloop) is a method of transportation that is currently being sought out as a new form of long-distance travel, preferably within cities. Immune to weather and self-powering sustainability, hyperloop was promoted by Tesla and SpaceX. Sounding very futuristic, hyperloop capsule designs generated by different corporations such as The Boring Company have made concepts of having aboveground and underground tubes that would stretch long distances. The engineering concept for hyperloop is sending pods through tubes using magnetic fields which would allow people to travel long distances at high speeds. The systems vary depending on the companies that are working on this, but the basic is a capsule design for passengers to sit, a tube structure where the capsule is set to travel from place to place, and solar panels can be used to power the system. The benefits of such technology are not just high-speed travel but also cutting down on emission of greenhouse gasses from using other transportation such as cars and airplanes. With new advances come factors to develop it, which remain to be cost, safety, and technology. Constructing a tube hundreds of miles long poses challenges to robustness and durability. The engineering factors affecting hyperloop are the pressure, spontaneous decomposition, and thermal expansion. Considering all the factors' pros and cons for hyperloop technology review suggests that the concept is highly innovative and that the design entails SoS approach.

9 MEMS and NEMS Technology

The technology that considers micromanufacturing of microscale transducers, actuators, probes, capacitors, inductors, valves, gears, pumps, gyroscope, mirrors, switches, and so on similar to the fabrication of VLSI (very-large-scale integrated circuits) components within the semiconductor chips is referred to as microelectromechanical systems or MEMS in short. MEMS is an advanced miniaturized product and equipment design concept that has already emerged in order to cater the development of miniaturized products. It has become the preferred choice for the development of next-generation products and equipment, which are to be used for many kinds of applications such as biomedical devices, sensing, actuation, communications, and space. In essence, MEMS and NEMS (nanoelectromechanical systems) are small and integrated devices, which combine electronics and electrical as well as mechanical elements to meet the control-related functional requirements. MEMS and NEMS technology is an extended form of traditional microelectronic IC fabrication techniques and an advanced process as far as micromanufacturing of microsystems is concerned. MEMS technology can fabricate capacitors and inductors as well as mechanical elements such as springs, gears, beams, diaphragms, and so on. MEMS and NEMS are a kind of multi-science, multi-engineering discipline, and its scope is vast. Ideally MEMS combines physics, microelectronics, micromechanics, materials science, and computer-aided design (CAD) technology. Multi-science-multi-engineering (in other words, system of systems) integration methods look at systems design that adheres to the basic principles of microscale and nanoscale integration phenomenon considering the physics, chemistry, biology, and several engineering backbones. As mentioned, the technology embeds computer-aided simulation methods for the optimization of design parameters at all length scales, i.e., from the nanoscale to microscale to macroscale. The microscale design adheres to the SoS approach that entails not only interdisciplinary knowledge base but also sustainability. The scope of microscale designs is (a) fabrication processes and methods; (b) system modeling and analysis; (c) characteristics of active and passive devices and their integration; (d) mechanical transducers and actuators; (e) thermal MEMS; (f) MOEMS (Micro Opto Electro Mechanical Sysyems) and QoS (quality of service); (f) RF, bio, fluidic, and chemical MEMS; (g) NEMS (nanoelectromechanical systems); (h) simulation-based design methodology; (h) multiscale design approach; and (i) performance indices and optimal design (Fig. 6).

10 Entertainment Technology

Entertainment has played a very crucial role in society. Technological advancements have changed the whole perception of the entertainment industry today. From traditional magic shows to gladiators to theater plays to the film industry to computer games, it was all a part of an evolution of technological advances. And the

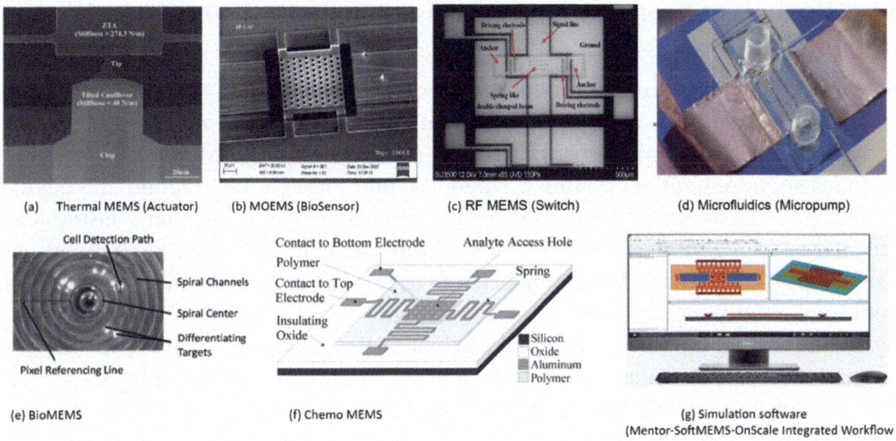

Fig. 6 Some MEMS designs that confirm SoS implications (Ouyang and Zhu 2012; Volkov et al. 2021; Shi and Shen 2022)

recent skyrocketing revolution would have been incomprehensible to humans 100 years ago. The chronological evolution in this field started in the early 1800s as individuals started traveling and performing entertainment programs such as rope rod, short play, animal show, etc., around the colonies. Some had a team of artists who performed plays or magic shows and circus-like tricks with animals. Later as the World War and the industrial revolution changed the scenario, inventions like electricity and radio were discovered. As the first radiobroadcaster in the world, the British Broadcasting Corporation (BBC) started using the technology to educate the public about the national and world news. Later, television systems ruined the entertainment programs in terms of comedy shows, serials, dramas, movies, etc. These advancements changed the living style. Radio and TV broadcasters had to employ artists. Hundred years later video game machines and mobile phones augmented the entertainment industry in a bigger way. The screen time for entertainment games has significantly increased in the recent past. In the present society, entertainment programs have become a means to many social activities including education. Entertainment programs and activities have impacted societal norms in the public that have been a huge part of our lives. It is convincing that not only the evolutionary scenarios of all aspects of technological advancements in entertainment are seen as dominant but also its impacts on the new generation are tremendous. Schools want kids to take classes online using pedagogically entertainment interfaces. There are pros and cons of modern-day entertainment programs and education where technology plays a crucial role and the setting undoubtedly entails SoS design (cognitive science, computer applications and software, ergonomics, human-machine interface, graphical user interface, cyber-physical systems, wireless technology and communication, diversity and inclusion technology, and sociology) and implementation strategy while considering sustainability.

11 Business Opportunity on Social Media and Networking

Sales, supply chain, storage, production, societal needs, and diversity within business have become a SoS approach, and all entail sustainable integration. In this context, the meaning of "networking" has changed recently. Modern networking includes sociology (ethics), digital, cloud, AI, and intuitive interfacing, and cognitive science. Networking via social media plays a critical role in any business in recent years. According to Forbes, 85% of job vacancies are filled via networking using apps such as LinkedIn and Indeed that help find people who are interested and could potentially create future transactions, and 60% of professionals worldwide agree that regular online interaction with their professionals can lead to possible job opportunities and an improved business and hence economy. Establishing a transaction with businesses that have a similar service or product can be bundled together to make profits for citizens and the business partners. The objective is to ensure a connection, customer satisfaction, targeted audience, partnership, and visibility. Using social media for business is how it attracts buyers from important parts of the demography. As an example, marketing utilizing Facebook, Instagram, Twitter, etc. has increased significantly over the past years. People are likely to find information on social media about the products due to it being much more convenient than going to the actual store and even to the companies' websites. Also, people are looking for reviews from other people using the social media platform. With over 50% of the world's population on social media, it's critical that the business has an effective social media strategy that should help reach the intended audience. Modern networking is also capable of executing comprehensive management policies of the entire platform by using cognitive science-based AI algorithms as they target specific audiences. These software intervention tactics are in place to ensure that business opportunity goals are met. In order to achieve greater business opportunity and connections, continuous development and applications of various AI tools, techniques, and methods into social media will facilitate implementation of futuristic Industry 5.0 standards and implications to next-generation networking.

12 Conclusions

The technological education and research scenario all over the world is converging towards a multidiscipline one. Sustainability plays an important role in engineering designs. The present scenario is different as compared to the recent past in the sense that the technical disciplines are now dilating instead of diverging. As such, the worldwide technological developments have been a reverse structure. The primary reason is the fact that the technological designs are of highly complex and interdisciplinary nature involving synergistic integration of many aspects of the knowledge base including sociology and STEM. The amalgamation of various mandates, theories, principles, phenomena, techniques, and methodologies to cater the pressing needs have long been emerging as new disciplines. For example, fundamentals to the state-of-the-art field devices (sensors and actuators are called field devices), subsystems, and modules must be understood along with not only the overview of underlying architectures and design approaches but also the design aesthetics and societal needs. Since this chapter covers aspects of interdisciplinary subjects synergistically in the areas of industrial technology, the importance is considered significant.

Acknowledgments The author acknowledges Jordan College of Agricultural Sciences and Technology, California State University, Fresno, for providing PRSCA award to develop this chapter. The author also acknowledges the graduate students in my Research Methodology (IT280) class in Spring 2022 such as Leen Almasri, Tarun Angrish, Chandramouli Balachandar, Mukesh Chinreddy, Heather Galindo, Enrique Gutierrez, Malvika Jagtap, Saikrishna Kalimi, Joseph Kryger, Vajidullah Molvizadah, Ramakrishna Parsana, Jingjing Piao, Shadrach Samson, Jonathan Sanchez, and Jagadesh Sriram as I was able to examine the currency on the topics cited above and then analyze the reviews for validation and consolidation.

References

Tabim, V.M., Ayala, N.F. & Frank, A.G. Implementing Vertical Integration in the Industry 4.0 Journey: Which Factors Influence the Process of Information Systems Adoption?. Inf Syst Front (2021).

Asok Ray (1989). Introduction to networking for integrated control systems, IEEE Control Systems Magazine, pp. 76–79.

https://www.techtarget.com/searchapparchitecture/definition/system-of-systems-SoS

Lirim Ashiku, Cihan H Dagli, System of Systems (SoS) Architecture for Digital Manufacturing Cybersecurity, Procedia Manufacturing, Volume 39, 2019, Pages 132–140, ISSN 2351-9789.

https://www.i-scoop.eu/industry-4-0/

Wang, B., Dehghanian, P., Wang, S., & Mitolo, M. (2019). Electrical Safety Considerations in Large-Scale Electric Vehicle Charging Stations. IEEE Transactions on Industry Applications, 55, 6603–6612.

Stacia Sherman Garr and Carole Jackson, Diversity & Inclusion Technology: The Rise of a Transformative Market, RedThread Research in partnership with Mercer, February 2019. Available at https://info.mercer.com/rs/521-DEV-513/images/Mercer_DI_Report_Digital.pdf

Akyazi T, Goti A, Oyarbide A, Alberdi E, Bayon F. A Guide for the Food Industry to Meet the Future Skills Requirements Emerging with Industry 4.0. Foods. 2020 Apr 14;9(4):492. doi: 10.3390/foods9040492. PMID: 32295238; PMCID: PMC7230995.

Increasing Efficiency of Building Systems and Technologies (An Assessment of Energy Technologies And Research Opportunities), Quadrennial Technology Review (retrieved on 9/5/2022) available at https://www.energy.gov/sites/prod/files/2017/03/f34/qtr-2015-chapter5.pdf

Christian Reinisch, Mario J. Kofler, Felix Iglesias, and Wolfgang Kastner, ThinkHome Energy Efficiency in Future Smart Homes, Hindawi Publishing Corporation, EURASIP Journal on Embedded Systems, Volume 2011, https://doi.org/10.1155/2011/104617

Micu, A.; Micu, A.-E.; Geru, M.; Capatina, A.; Muntean, M.-C. The Challenge for Energy Saving in Smart Homes: Exploring the Interest for IoT Devices Acquisition in Romania. Energies 2021, 14, 7589. https://doi.org/10.3390/en14227589

Jing Ouyang, and Yong Zhu, Z-Shaped MEMS Thermal Actuators: Piezoresistive Self-Sensing and Preliminary Results for Feedback Control, Journal of Microelectromechanical Systems 21(3):596–604, 2012

Volkov, Petr, Andrey Lukyanov, Alexander Goryunov, Daniil Semikov, Evgeniy Vopilkin, Stanislav Kraev, Andrey Okhapkin, Anatoly Tertyshnik, and Ekaterina Arkhipova. 2021. "Wideband MOEMS for the Calibration of Optical Readout Systems" Sensors 21, no. 21: 7343.

Shi, Y.; Shen, Z. Recent Advances in Flexible RF MEMS. Micromachines, 2022, 13, 1088.

Correction to: Driving Effective Sustainable Housing Infrastructure Delivery in South Africa Through Incorporation of Socioeconomic Development factor

Oluwatobi Mary Owojori and Emem O. Anwana

Correction to:
Chapter 16 in: I. S. Dunmade et al. (eds.),
***Sustainable Engineering*, Green Energy and Technology,**
https://doi.org/10.1007/978-3-031-47215-2_16

This book was inadvertently published with incorrect affiliation of the author Emem O. Anwana in Chapter 16. It has been corrected. The correct affiliation of the author is:

Department of Applied Law, Faculty of Management Sciences, Durban University of Technology, Durban, South Africa

The updated version of this chapter can be found at
https://doi.org/10.1007/978-3-031-47215-2_16

I. S. Dunmade et al. (eds.), *Sustainable Engineering*, Green Energy and Technology, https://doi.org/10.1007/978-3-031-47215-2_29

Correction to: Digitalization for Sustainable Agriculture: Enabling Farm Digitalization Through Decentralized Control and Ownership

Alvaro Romera, Glenn Parry, James Turner, Martin Espig, Michael Rogerson, and Munir Shah

Correction to:
Chapter 1 in: I. S. Dunmade et al. (eds.), *Sustainable Engineering*, Green Energy and Technology, https://doi.org/10.1007/978-3-031-47215-2_1

This book was inadvertently published with incorrect affiliation of author, Michael Rogerson, in Chapter 1. It has been corrected. The correct affiliation of the author is:

University of Sussex Business School, University of Sussex, Brighton, UK

The updated version of this chapter can be found at
https://doi.org/10.1007/978-3-031-47215-2_1

I. S. Dunmade et al. (eds.), *Sustainable Engineering*, Green Energy and Technology, https://doi.org/10.1007/978-3-031-47215-2_30

Index

I. S. Dunmade et al. (eds.), *Sustainable Engineering*, Green Energy and Technology, https://doi.org/10.1007/978-3-031-47215-2

The manufacturer's authorised representative in the EU is Springer Nature Customer Service Centre GmbH, Europaplatz 3, 69115 Heidelberg, Germany. If you have any concerns regarding our products, please contact ProductSafety@springernature.com

Printed and bound by CPI Group (UK) Ltd, Croydon, CR0 4YY
01/10/2026
02225896-0018